THE THERMODYNAMICS OF FLUID SYSTEMS

BY

L. C. WOODS

CLARENDON PRESS · OXFORD
1975

Oxford University Press, Ely House, London W.1

GLASGOW NEW YORK TORONTO MELBOURNE WELLINGTON
CAPE TOWN IBADAN NAIROBI DAR ES SALAAM LUSAKA ADDIS ABABA
DELHI BOMBAY CALCUTTA MADRAS KARACHI LAHORE DACCA
KUALA LUMPUR SINGAPORE HONG KONG TOKYO

ISBN 0 19 856125 3

PRINTED IN GREAT BRITAIN BY
J. W. AROWSMITH LTD., BRISTOL, ENGLAND

PREFACE

A SUITABLE sub-title for this book would have been 'The time-variable in thermodynamics', for throughout the emphasis is placed on the role of time scales, in determining both the state coordinates of equilibrium thermodynamics (Part I) and the affinities or thermodynamic forces of irreversible processes (Part II). The notion that the entropy of a system depends *inter alia* on the time scale on which it is observed (or equivalently on how detailed is the observer's information about it) is employed in both parts to make clear the unity of classical, kinetic, statistical, and process thermodynamics. This practical approach also enables us to resolve the occasional paradoxes that arise with entropy and its increase in isolated systems, for these can usually be traced to the false belief in the existence in an absolute entropy, a belief unfortunately encouraged by the currently-popular axiomatic treatments of thermodynamics. Thermodynamics is, above all, an approximating science in which it is necessary not only to understand the various limit forms it can take but also the physics of the asymptotics involved.

My aim initially was to write a concise account of the thermodynamics of irreversible processes in gases, plasmas, and other fluids, placing a greater emphasis on the underlying assumptions and less on chemical applications than the few reliable texts presently available. But it became clear at an early stage of writing that the key and unavoidable hypothesis of 'local thermodynamic equilibrium' needs much more elaboration than it usually receives. The concept of thermodynamic equilibrium is not nearly as simple and precise as most accounts imply; and without a clear idea of its meaning, it is pointless for the student to advance to process thermodynamics. Thus Part I (on equilibrium thermodynamics) was written to serve both as a survey of the principal ideas of the subject and, because of its emphasis on the relative nature of 'equilibrium', as a suitable introduction to Part II.

Part I starts with what may be called 'engineering' thermodynamics, for it deals with long-time-scale external processes; it is based on the first three laws of thermodynamics, which combine to define an entropy S via the fundamental formula $T\,\mathrm{d}S = \mathrm{d}U + p\,\mathrm{d}V$. The observer's time scale is then progressively shortened, and at each step terms added to the fundamental equation to represent the increasing knowledge assumed of the system; and so we proceed through 'chemical' thermodynamics and 'kinetic' thermodynamics to the ultimate description contained in statistical thermodynamics. By introducing an internal coordinate ξ_y defined over an 'interior' space y, we are able to transform the fundamental expression for $T\,\mathrm{d}S$ into the prescription $S = -k\sum_y \xi_y \ln \xi_y$, which, unconstrained, holds on all time scales. It is the constraints that are added to turn the prescription into a genuine definition that are time-scale dependent, and the shorter the observer's time scale the greater the number of constraints that must be imposed.

I hope this novel approach will provide a useful skeletal survey of basic thermodynamics suitable for final-year students in the physical sciences.

In Part II the concepts of process thermodynamics (often confusingly termed 'non-equilibrium' thermodynamics, despite the fact that local thermodynamic *equilibrium* is a *sine qua non* ingredient) are explained and applied to a variety of fluid phenomena. There are four basic ideas.

First there is the separation of processes into reversible and irreversible elements, a division that depends on the physical model assumed to represent the system. And when the reversible changes of entropy are known, the rate σ at which entropy is irreversibly created can be expressed as a bilinear form, say, $\sigma = \mathbf{JX}$, where $\mathbf{J}$ and $\mathbf{X}$ are vectors representing the various processes and thermodynamic forces in the system. In all this time scales play an important part. As σ is non-negative, we have the basic inequality $\sigma = \mathbf{JX} \geqslant 0$, which has much the same role in process thermodynamics as does the fundamental equation $\mathrm{d}S = T^{-1}\,\mathrm{d}V + T^{-1}p\,\mathrm{d}V + \ldots = Z\,\mathrm{d}Y$, say, for equilibrium thermodynamics. From the latter we use the fact that $\mathrm{d}S$ is the exact differential of a function $S = S(Y)$ to infer the existence of *state* equations $Z = Z(Y)$; and likewise as $\sigma \geqslant 0$ requires $\mathbf{J}$ and $\mathbf{X}$ to vanish together, we may infer *constitutive* relations $\mathbf{J} = \mathbf{J}(\mathbf{X})$ from the bilinear form.

With the assumption of linear constitutive relations, say $\mathbf{J} = L\mathbf{X}$, where L is a phenomenological matrix, we come to the second basic idea of process thermodynamics, namely, the existence of reciprocal relations between the coefficients L_{ij} of L, which in the simplest case are $L_{ij} = L_{ji}$. In essence, these relations are due to W. Thomson (Lord Kelvin), although it is usual to dismiss his arguments as being too intuitive to offer a convincing proof. On the other hand, the accepted treatment, due to L. Onsager, relies too much on the analogy taken to exist between the regression of large thermodynamic fluctuations and macroscopic processes like heat or current flux that persist not only in stationary flow conditions but, more importantly, in systems pressed to the (thermodynamic) limit at which fluctuations strictly vanish. The new approach offered in this book (along with accounts of the works of Thomson and Onsager) resembles Thomson's in being genuinely 'macroscopic'. By the time-scale arguments required to draw the distinction between reversibility and irreversibility, it is shown that the essential nature of the reciprocal relations is that they ensure that the entropy production rate $\sigma = \mathbf{X}L\mathbf{X}$ is due only to irreversible processes.

The two remaining basic concepts are common to all branches of macroscopic physics. They are 'reference-frame indifference', which requires the constitutive relations to be expressible in covariant form, and 'material invariance', by which spatial symmetries known to exist in the system enable us to reduce substantially the number of independent elements appearing in

L. Particular attention has been given to the lateral isotropy of magneto-plasmas, in which the magnetic field imposes a characteristic direction on the medium.

The use of internal coordinates ξ permits a great extension in the scope of process thermodynamics. And when constitutive relations for $\dot{\xi}(t)$ are known or assumed, some or all of these coordinates may be eliminated with the effect of replacing the instantaneous constitutive relation $\mathbf{J} = L\mathbf{X}$ by the integral relation $\mathbf{J}(t) = \int_{-\infty}^{t} \psi(t-t')\mathbf{X}(t')\,dt'$, where $\psi(t)$ is a phenomenological response function. If, further, owing to thermodynamic fluctuations, the $\dot{\xi}(t)$ are assumed to be stochastic processes, the theory leads on to the important fluctuation–dissipation theorem, of which the earliest examples are contained in Einstein's work on Brownian motion and Nyquist's on Johnson noise.

Fluid mixtures are systems for which the systematic approach of process thermodynamics is particularly useful, for the number of possible processes and cross-couplings between them may be quite large. A general account of mixture theory, which includes the effects of mass transfer between the components, is followed by applications to two important examples, namely, magneto-plasmas and helium II. A magneto-plasma can be described adequately by a two-fluid theory on a certain time scale and is a good example of an anisotropic medium. Helium II, on the other hand, has the interesting feature (rather neglected in standard treatments) that its normal and superfluid components can exchange mass elements. These aspects have been pursued to the stage of obtaining some new results for each mixture. Rotating helium II, which is an anisotropic medium, is also given some attention.

Kinetic constitutive relations are described in the final chapter. The set $S_1, S_2, \ldots S_r$ of entropies for the one-particle (F_1), two-particle (F_2), ... r-particle (F_r) distribution functions yields a sequence of non-negative entropy production rates $\dot{S}_1, \dot{S}_2, \ldots \dot{S}_r$ from which may be inferred constitutive relations for $F_1, F_2 \ldots$ of increasing accuracy. This approach yields the kinetic equations of Boltzmann, Bogoliubov, and others, and may have some merit over the usual methods of closing the BBGKY hierarchy of equations. In any case it reveals the close similarity between kinetic equations and the constitutive relations of macroscopic fluid dynamics.

Part II will, I hope, be of interest to graduates working in continuum mechanics and other branches of macroscopic physics.

Among the many relevant texts and articles I have studied—of which only those essential to the textual argument have been cited—I am particularly indebted to the work of S. R. de Groot and P. Mazur (1962), which first engaged my attention in the thermodynamics of irreversible processes. And perhaps it is right to admit too that a growing aversion to axiomatic thermodynamics, with its *penchant* for substituting deductive mathematical constructions in place of physics, has stimulated me to write this book, and in it to emphasize the approximations inherent in thermodynamics, or at least those approximations associated with time scales. It is, of course, the existence of such approximations that makes thermodynamics such a rich and diverse branch of physics.

I am grateful to both H. Troughton and Stanley Morris for help with checking the manuscript and proofs, to Mrs. Ina Godwin for her excellent typing of rather complicated material, and finally to the Delegates of the Oxford University Press for accepting my biased recommendation as a series editor that the work was suitable for publication in the Oxford Engineering Science Series.

Mathematical Institute L.C.W.
Oxford University
August, 1974

CONTENTS

PART I

EQUILIBRIUM THERMODYNAMICS

1

THERMODYNAMIC VARIABLES AND PROCESSES

1. Thermodynamic systems

1.1. Macroscopic variables

BEFORE describing a thermodynamic system, we shall consider some of the general properties of a real physical system consisting of matter and radiation lying in a finite region of space. Because of the atomic constitution of matter, there are small lengths and times, termed 'microscopic', in which substantial changes in the properties of individual particles (molecules, atoms, protons, electrons, photons, etc.) can occur. Thus, for example, under STP conditions in air the momentum $m\mathbf{w}$ of a typical molecule of mass m and velocity $\mathbf{w}$ is significantly changed over a distance of a mean free path λ ($\approx 10^{-5}$ cm) and in a mean free time $\tau = \lambda/|\mathbf{w}|$ ($\approx 10^{-10}$ s). On the other hand, the *averages* of particle properties like the momentum per unit mass $\mathbf{v} = \langle m\mathbf{w}\rangle/\langle m\rangle$ and the energy per unit mass $u = \langle \frac{1}{2}mw^2\rangle/\langle m\rangle$ taken over the enormous number of particles ($\approx 10^7$) in a microscopic volume element ($\approx 10^{-12}$ cm^3) usually change over much larger length L and time $\mathscr{T}$ scales. Such averages are termed 'macroscopic' properties and L and $\mathscr{T}$ are the macroscopic length and time scales. Provided $L \gg \lambda$ and $\mathscr{T} \gg \tau$, we can introduce macroscopic infinitesimals $|\mathrm{d}\mathbf{r}|$ and $\mathrm{d}t$ of length and time that are large on a microscopic scale but small on a macroscopic scale. Then, save in exceptional regions, macrosopic properties like $\mathbf{v}$ and u will be smooth functions of the position vector $\mathbf{r}$ and of the time variable t, changing by small increments $|\mathrm{d}\mathbf{v}|$ and $\mathrm{d}u$ over $|\mathrm{d}\mathbf{r}|$ and $\mathrm{d}t$.

The study of the spatial and time fields of macroscopic variables like $u(\mathbf{r}, t)$ is termed 'continuum mechanics' if $|\mathrm{d}\mathbf{r}| \gg \lambda$, $\mathrm{d}t \gg \tau$, for in this case the particulate nature of matter is completely submerged and, further, the scale lengths are such that particle behaviour is aptly described as being 'collision-dominated'. This has the consequence that many physical effects—like, for example, the mechanical stress—are transmitted *locally*, and so can be described by differential equations. As the average distance between particles ($\approx 10^{-7}$ cm for air at STP) is considerably less than λ, smooth functions $u(\mathbf{r}, t)$ *can* be defined for $|\mathrm{d}\mathbf{r}|$ and $\mathrm{d}t$ much less than λ and τ. Under these circumstances the particles are 'collision-free', physical effects are transmitted on a global scale, and a more detailed description of the behaviour of the macroscopic fields is necessary. This is the province of 'kinetic theory', which will be discussed in a later chapter. In both kinetic theory and continuum mechanics the dependent variables are macroscopic and the independent variables include **r** and t. Equilibrium thermodynamics is also concerned with macroscopic variables, its principal results being relationships between certain of the variables occurring in the field theories; such relationships are independent of **r** and t, which may, however, play parametric roles.

1.2. Equilibrium on the observer's length and time scales

An omnipresent observer could specify the macroscopic state of a physical system P, say, at a given time t, by giving all the macroscopic properties of P, both physical and chemical, as functions of **r**. A human observer must be content with a greatly reduced set of macroscopic variables, and the problem of defining a thermodynamic system lies in how to specify this set, say $\{x_1, x_2, ...\}$, for a given system P and a given observer Q. The observer's time and length scales, say $\mathscr{T}$ and $\mathscr{L}$, are the most important elements in the problem. These scales, which depend on the observer's objectives and knowledge, must be compared with the variety of time and length scales occurring naturally in P, due either to purely internal changes or to interactions between P and its environment. For example, if the variable x_r has a maximum gradient x_r', the length $L = x_r/x_r'$ occurs naturally in P and it is reasonable to assume that over a length $\mathscr{L}$, where $\mathscr{L} \ll L$, it is permissible for Q to ignore gradients in x_r, that is, to take x_r as uniform in P. The variables in equilibrium thermodynamics are always assumed to be uniform; departures from uniformity lie within the ambit of 'process thermodynamics' to be described in Part II.

Similarly, a further reduction in the set $\{x_r\}$ defining the state of P results if 'temporal equilibrium' may be assumed by Q for some of the variables. The role of the time variable in thermodynamics is particularly important, although not easily explained at the beginning of the subject; the following description is amplified in a later section (§ 19.1). There are two quite distinct kinds of temporal equilibrium in thermodynamics. To explain these we shall introduce the natural time scales

$$\tau_r = (x_r^{\text{eq}} - x_r)/\dot{x}_r \qquad (r = 1, 2, ...), \tag{1.1}$$

where x_r^{eq} denotes the value x_r would attain as $t/\tau_r \to \infty$. Now order the time scales so that $\tau_1 \geqslant \tau_2 \geqslant \tau_3 ...$ and suppose that Q's time scale $\mathscr{T}$ satisfies

$$\tau_m \gg \mathscr{T} \gg \tau_{m+n}, \tag{1.2}$$

where τ_m is the least scale and τ_{m+n} is the greatest scale that will satisfy the gross inequalities in (1.2).† On Q's time scale the variables $x_r, r = 1, 2, ...$ $m-1$, are said to be in 'frozen' equilibrium, and Q observes them to be

† It is impossible to give such inequalities precision out of physical context; the reader should think in terms of factors of 10^2 or more.

constants in P. On the other hand, Q observes that the variables x_r, $r > m+n$, can change so rapidly that they never deviate measureably from their equilibrium values x_r^{eq}; such variables are in 'relaxed' equilibrium. Stability requires that x_r^{eq} depend only on the variables x_s, $s < r$, that change more slowly than x_r. Thus for the system (P, Q) (P observed by Q), the relevant thermodynamic variables are the set $\{x_r\}$, $m \leqslant r \leqslant m+n$, which change more or less on Q's time scale.

1.3. Thermodynamic systems

The time scale $\mathscr{T}$ is not the only observer-dependent element in thermodynamics: two observers may have the same $\mathscr{T}$ but different views on what is important in P. For example, Q_1 may hypothesise that the ionization level χ is an important variable in a strong shock wave and thus include χ as a thermodynamic variable, whereas Q_2 may consider χ unimportant. And with different objectives they could both be correct. Such subjective aspects are, of course, present in all idealized models of real physical systems and can never be completely removed.

The system (P, Q) is called a *thermodynamic system* and $\{x_r\}$, $m \leqslant r \leqslant m+n$, are its *thermodynamic variables*. The principal objective of classical thermodynamics is to establish relationships involving just these variables and their derivatives. A subset of these variables, termed thermodynamic *coordinates*, serves to define a state-space, and the remaining variables are then functions of state. The coordinates may be chosen in a variety of ways, and it is clearly convenient to include among them any of the x_r known to be constant (on Q's time scale). Those (thermodynamic) variables that are already familiar from mechanics and electromagnetism like pressure p, volume V, mass M, magnetic flux $\mathbf{B}$, and electric field $\mathbf{E}$ are accepted as primitive concepts, and the derivation of additional variables, entirely thermodynamic in origin, is the first task in the development of thermodynamics.

To derive new thermal variables it is necessary to assign some physical properties to the thermodynamic system, distinct from those just mentioned. There are several ways of going about this, all equally valid but with distinct metaphysics. The usual textbook treatment is to introduce the additional physical properties implicitly via the so-called 'laws' of thermodynamics, and with the aid of 'walls' that separate the system from its immediate surroundings and that limit in various ways the nature of the exchanges between them. In this axiomatic approach, of which there are several variants, the thermal variables are defined in terms of the non-thermal ones, so that their intrinsic physical significance is not clarified. This is possible because on a long time scale the state may be defined by coordinates chosen from the list of variables accepted as primitive or non-thermal. For example, the simplest thermodynamic system is a gas of a single chemical constituent in a container, and on a long time scale its state coordinates may be taken to be its pressure p

and volume V (in what follows such a system will be termed a 'simple' gas system). Alternatively we can accept the atomic nature of matter and directly attribute macroscopic thermal variables to the system, such as the internal energy u defined above, with a consequent reduction in the 'laws' required. In the prevailing philosophy of continuum mechanics it is usual to eschew deductions from a knowledge of the existence of atoms, and this fashion will be followed for the present.

2. The empirical temperature

2.1. *Adiabatic walls*

'Temperature' is the most important of the entirely thermodynamic properties. To define it we first need to introduce the concept of a 'wall' C that separates a system P from its surroundings P*. At one extreme C need have no physical properties at all, save that of merely defining the extent of P; at the opposite end of the scale C may isolate P completely from P*, by which is meant that there are no changes possible in P* able to affect the state of P, and vice versa. Such isolation is an idealized concept for it is a well-established principle that no part of the universe can be completely isolated from the rest of the (visible) universe. However if the observer's time scale $\mathscr{T}$ is very small compared with the transit times of various phenomena through C, we may assert that P is effectively isolated from the effects of such phenomena. Clearly $\mathscr{T}$ must be sufficiently small for C to appear to Q as being impermeable, rigid, and of infinite electrical conductivity (to exclude electromagnetic fields). These properties will exclude mass and work transfers across (C, Q) (wall C observed on Q's time scale), but experimentally it is found that they are insufficient to isolate P completely. In addition, C must possess a special thermal property that cannot be defined in terms of primitive concepts. This property is called *adiabaticity* and is found to be imperfectly present in varying degrees as the chemical composition and physical structure of C is changed.

Suppose (C, Q) isolates P from mass and work transfers, and let τ_q be the least time for a measurable change to occur in the state of P after all possible changes in P*.† Then (C, Q) is characterized as follows:‡

if $\mathscr{T} \ll \tau_q$, (C, Q) is *adiabatic*;

if $\mathscr{T} \gg \tau_q$, (C, Q) is *diathermic*.

† Of course the important change we have in mind here is the placing of a heat source in contact with C, but the concept of heat follows later from the first law of thermodynamics, hence our circumlocution.

‡ Similar inequalities may be adopted to define 'open' and 'closed' systems

$\mathscr{T} \ll \tau_D$, (C, Q) is *closed*; $\mathscr{T} \gg \tau_D$, (C, Q) is *open*,

where τ_D is a mass diffusion time. Consideration of open systems will be deferred to § 14.

An impermeable wall that possesses adiabaticity, but which may be flexible, may admit electromagnetic fields and currents to P, and could allow mechanical devices to stir the medium of P, will be called an *adiabatic* wall. Armed with the concept of adiabatic walls, we can now introduce temperature; in the sequel the role of the observer will be supposed understood and omitted from the account.

2.2. *Thermal equilibrium*

Consider two adjacent systems, P_1 in state X_1 defined by m_1 coordinates and P_2 in a state X_2 defined by m_2 coordinates, isolated from their surroundings but not necessarily from each other. If P_1 and P_2 are isolated from each other, the state of the composite system (X_1, X_2) will require (m_1+m_2) coordinates to define it, whereas, if P_1 and P_2 are in some form of contact, fewer than (m_1+m_2) coordinates will be required to define (X_1, X_2). This contact will permit an exchange between P_1 and P_2. For example, if P_1 and P_2 are two simple gas systems, $X_1 = (p_1, V_1)$, $X_2 = (p_2, V_2)$, and if they are separated by a perfectly flexible membrane, p_1 will equal p_2 and the composite system is completely defined by (p_1, V_1, V_2). Should p_1 not equal p_2 the membrane will move to make the pressures equal and P_1 will do work on P_2 or vice versa.

Again, if P_1 and P_2 are first isolated from one another by a rigid adiabatic wall, then $(X_1, X_2) = (p_1, V_1, p_2, V_2)$; if this partition is now replaced by a diathermic wall it is found that P_1 and P_2 move to new equilibrium states in which any three of p_1, V_1, p_2, V_2 are sufficient to define the composite state. Hence there must exist a definite relationship

$$F(p_1, V_1, p_2, V_2) = 0, \tag{2.1}$$

and P_1 and P_2 are said to be in *thermal equilibrium* with each other. Similarly, in the general composite state space of (m_1+m_2) coordinates, the thermal equilibrium states will lie on a (m_1+m_2-1) surface.

2.3. *Zeroth law and empirical temperature*

The first of the four fundamental laws of thermodynamics, called the 'zeroth law' (being discovered last), states

0. *Two systems in thermal equilibrium with a third are in thermal equilibrium with each other.*

Actual thermal contact via a diathermic partition is not necessary here; we mean rather that if such contact were established, there would be no change of state for the two systems involved. It is sufficient to consider three simple systems P_1, P_2, and P_3. Let P_3 be in thermal equilibrium with each of P_1 and P_2, then functional relations

$$F_{13}(p_1, V_1, p_3, V_3) = 0 \quad \text{and} \quad F_{23}(p_2, V_2, p_3, V_3) = 0 \tag{2.2}$$

must hold. And provided each relation can be solved to give a *unique* value of p_3 (or V_3), then p_3 may be eliminated to yield a unique relation of the form

$$f_{13}(p_1, V_1, V_3) = f_{23}(p_2, V_2, V_3). \tag{2.3}$$

Further, from the zeroth law P_1 and P_2 are also in thermal equilibrium, so that there is a relation

$$F_{12}(p_1, V_1, p_2, V_2) = 0, \tag{2.4}$$

that is, (2.2) imply (2.4). This means that it must be possible to eliminate V_3 from (2.3). All the consequences of (2.3) are contained in

$$\Phi[f_{13}(p_1, V_1, V_3), V_3] = \Phi[f_{23}(p_2, V_2, V_3), V_3], \tag{2.5}$$

where $\Phi[\alpha, \beta]$ is an arbitrary function, and hence a particular function $\varphi[\alpha, \beta]$ must exist such that for $\Phi = \varphi$, V_3 drops out of (2.5) and (2.4) is the result. This requires that

$$\varphi[f_{13}(p_1, V_1, V_3), V_3] = G(V_3)[F_1(p_1, V_1) - H(V_3)]$$

$$\varphi[f_{23}(p_2, V_2, V_3), V_3] = G(V_3)[F_2(p_2, V_2) - H(V_3)],$$

the right-hand sides having the most general functional form that permits the cancellation property required; here only the function $H(V_3)$ may possibly be zero. Then (2.5) becomes

$$F_1(p_1, V_1) = F_2(p_2, V_2). \tag{2.6}$$

Thus the zeroth law restricts the condition of thermal contact (2.4) to have the particular form given by (2.6), and of course a similar result applies to the functional forms of (2.2). Hence

$$F_1(p_1, V_1) = F_2(p_2, V_2) = F_3(p_3, V_3) = \theta, \quad \text{say}, \tag{2.7}$$

so that θ is a state variable of each of the three systems. It is called the *empirical temperature*.

2.4. *Thermal equation of state*

The equation

$$F_1(p_1, V_1) = \theta$$

is called the *thermal equation of state* for P_1, and the curves θ = constant in the (p_1, V_1)-plane are called the *isotherms* of P_1; similarly for P_2 and P_3. All the states lying on those isotherms in each of the three state planes that share the same constant value for θ will be in thermal equilibrium with each other. Thus the temperature of a system is a property that determines whether or not the system is in thermal equilibrium with other systems.

The actual numbers θ attached to the isotherms for one particular system, say P_1, may be arbitrary provided only that if $\theta(p_1, V_1)$ is taken to be a continuous function of each of its arguments—it must have no turning points or finite lengths of zero slope. Further, if two states of P_1 have different values of θ, then at least one of the thermodynamic coordinates p_1, V_1 must have different values for the two states. These restrictions are necessary to secure the *uniqueness* of θ as a function of state, and they permit us to write the thermal equation of state for a system P in the form

$$p = p(\theta, V). \tag{2.8}$$

Of course, once $\theta(p, V)$ has been established empirically for a given system, say P_1 (termed the thermometric body) then for other systems $P_2, P_3, \ldots$ thermal equilibrium with P_1 will determine the appropriate values of θ for the isotherms of $P_2, P_3, \ldots$. If a different temperature scale $\hat{\theta}$ is chosen, then clearly there must exist strictly increasing or strictly decreasing functions h and h^{-1} such that $\theta = h(\hat{\theta})$ and $\hat{\theta} = h^{-1}(\theta)$.

A convenient thermometric body, such as the mercury-in-glass thermometer, has almost constant values for all but one of its thermodynamic variables, the change in which provides a convenient scale for θ.

2.5. *Perfect gas*

Boyle's law states that as the pressure of a real gas tends to zero, the product pV tends to a finite limit, and this behaviour in the limit of vanishing pressure is taken to be the defining attribute of a *perfect gas*. Thus a perfect-gas thermometer has a thermal equation of state

$$pV = F(\theta), \tag{2.9}$$

where the function F depends on the nature of the empirical temperature scale. With θ taken to be proportional to the volume of the mercury in a mercury-in-glass thermometer, $F(\theta)$ is found to be linear over a wide range of temperature and, as p tends to zero, (pV/θ) tends to a constant value nR, where n is the number of moles in the system and R is known as the *universal gas constant*, having the same value for all gases. Thus with this choice of θ scale,

$$pV = nR\theta. \tag{2.10}$$

The sense of the scale, that is the physical distinction between increasing and decreasing θ, will be assigned in § 4.

3. Work processes

3.1. *Work*

A mechanical system increases its potential and kinetic energy when work is performed on it and—neglecting losses—this energy may be recovered subsequently in the form of work. It is usually accepted as a primitive concept that work may also be transferred to and from a thermodynamic system, and that such transfers change the state of the system. Of course, the 'state' here is the internal thermodynamic state and excludes account of any purely mechanical energy that the system may acquire qua rigid mechanical body. Work may be applied to a system P in several ways; these include deforming it against internal pressures or stresses either by mechanical or electromagnetic means, stirring P mechanically, or passing electric currents through it.

Suppose that P is initially in a state X_1 at $t = t_1$ and work is applied to it continuously to bring it to a final state X_2 at $t = t_2$. The change is called a *process* and will be denoted by $X_1 \to X_2$. An obvious question is: What can be said about the state of P *during* the process, that is, for times $t_1 < t < t_2$? As we shall see the answer depends both on how the work is supplied to P and the rate at which this happens.

3.2. *Equilibrium process*

Let P be enclosed by flexible walls or by walls containing moving sections like pistons and suppose that at some time during the process $X_1 \to X_2$ the volume V is changed by a macroscopically infinitesimal amount $\mathrm{d}V$ in a time $\mathrm{d}t$. As $\mathrm{d}V$ is macroscopic, $\mathrm{d}t$ must be very large on a microscopic time scale but, of course, infinitesimal on a macroscopic time scale. An observer Q would describe the change $\mathrm{d}V$ as being 'rapid' or not depending whether the time $\mathrm{d}t$ it takes is small or not compared with his time scale $\mathscr{T}$. And as described in § 1.2 the state X of P also depends on $\mathscr{T}$.

A process $X_1 \to X_2$ such that $\mathrm{d}t \sim \mathscr{T}$ at each instant t in (t_1, t_2) is termed (by Q) an *equilibrium process*. Variables that are either relaxed or frozen on the time scale $\mathscr{T}$ are also relaxed or frozen during the change in P occurring in $\mathrm{d}t$, consequently the state of P can be represented by a point in the state space X throughout the process, which thus corresponds to a continuous curve $X(t)$ starting at $X(t_1)$ and finishing at $X(t_2)$. On the other hand, if, at some instant t in (t_1, t_2), $\mathrm{d}t \ll \mathscr{T}$, some variables, relaxed for Q and not represented in X-space, are removed from equilibrium and become important during the process. Such *non-equilibrium processes* cannot be represented in X-space by a continuous curve. The various possibilities are set out in Table 1.

If Q wishes to reduce $\mathscr{T}$, variables previously in relaxed equilibrium must be added to the list of thermodynamic coordinates, and perhaps a variable or two at the other end of the time scale may be assumed frozen. A central problem in thermodynamics is that while frozen variables are easily identifiable, being the most permanent features of the system, the relaxed variables are often elusive. Amongst the possible relaxed descriptors of the internal state Q needs to identify the one with the longest relaxation time, so that,

TABLE 1

Intrinsic time scales (see § 1.2) Types of variable	$\tau_r \ll \mathcal{T}$ relaxed	$\tau_r \sim \mathcal{T}$ state	$\tau_r \gg \mathcal{T}$ frozen
Process time scales Types of process	$dt \ll \mathcal{T}$ non-equilibrium (irreversible)	$dt \sim \mathcal{T}$ equilibrium (reversible or irreversible)	$dt \gg \mathcal{T}$ static

on reducing $\mathcal{T}$, it may be added to the list of thermodynamic coordinates. Thus $\mathcal{T}$ can be reduced only if Q has a greater physical insight into the behaviour of P than represented in the original state-space. Suppose for a given $\mathcal{T}$ the state space is $X(\mathcal{T})$, then if we agree to retain in X any variables that could be frozen by reducing $\mathcal{T}$, the reduction of $\mathcal{T}$ to $\mathcal{T}'$ requires an extended state space $X'(\mathcal{T}')$ containing $X(\mathcal{T})$ as a sub-space. Then if $\mathcal{T}' \sim dt \ll \mathcal{T}$, a non-equilibrium process in $X(\mathcal{T})$ will appear as an equilibrium process in $X'(\mathcal{T}')$, and the states traversed by this process in $X'(\mathcal{T}')$ would be termed *non-equilibrium states* by an observer on the longer time scale $\mathcal{T}$.

Equilibrium processes are usually described as being 'quasi-static', but this is a misleading term, often being misinterpreted as meaning 'static' in the sense indicated in the table above.

As the thermodynamic coordinates specifying the state–space X are independent variables, hypothetically at least within the observer's control, an equilibrium process $X_1 \underset{\text{eq}}{\rightarrow} X_2$ represented by a curve in X may be traced in reverse. On the other hand, in a non-equilibrium process $X_1 \rightarrow X_2$ the observer has incomplete knowledge of the details of the change itself and hence is unable to regain X_1 from X_2 by a *reverse* process, although he may be able to effect a quite different return process $X_2 \rightarrow X_1$. Because of the speed at which it proceeds, a non-equilibrium process may be said to be 'out of control', and it is certainly *irreversible*. A process is said to be *reversible* when it is an equilibrium process in which the thermodynamic coordinates may be controlled by the observer without altering any of the physical

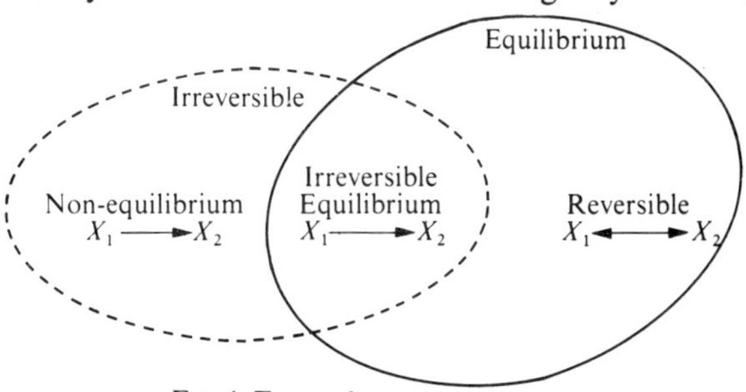

FIG. 1. Types of processes.

properties of the system or of its surroundings. Most of the equilibrium processes we shall meet in the next few sections are controllable in direction, albeit with some ingenuity in particular cases; later we shall meet some irreversible equilibrium processes. For the controlled or reversible equilibrium processes we shall adopt the notation $X_1 \leftrightarrow X_2$, with X_1 the initial state and X_2 the final state. Irreversible processes, whether equilibrium or non-equilibrium, will be indicated by $X_1 \rightarrow X_2$ (see Fig. 1).

3.3. Reversible and irreversible work

Consider a simple gas system P of pressure p and volume V enclosed by adiabatic walls. In an infinitesimal reversible process in which the volume changes by $\mathrm{d}V$, the work $\mathrm{d}'W$ done on P is $-p\,\mathrm{d}V$; such an expression cannot hold for non-equilibrium processes, for which we shall write

$$\mathrm{d}'W = -p\,\mathrm{d}V + \mathrm{d}'W_{\mathrm{i}}, \tag{3.1}$$

where $\mathrm{d}'W_{\mathrm{i}}$ is the additional work done on P owing to the time interval $\mathrm{d}t$ during which the work is transferred, being too small to preserve equilibrium (*note*. The notation d′ is used to denote an infinitesimal that is not the differential of a state variable or coordinate).

Suppose, for example, that the work is transferred to P via a piston on which there acts an external pressure p_{s}, say. Then p_{s} must differ from the opposing internal pressure p in order that the work can be transferred in a short time. Clearly if the system P is to be compressed ($\mathrm{d}V < 0$) we need $p_{\mathrm{s}} > p$, and if P is to expand ($\mathrm{d}V > 0$) we need $p_{\mathrm{s}} < p$. For a non-equilibrium process, work $\mathrm{d}'W = -p_{\mathrm{s}}\,\mathrm{d}V$ applied outside P is transmitted into P as $-p\,\mathrm{d}V + \mathrm{d}'W_{\mathrm{i}}$; hence

$$\mathrm{d}'W_{\mathrm{i}} = (p - p_{\mathrm{s}})\,\mathrm{d}V, \tag{3.2}$$

which is *always* positive.†

This means that only part of the work done on a system can be recovered by simply reversing the motion, a result that anticipates the content of the second law of thermodynamics (p. 17). The work $\mathrm{d}'W_{\mathrm{i}}$ lost in this sense is termed 'dissipated' work and it renders the non-equilibrium process quite irreversible. A more obvious form of irreversible work is work transmitted to P via a mechanical stirrer; such stirrers never return the work to the surroundings by spontaneous or even induced self-movements.

3.4. General expressions for reversible work

Suppose that a thermodynamic system is doubled in size by taking an adjacent replica and removing the interface wall. Those variables like p

† Incidentally with $p > p_{\mathrm{s}}$ and $\mathrm{d}V > 0$, stability requires the expansion to reduce the internal pressure, bringing it closer in value to p_{s}, until eventually mechanical equilibrium is attained at $p = p_{\mathrm{s}}$. Thus if $\mathrm{d}V > 0$ then $\mathrm{d}p < 0$. Similarly if $\mathrm{d}V < 0$ then $\mathrm{d}p > 0$. Hence we have the *mechanical stability condition* $(\mathrm{d}V/\mathrm{d}p)_{\mathrm{a}} < 0$ (a denotes adiabatic).

and θ that are unchanged by the operation are termed *intensive* variables, whereas variables that are changed, such as the volume V, are described as being *extensive*. Other extensive variables will be introduced later, and we shall find that, like the volume, they are all *additive*, that is, if X_α is such a variable for a sub-system P_α, then $X = \sum_\alpha X_\alpha$ is its value for the composite system $\mathrm{P} = \sum_\alpha \mathrm{P}_\alpha$. Work has this additive property and hence the expressions for reversible work are always the product of an intensive variable (generalized force) and an extensive variable (a generalized displacement).

Consider, for example, the composite system consisting of two simple systems in thermal contact shown in Fig. 2. The state may be specified by (θ, V_1, V_2) and the expression for the reversible work is $-(p_1\,\mathrm{d}V_1 + p_2\,\mathrm{d}V_2)$. The extensive variables V_1, V_2 appearing in the form for the reversible work are known as *deformation* coordinates, and have the special virtue of having clearly defined meanings even for non-equilibrium processes.

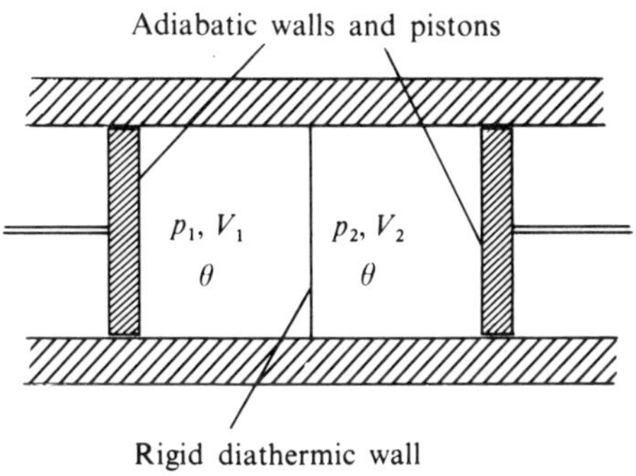

FIG. 2. A three-dimensional thermodynamic system.

Later we shall need an expression for the reversible electromagnetic work $\mathrm{d}W^*$ done on a system of volume V and volume element $\mathrm{d}\tau$. In standard notation and SI units this expression is (see § 11 for further discussion)

$$\mathrm{d}'W = \int_V (\mathbf{B}_\mathrm{e}\,.\,\mathrm{d}\mathbf{M} + \mathbf{E}_\mathrm{e}\,.\,\mathrm{d}\mathbf{P})\,\mathrm{d}\tau. \tag{3.3}$$

When the magnetic induction $\mathbf{B}_\mathrm{e}$ and electric field $\mathbf{E}_\mathrm{e}$ are uniform over the volume, then in terms of the total magnetic and electric polarizations for the system, namely,

$$\mathscr{M} \equiv \int_V \mathbf{M}\,\mathrm{d}\tau, \qquad \mathscr{P} \equiv \int_V \mathbf{P}\,\mathrm{d}\tau,$$

we have

$$\mathrm{d}'W = \mathbf{B}_\mathrm{e}\,.\,\mathrm{d}\mathscr{M} + \mathbf{E}_\mathrm{e}\,.\,\mathrm{d}\mathscr{P}.$$

which involves six scalar deformation coordinates in $\mathscr{M}$ and $\mathscr{P}$.

More generally, with infinitesimal changes of a thermodynamic system, whose state $X(\theta, \mathbf{V})$ depends on the temperature θ and n deformation co-ordinates $\mathbf{V} = (V_1, V_2, \dots V_n)$, the reversible work has the form

$$d'W_r = -\sum_{j=1}^{n} p_j \, dV_j = -\mathbf{p} \cdot d\mathbf{V}, \tag{3.4}$$

where p_j is the (intensive) generalized force associated with the (extensive) generalized displacement V_j, and $\mathbf{p}$ is the vector $(p_1, p_2, \dots p_n)$. The total work is

$$d'W = -\mathbf{p} \cdot d\mathbf{V} + d'W_i, \tag{3.5}$$

where $d'W_i$ is the irreversible work transferred by stirring, electric currents, absence of mechanical equilibrium, etc. Each of the generalized forces will satisfy a thermal state equation similar to (2.8):

$$p = p(\theta, \mathbf{V}). \tag{3.6}$$

2

THE BASIC VARIABLES OF THERMODYNAMICS

So far we have introduced just one variable entirely thermal in origin, namely, the empirical temperature θ. Its appearance depended on the zeroth law and on the use of time scales short enough for the existence of adiabatic walls. The transfer of work to a system takes a finite time, but by choosing the type of wall enclosing it, this transfer can be done in a much shorter time than required for adiabaticity, even when the transfer is undertaken by pistons moving slowly enough to make the process reversible. This fact enables us to introduce further purely thermal variables known as the internal energy U, the empirical entropy φ, the absolute temperature T (replacing θ), and the entropy S (replacing φ).

From a theoretical view the most convenient set of independent coordinates for an observer with a time scale $\mathscr{T}_R$ large enough to admit reversible processes is $\{U, V_1, V_2, .. V_n\}$ or $(U, \mathbf{V})$, where $\mathbf{V}$ is the vector of displacement co-ordinates, hereafter termed the 'deformation vector'. These variables can be controlled in the sense described in § 3.2. Processes occurring on a shorter time scale than $\mathscr{T}_R$ are irreversible, although it may be possible to describe them as 'equilibrium' processes in some extended state space $(U, \mathbf{V}, \mathbf{Y})$. Such processes will be discussed in the next chapter. Much of the subject matter of the present chapter could be termed 'reversible, long-time-scale thermodynamics'; the term 'classical thermodynamics' is often used with essentially this meaning.

4. Internal energy

4.1. The first law of thermodynamics

The existence of a function of state called the internal energy is a direct consequence of the *first law of thermodynamics*, of which there are several equivalent forms. The one adopted here is an immediate induction from Joule's famous paddle-wheel experiments. First we must remember that $X_i \rightarrow X_j$ indicates an irreversible process that commences at the equilibrium state X_i and finishes at the equilibrium state X_j; a reversible process proceeding in the same direction is written $X_i \leftrightarrow X_j$. We shall say that X_j is 'attainable' from X_i; in general all points in state-space are attainable from all other points unless the system P is subject to constraints. The first law can now be stated:

I. *Let* X_1 *and* X_2 *be two arbitrary equilibrium states of an adiabatically enclosed thermodynamic system P, then by the performance of work, one and only one of the processes* $X_1 \rightarrow X_2$, $X_1 \leftrightarrow X_2$, $X_2 \rightarrow X_1$ *is possible; the amount of work transferred is independent of the type of work involved and any other details of the process.*

Work performed on an adiabatically enclosed system is termed 'adiabatic work'.

Suppose, for definiteness, that the process is $X_1 \rightarrow X_2$ and consider a particular case $X_1 \rightarrow X_j \rightarrow X_2$ that includes an intermediate state attainable from X_1 and from which X_2 may be reached. Now the adiabatic work, say W^a_{12} consists of two parts, namely, W^a_{1j} and W^a_{j2}, and from I $W^a_{12} = W^a_{1j} + W^a_{j2}$ must be independent of the intermediate state X_j. This is possible only if W^a_{rs} can be written as $U_s - U_r$, where U_α, $\alpha = r, s$, depends only on X_α. With U_1 now denoting a suitable reference value, the new state variable is defined by

$$U_j = U_1 + W^a_{1j}. \tag{4.1}$$

The same discussion also holds for a reversible process. That the *internal energy* U is an additive function follows immediately from the additivity of work. The functional relation

$$U = U(\theta, \mathbf{V}) \tag{4.2}$$

is termed the *caloric equation of state* of the system P.

4.2. *Empirical temperature as a function of internal energy*

In the second paragraph of § 1.3 it was stated that thermodynamic coordinates may be chosen freely from the state variables, for example, a state $X(\theta, \mathbf{V})$ defined by the empirical temperature θ and deformation coordinates $\mathbf{V} = (V_1, V_2, \ldots V_n)$ could equally well be defined by $\mathbf{V}$ and the internal energy U. For such an interchange of the roles of θ and U to be valid over a wide domain of state-space, we must be able to solve $U = U(\theta, \mathbf{V})$ to obtain a single-valued temperature function $\theta = \theta(U, \mathbf{V})$. That $U(\theta, \mathbf{V})$ is itself single-valued is ensured, of course, by I, but to establish the same property for $\theta(U, \mathbf{V})$ we need to appeal to the *second law of thermodynamics.*

The second law manifests itself in a remarkably large number of forms, and discussion still continues as to which of these forms is 'optimal' in some sense or other. The author's preference is to take a form for the law that is a direct induction from experiment, and then to deduce as consequences other forms commonly adopted. The law that follows is an inference from Joule's experiments, and should be read in conjunction with I.

II. *If the states X_1 and X_2 in I have the same deformation coordinates, then the adiabatic work required for the process $X_1 \rightarrow X_2$ (or $X_2 \rightarrow X_1$) must be strictly positive, that is, work must be done on P to effect the change of state.* (We need not consider $X_1 \leftrightarrow X_2$ for the reversible work is zero, so that P's state cannot be changed by such a process.) By eqn (3.5) the law may be written

$$d'W_{\mathrm{i}}^{\mathrm{a}} > 0, \quad \mathbf{V} = \text{constant}, \tag{4.3}$$

the superscript (a) denoting adiabatic. The constraint $\mathbf{V}$ = constant applies only to the initial and final states, so that intermediate pumping actions in which the deformation coordinates change are permitted.

From (4.1) and (4.3) it follows that if $X_1 \rightarrow X_2$ is a possible process and both X_1 and X_j are on a line L defined by $\mathbf{V}$ = constant, then $W_{1j}^{\mathrm{a}} = U_j - U_1 > 0$, that is, $U_j > U_1$. Hence all points on L attainable from X_1 have greater internal energy. To this stage the scale of θ is not defined in 'sense', so we are free to choose the direction of increasing θ along L to be in the direction of the natural process $X_1 \rightarrow X_j$. With this we have

$$\text{on L:} \quad X_1 \rightarrow X_j \Rightarrow W_{1j}^{\mathrm{a}} = U_j - U_1 > 0, \qquad \theta_j - \theta_1 > 0. \tag{4.4}$$

Notice from (4.4) that two distinct states on L cannot have the same internal energy, for the first law requires one of the states to be attainable from the other, and the second law requires that positive (and not zero) work be transferred in this process. We conclude that, on L, $U(\theta)$ is a strictly increasing function and therefore—save in physically singular regions where the partial derivative may not exist—

$$\left(\frac{\partial U}{\partial \theta}\right)_{\mathbf{V}} > 0. \tag{4.5}$$

It follows that $\theta(U, \mathbf{V})$ is single-valued, so that U may replace θ as a thermodynamic coordinate.

As remarked in § 2.5, a *perfect gas* is one that has the limit properties of real gases as the pressure tends to zero; near $p = 0$ it is found from experiment that U depends only on θ. Thus a perfect gas has the caloric equation of state

$$U = U(\theta). \tag{4.6}$$

Later (§ 10.6) we shall find that the thermal state eqn (2.9) implies a caloric state equation of the type given in (4.6).

4.3. *Internal energy in non-equilibrium processes*

By considering the nature of the adiabatic work W_{12}^{a} required to take a system from a state X_1 to a state X_2, we have been able to introduce the internal energy $U_2 = U_1 + W_{12}^{\mathrm{a}}$ as a function in the state space $X(\mathscr{T})$, where

$\mathscr{T}$ is the observer's time scale. The nature of the process linking X_1 and X_2 plays no role, which raises the important question: Can U be defined for non-equilibrium states?

Suppose $X'(\mathscr{T}')$ is a suitably extended state-space, as described in § 3.2, then the non-equilibrium process $X_1 \to X_2$ may have an intermediate state X'_0, which is non-equilibrium on the time scale $\mathscr{T}(\mathscr{T}' \ll \mathscr{T})$. Let $W^a{}_{10}$ be the work transferred in $X_1 \to X'_0$, and suppose no work is transferred in $X'_0 \to X_2$. Then $W^a_{10} = U_0 - U_1 = W^a_{12} = U_2 - U_1$, so that $U_0 = U_2$. Thus reverting to the longer time scale, there is no need to restrict the concept of internal energy to equilibrium states. Like the deformation vector $\mathbf{V}$, U is a clearly defined physical quantity throughout a non-equilibrium process. This property makes U a more generally useful coordinate than θ, which in general cannot be liberated from the concept of equilibrium.

5. Heat

5.1. Energy transfer through diathermic walls

For an adiabatic process $X_1 \to X_2$ the first law yields $W^a_{12} = U_2 - U_1$. Now let the adiabatic walls surrounding the system P be replaced by diathermic walls, thus placing P in thermal contact with its surroundings P*. In general we can expect the non-adiabatic work W_{12} to differ from W^a_{12}. With the definition

$$Q_{12} = W^a_{12} - W_{12}, \tag{5.1}$$

the first law is generalized to

$$Q_{12} + W_{12} = U_2 - U_1. \tag{5.2}$$

This quantity Q_{12} is the energy that is transferred through the diathermic walls during the process; it is termed the *heat*, and like work is reckoned to be positive if the energy is transferred from P* to P.

For infinitesimal amounts of heat $\mathrm{d}'Q$ and work $\mathrm{d}'W$ transmitted across the boundary of a closed system (5.2) takes the form

$$\mathrm{d}'Q + \mathrm{d}'W = \mathrm{d}U, \tag{5.3}$$

where $\mathrm{d}U$ is the differential of the state variable U. As Q and W depend on interactions between P and P*, unlike U, they are not functions of state and so in general do not have differentials $\mathrm{d}Q$ and $\mathrm{d}W$ in state-space.

5.2. Direction of heat transfer

Let two adiabatically enclosed systems P′ and P″ with initial internal energies U'_1 and U''_1 and different temperatures θ'_1 and θ''_1 be placed in thermal contact in such a way that the combined system P′ + P″ remains adiabatic.

Let the internal energies become U'_2 and U''_2 when thermal equilibrium is recovered. No work is transferred, the only change being a transfer of heat from P′ to P″, which is internal for the combined system P′+P″. Applying (5.2) to P′, P″, and P′+P″ we have

$$Q'_{12} = U'_2 - U'_1, \qquad Q''_{12} = U''_2 - U''_1, \qquad 0 = (U'_2 + U''_2) - (U'_1 + U''_1),$$

on using the additivity property of internal energy. Suppose that P′ receives heat from P″, that is that $Q'_{12} > 0$, then $U'_2 > U'_1$, so that the internal energy of P′ is increased. From eqn (4.5) this means that $\theta'_2 > \theta'_1$, and by a similar argument $\theta''_2 < \theta''_1$. In the final state $\theta'_2 = \theta''_2$, whence in the initial state $\theta'_1 < \theta''_1$ and the heat has been transferred from the system of higher initial temperature (P″) to the system of lower initial temperature (P′). At the initial state we say that P″ is 'hotter' than P′, or that P′ is 'cooler' than P″. The process just described is a non-equilibrium process, for when P′ and P″ are initially placed in thermal contact, the combined system P′+P″ is not in thermodynamic equilibrium. Also the process is certainly not reversible, for if P′ and P″ are separated by an adiabatic wall there is no way in which their original temperature difference can be restored without assistance from their surroundings.

The concept of reversible processes, described in § 3.2 for work transfer, may also be applied to heat transfer. Suppose heat is to be transferred from P* to P in such a way that P remains in thermodynamic equilibrium throughout the process. Then P* should be only infinitesimally hotter than P, so that the direction of heat flow could be reversed by an infinitesimal temperature reduction of P* to make it cooler than P. Thus in a reversible process involving both heat and work transfer between P and P*, it is necessary for P and P* to be in *both* mechanical and thermal equilibrium throughout the process. We shall denote by $d'Q_r$ an element of heat transferred reversibly between P and P*.

Eqns (3.4), (3.5), and (5.3) yield

$$d'Q + d'W_i = dU - d'W_r = dU + \mathbf{p} \, . \, d\mathbf{V}, \tag{5.4}$$

and in particular for a simple system

$$d'Q + d'W_i = dU + p\,dV, \tag{5.5}$$

where $d'W_i = 0$, $d'Q = d'Q_r$, for a reversible process.

5.3. The distinction between work and heat

The state of a thermodynamic system P may be changed by a transfer of work or heat; indeed we have defined heat to be all these effects (like conduction, convection, and radiation) able to change the state of P that are not *recognized* as being work, and the task of precise definition has been diverted from the concept of 'heat' to the concept of 'adiabaticity'. This indirect approach, due to Carathéodory, has largely superseded an earlier treatment due to Joule and Thomson in which heat as a form of energy is accepted as a primitive concept and in which eqn (5.2) simply expresses the law of conservation of energy. Carathéodory's method is precise for closed systems, where impermeable adiabatic walls have some relation to physical reality, but, as we shall see later (§ 53.2) for open systems consisting of several diffusing phases, the distinction between work and heat cannot be drawn so clearly. This would not matter if heat were merely equivalent to work, as appears to be the case from the first law. However, the second law, in one of its important formulations, is concerned with the qualitative *non-equivalence* of heat and work, and so the distinction is very important. Before explaining this further it is convenient to introduce a new function of state called the 'entropy'.

6. The empirical entropy

6.1. *Isentropic surfaces*

In § 4 we were concerned with states X_2 (or X_j) attainable from a given state X_1 by adiabatic processes that could be either irreversible or reversible. Here we shall restrict attention to *reversible* adiabatic processes, and *define* states that can be attained from X_1 by such processes to be *isentropic* with it.

Suppose that the state X_2 is isentropic with X_1 and let L_1, L_2 denote the lines $\mathbf{V}$ = constant in $(U, \mathbf{V})$-state-space passing through X_1, X_2. First we shall prove that there are no other points on L_2 isentropic with X_1. For suppose X'_2 on L_2 is also isentropic with X_1, and for definiteness take U'_2 to be greater than U_2 as shown in Fig. 3. Then the reversible path $X'_2 \leftrightarrow X_1 \leftrightarrow X_2$ results in a reduction of the internal energy of the system P and, by

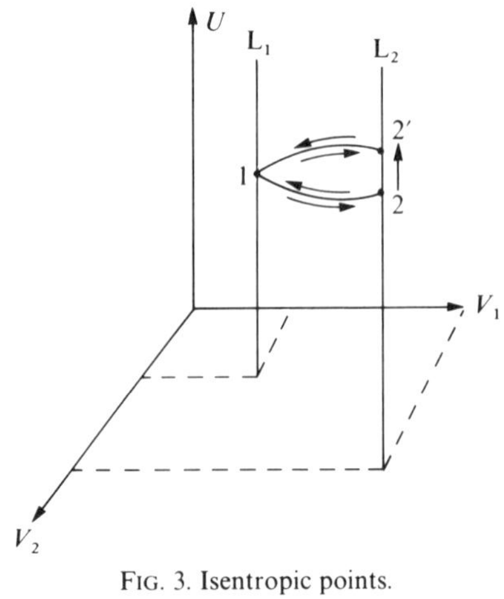

FIG. 3. Isentropic points.

(4.1), the performance of *negative* work. As the deformation coordinates are the same at the initial and final states, we have a contradiction of the second law of thermodynamics and the hypothesis that X'_2 is also isentropic with X_1 is false.

Now suppose that the line L_2 is displaced parallel to itself so that its intersection with the hyperplane U = constant generates a connected region lying in that plane. By virtue of the continuity to be expected of the physical behaviour of the system—if not everywhere, then at least over wide regions of state-space—the single point on L_2 isentropic with X_1 will also generate a coherent hypersurface. It follows that with a set of states $X_1, X'_1, X''_1, \ldots$ along L_1 we may associate a set of isentropic hypersurfaces φ_1 = constant, φ'_1 = constant, φ''_1 = constant, ... , where φ is a conveniently chosen isentropic 'potential'. These surfaces of constant potential cannot intersect, for if so it would be possible to find two or more points on a line parallel to the U-axis isentropic with a single point on another similar line passing through the intersection. The actual numbers φ attached to the isentropic surfaces may be arbitrary, provided, of course, that φ is a unique function of position $X = (U, \mathbf{V})$ in state-space, just as described on p. 9 for the empirical temperature.

6.2. *Differential of the empirical entropy*

The potential $\varphi = \varphi(U, \mathbf{V})$ is a new state function, often called the *empirical entropy* to distinguish it from the entropy function $S(\varphi)$ to be defined in the next article. To move from one value of φ to another it is necessary to violate either the irreversibility or the adiabaticity assumed in the generation of the hypersurfaces φ = constant. This point is manifest in the eqn (5.4), which for an isentropic surface reduces to the differential form—known as a 'Pfaffian'—

$$\mathrm{d}U + \mathbf{p} \cdot \mathrm{d}\mathbf{V} = 0. \tag{6.1}$$

It is important to distinguish clearly between adiabatic processes ($\mathrm{d}'Q = 0$), reversible processes ($\mathrm{d}'W_\mathrm{i} = 0$, $\mathrm{d}'Q = \mathrm{d}'Q_\mathrm{r}$), and isentropic processes ($\mathrm{d}'Q = 0$, $\mathrm{d}'W_\mathrm{i} = 0$).

With vectors defined by $\mathbf{R} = (1, \mathbf{p})$, $\mathbf{r} = (U, \mathbf{V})$, $\mathrm{d}\mathbf{r} = (\mathrm{d}U, \mathrm{d}\mathbf{V})$, (6.1) can be written

$$\mathbf{R} \cdot \mathrm{d}\mathbf{r} = 0. \tag{6.2}$$

As the intensive variables $\mathbf{p}$ are state functions, $\mathbf{R}$ is a vector function of $\mathbf{r}$. All curves satisfying (6.2) lie on isentropic surfaces φ = constant, on which $\mathrm{d}\varphi = \nabla\varphi \cdot \mathrm{d}\mathbf{r} = 0$. It follows that $\mathbf{R}$ and $\nabla\varphi$ are parallel, that is, there exists a scalar function $\lambda(\mathbf{r})$ such that $\mathbf{R} = \lambda\,\nabla\varphi$. Hence

$$\mathbf{R} \cdot \mathrm{d}\mathbf{r} = \lambda\,\mathrm{d}\varphi = \mathrm{d}U + \mathbf{p} \cdot \mathrm{d}\mathbf{V}, \tag{6.3}$$

which provides an expression for the change in φ between neighbouring isentropic surfaces. For a non-adiabatic irreversible process $\mathrm{d}U + \mathbf{p}\,.\,\mathrm{d}\mathbf{V} = \mathrm{d}'Q + \mathrm{d}'W_{\mathrm{i}}$, whence

$$\lambda\,\mathrm{d}\varphi = \mathrm{d}'Q + \mathrm{d}'W_{\mathrm{i}} = \mathrm{d}U + \mathbf{p}\,.\,\mathrm{d}\mathbf{V}. \tag{6.4}$$

6.3. Empirical entropy as a thermodynamic coordinate

Now consider an adiabatic path in $(U, \mathbf{V})$-state-space from X_2 to X_1 (Fig. 3) and then back to a point X_2'' on L_2 by an irreversible adiabatic process. As $\mathrm{d}'W_{\mathrm{i}} = 0$ on $X_2 \leftrightarrow X_1$ and $\mathrm{d}'W_{\mathrm{i}} \neq 0$ on $X_1 \rightarrow X_2''$, then $\mathrm{d}'W_{\mathrm{i}} \neq 0$ for the combined process. By the second law of thermodynamics (p. 17) $\mathrm{d}'W_{\mathrm{i}} > 0$, but by (6.4) in an adiabatic process $\mathrm{d}'W_{\mathrm{i}} = \lambda\mathrm{d}\varphi > 0$. If λ were to vanish, so would $\mathrm{d}'W_{\mathrm{i}}$ in contradiction to law II, so either $\lambda > 0$ or $\lambda < 0$. We shall adopt $\lambda > 0$, then by (6.4)

$$\left(\frac{\partial\varphi}{\partial U}\right)_{\mathbf{V}} = \frac{1}{\lambda} > 0; \qquad \left(\frac{\partial\varphi}{\partial V_j}\right)_{U, V_{i \neq j}} = \frac{p_j}{\lambda} \qquad (i, j = 1, 2, \dots n). \tag{6.5}$$

Thus at constant $\mathbf{V}$, φ is a strictly increasing function of U and could replace U (or θ) as a thermodynamic coordinate. Also, for deformable systems φ will depend on at least one of the V_j, say V_1, and as φ is a unique function of $(U, V_1, V_2, \dots)$, the independent variable V_1 can be freely replaced by φ, for example, the coordinates $(\theta, V_1, \dots V_n)$ may be replaced by $(\theta, \varphi, V_2, \dots V_n)$.

6.4. Second law for non-adiabatic processes

The content of the second law contained in eqn (4.3)—namely, $\mathrm{d}'W_{\mathrm{i}}^{\mathrm{a}} > 0$ if $\mathbf{V} = $ constant—can be extended to non-adiabatic processes as follows. Let the system P be in thermal contact with another system R, and let the composite system P+R be adiabatically enclosed. Consider a reversible process $X_1 \leftrightarrow X_2'$ and an irreversible process $X_1 \rightarrow X_2$ for system P, in each of which the same heat $\mathrm{d}'Q$ is transferred from R to P and with X_2, X_2' having the same deformation coordinates. No work is transferred to R in these processes. Now reverse the first process and consider the combined process $X_2' \leftrightarrow X_1 \rightarrow X_2$. Eqn (6.4) gives $\mathrm{d}'Q + \mathrm{d}'W_{\mathrm{i}} = \mathrm{d}U + \mathbf{p}\,.\,\mathrm{d}\mathbf{V}$ for P in $X_1 \rightarrow X_2$. However, for R+P in the *adiabatic* process $X_2' \leftrightarrow X_1 \rightarrow X_2$, $\mathrm{d}'W_{\mathrm{i}} > 0$, and as this is the same dissipated work experienced by P in the non-adiabatic process $X_1 \rightarrow X_2$, it follows from (6.4) that for P

$$\mathrm{d}'Q < \lambda\,\mathrm{d}\varphi, \qquad \mathrm{d}'W_{\mathrm{i}} > 0. \tag{6.6}$$

7. Absolute temperature and entropy

7.1. Absolute temperature

The arbitrariness in the functions θ and φ can be removed by considering the thermal coupling between two sub-systems P′, P″ of a system P. Let the sub-systems have state coordinates $(\theta, \mathbf{V}')$ and $(\theta, \mathbf{V}'')$, then P has the coordinates $(\theta, \mathbf{V}', \mathbf{V}'')$. By the theory given in § 6 empirical entropies φ, φ', and φ'' exist for P, P′, and P″, where the states of P′, P″ may be written as $X'(\theta, \varphi', V'_2, \ldots V'_n)$, $X''(\theta, \varphi'', V''_2, \ldots V''_n)$. For an infinitesimal reversible process it follows from (6.4) that

$$\lambda'\,\mathrm{d}\varphi' = \mathrm{d}'Q'_{\mathrm{r}}, \qquad \lambda''\,\mathrm{d}\varphi'' = \mathrm{d}'Q''_{\mathrm{r}},$$

$$\lambda\,\mathrm{d}\varphi = \mathrm{d}'Q_{\mathrm{r}} = \mathrm{d}'Q'_{\mathrm{r}} + \mathrm{d}'Q''_{\mathrm{r}} = \lambda'\,\mathrm{d}\varphi' + \lambda''\,\mathrm{d}\varphi''.$$

Hence

$$\mathrm{d}\varphi = (\lambda'/\lambda)\,\mathrm{d}\varphi' + (\lambda''/\lambda)\,\mathrm{d}\varphi'', \tag{7.1}$$

where λ', λ'' depend on X', X'' and λ, φ depend on the state $X(\theta, \varphi', \varphi'', V'_2, V''_2, \ldots)$. The absence of differentials $\mathrm{d}\theta, \mathrm{d}V'_2, \mathrm{d}V''_2, \ldots$ from the right-hand side of (7.1) means that in fact φ depends only on φ' and φ''. Thus $\partial\varphi/\partial\theta = 0$ and

$$\varphi = \varphi(\varphi', \varphi''). \tag{7.2}$$

Now

$$\frac{\partial\varphi}{\partial\varphi'} = \frac{\lambda'}{\lambda}, \qquad \frac{\partial^2\varphi}{\partial\varphi'\partial\theta} = \frac{\partial}{\partial\theta}\left(\frac{\lambda'}{\lambda}\right) = 0,$$

and similarly

$$\frac{\partial}{\partial\theta}\left(\frac{\lambda''}{\lambda}\right) = 0.$$

Hence

$$\frac{\partial}{\partial\theta}(\ln\lambda) = \frac{\partial}{\partial\theta}(\ln\lambda') = \frac{\partial}{\partial\theta}(\ln\lambda''),$$

which expressions can be equal only if they depend on θ alone. Therefore we have

$$\left.\begin{aligned} \ln\lambda &= \int f(\theta)\,\mathrm{d}\theta + \ln\Phi(\varphi', \varphi'', V'_2, V''_2, \ldots),\\ \ln\lambda' &= \int f(\theta)\,\mathrm{d}\theta + \ln\Phi'(\varphi', V'_2, \ldots),\\ \ln\lambda'' &= \int f(\theta)\,\mathrm{d}\theta + \ln\Phi''(\varphi'', V''_2, \ldots), \end{aligned}\right\} \tag{7.3}$$

where Φ, Φ', and Φ'' are functions of all the independent variables of P, P′, and P″ respectively, *except* the temperature θ. The function $f(\theta)$ is a *universal* function, numerically the same for the three systems and the integrations are from some standard empirical temperature θ_0 assumed to be the same for all physical systems.

The *absolute temperature* $T(\theta)$ is now defined by

$$T(\theta) = C \exp\left[\int_{\theta_0}^{\theta} f(\theta)\,\mathrm{d}\theta\right], \tag{7.4}$$

where C is a scale factor that can be assigned by prescribing the value of $(T_1 - T_2)$ for two reproducible states X_1 and X_2 of a convenient thermometric body.

7.2. Entropy

The entropy for P′ is defined by

$$S' = \frac{1}{C}\int \Phi'\,\mathrm{d}\varphi' + \text{constant}, \tag{7.5}$$

with a corresponding formula for S'', and it remains to be shown that the entropy S for the complete system can be defined similarly. From (7.3) it follows that consistency requires $\Phi(\varphi', \varphi'', V_2', V_2'', \dots)$ to be reducible to $\Phi(\varphi, V_2', V_2'', \dots)$, so that by (7.2) Φ depends on φ', φ'' only through φ. From (7.1) and (7.3)

$$\Phi\,\mathrm{d}\varphi = \Phi'\,\mathrm{d}\varphi' + \Phi''\,\mathrm{d}\varphi'',$$

hence

$$\Phi\varphi_{\varphi'} = \Phi', \qquad \Phi\varphi_{\varphi''} = \Phi'',$$

$$\Phi_{\varphi''}\varphi_{\varphi'} + \Phi\varphi_{\varphi'\varphi''} = 0, \qquad \Phi_{\varphi'}\varphi_{\varphi''} + \Phi\varphi_{\varphi'\varphi''} = 0,$$

that is,

$$\Phi_{\varphi'}/\varphi_{\varphi'} = \Phi_{\varphi''}/\varphi_{\varphi''} = \Phi_{\varphi},$$

which shows that Φ depends on φ' and φ'' only in the combination $\varphi(\varphi', \varphi'')$. For the complete system we can therefore take S to be defined by

$$S = \frac{1}{C}\int \Phi\,\mathrm{d}\varphi + \text{constant}. \tag{7.6}$$

7.3. S and T independent of φ and θ

That the absolute temperature is independent of the thermometric body used to define the empirical temperature scale is easily shown. For suppose $\hat{\theta}$ is a different scale, then the theory leading to (7.3) and (7.4) is unchanged, save that $f(\theta)$ is replaced by another function of $\hat{\theta}$, say $g(\hat{\theta})$. Now (see § 2.4)

θ is a non-decreasing function of $\hat{\theta}$, so that

$$f(\theta)\,\mathrm{d}\theta = \frac{\partial(\ln\lambda)}{\partial\theta}\frac{\partial\theta}{\partial\hat{\theta}}\,\mathrm{d}\hat{\theta} = \frac{\partial(\ln\lambda)}{\partial\hat{\theta}}\,\mathrm{d}\hat{\theta} = g(\hat{\theta})\,\mathrm{d}\hat{\theta},$$

which when substituted in (7.4) yields $T(\theta) = T(\hat{\theta})$. Notice that if T is a non-decreasing function of a given empirical temperature scale θ, it will have this property for any other scale $\hat{\theta}$; and we need only consider the perfect gas thermometer, for which it is possible to select a scale such that $T = \theta$ (see § 8), to conclude that T is a single-valued function of all scales θ. Thus T may be used freely in place of θ throughout the theory.

Now suppose that a different empirical entropy function $\hat{\varphi}$ is used in place of φ in (7.6). As $\mathrm{d}'Q_\mathrm{r} = \lambda\,\mathrm{d}\varphi$ for a reversible process, we have $\mathrm{d}'Q_\mathrm{r} = \lambda\,\mathrm{d}\varphi = \hat{\lambda}\,\mathrm{d}\hat{\varphi}$, where $\hat{\lambda}$ is a new factor. Now $\mathrm{d}\varphi/\mathrm{d}\hat{\varphi}$ is independent of the empirical temperature, and hence the new universal function $\hat{f}(\theta)$ satisfies

$$\hat{f}(\theta) = \frac{\partial(\ln\hat{\lambda})}{\partial\theta} = \frac{\partial}{\partial\theta}\left(\ln\lambda + \ln\frac{\mathrm{d}\varphi}{\mathrm{d}\hat{\varphi}}\right) = \frac{\partial(\ln\lambda)}{\partial\theta} = f(\theta).$$

Thus, provided C and θ_0 in (7.4) are unchanged, the temperature T is unaffected by the change. Furthermore, $\lambda = \Phi T/C$, $\hat{\lambda} = \hat{\Phi}T/C$, so that $\Phi\,\mathrm{d}\varphi = \hat{\Phi}\,\mathrm{d}\hat{\varphi}$ and provided the constant in (7.6) is unchanged, the entropy S is independent of the choice of the empirical entropy scale. Unlike T, however, it does depend on the system under consideration.

7.4. *Basic thermodynamic equations for closed systems*

With the above definitions of temperature and entropy it follows that $\lambda'\,\mathrm{d}\varphi' = T\,\mathrm{d}S'$, $\lambda''\,\mathrm{d}\varphi'' = T\,\mathrm{d}S''$, $\lambda\,\mathrm{d}\varphi = T\,\mathrm{d}S$, and as $\lambda\,\mathrm{d}\varphi = \lambda'\,\mathrm{d}\varphi' + \lambda''\,\mathrm{d}\varphi'' = T\,\mathrm{d}S' + T\,\mathrm{d}S''$, we have established the additive property of entropy, $\mathrm{d}S = \mathrm{d}S' + \mathrm{d}S''$, for systems at the same temperature.

Eqn (6.4) can now be written

$$\mathrm{d}'Q + \mathrm{d}'W_\mathrm{i} = T\,\mathrm{d}S = \mathrm{d}U + \mathbf{p}\,.\,\mathrm{d}\mathbf{V}, \tag{7.7}$$

in which by (6.6)

$$T\,\mathrm{d}S \geqslant \mathrm{d}'Q, \qquad \mathrm{d}'W_\mathrm{i} \geqslant 0, \tag{7.8}$$

the equality holding only for reversible processes.

The first of eqns (6.5) takes the form

$$\left(\frac{\partial S}{\partial U}\right)_\mathbf{V} = \frac{1}{T} > 0, \tag{7.9}$$

and the inequality in (4.5) gives

$$\left(\frac{\partial U}{\partial T}\right)_\mathbf{V} > 0. \tag{7.10}$$

This inequality is often termed the *thermal stability* condition, for the supply of heat to a colder region P from a region at temperature T_s, by raising the internal energy of P, also *increases* its temperature T by virtue of (7.10) and brings it closer in value to T_s. Were the inequality reversed the equilibrium at $T = T_s$ would clearly be unstable.

The thermal equation of state (3.6) and the caloric equation of state (4.2) may now be written as

$$\mathbf{p} = \mathbf{p}(T, \mathbf{V}) \tag{7.11}$$

and

$$U = U(T, \mathbf{V}). \tag{7.12}$$

7.5. Entropy in non-equilibrium states

In § 4.3 the internal energy U was shown to have a precise physical value even for non-equilibrium states. This result followed from the fact that the energy difference $(U_2 - U_1)$ is (1) independent of the nature of the process $X_1 \rightarrow X_2$,† and (2) equal to the adiabatic work transferred during the process. The question now arises: Can entropy also be defined precisely for non-equilibrium states? As we shall see, the answer is *no*, which implies that S is not a suitable *independent* variable in general accounts of thermodynamics, which may include non-equilibrium situations.

The entropy function S is introduced into state-space X via reversible adiabatic processes $X_1 \leftrightarrow X_2$, for which $S_2 - S_1 = 0$. As $S = S(X)$, then for any two given states X_1 and X_2, the difference $S_2 - S_1$ is precisely determined, and thus, like $(U_2 - U_1)$, is independent of the nature of the process $X_1 \rightarrow X_2$. However, unlike U, S has no physical significance separable from its being a state variable in $X(\mathscr{T})$. If, as in § 4.3, we introduce an intermediate state $X'_0(\mathscr{T}')$, which is non-equilibrium on the longer time scale $\mathscr{T}$, we may certainly envisage an entropy state variable $S'(X')$, but as X' is undefined on the time scale $\mathscr{T}$, the value of S' will be unknown to the observer Q. In other words the actual *value* of S depends on Q's time scale $\mathscr{T}$, whereas this is not true of U. Of course, when U is taken to be a dependent variable, the actual form of its functional dependence will change with $\mathscr{T}$, but this is quite a different matter. Further remarks on the question of entropy in non-equilibrium states will be found in § 26.4.

8. Degeneracy of simple systems

From eqn (5.5) the isentropes $(\mathrm{d}'Q, \mathrm{d}'W_\mathrm{i} = 0)$ of a simple system satisfy

$$\mathrm{d}U + p(U, V)\,\mathrm{d}V = 0,$$

‡ Used here to denote any of the processes shown in Fig. 1.

that is, an ordinary differential equation of which the solution is a one-parameter set of isentropic curves $\varphi(U, V) = \text{constant}$. We conclude that in the two-dimensional state-space of a simple system an empirical entropy exists despite the fact that we have made no use of the second law of thermodynamics (cf. § 6.1). However, the second law is still required to establish the existence of the absolute temperature T and the entropy S.

To illustrate, take the case of a perfect gas, for which U depends only on θ and the isotherms are given by Boyle's law $pV = F(\theta)$, where $F(\theta)$ is an experimentally determined function. Let

$$\psi = \ln V + \int \frac{\mathrm{d}U}{\mathrm{d}\theta}\frac{\mathrm{d}\theta}{F(\theta)}, \qquad \varphi = f(\psi), \qquad \lambda = \frac{F(\theta)}{f'(\psi)}, \tag{8.11}$$

where f is an arbitrary function, then $\mathrm{d}U + p\,\mathrm{d}V = F(\theta)\,\mathrm{d}\psi = \lambda\,\mathrm{d}\varphi$. The arbitrariness in λ and φ can be removed only by applying the second law of thermodynamics to a compound system of two simple systems in thermal contact (as in Fig. 2) then reaching the conclusion, as in § 7, that it must be possible to choose f so that $\lambda(\theta, \psi)\,\mathrm{d}\varphi(\theta, \psi) = T(\theta)\,\mathrm{d}S(\psi)$. In (8.1) take $f = nR\psi$, where n is the number of moles of the gas and R is a constant, then

$$\lambda = F(\theta)/nR = T(\theta), \qquad S = nR\psi$$

or

$$S = R\ln V + \int^{T} \frac{\mathrm{d}U}{\mathrm{d}T}\frac{\mathrm{d}T}{T} + \text{constant} \tag{8.2}$$

and

$$pV = nRT. \tag{8.3}$$

In particular when (2.10) applies, it follows that $T = \theta$.

It may be thought that, just as in the two-dimensional case, the existence of empirical entropy for the n-dimensional state-space ($n > 2$) could be established without reference to the second law (II). This would mean that an 'integrating divisor' λ would always exist to convert the Pfaffian $\mathbf{R}\,.\,\mathbf{dr}$ defined in § 6.2 into an exact differential $\mathrm{d}\varphi$. That this is not so can be shown as follows. Consider an infinitesimal change in which all but three scalar components of $\mathbf{dr}$ are zero. Denote by $\mathbf{r}'$ the three-dimensional sub-space of state-space in which the change takes place and let ∇' be the gradient operator in $\mathbf{r}'$-space. Then if λ exists such that $\lambda\,\mathrm{d}\varphi = \mathbf{R}'\,.\,\mathbf{dr}'$, then $\mathbf{R}' = \lambda\nabla'\varphi$, where $\mathbf{R}'$ is a three-dimensional vector in $\mathbf{r}'$-space. Now $\nabla' \times \mathbf{R}' = \nabla'\lambda \times \nabla'\varphi$ and

$$\mathbf{R}'\,.\,\nabla' \times \mathbf{R}' = 0, \tag{8.4}$$

a constraint that must hold for all possible choices of the $\mathbf{r}'$-space. Eqn (8.4) is a necessary condition for $\mathbf{R}\,.\,\mathrm{d}\mathbf{r}$ to possess an integrating divisor λ; it may also be shown to be sufficient.

In a three-dimensional state-space $\mathbf{r}' = (x, y, z)$ say, in which $\mathbf{R}' = (X, Y, Z)$, where X, Y, Z each depend on (x, y, z), (8.4) takes the form

$$X(Z_y - Y_z) + Y(X_z - Z_x) + Z(Y_x - X_y) = 0. \tag{8.5}$$

When the Pfaffian expression, namely, $X\,\mathrm{d}x + Y\,\mathrm{d}y + Z\,\mathrm{d}z$, has one term absent, say the last, then (8.5) reduces to $XY_z = YX_z$, that is,

$$\frac{\partial}{\partial z}(\ln X) = \frac{\partial}{\partial z}(\ln Y). \tag{8.6}$$

It was this case, or rather a generalization of it, that occurred in § 7.1. Finally, the degeneracy of the two-dimensional case is clear from (8.6), for when X and Y are independent of Z this necessary condition is always satisfied.

9. The second law of thermodynamics

9.1. The Kelvin–Planck form

The form of the second law given in § 4 is seldom the one adopted in textbooks, so it is useful to consider other forms of the law at this stage. The Kelvin–Planck statement, namely,

> *No cyclic process is possible whose sole result is the absorption of heat from a reservoir and the conversion of this heat into work,*

is the one most often encountered. What always happens in fact is that the system undergoing the cyclic process *rejects* heat during part of the cycle.

The first law may be regarded from one standpoint as expressing the quantitative equivalence of heat and work. On the other hand, the second law is concerned with the qualitative *non*-equivalence of heat and work, namely that, whereas it is possible to convert all of the work transferred into a system into heat, it is not possible to convert all of the heat transferred into a system into work.

The equivalence of the Kelvin–Planck (KP) form of the law with that labelled II on p. 17 is easily established. Suppose at first that law II is true for some system P, then the consequences $\lambda\,\mathrm{d}\varphi = \mathrm{d}'Q + \mathrm{d}'W_{\mathrm{i}}$, $\mathrm{d}'W_{\mathrm{i}} > 0$, $\lambda > 0$ (see § 6) of this assumption show that, while φ can be increased by supplying *either* heat *or* irreversible work to P, it can be reduced only by removing heat from P. As φ must regain its initial value after a cyclic process, it is clearly impossible to avoid the rejection of heat from P. Conversely suppose that the KP law is true and consider a cyclic process $X_1 \to X_1' \to X_1$ on a line L without change in deformation coordinates. If heat is absorbed by P in $X_1 \to X_1'$ then the law prevents adiabatic work being performed in $X_1' \to X_1$, whereas if positive adiabatic work is done on P in $X_1 \to X_1'$, heat can be *rejected* in $X'_1 \to X_1$. For the classical development of thermodynamics on the basis of the KP law the reader is referred to a text such as Zemansky's (1968) excellent account.

9.2. *Carathéodory's principle*

Another formulation of the second law, due to Carathéodory (1909), is more abstract in form:

In the neighbourhood of each equilibrium state of a thermodynamic system there exist states that are inaccessible by adiabatic processes.

This is an immediate consequence of law II because the process $X_1 \to X'_1$ mentioned above cannot be reversed adiabatically, however close X_1 and X'_1 are, that is, X'_1 is a point in the neighbourhood of X_1 that is adiabatically inaccessible from it.

Although Carathéodory's 'principle' has had a great influence on the metaphysics of thermodynamics, it has not replaced the KP law to any extent. There seem to be two main reasons for this. First, it is neither intuitively obvious nor much supported by *direct* experimental evidence, and secondly the analysis required to show that it restricts the Pfaffian $\mathbf{R} \cdot \mathbf{dr}$ defined in § 6 to the form $\lambda \, d\varphi$ is difficult for the average undergraduate. The treatment given in § 6 avoids these difficulties. Once the existence of empirical entropy is established, the further analysis leading to S and T, given in § 7 and also due to Carathéodory, is straightforward.

9.3. *The status of the equation for* $T \, dS$

The basic relation between S, U, $\mathbf{V}$, $\mathbf{p}$, and T, namely,

$$T \, dS = dU + \mathbf{p} \cdot d\mathbf{V}, \tag{9.1}$$

is termed the 'Gibbs relation' or sometimes the 'fundamental equation of thermodynamics'. It and various generalizations given later play a central role in thermodynamics, and it is therefore important to understand the sort of relation it is.

The term $\mathbf{p} \cdot d\mathbf{V}$ is a general term for reversible work; to give it physical substance we must specify the nature of the physical system P, that is, make a constitutive assumption about P based on experimental evidence. With such an assumption $\mathbf{p}$ and $\mathbf{V}$ are primitive quantities (relative to the subject of thermodynamics) that may be obtained at any instant by measurements made on P. Of course, $\mathbf{p}$ and $\mathbf{V}$ need not be directly measured, but may be inferred from measurements of other quantities. The variable U is not a primitive quantity but is defined in terms of such a quantity, namely, the adiabatic work transferred. Thus the value of the right-hand side of (9.1) may be inferred from measurements on P, and it is irrelevant as to whether P is in a temporal equilibrium state or not, although spatial equilibrium (uniformity) is necessary.

The term $T\,dS$ is quite different in character from those on the right-hand side of (9.1). If this were not so, that is, if $T\,dS$ could be measured as a physical quantity, then (9.1) would be a 'law of nature', albeit not a general one because of the constitutive assumption about P. But certainly S cannot be measured for a physical system, so (9.1) is not a physical law. The temperature T, however, can be inferred from various measurements, the theories of which rely on the concept of equilibrium state, short of which T has no meaning. Of course we are using 'equilibrium' here in the general sense introduced in §§ 1.2 and 3.2, that is, to mean a state-space adequate to describe the evolution of the system for a given process. And, as we shall see, this concept can be pushed to shorter and shorter time scales, so that in fact the equilibrium requirement for T is not very restrictive, although it cannot be ignored. In the kinetic theory of a dilute gas (§ 23) it emerges that T is proportional to U, that is, is apparently independent of the concept of equilibrium, but this is a very degenerate case.

Even if we could talk of a temperature *value*, without necessarily being clear about the associated equilibrium state, this is not possible with entropy, a point that will be developed in the next chapter. Here we shall just remark that the status of (9.1) is that it defines the *value* of S, and that as its right-hand side is generalized to more complicated state-spaces (see §§ 16.1, 21.3) the definition of S is accordingly changed. The dimensionality of state-space increases as the observer's time scale $\mathscr{T}$ is reduced and, as we shall see later, this results in a reduction in the actual value of S. The *existence* of S as a function of state is of course a consequence of the second law of thermodynamics.

10. Heat capacities, compressibility, and other derivatives

10.1. The basic partial derivatives

Partial derivatives play a dominant role in practical applications of thermodynamics, and it is useful at this stage to give an account of these for a simple system P. In this system

$$T\,dS = dU + p\,dV \tag{10.1}$$

or

$$T\,dS = dH - V\,dp, \tag{10.2}$$

where $H = U + pV$, is a new state function called the *enthalpy*. There are also the two state equations

$$p = p(T, V), \qquad U = U(T, V), \tag{10.3}$$

although, as we shall see, it is unnecessary to be given both relations completely to fully specify the thermodynamics of P.

Let $d'Q_r = T\,dS$ be the heat reversibly supplied to P while its temperature increases by dT, with either p or V held constant during the process, then the *heat capacities* C_X are defined by

$$C_X = (d'Q_r/dT) = T(dS/dT)_X \qquad (X = p \text{ or } V).$$

From (10.1) and (10.2) it follows that

$$C_V = T\left(\frac{\partial S}{\partial T}\right)_V = \left(\frac{\partial U}{\partial T}\right)_V; \qquad C_p = T\left(\frac{\partial S}{\partial T}\right)_p = \left(\frac{\partial H}{\partial T}\right)_p. \tag{10.4}$$

Three other derivatives that commonly appear in applications are the *isothermal* and *isentropic compressibilities* κ_T and κ_S, and the *coefficient of thermal expansion* β, which are defined by

$$\kappa_T \equiv -\frac{1}{V}\left(\frac{\partial V}{\partial p}\right)_T, \qquad \kappa_S \equiv -\frac{1}{V}\left(\frac{\partial V}{\partial p}\right)_S, \qquad \beta \equiv \frac{1}{V}\left(\frac{\partial V}{\partial T}\right)_p. \tag{10.5}$$

Only three of the five quantities just defined are independent because of the relations

$$\kappa_T/\kappa_S = C_p/C_V, \tag{10.6}$$

and

$$C_p - C_V = TV\beta^2/\kappa_T, \tag{10.7}$$

which will be established shortly. We shall find in fact that *all* the partial derivations can be expressed in terms of a standard set of just three independent partial derivatives.

10.2. Maxwell relations

The following mathematical identities are used to manipulate partial derivations into alternative forms,†

$$\left(\frac{\partial x}{\partial y}\right)_z = 1\bigg/\left(\frac{\partial y}{\partial x}\right)_z, \qquad \left(\frac{\partial x}{\partial y}\right)_z = \left(\frac{\partial x}{\partial w}\right)_z\bigg/\left(\frac{\partial y}{\partial w}\right)_z,$$
$$\left(\frac{\partial x}{\partial y}\right)_z = -\left(\frac{\partial z}{\partial y}\right)_x\bigg/\left(\frac{\partial z}{\partial x}\right)_y. \tag{10.8}$$

† Using an abbreviated notation $(\partial a/\partial b)_c = a_{b/c}$, we have for functions of two variables, $dx = x_{y/z}\,dy + x_{z/y}\,dz$, $dy = y_{x/z}\,dx + y_{z/x}\,dz$, $dw = w_{z/y}\,dz + w_{y/z}\,dy$. Eliminating dy from the first pair, $(1 - x_{y/z}y_{x/z})\,dx = (x_{y/z}y_{z/x} + x_{z/y})\,dz$, and the differentials dx, dz set equal to zero in turn yield the first and third of eqns (10.8) respectively. Eliminating dy from the second pair, $dw = w_{y/z}y_{x/z}\,dx + (w_{z/y} + w_{y/z}y_{z/x})\,dz$, the first coefficient of which gives $w_{x/z} = w_{y/z}y_{x/z}$ or by the first of eqns (10.8) $x_{w/z} = y_{w/z}x_{y/z}$, which is the same as the second of eqns (10.8). Eqn (10.9) follows directly on noting that $\varphi_{x/y} = X$, $\varphi_{y/z} = Y$, and $\varphi_{xy} = \varphi_{yx}$.

We shall also use the result that $X\,dx + Y\,dy$ is an exact differential $d\varphi$ if and only if

$$\left(\frac{\partial X}{\partial y}\right)_x = \left(\frac{\partial Y}{\partial x}\right)_y, \qquad (d\varphi = X\,dx + Y\,dy). \tag{10.9}$$

The Maxwell relations

$$\begin{aligned}\left(\frac{\partial V}{\partial S}\right)_p &= \left(\frac{\partial T}{\partial p}\right)_S, & \left(\frac{\partial S}{\partial p}\right)_T &= -\left(\frac{\partial V}{\partial T}\right)_p,\\ \left(\frac{\partial p}{\partial T}\right)_V &= \left(\frac{\partial S}{\partial V}\right)_T, & \left(\frac{\partial T}{\partial V}\right)_S &= -\left(\frac{\partial p}{\partial S}\right)_V,\end{aligned} \tag{10.10}$$

can be derived by writing (10.1) in the various exact differential forms

$$\begin{aligned}&(\text{internal energy } U) & dU &= T\,dS - p\,dV, &&\\ &(\text{enthalpy } H) & dH &= T\,dS + V\,dp, & H &= U + pV,\\ &(\text{free energy } F) & dF &= -S\,dT - p\,dV, & F &= U - TS,\\ &(\text{Gibbs function } G) & dG &= -S\,dT + V\,dp, & G &= U + pV - TS,\end{aligned} \tag{10.11}$$

and applying (10.9). They may also be deduced from eqns $(10.11)_1$, $(10.8)_2$, and $(10.8)_3$. The functions $U(S, V)$, $H(S, p)$, $F(T, V)$, and $G(T, p)$ are called *thermodynamic potentials*, and the pair of independent variables in each case are said to be the *natural* variables for the function in question.†

10.3. Use of thermodynamic potentials

Knowledge of a thermodynamic potential for a system is sufficient to determine *all* the properties of that system. Suppose, for example, that in place of the caloric equation of state $U(T, V)$—which is not a potential—we

† The mnemonic diagram aids the recovery of all the results in eqns (10.10) and (10.11). The potentials are found between their natural variables and the diagonals indicate the appropriate coefficients for the differentials of these variables, the arrow direction yielding the

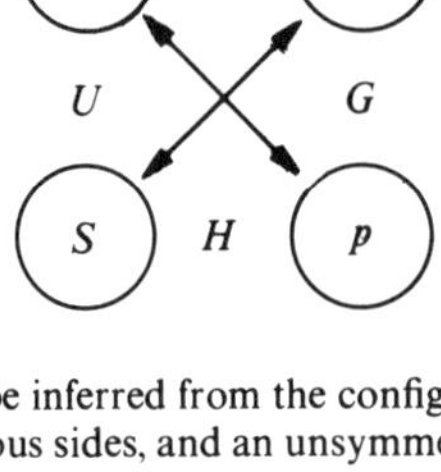

signs. The Maxwell relations can be inferred from the configurations of the corner elements, the square being rotated on to its various sides, and an unsymmetric placing of the arrows denotes a minus sign.

are given $U(S, V)$, then by $(10.11)_1$ the first derivatives yield the state equations

$$\left(\frac{\partial U}{\partial S}\right)_V = T, \qquad \left(\frac{\partial U}{\partial V}\right)_S = -p. \tag{10.12}$$

By eqns (10.4)–(10.8) we find for the second derivatives

$$\frac{\partial^2 U}{\partial S^2} = \left(\frac{\partial T}{\partial S}\right)_V = \frac{T}{C_V},$$

$$\frac{\partial^2 U}{\partial S \partial V} = -\left(\frac{\partial p}{\partial S}\right)_V = -\left(\frac{\partial T}{\partial S}\right)_V \bigg/ \left(\frac{\partial T}{\partial p}\right)_V = -\frac{T}{C_V}\left(\frac{\partial p}{\partial T}\right)_V$$

$$= \frac{T}{C_V}\left(\frac{\partial V}{\partial T}\right)_p \bigg/ \left(\frac{\partial V}{\partial p}\right)_T = -\frac{T\beta}{C_V \kappa_T},$$

$$\frac{\partial^2 U}{\partial V^2} = -\left(\frac{\partial p}{\partial V}\right)_S = \frac{1}{V\kappa_S};$$

and by (10.6) and (10.7) we can express these results in terms of C_p, κ_T, and β, which are a convenient standard set of partial derivatives. This set is based on the choice of $G(T, p)$ as the thermodynamic potential, for by $(10.4)_2$, (10.5), and $(10.11)_4$,

$$\left(\frac{\partial G}{\partial T}\right)_p = -S, \qquad \left(\frac{\partial G}{\partial p}\right)_T = V$$

and (10.13)

$$\frac{\partial^2 G}{\partial T^2} = -\frac{C_p}{T}, \qquad \frac{\partial^2 G}{\partial T \partial p} = V\beta, \qquad \frac{\partial^2 G}{\partial p^2} = -V\kappa_T.$$

It should now be apparent why just three partial derivatives are sufficient for a standard set—potential functions have only three independent second derivatives.

10.4. Proof of relations between heat capacities

It remains to prove relations (10.6) and (10.7). By (10.4), (10.5), and (10.8) we have

$$\kappa_S = -\frac{1}{V}\left(\frac{\partial V}{\partial p}\right)_S = \frac{1}{V}\left(\frac{\partial S}{\partial p}\right)_V \bigg/ \left(\frac{\partial S}{\partial V}\right)_p = \frac{1}{V}\left(\frac{\partial S}{\partial T}\right)_V \left(\frac{\partial T}{\partial p}\right)_V \quad \left(\frac{\partial S}{\partial T}\right)_p \left(\frac{\partial T}{\partial V}\right)_p$$

$$= \frac{C_V}{C_p}\frac{1}{V}\left(\frac{\partial T}{\partial p}\right)_V \bigg/ \left(\frac{\partial T}{\partial V}\right)_p = -\frac{C_V}{C_p}\frac{1}{V}\left(\frac{\partial V}{\partial p}\right)_T = \kappa_T \frac{C_V}{C_p},$$

which proves (10.6).

To prove (10.7) we return to eqns (10.1) and (10.2) written in the forms

$$T\,\mathrm{d}S = C_V\,\mathrm{d}T + \left[\left(\frac{\partial U}{\partial V}\right)_T + p\right]\mathrm{d}V = C_p\,\mathrm{d}T + \left[\left(\frac{\partial H}{\partial p}\right)_T - V\right]\mathrm{d}p.$$

By (10.1) and $(10.10)_3$

$$p + \left(\frac{\partial U}{\partial V}\right)_T = p + T\left(\frac{\partial S}{\partial V}\right)_T - p\left(\frac{\partial V}{\partial V}\right)_T = T\left(\frac{\partial p}{\partial T}\right)_V$$

$$= -T\left(\frac{\partial V}{\partial T}\right)_p \Big/ \left(\frac{\partial V}{\partial p}\right)_T = \frac{T\beta}{\kappa_T},$$

similarly

$$\left(\frac{\partial H}{\partial p}\right)_T - V = -T\left(\frac{\partial V}{\partial T}\right)_p = -TV\beta.$$

Thus

$$T\,\mathrm{d}S = C_V\,\mathrm{d}T + \frac{T\beta}{\kappa_T}\mathrm{d}V = C_p\,\mathrm{d}T - TV\beta\,\mathrm{d}p, \tag{10.14}$$

so that

$$(C_p - C_V)\,\mathrm{d}T = \frac{T\beta}{\kappa_T}\mathrm{d}V + TV\beta\,\mathrm{d}p.$$

Eqn (10.7) now follows on dividing through by $\mathrm{d}T$ and taking p constant ($\mathrm{d}p = 0$).

Notice that κ_T and β can be calculated from the thermal state equation (the first of (10.3)), and that to calculate C_V only the temperature dependence of $U(T, V)$ is required; the volume dependence, in fact, is already determined through the relation

$$\left(\frac{\partial U}{\partial V}\right)_T = \frac{T\beta}{\kappa_T} - p \tag{10.15}$$

just derived.

The mechanical stability condition given in the footnote on p. 12 may be expressed $(\partial V/\partial p)_S < 0$ or $\kappa_S > 0$, while the thermal stability condition (7.10) is $C_V > 0$. Thus by (10.6) and (10.7)

$$C_p > C_V > 0, \qquad \kappa_T > \kappa_S > 0. \tag{10.16}$$

(thermal stability) (mechanical stability)

10.5. General method for reduction of derivatives

One technique for reducing derivatives to a combination of the standard set is to take the following steps:

1. If the derivative contains a potential, bring it into the numerator by (10.8) and then eliminate it in favour of T, S, p, V by (10.11). The expression may now contain several derivatives each of which is treated as follows.
2. Bring the entropy into the numerator and eliminate it, using either Maxwell's relations (eqns $(10.10)_2$ and $(10.10)_3$) or the definitions (10.4) of C_p and C_V. Use (10.7) to eliminate C_V.
3. The remaining derivatives will contain only p, T, and V, and can be removed in favour of κ_T, β or $\beta/\kappa_T = (\partial p/\partial T)_V$.

As a simple example consider the Joule–Thomson coefficient (§ 10.7) $(\partial T/\partial p)_H$. Following the steps given above, we have

$$\left(\frac{\partial T}{\partial p}\right)_H = -\left(\frac{\partial H}{\partial p}\right)_T \Big/ \left(\frac{\partial H}{\partial T}\right)_p = -\left[T\left(\frac{\partial S}{\partial p}\right)_T + V\left(\frac{\partial p}{\partial p}\right)_T\right] \Big/ C_p$$

$$= \left[T\left(\frac{\partial V}{\partial T}\right)_p - V\right] \Big/ C_p = V(T\beta - 1)/C_p. \tag{10.17}$$

A more complicated example is offered by $\left(\frac{\partial(p/T)}{\partial V}\right)_U$. The procedure yields

$$\left(\frac{\partial(p/T)}{\partial V}\right)_U = \frac{1}{T}\left(\frac{\partial p}{\partial V}\right)_U - \frac{p}{T^2}\left(\frac{\partial T}{\partial V}\right)_U$$

$$= -\frac{1}{T}\left(\frac{\partial U}{\partial V}\right)_p \Big/ \left(\frac{\partial U}{\partial p}\right)_V + \frac{p}{T^2}\left(\frac{\partial U}{\partial V}\right)_T \Big/ \left(\frac{\partial U}{\partial T}\right)_V$$

$$= -\frac{1}{T}\left[T\left(\frac{\partial S}{\partial V}\right)_p - p\right] \Big/ T\left(\frac{\partial S}{\partial p}\right)_V + \frac{p}{T^2 C_V}\left(\frac{T\beta}{\kappa_T} - p\right)$$

on using (10.15) to evaluate the last term. Next substituting

$$\left(\frac{\partial S}{\partial V}\right)_p = \left(\frac{\partial S}{\partial T}\right)_p \left(\frac{\partial T}{\partial V}\right)_p = \frac{C_p}{T\beta V}, \qquad \left(\frac{\partial S}{\partial p}\right)_V = \frac{C_V \kappa_T}{T\beta},$$

and using (10.7) to eliminate C_p, we arrive at

$$\left(\frac{\partial(p/T)}{\partial V}\right)_U = -\left[\frac{1}{TV\kappa_T} + \frac{1}{C_V}\left(\frac{\beta}{\kappa_T} - \frac{p}{T}\right)^2\right] < 0, \tag{10.18}$$

the inequality here being a consequence of (10.16).

10.6. State equations for a perfect gas

A perfect gas is defined by the thermal equation of state

$$pV = nRT, \tag{10.19}$$

where n is the number of moles of gas and R is the universal gas constant.† From (10.19) we find that

$$\kappa_T = 1/p, \qquad \beta = 1/T, \tag{10.20}$$

and $(\partial U/\partial V)_T = 0$ by (10.15). This gas therefore has a caloric equation of state

$$U = U(T),$$

and so

$$H = H(T), \qquad (H = U + nRT).$$

By (10.4), (10.7), and (10.14) it follows that

$$\left.\begin{aligned} \mathrm{d}U = C_V\,\mathrm{d}T, \qquad \mathrm{d}H = C_p\,\mathrm{d}T,\\ C_p - C_V = nR,\\ \text{and} \qquad T\,\mathrm{d}S = C_V\,\mathrm{d}T + p\,\mathrm{d}V = C_p\,\mathrm{d}T - V\,\mathrm{d}p. \end{aligned}\right\} \tag{10.21}$$

In general, C_V, C_p, and their ratio,

$$\gamma \equiv C_p/C_V,$$

are slowly varying functions of temperature.

A calorically perfect gas is one for which C_p and C_V are constant. For such a gas let $U = 0$ at $T = 0$ and $S = S_0$ at $TV^{\gamma-1} = 1$, then by (10.19) and (10.21),

$$U = C_V T = \frac{1}{\gamma-1}pV, \qquad H = C_p T = \frac{\gamma}{\gamma-1}pV, \tag{10.22}$$

$$S - S_0 = C_V \ln T + nR \ln V = C_p \ln T - nR \ln p,$$

or

$$S - S_0 = C_V \ln p + C_p \ln V - C_V \ln nR,$$

that is,

$$pV^{\gamma} = nR \exp\,[(S - S_0)/C_V].$$

In an isentropic process, we conclude that

$$\frac{p}{p_0} = \left(\frac{V_0}{V}\right)^{\gamma} = \left(\frac{T}{T_0}\right)^{\gamma/(\gamma-1)},$$

where p_0, V_0, and T_0 are reference values.

† $8{\cdot}314 \times 10^3\ \mathrm{J\,K^{-1}\,mol^{-1}}$.

10.7. The Joule–Thomson effect

Deviations from perfect gas behaviour are particularly important in the production of low temperatures. The Joule–Thomson *throttling process*, in which gas at high pressure is expanded through a porous plug or a throttling valve in an adiabatic pipe to a low pressure, is often used for this purpose. The whole system can be treated as being adiabatically enclosed by adding suitable pistons that move so as to maintain constant gas pressures on each side of the porous plug. We shall assume that the pistons move slowly enough for the viscous work done by the gas on the walls and piston to be negligible.

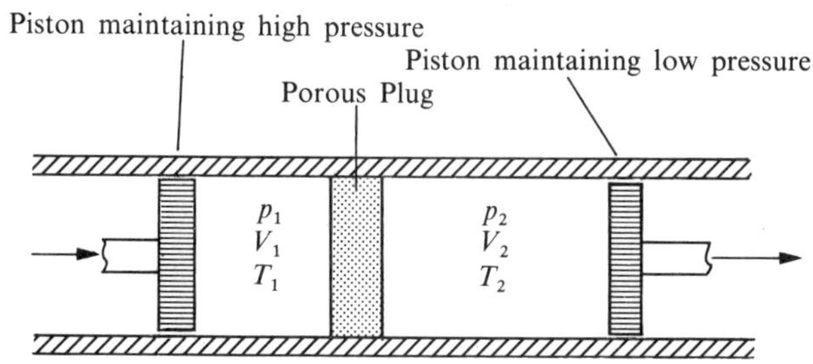

FIG. 4. The Joule–Thomson throttling process.

Treating this as an equilibrium process, so that thermodynamic properties may be assigned on each side of the plug, we observe that in a change in which the high-pressure piston sweeps through a volume V_1 towards the plug and the low-pressure piston sweeps through a volume V_2 away from the plug, the net work done on the gas is $p_1V_1 - p_2V_2$. No heat is transferred, hence $U_2 - U_1 = p_1V_1 - p_2V_2$, that is, $U_1 + p_1V_1 = U_2 + p_2V_2$ or

$$H = \text{constant in a throttling process.}$$

In such a process it is generally found that $T_2 \neq T_1$ (cf. the second of eqns (10.22)) which fact is termed the Joule–Thomson effect. For an infinitesimal change $\mathrm{d}H = 0$, so by (10.17)

$$\mathrm{d}T = \left(\frac{\partial T}{\partial p}\right)_H \mathrm{d}p = \frac{V(T\beta - 1)}{C_p}\,\mathrm{d}p.$$

By (10.20) this effect vanishes for a perfect gas. For real gases it is found that $\mathrm{d}T$ is negative at sufficiently low temperatures, hence by a continuous process in which the throttled gas is used to pre-cool the newly arriving gas, very low temperatures can be attained.

11. Systems in electromagnetic fields

11.1. Maxwell's equations

The four basic field vectors of electromagnetic theory (see, for example, Stratton (1941)) are the *electric field* **E**, the *electric displacement* **D**, the *magnetic field* **H**, and the *magnetic induction* **B**. The source of an electromagnetic field is a distribution of electric charge of density q and an electric current of current density **j**. These vectors are related by the Maxwell equations

$$\nabla \times \mathbf{E} + \frac{\partial \mathbf{B}}{\partial t} = 0, \qquad \nabla \cdot \mathbf{B} = 0, \tag{11.1}$$

$$\nabla \times \mathbf{H} - \frac{\partial \mathbf{D}}{\partial t} = \mathbf{j}, \qquad \nabla \cdot \mathbf{D} = q. \tag{11.2}$$

In free space

$$\mathbf{B} = \mu_0 \mathbf{H}, \qquad \mathbf{D} = \varepsilon_0 \mathbf{E}, \tag{11.3}$$

where μ_0 and ε_0 are constants known as the free-space *permeability* and free-space *permittivity* respectively.

In a material body (11.3) are generalized to

$$\mathbf{B} = \boldsymbol{\mu} \cdot \mathbf{H}, \qquad \mathbf{D} = \boldsymbol{\varepsilon} \cdot \mathbf{E}, \tag{11.4}$$

where $\boldsymbol{\mu}$, $\boldsymbol{\varepsilon}$ are second-order tensors, the elements of which are functions of the local thermodynamic state and the electromagnetic field strengths. In an isotropic medium scalar values μ, ε suffice,

$$\mathbf{B} = \mu \mathbf{H}, \qquad \mathbf{D} = \varepsilon \mathbf{E}. \tag{11.5}$$

An alternative representation to describe the electromagnetic state of a sample of matter is to use the magnetic and electric polarization vectors,

$$\mathbf{M} = \frac{1}{\mu_0}\mathbf{B} - \mathbf{H}, \qquad \mathbf{P} = \mathbf{D} - \varepsilon_0 \mathbf{E}, \tag{11.6}$$

which vanish in free space. By (11.4)

$$\mathbf{M} = (\boldsymbol{\mu} - \mu_0 \mathbf{1}) \cdot \mathbf{H}/\mu_0, \qquad \mathbf{P} = (\boldsymbol{\varepsilon} - \varepsilon_0 \mathbf{1}) \cdot \mathbf{E}, \tag{11.7}$$

where **1** is the second-order unit tensor.

11.2. Electromagnetic work

The rate at which electromagnetic work is performed within a thermodynamic system P and its environment—which may extend to infinity—may be calculated as follows. Let the bounded domain $\tilde{\mathrm{P}}$ contain P plus all electric charge and current distributions able to generate electromagnetic fields within P. Within $\tilde{\mathrm{P}}$ there may be several current carrying components

P_i, $i = 1, 2, \ldots$, (electrons and different species of ions) moving with velocities $\mathbf{v}_i$ and carrying electric charges q_i per unit volume. The force per unit volume on component P_i moving in an electric field $\mathbf{E}$ and a magnetic induction $\mathbf{B}$ is given by

$$\mathbf{F}_i = q_i(\mathbf{E}+\mathbf{v}_i \times \mathbf{B}), \tag{11.8}$$

and so the rate at which P_i does work against this force is $\mathbf{F}_i \cdot \mathbf{v}_i = q_i \mathbf{v}_i \cdot \mathbf{E}$ per unit volume. The rate at which all the components P_i within $\tilde{P}$ do work is $\mathbf{j} \cdot \mathbf{E}$ per unit volume,

$$\mathbf{j} = \sum_i q_i \mathbf{v}_i, \tag{11.9}$$

being the electric current density. Hence the rate at which electromagnetic work is done *on* $\tilde{P}$ is

$$\frac{d'\tilde{W}}{dt} = -\int \mathbf{j} \cdot \mathbf{E}\, d\tau, \tag{11.10}$$

where the volume integral extends at least over $\tilde{P}$ and may be considered to extend over the whole of space as $\mathbf{j}$ is zero outside $\tilde{P}$.

If the scalar product of the first of eqns (11.2) with $\mathbf{E}$ is subtracted from the scalar product of the first of eqns (11.1) with $\mathbf{H}$, and use is made of the vector identity

$$\mathbf{H} \cdot \nabla \times \mathbf{E} - \mathbf{E} \cdot \nabla \times \mathbf{H} = \nabla \cdot \mathbf{S},$$

where

$$\mathbf{S} \equiv \mathbf{E} \times \mathbf{H}, \tag{11.11}$$

there results

$$-\mathbf{j} \cdot \mathbf{E} = \nabla \cdot \mathbf{S} + \mathbf{H} \cdot \dot{\mathbf{B}} + \mathbf{E} \cdot \dot{\mathbf{D}}, \tag{11.12}$$

time derivatives being denoted by dots. This result is to be substituted into (11.10). The volume integral of $\nabla \cdot \mathbf{S}$ can be replaced by a surface integral over the sphere at infinity and with the boundary conditions

$$\lim_{r\to\infty} r\mathbf{E} = 0, \qquad \lim_{r\to\infty} r\mathbf{H} = 0 \tag{11.13}$$

(a consequence of $\tilde{P}$ being bounded in extent), this surface integral is zero. Let $d\mathbf{B}$, $d\mathbf{D}$ denote the change in $\mathbf{B}$ and $\mathbf{D}$ in a time dt, then (11.10) may now be written

$$d'\tilde{W} = \int (\mathbf{H} \cdot d\mathbf{B} + \mathbf{E} \cdot d\mathbf{D})\, d\tau, \tag{11.14}$$

the integral being taken over all of space.

11.3. Work done on a material system

Some of the work included in (11.14) has no thermodynamic interest, for it is expended in establishing field changes throughout space. Denote by the subscript e (external) the field values that would be produced by the given charge and current distributions were the thermodynamic system P replaced by empty space. Then, by (11.3),

$$\mathbf{B}_e = \mu_0 \mathbf{H}_e, \qquad \mathbf{D}_e = \varepsilon_0 \mathbf{E}_e, \tag{11.15}$$

and (11.14) gives

$$\mathrm{d}'\tilde{W}_e = \mathrm{d}U_e, \qquad U_e \equiv \int (\tfrac{1}{2}\mu_0 H_e^2 + \tfrac{1}{2}\varepsilon_0 E_e^2)\,\mathrm{d}\tau, \tag{11.16}$$

in empty space. As this work does not involve P at all, it is convenient to exclude it and to regard

$$\mathrm{d}'W = \mathrm{d}'\tilde{W} - \mathrm{d}'\tilde{W}_e = \int [\mathbf{H}\,.\,\mathrm{d}\mathbf{B} + \mathbf{E}\,.\,\mathrm{d}\mathbf{D} - (\mathbf{H}_e\,.\,\mathrm{d}\mathbf{B}_e + \mathbf{E}_e\,.\,\mathrm{d}\mathbf{D}_e)]\,\mathrm{d}\tau \tag{11.17}$$

as being the electromagnetic work done on P. Suppose that the only other type of reversible work is due to volume changes, then $-p\,\mathrm{d}V + \mathrm{d}'W$ is the total work done on P in a reversible change, and (7.7) may be written

$$\mathrm{d}(U - U_e) = T\,\mathrm{d}S - p\,\mathrm{d}V + \mathrm{d}'W. \tag{11.18}$$

However, the expression for $\mathrm{d}'W$ still has the disadvantage of containing an integral over the whole of space.

We define internal fields, indicated by the subscript i, by

$$\mathbf{B}_i = \mathbf{B} - \mathbf{B}_e, \qquad \mathbf{H}_i = \mathbf{H} - \mathbf{H}_e, \qquad \mathbf{D}_i = \mathbf{D} - \mathbf{D}_e, \qquad \mathbf{E}_i = \mathbf{E} - \mathbf{E}_e, \tag{11.19}$$

then by (11.6) and (11.15)

$$\left.\begin{aligned} \mathbf{H}\,.\,\mathrm{d}\mathbf{B} - \mathbf{H}_e\,.\,\mathrm{d}\mathbf{B}_e &= \mathbf{B}_e\,.\,\mathrm{d}\mathbf{M} + \mathbf{H}_i\,.\,\mathrm{d}\mathbf{B} + \mathbf{B}_e\,.\,\mathrm{d}\mathbf{H}_i \\ \mathbf{E}\,.\,\mathrm{d}\mathbf{D} - \mathbf{E}_e\,.\,\mathrm{d}\mathbf{D}_e &= \mathbf{E}_e\,.\,\mathrm{d}\mathbf{P} + \mathbf{E}_i\,.\,\mathrm{d}\mathbf{D} + \mathbf{D}_e\,.\,\mathrm{d}\mathbf{E}_i. \end{aligned}\right\} \tag{11.20}$$

As $\mathbf{j}$ and q in (11.2) are the same for the actual fields $\mathbf{H}$, $\mathbf{D}$ as for the external fields $\mathbf{H}_e$, $\mathbf{D}_e$ it follows that $\mathbf{j}$ and q vanish for the fields $\mathbf{H}_i$, $\mathbf{D}_i$. Hence by (11.12) $\mathbf{H}_i\,.\,\mathrm{d}\mathbf{B}_i + \mathbf{E}_i\,.\,\mathrm{d}\mathbf{D}_i = -\nabla\,.\,\mathbf{S}_i\,\mathrm{d}t$, and when this is integrated over all space, the surface integral of $\mathbf{S}_i$ will vanish as previously explained for $\mathbf{S}$. By this last result and (11.20)

$$\begin{aligned} &\mathbf{H}\,.\,\mathrm{d}\mathbf{B} + \mathbf{E}\,.\,\mathrm{d}\mathbf{D} - (\mathbf{H}_e\,.\,\mathrm{d}\mathbf{B}_e + \mathbf{E}_e\,.\,\mathrm{d}\mathbf{D}_e) \\ &\qquad = \mathbf{B}_e\,.\,\mathrm{d}\mathbf{M} + \mathbf{E}_e\,.\,\mathrm{d}\mathbf{P} + \mathrm{d}(\mathbf{H}_i\,.\,\mathbf{B}_e + \mathbf{E}_i\,.\,\mathbf{D}_e) \\ &\qquad = \mathbf{B}_e\,.\,\mathrm{d}\mathbf{M} - \mathbf{P}\,.\,\mathrm{d}\mathbf{E}_e + \mathrm{d}(\mathbf{H}_i\,.\,\mathbf{B}_e + \mathbf{E}_e\,.\,\mathbf{D}_i), \end{aligned} \tag{11.21}$$

where the second equation follows from the definition of $\mathbf{P}$ and the volume integration has been anticipated.

11.4. Work in quasi-stationary processes

In deducing the general expression for $d'W$ in (11.17), we imposed no limits on the time rates of change, although reversibility would certainly require quasi-stationary processes within the material body P. We shall now assume such processes, then, as $\mathbf{j}$, q vanish for the internal fields, (11.1) and (11.2) show that $\mathbf{E}_i$, $\mathbf{E}_e$, and $\mathbf{H}_i$ are irrotational vector fields and $\mathbf{B}_e$, $\mathbf{D}_i$ are solenoidal vector fields. The vector $\mathbf{D}_e$ is a solenoid field when there is no charge distribution.

The further reduction of (11.21) relies on the following theorem:

The integral over all space of the scalar product $\mathbf{X}.\mathbf{Y}$ of an irrotational vector $\mathbf{X}$ and a solenoidal vector $\mathbf{Y}$ is zero provided $\mathbf{X}$ and $\mathbf{Y}$ are continuous and vanish at infinity as rapidly as r^{-2}. For, writing $\mathbf{X} = \nabla\varphi$, we have $\mathbf{X}.\mathbf{Y} = \nabla\varphi.\mathbf{Y} = \nabla.(\varphi\mathbf{Y})$, and the resulting surface integral over the sphere at infinity is zero.

By this theorem the total differentials in (11.21) vanish on integration over all space. Hence by (11.17) and (11.21) and the fact that $\mathbf{M}$ and $\mathbf{P}$ are zero outside the thermodynamic system, we arrive finally at

$$d'W = \int_V (\mathbf{B}_e.d\mathbf{M} - \mathbf{P}.d\mathbf{E}_e)\,d\tau \tag{11.22}$$

or

$$d'W = \int_V (\mathbf{B}_e.d\mathbf{M} + \mathbf{E}_e.d\mathbf{P})\,d\tau \quad \text{(zero charge distribution).} \tag{11.23}$$

When the external fields are uniform and there is no charge distribution,

$$d'W = \mathbf{B}_e.d\mathcal{M} + \mathbf{E}_e.d\mathcal{P}, \tag{11.24}$$

where

$$\mathcal{M} \equiv \int_V \mathbf{M}\,d\tau, \qquad \mathcal{P} \equiv \int_V \mathbf{P}\,d\tau$$

are termed the *total* magnetic and electric polarizations of the system. The fundamental equation (11.18) now appears as

$$dU' = T\,dS - p\,dV + \mathbf{B}_e.d\mathcal{M} + \mathbf{E}_e.d\mathcal{P}, \tag{11.25}$$

where $U' = U - U_e$. Generalized thermodynamic potentials can now be written down and the Maxwell relations immediately deduced (see § 15.2).

11.5. Systems separable into material and electromagnetic components

Suppose now that a closed system P, comprised of a single *material* component P^0 and an electromagnetic component P^*, has uniform electromagnetic fields over its volume V. By (11.14) the electromagnetic work term becomes

$$d'\tilde{W} = V\mathbf{H}\,.\,d\mathbf{B} + V\mathbf{E}\,.\,d\mathbf{D}. \tag{11.26}$$

For this case the free energy F satisfies

$$dF = -S\,dT - p\,dV + V\mathbf{H}\,.\,d\mathbf{B} + V\mathbf{E}\,.\,d\mathbf{D}, \tag{11.27}$$

and the four equations of state prescribe S, p, $\mathbf{H}$, $\mathbf{E}$ as functions of the state $X(T, V, \mathbf{B}, \mathbf{D})$. We shall restrict attention to isotropic systems, so that

$$\mathbf{B} = \mu\mathbf{H}, \qquad \mathbf{D} = \varepsilon\mathbf{E},$$

where μ and ε are functions of state, knowledge of which is equivalent to being given the two magnetic equations of state.†

For many substances and conditions commonly experienced, the dependence of μ and ε on the electromagnetic fields may be neglected, leaving

$$\mu = \mu(T, V), \qquad \varepsilon = \varepsilon(T, V). \tag{11.28}$$

It will now be shown that with such equations the thermodynamics of the material component P^0 and the electromagnetic component P^* can be separated. On substituting (11.28) into (11.26) we find that the result can be cast into the form

$$d'\tilde{W} = dF^* + S^*\,dT - p^*\,dV, \tag{11.29}$$

where

$$\left.\begin{aligned} F^* &\equiv V(\tfrac{1}{2}\mu H^2 + \tfrac{1}{2}\varepsilon E^2) \\ S^* &\equiv V(\tfrac{1}{2}\mu_T H^2 + \tfrac{1}{2}\varepsilon_T E^2), \\ p^* &\equiv \tfrac{1}{2}\mu H^2(1 - V\mu_V/\mu) + \tfrac{1}{2}\varepsilon E^2(1 - V\varepsilon_V/\varepsilon), \end{aligned}\right\} \tag{11.30}$$

and the subscripts denote partial derivatives. Hence from (11.26), (11.27), and (11.29)

$$dF^0 = -S^0\,dT - p^0\,dV, \tag{11.31}$$

where

$$F^0 = F - F^*, \qquad S^0 = S - S^*, \qquad p^0 = p + p^*.$$

Eqn (11.31) shows that F^0 is independent of the electromagnetic field, and the same is true of its partial derivatives $-S^0$, $-p^0$. Thus (11.31) refers only to the material component P^0, and the separation earlier mentioned is established.

† The phenomenon known as *hysteresis*, in which μ and ε are not functions of state alone but are also functions of the history of the system, is excluded from consideration.

The quantity p^* is known as the *electromagnetic pressure*. The internal energy associated with P^* is $F^* + TS^*$, that is,

$$U^* = V[\tfrac{1}{2}\mu H^2(1 + T\mu_T/\mu) + \tfrac{1}{2}\varepsilon E^2(1 + T\varepsilon_T/\varepsilon)]. \tag{11.32}$$

When it is possible, as for example in a plasma (§ 56), to consider μ and ε as constants, then by $(11.30)_3$ and (11.32)

$$U^* = p^* V, \tag{11.33}$$

and comparing this with the first result in (10.22) for a perfect gas, namely, $U = pV/(\gamma - 1)$, we see that the magnetic field behaves as if it were a perfect gas with $\gamma = 2$.

3

SYSTEMS INVOLVING INTERNAL VARIABLES

THE equilibrium processes considered in Chapter 2 had time scales $\mathrm{d}t$ comparable with the time scale $\mathscr{T}_{\mathrm{R}}$ on which the internal energy U and the deformation vector $\mathbf{V}$ are adequate to define the state of a system P. And as—by changes external to P—both U and $\mathbf{V}$ may be independently controlled, reversible processes are possible in the space $X_{\mathrm{R}} = (U, \mathbf{V})$. Our present aim is to study processes on time scales $\mathrm{d}t$ somewhat shorter than $\mathscr{T}_{\mathrm{R}}$.

It is often possible to find a thermodynamic state-space $X_{\mathrm{I}} = (U, \mathbf{V}, \mathbf{Y})$ that includes X_{R} as a sub-space, in which such processes can be fully described, that is, appear as equilibrium processes. Let $\mathscr{T}_{\mathrm{I}}$ be the observer's time scale appropriate to X_{I}, then

$$\mathrm{d}t \sim \mathscr{T}_{\mathrm{I}} \ll \mathscr{T}_{\mathrm{R}}.$$

The additional variables $\mathbf{Y}$ are always *internal* in the sense that, unlike U and $\mathbf{V}$, they can alter without any corresponding changes occurring outside P. Further, the observer is not usually able to control $\mathbf{Y}$, so that the equilibrium processes in $X_{\mathrm{I}}(U, \mathbf{V}, \mathbf{Y})$ are often irreversible.

12. Compound systems not in thermal equilibrium

12.1. Entropy of compound systems

A compound system P is one that consists of two or more distinct sub-systems, in some form of contact with each other so that interactions are possible between the sub-systems. These may be transfers of work or heat or mass, and so on, but in this article we shall restrict attention to *closed* sub-systems, that is, sub-systems of fixed content. There are two distinct types of compound system, namely, (1) the *heterogenous* system consisting of spatially distinct sub-systems P_α, $\alpha = 1, 2, \ldots n$, termed *phases*, and (2) the *homogeneous* system consisting of spatially over-lapping sub-systems P_i, $i = 1, 2, \ldots m$, termed *components*. Thus phases occupy distinct volumes (usually contiguous) while components occupy the same volume. The transfer of work and heat is a surface interaction with phases and a volume distributed phenomenon with components. Distinct temperatures T_α are certainly possible for phases, and even components may have different temperatures

T_i, although in the latter case this is a sensible model only on an observer's time scale $\mathcal{T}$ comparable with or less than the temperature equilibration time between the components. To denote sub-systems in general, whether they be phases or components, we shall use the subscript σ.

Now suppose the system P has internal rigid adiabatic partitions isolating its sub-systems P_σ; for phases P_α such partitions will physically separate the P_α from each other, whereas for components they can only be conceptual inhibitors of heat and work transfer between the P_i. When these partitions are replaced by diathermic partitions, heat will flow irreversibly between the phases or components at different temperatures. It will be assumed that these heat exchanges are much slower than heat equilibration within each sub-system P_σ, so that during the irreversible process each P_σ is in thermodynamic equilibrium, although not in thermal equilibrium with its neighbours. This assumption about time scales enables us to attribute thermodynamic variables to P_σ during the heat exchanges; these processes can thus be traced in a state-space, and so are examples of the irreversible equilibrium process described at the end of § 3.2.

By the additive property of internal energy, the energy of the compound system P is

$$U = \sum_\sigma U_\sigma, \qquad (12.1)$$

but the corresponding result for entropy has been established only for sub-systems at the *same temperature*. This limited additivity property follows directly from (7.7),

$$T\,\mathrm{d}S = \mathrm{d}'Q + \mathrm{d}'W_i = \mathrm{d}U + \mathbf{p}\,.\,\mathrm{d}\mathbf{V}, \qquad (\mathrm{d}'W_i \geqslant 0), \qquad (12.2)$$

written for each P_σ and the additivity of internal energy and work. We now extend the definition of entropy by postulating that it is also an additive function for compound systems whose sub-systems are at different temperatures T_σ. Thus the entropy S of P is given by

$$S = \sum_\sigma S_\sigma; \qquad \mathrm{d}S = \sum_\sigma \mathrm{d}S_\sigma = \sum_\sigma (\mathrm{d}U_\sigma + \mathbf{p}_\sigma\,.\,\mathrm{d}\mathbf{V}_\sigma)/T_\sigma. \qquad (12.3)$$

This very important postulate is fundamental for process thermodynamics as it liberates thermodynamics from the striaghtjacket of thermal equilibrium. However, for all its importance, it is not a new law but merely extends a definition to a new situation, which is always possible provided the extended definition is consistent with the original meaning.

12.2. The Clausius inequality

For an isolated homogeneous system of just one component $\mathrm{d}U = 0$, $\mathrm{d}\mathbf{V} = 0$, and it follows from (12.2) that $\mathrm{d}S = 0$. However, for an isolated compound system, heat may be transferred internally from one phase or component to another, and it is easily shown that this leads to an increase in entropy. Let a rigid adiabatic wall, separating two sub-systems P_1, P_2 of an isolated system P, be replaced by a diathermic wall, and consider the process of heat transfer from P_1 and P_2 that results. If this is an equilibrium process, as described in the second paragraph of § 12.1, then for an infinitesimal step of the process during which heat $-d'Q_1 = d'Q_2 > 0$ flows from P_1 to P_2, $T_1\,dS_1 = d'Q_1$, $T_2\,dS_2 = d'Q_2$. The thermal stability condition $C_V > 0$ (see eqn (10.16)) requires that $T_1 > T_2$. Hence

$$dS = dS_1 + dS_2 = d'Q_2\left(\frac{1}{T_2} - \frac{1}{T_1}\right) > 0. \tag{12.4}$$

As the initial and final states of P are independent of the details of the process involved, we conclude that the entropy is always increased by internal heat transfers.

More generally if a system P at temperature T receives elements of heat $d'Q_1, d'Q_2, \ldots$ from systems $P_1, P_2, \ldots$ at temperatures $T_1, T_2, \ldots$, then the change of entropy of P is

$$dS = \frac{d'Q}{T} = \sum_\sigma \frac{d'Q_\sigma}{T_\sigma} + \sum_\sigma d'Q_\sigma\left(\frac{1}{T} - \frac{1}{T_\sigma}\right) \qquad (\sigma = 1, 2, \ldots), \tag{12.5}$$

where $d'Q = \sum_\sigma d'Q_\sigma$. As $T_\sigma \geqslant T$ implies $d'Q_\sigma \geqslant 0$ and $T_\sigma \leqslant T$ implies $d'Q_\sigma \leqslant 0$, the last term in (12.5) is non-negative, whence

$$dS \geqslant \sum_\sigma \frac{d'Q_\sigma}{T_\sigma}. \tag{12.6}$$

The inequality is one form of *Clausius' inequality*; the equals sign holds only when the heat transfer is reversible, that is, when $T = T_\sigma$ (see § 5.2). Notice that as $-d'Q_\sigma/T_\sigma$ is the entropy change of P_σ, (12.6) can be interpreted as the statement that irreversible *internal* heat transfers in an adiabatic system $\mathscr{P} = P + \sum_\sigma P_\sigma$ always result in an increase in the entropy of $\mathscr{P}$. As the inequality is often written in the condensed and misleading form $dS > d'Q/T$, even in the absence of work transfer, it is important to bear in mind that the temperatures in (12.6) are the temperatures of the systems contiguous to P, that is, the temperatures of the *sources* of heat and not the temperature of P itself.

12.3. *Clausius inequality for a continuum*

Finally let the homogeneous system P be contained within a surface Σ, with an outward directed normal $\mathbf{n}$, and allow the number of phases surrounding P to be increased indefinitely to form a continuum. Also suppose that P has components P_i, $i = 1, 2, \ldots$, at distinct temperatures T_i. The heat passing inwards to the ith component P_i of P through the surface element

$\mathbf{n}\,\mathrm{d}\Sigma$ in time $\mathrm{d}t$ may be written

$$\mathrm{d}'Q_i = -\mathbf{n}\,.\,\mathbf{q}_i\,\mathrm{d}\Sigma\,\mathrm{d}t, \tag{12.7}$$

where $\mathbf{q}_i$ is termed the ith component *heat flux vector*. Suppose that in a time $\mathrm{d}t$ unit volume within P_i receives heat $\varphi_{ij}\,\mathrm{d}t$ from component P_j and heat $\psi_i\,\mathrm{d}t$ by radiation from a source $\mathscr{S}$ external to P. The heat source P_j is at a temperature T_j, while—on the assumption of radiative equilibrium—$\mathscr{S}$ is at an effective temperature T_i as far as P_i is concerned. Now splitting the right-hand side of (12.6) into contributions from the continuum of phases and the internal components, we obtain

$$\frac{\mathrm{d}S}{\mathrm{d}t} \geqslant -\int_\Sigma \sum_i \frac{\mathbf{n}\,.\,\mathbf{q}_i}{T_i}\,\mathrm{d}\Sigma + \int_V \sum_i \left(\sum_j \frac{\varphi_{ij}}{T_j} + \frac{\psi_i}{T_i}\right)\mathrm{d}\tau, \tag{12.8}$$

which is a fundamental result for the theory of process thermodynamics.

13. The law of entropy increase

13.1. Mechanical and thermal stability

It has been shown above that internal transfers of heat in an isolated system P always results in an increase in the entropy of P, and that this is a direct consequence of the thermal stability condition. This law of entropy increase also applies when there is an internal transfer of work resulting from a relaxation of rigid internal barriers across which there is a pressure discontinuity.

Suppose that a rigid adiabatic wall separating two phases P_1 and P_2 of an isolated system P is replaced by a perfectly flexible diathermic wall, and that the heat and work transfers that ensue are equilibrium processes on a small enough time scale. Then for an infinitesimal step $T_1\,\mathrm{d}S_1 = \mathrm{d}U_1 + p_1\,\mathrm{d}V_1$, $T_2\,\mathrm{d}S_2 = \mathrm{d}U_2 + p_2\,\mathrm{d}V_2$, where we have taken the phases to be simple systems. By the isolation of P, $U_1 + U_2 = U = \text{constant}$, $V_1 + V_2 = V = \text{constant}$, so that $\mathrm{d}U_1 + \mathrm{d}U_2 = 0$, and $\mathrm{d}V_1 + \mathrm{d}V_2 = 0$. Thus

$$\mathrm{d}S = \mathrm{d}(S_1 + S_2) = \left(\frac{1}{T_1} - \frac{1}{T_2}\right)\mathrm{d}U_1 + \left(\frac{p_1}{T_1} - \frac{p_2}{T_2}\right)\mathrm{d}V_1; \tag{13.1}$$

thermal and mechanical equilibrium between P_1 and P_2 are achieved at $T_1 = T_2$ and $p_1 = p_2$.

The criterion† for stability is that the spontaneous processes induced by a deviation from thermal and mechanical equilibrium must be in a direction to restore the system to equilibrium.‡ Thus considering the term

† A principle usually associated with the name Le Châtelier.

‡ The word 'equilibrium' is somewhat overworked in thermodynamics. In this article we have used it in three distinct senses, namely, 'equilibrium process', 'thermal equilibrium', and 'mechanical equilibrium'.

$(T_1^{-1} - T_2^{-1})\,\mathrm{d}U_1$, we conclude that if $T_2 > T_1, \mathrm{d}U_1$ at $\mathrm{d}V_1 = 0$ must be such as to increase T_1, so bringing P_1 closer to thermal equilibrium. As $(\partial U/\partial T)_V = C_V > 0$, then $\mathrm{d}U_1 > 0$. Likewise, if $T_2 < T_1$ we find that $\mathrm{d}U_1 < 0$, and so in both cases $(T_1^{-1} - T_2^{-1})\,\mathrm{d}U_1 > 0$. This is just a variant of the argument used to deduce the inequality in (12.4).

Again considering the term $(p_1 T_1^{-1} - p_2 T_2^{-1})\,\mathrm{d}V_1$, we see that if $p_1 T_1^{-1} > p_2 T_2^{-1}$, $\mathrm{d}V_1$ at $\mathrm{d}U_1 = 0$, must be such as to reduce $p_1 T_1^{-1}$, and as $(\partial V/\partial(p/T))_U < 0$ by (10.18), it follows that $\mathrm{d}V_1 > 0$. Similarly $\mathrm{d}V_1 < 0$ if $p_1 T_1^{-1} < p_2 T_2^{-1}$, so always $(p_1 T_1^{-1} - p_2 T_2^{-1})\,\mathrm{d}V_1 > 0$. Hence both of the terms in (13.1) are non-negative, that is,†

$$\mathrm{d}S(U, V, U_1, V_1) \geqslant 0, \tag{13.2}$$

the equality applying when the system $P = P_1 + P_2$ is in internal equilibrium. Eqn (13.1) is an example of the irreversible equilibrium process described in § 3.2.

13.2. *Internal coordinates*

The theory easily generalizes to many phases P_α, $\alpha = 1, 2, \ldots$, to generalized forces $\mathbf{p} = (p_1, p_2, \ldots p_n)$ with their associated deformations $\mathbf{V} = (V_1, V_2, \ldots V_n)$ and to changes $\mathrm{d}U = \sum_\alpha \mathrm{d}U_\alpha, \mathrm{d}\mathbf{V} = \sum_\alpha \mathrm{d}\mathbf{V}_\alpha$ in the total internal energy and total deformations of the compound system $P = \sum_\alpha P_\alpha$. In place of (13.1) we find

$$\mathrm{d}U = T\,\mathrm{d}S - \mathbf{p}\,.\,\mathrm{d}\mathbf{V} + \sum_\alpha [(T_\alpha - T)\,\mathrm{d}S_\alpha + (\mathbf{p} - \mathbf{p}_\alpha)\,.\,\mathrm{d}\mathbf{V}_\alpha],$$

where $\mathbf{p}$ and T denote the values of $\mathbf{p}_\alpha$ and T_α when internal equilibrium prevails. By the thermal stability condition, and the mechanical stability conditions $-(\partial p_j/\partial V_j)_S > 0$ (cf. discussion in footnote on p. 12 applied to each phase, we find that the last term must be non-positive. Hence collecting them into single vector form $\sum_\sigma z_\sigma\,\mathrm{d}Y_\sigma = \mathbf{z}\,.\,\mathrm{d}\mathbf{Y}$, we have

$$T\,\mathrm{d}S\,(U, \mathbf{V}, \mathbf{Y}) = \mathrm{d}U + \mathbf{p}\,.\,\mathrm{d}\mathbf{V} - \mathbf{z}\,.\,\mathrm{d}\mathbf{Y} \qquad (-\mathbf{z}\,.\,\mathrm{d}\mathbf{Y} \geqslant 0). \tag{13.3}$$

At internal equilibrium (13.3) becomes

$$T\,\mathrm{d}S\,(U, \mathbf{V}) = \mathrm{d}U + \mathbf{p}\,.\,\mathrm{d}\mathbf{V} \qquad (\mathbf{z}\,.\,\mathrm{d}\mathbf{Y} = 0). \tag{13.4}$$

The new thermodynamic coordinates $\{Y_\sigma\}$ introduced here are called *internal variables*, as they describe the internal structure of the system during its progress between states defined only by the 'external' and controllable variables U and $\mathbf{V}$. The negative sign for $\mathbf{z}\,.\,\mathrm{d}\mathbf{Y}$ in (13.3) emphasizes the

†One of the most confusing aspects of entropy is that quite often the appropriate set of independent variables is not specified, and statements are made about 'entropy' without classifying which function is intended. Here we have changed attention from a long time scale entropy function $S(U, V)$ for which $\mathrm{d}S$ is zero, to a short time scale entropy function $S(U, V, U_1, V_1)$, which by (13.2) has a smaller value.

essential difference between the work $-\mathbf{p}\,.\,\mathrm{d}\mathbf{V}$ transferred from outside P and the work and heat transfers $\mathbf{z}\,.\,\mathrm{d}\mathbf{Y}$ within P. In (13.3) T, $\mathbf{p}$, and $\mathbf{z}$ are functions of U, $\mathbf{V}$, and $\mathbf{Y}$. If the coordinates Y_σ are independent, the equilibrium condition $\mathbf{z}\,.\,\mathrm{d}\mathbf{Y} = 0$ yields

$$\mathbf{z} = 0, \qquad \mathbf{Y}^{\mathrm{eq}} = \mathbf{Y}^{\mathrm{eq}}(S, \mathbf{V}), \tag{13.5}$$

whence $S(U, \mathbf{V}, \mathbf{Y})$ reduces to $S(U, \mathbf{V})$, and similarly for the functional dependence of T and $\mathbf{p}$.

Later we shall meet other examples of internal variables, which allow us to describe systems on short time scales and in states that on the longer time scale appropriate to (13.4) would be termed 'non-equilibrium'.

13.3. Thermodynamic inequalities

It is, of course, the inequality in (13.3) that renders irreversible the equilibrium process described by eqn (13.3). The variables $\{Y_\sigma\}$ cannot be completely controlled by the observer for the sign of Y_σ is determined by that of z_σ. Eqns (13.3) and (13.4) yield the inequality

$$T\,\mathrm{d}S\,(U, \mathbf{V}, \mathbf{Y}) \geqslant T\,\mathrm{d}S\,(U, \mathbf{V}) = \mathrm{d}U + \mathbf{p}\,.\,\mathrm{d}\mathbf{V}, \tag{13.6}$$

where T has the same functional dependence as the S with which it is associated. As this inequality does not depend on the particular sub-division of P into phases P_α, it is true for all possible sub-divisions,† so we may regard $\mathbf{Y}$ as unspecified and write (13.6) in the general form

$$T\delta S \geqslant \mathrm{d}U + \mathbf{p}\,.\,\mathrm{d}\mathbf{V} \tag{13.7}$$

where δ is a differential operator in some state-space that includes $(U, \mathbf{V})$ as a sub-space. Of course T and S are now unspecified functions, with undetermined values; nevertheless useful conclusions may be drawn from (13.7).

A process that satisfies (13.7), in which δS may be any possible change in S, is termed a *natural* process; the reversible process in the second relation (13.6) is included as a limiting case. As $\mathrm{d}U$ and $\mathrm{d}\mathbf{V}$ merely denote independent differentials in (13.7), they may be written as δU and $\delta\mathbf{V}$. Thus we arrive at the basic inequality of thermodynamics,

$$T\,\delta S \geqslant \delta U + \mathbf{p}\,.\,\delta\mathbf{V}; \tag{13.8}$$

alternatively

$$\delta U \leqslant T\,\delta S - \mathbf{p}\,.\,\delta\mathbf{V}, \tag{13.9}$$

for natural processes. It must be remembered that these inequalities involve an entropy that has no determinate value so long as the appropriate Gibbs relation (eqn (13.3) is one such relation) is absent.

† Subject to the limitations imposed by quantum theory—see § 25.3.

With an isolated system the right-hand side of (13.8) is zero, and the only possible changes are variations in any existing internal constraints. Hence the entropy law may be expressed:

However the internal constraints of an isolated system are varied, this can never result in a decrease in its entropy.

When constraints are relaxed in any way, the system is removed from equilibrium and natural processes set in, during each infinitesimal step of which $\delta S > 0$; when these processes cease and equilibrium is re-established, the entropy of the system will then be the maximum value permitted by the new constraints. Conversely if the entropy is a maximum with respect to all imaginary small internal disturbances, then $\delta S \leqslant 0$, but changes in which $\delta S < 0$, while conceptually possible are certainly physically impossible, so $\delta S = 0$ and the system is in equilibrium.

Equilibrium occurs when there are no natural processes available to take P to another state. Thus when P is not isolated it follows from (13.9) that the general equilibrium condition is

$$\delta U \geqslant T\,\delta S - \mathbf{p}\,.\,\delta \mathbf{V} \quad \text{(equilibrium)}, \tag{13.10}$$

for all virtual displacements $\delta U, \delta \mathbf{V}, \delta S$ consonant with the general interaction conditions between P and its surroundings. We shall return to this topic in § 17.

14. Homogeneous open systems

14.1. Specific values

To this stage we have considered only *closed* thermodynamic systems, that is, systems containing a constant mass in each phase or component. In § 13 we were concerned with the space-wise sub-division of a system P into *phases* and arrived at an extended state-space $X(S, \mathbf{V}, \mathbf{Y})$ suitable for irreversible process; here by analysing P into overlaying *component* systems P_i, $i = 1, 2, \ldots l$, we shall obtain a similar extension of state-space permitting the description of open systems. The components P_i may be different molecular species, or ions, electrons, etc., which are able to interact physically or chemically and change their masses M_i. The additivity property of energy and entropy provides the means of extending thermodynamics to such systems; when this property is applied to conceptually distinct neighbouring sub-systems that are not physically different, it is termed the 'extensive' nature of U and S, and it permits the introduction of 'specific' quantities—that is, quantities per unit mass—in place of extensive ones. Lower-case letters are used to denote specific quantities, for example, $u_i = U_i/M_i$ is the specific energy of the ith component.

The additivity properties may be written

$$M = \sum_i M_i, \qquad \begin{aligned} U &= \sum_i U_i = \sum_i M_i u_i, & u &= \sum_i c_i u_i, \\ S &= \sum_i S_i = \sum_i M_i s_i, & s &= \sum_i c_i s_i, \end{aligned} \tag{14.1}$$

where $c_i = M_i/M$ is termed the 'concentration' of the ith component. Notice that

$$\sum_i c_i = 1. \tag{14.2}$$

Later we shall need Dalton's law of partial pressures for the total pressure p in a compound system, namely,

$$p = \sum_i p_i. \tag{14.3}$$

14.2. Gibbs relation for an open system

The analysis to follow applies to the ith component system P_i and for convenience we shall omit the subscript i from all variables. Let a system of fixed mass M have an entropy $S(U, \mathbf{V}; M)$, where the mass M is merely a parameter at this stage. And suppose this system to be replicated λ times to produce a different system having an entropy $S'(\lambda U, \lambda \mathbf{V}; \lambda M)$. Then by the extensivity of entropy, $S'(\lambda U, \lambda \mathbf{V}; \lambda M) = \lambda S(U, \mathbf{V}; M)$. Now take λ^{-1} to be the number of kilograms of material present in the original system, then $S'(u, \mathbf{v}; 1) = S(U, \mathbf{V}; M)/M$, that is, $S'(u, \mathbf{v}; 1)$ is the specific entropy s. Hence

$$S(U, \mathbf{V}; M) = Ms(u, \mathbf{v}).$$

But $S = Ms$ is true whether M is fixed in value or not, and as s is independent of M, it follows that for changes dS during which M is permitted to vary

$$\mathrm{d}S = M\,\mathrm{d}s\,(u, \mathbf{v}) + s\,\mathrm{d}M.$$

From the basic Gibbs relation (7.7) written in the form

$$T(\mathrm{d}S)_M = \mathrm{d}U + \mathbf{p}\,.\,\mathrm{d}\mathbf{V}, \tag{14.4}$$

and $S = Ms$, $U = Mu$, it follows that

$$\left(\frac{\partial S}{\partial U}\right)_{\mathbf{V},M} = \left(\frac{\partial s}{\partial u}\right)_{\mathbf{v}} = \frac{1}{T}, \qquad \left(\frac{\partial S}{\partial \mathbf{V}}\right)_{U,M} = \left(\frac{\partial s}{\partial \mathbf{v}}\right)_{u} = \frac{\mathbf{p}}{T},$$

so

$$T\,\mathrm{d}s = \mathrm{d}u + \mathbf{p}\,.\,\mathrm{d}\mathbf{v}.$$

Hence

$$T\,\mathrm{d}S = M(\mathrm{d}u+\mathbf{p}\,.\,\mathrm{d}\mathbf{v})+Ts\,\mathrm{d}M$$
$$= \mathrm{d}(Mu)+\mathbf{p}\,.\,\mathrm{d}(M\mathbf{v})-(u+\mathbf{p}\,.\,\mathbf{v}-Ts)\,\mathrm{d}M,$$

or

$$T\,\mathrm{d}S = \mathrm{d}U+\mathbf{p}\,.\,\mathrm{d}\mathbf{V}-g\,\mathrm{d}M, \tag{14.5}$$

where g is the generalized specific Gibbs function

$$g = u+\mathbf{p}\,.\,\mathbf{v}-Ts$$

(cf. the fourth relation of (10.11)). Eqn (14.5) is the required extension of the Gibbs relation.

For the component P_i then

$$\left.\begin{aligned} T_i\,\mathrm{d}s_i &= \mathrm{d}u_i+\mathbf{p}_i\,.\,\mathrm{d}\mathbf{v}_i, \\ T_i\,\mathrm{d}S_i &= \mathrm{d}U_i+\mathbf{p}_i\,.\,\mathrm{d}\mathbf{V}-g_i\,\mathrm{d}M_i, \end{aligned}\right\} \tag{14.6}$$

where

$$g_i = u_i+\mathbf{p}_i\,.\,\mathbf{v}_i-T_is_i \tag{14.7}$$

and the overlapping component systems are assumed to have the same deformation vector $\mathbf{V}$.

14.3. Components at the same temperature

To carry the theory further it is necessary to assume that the component temperatures T_i are the same, equal to T say; this will be so in the remainder of this chapter. This assumption enables us to sum the second of eqns (14.6) thus

$$T\,\mathrm{d}S = \mathrm{d}U+\mathbf{p}\,.\,\mathrm{d}\mathbf{V}-\sum_i g_i\,\mathrm{d}M_i, \tag{14.8}$$

where, analogously with (14.3), $\mathbf{p}$ is the total intensity vector

$$\mathbf{p} = \sum_i \mathbf{p}_i.$$

Let $\mathbf{M}$ denote the vector $(M_1, M_2, \ldots M_l)$ and $\mathbf{g}$ the vector $(g_1, g_2, \ldots g_l)$, then (14.8) may be written

$$T\,\mathrm{d}S(U,\mathbf{V},\mathbf{M}) = \mathrm{d}U+\mathbf{p}\,.\,\mathrm{d}\mathbf{V}-\mathbf{g}\,.\,\mathrm{d}\mathbf{M}; \tag{14.9}$$

the state-space $X(U, \mathbf{V})$ of eqn (14.4) has been now enlarged to $X(U, \mathbf{V}, \mathbf{M})$ with the component masses $M_1, M_2, \ldots$ as internal coordinates (cf. (13.3)).

In many applications the mass M of the compound system is constant, so that

$$\sum_{i=1}^{l} \mathrm{d}M_i = 0,$$

and (14.8) may be written

$$T\,\mathrm{d}S = \mathrm{d}U + \mathbf{p}\,.\,\mathrm{d}\mathbf{V} - \sum_{i=1}^{l-1} (g_i - g_l)\,\mathrm{d}M_i$$

or alternatively, on dividing through by the constant M,

$$T\,\mathrm{d}s = \mathrm{d}u + \mathbf{p}\,.\,\mathrm{d}\mathbf{v} - \sum_{i=1}^{l} g_i\,\mathrm{d}c_i. \tag{14.10}$$

The concept of negative catalysts or *inhibitors* to close the component system P_i is quite useful, for we can think of the term $\mathbf{g}\,.\,\mathrm{d}\mathbf{M}$ as being 'chemical work' performed within P during the removal of the inhibitors for a short period in which an infinitesimal amount of reaction can take place. Inhibitors for particle reactions have a parallel role to that of adiabatic walls for heat transfer. However, there is an important difference between the work term $-\mathbf{p}\,.\,\mathrm{d}\mathbf{V}$ and $\mathbf{g}\,.\,\mathrm{d}\mathbf{M}$ in that, while the sign of $\mathrm{d}\mathbf{V}$ may be reversed, this is not true in general of $\mathrm{d}\mathbf{M}$. This means that the equilibrium process described by (14.9) is irreversible. And, of course, another difference is that on a long enough time scale chemical equilibrium is maintained and the term $\mathbf{g}\,.\,\mathrm{d}\mathbf{M}$ vanishes.

From (14.9) and a rearrangement of it with $U(S, \mathbf{V}, \mathbf{M})$ as the dependent variable, we find that

$$g_i = -T\left(\frac{\partial S}{\partial M_i}\right)_{U,\mathbf{V},\mathbf{M}(i)} = \left(\frac{\partial U}{\partial M_i}\right)_{S,\mathbf{V},\mathbf{M}(i)} \qquad (i = 1, 2, \ldots l \ldots), \tag{14.11}$$

where $\mathbf{M}(i)$ means that each of $M_1, M_2 \ldots M_l$ except M_i is to be held constant. The quantity g_i is known as the 'chemical potential of the ith component' or sometimes as the 'partial specific Gibbs function'. Notice that g_i has the dimensions of energy/amount and is therefore a positive intensive quantity. As shown later (§ 18) chemical stability requires mass to tend to move out of components with high values of g_i into components with low values of g_i. And later (§ 54.1) it is shown that substance i diffuses down g_i gradients just as heat diffuses down temperature gradients.

15. Characteristic functions

15.1. The Gibbs–Duhem relation

Eqn (14.9), in which S has the functional dependence

$$S = S(U, \mathbf{V}, \mathbf{M}), \tag{15.1}$$

may be written

$$\mathrm{d}S = \frac{1}{T}\mathrm{d}U + \frac{\mathbf{p}}{T}\,.\,\mathrm{d}\mathbf{V} - \frac{\mathbf{g}}{T}\,.\,\mathrm{d}\mathbf{M}, \tag{15.2}$$

which shows that

$$T^{-1} = \left(\frac{\partial S}{\partial U}\right)_{\mathbf{V},\mathbf{M}}, \quad p_j T^{-1} = \left(\frac{\partial S}{\partial V_j}\right)_{U,\mathbf{V}(j),\mathbf{M}}, \quad g_i T^{-1} = -\left(\frac{\partial S}{\partial M_i}\right)_{U,\mathbf{V},\mathbf{M}(i)}. \tag{15.3}$$

These are the fundamental equations of what is called 'the entropy representation': $S(U, \mathbf{V}, \mathbf{M})$ is called a *characteristic function*, as it contains *all* thermodynamic information about the system. Its first derivatives given in (15.3), when written in the functional forms $T^{-1} = T^{-1}(U, \mathbf{V}, \mathbf{M})$, etc. are called *equations of state.*

By the extensive property of entropy $S(U, \mathbf{V}, \mathbf{M})$ is a homogeneous function of the first degree, that is, for any λ

$$\lambda S(U, \mathbf{V}, \mathbf{M}) = S(\lambda U, \lambda \mathbf{V}, \lambda \mathbf{M}). \tag{15.4}$$

Let us assume *pro tempore* that U, $\mathbf{V}$, $\mathbf{M}$ are the only extensive quantities on which S can depend, so that the parameter λ is not concealed as the coefficient of some further extensive parameter X say $-\lambda S(U, \mathbf{V}, \mathbf{M}; X) = S(\lambda U, \lambda \mathbf{V}, \lambda \mathbf{M}; \lambda X)$—then it is permissible to differentiate (15.4) with respect to λ to find

$$S = \frac{\partial S}{\partial U}U + \frac{\partial S}{\partial \mathbf{V}}\,.\,\mathbf{V} + \frac{\partial S}{\partial \mathbf{M}}\,.\,\mathbf{M},$$

which we shall term the 'Euler relation', being a special case of his theorem for homogeneous functions of the nth degree. Comparison with

$$\mathrm{d}S = \frac{\partial S}{\partial U}\mathrm{d}U + \frac{\partial S}{\partial \mathbf{V}}\,.\,\mathrm{d}\mathbf{V} + \frac{\partial S}{\partial \mathbf{M}}\,.\,\mathrm{d}\mathbf{M}$$

yields the simple rule 'to obtain the Euler relation merely suppress the differential operator d in the Gibbs relation'. By (15.2) the Euler relation may be written

$$S = \frac{1}{T}(U + \mathbf{p}\,.\,\mathbf{V} - \mathbf{g}\,.\,\mathbf{M}), \tag{15.5}$$

or by (14.7)

$$S = \frac{1}{T}\left(U + \mathbf{p}\,.\,\mathbf{V} - \sum_{i=1}^{l}\,[U_i + \mathbf{p}_i\,.\,\mathbf{V} - TS_i]\right),$$

revealing it to be an identity, which thus justifies our initial assumption concerning (15.4). When the differential of (15.5) is subtracted from (15.2) the result is the *Gibbs–Duhem relation*, namely,

$$S\,\mathrm{d}T = \mathbf{V}\,.\,\mathrm{d}\mathbf{p} - \mathbf{M}\,.\,\mathrm{d}\mathbf{g}. \tag{15.6}$$

As $T > 0$, the first of (15.3) shows S to be a monotonic function of U, so that (15.1) may be rewritten to yield the characteristic function

$$U = U(S, \mathbf{V}, \mathbf{M});$$

by (15.2) and (15.5)

$$\mathrm{d}U = T\,\mathrm{d}S - \mathbf{p}\,.\,\mathrm{d}\mathbf{V} + \mathbf{g}\,.\,\mathrm{d}\mathbf{M}, \qquad U = TS - \mathbf{p}\,.\,\mathbf{V} + \mathbf{g}\,.\,\mathbf{M}. \tag{15.7}$$

This is the 'energy representation' in which the state equations take the more convenient forms

$$T = \left(\frac{\partial U}{\partial S}\right)_{\mathbf{V},\mathbf{M}}, \qquad p_j = -\left(\frac{\partial U}{\partial V_j}\right)_{S,\mathbf{V}(j),\mathbf{M}}, \qquad g_i = \left(\frac{\partial U}{\partial M_i}\right)_{S,\mathbf{V},\mathbf{M}(i)}.$$

15.2. Thermodynamic potentials

In both the entropy and energy representations extensive parameters occur as independent variables and the intensive parameters arise as derived quantities. However, in experiments these roles are usually reversed, for the intensive quantities are often the more easily measured and controlled—for example no instrument exists for the measurement and control of entropy, whereas thermometers and thermostats are standard equipment. The roles of dependent and independent variables are interchanged by Legendre's transformation, which for the energy representation is the following. Let the scalar products in (15.7) be expanded in the forms

$$\mathrm{d}U = \sum_{r=1}^{K} P_r\,\mathrm{d}X_r, \qquad U = \sum_{r=1}^{K} P_r X_r,$$

then the transformation

$$\Psi_k = U - \sum_{r=1}^{k} P_r X_r = \sum_{s=k+1}^{K} P_s X_s \qquad (k < K)$$

yields

$$\mathrm{d}\Psi_k = -\sum_{r=1}^{k} X_r\,\mathrm{d}P_r + \sum_{s=k+1}^{K} P_s\,\mathrm{d}X_s.$$

The characteristic functions $\Psi_k = \Psi_k(P_1, P_2, \ldots P_k, X_{k+1}, \ldots X_K)$ are called *thermodynamic potentials.* Three types of Maxwell relations (10.10) can be written down from the expression for $\mathrm{d}\Psi_k$, namely,

$$\frac{\partial X_r}{\partial P_n} = \frac{\partial X_n}{\partial P_r}, \quad \frac{\partial P_s}{\partial X_m} = \frac{\partial P_m}{\partial X_s}, \quad \frac{\partial X_r}{\partial X_s} = -\frac{\partial P_s}{\partial P_r} \quad (r, n \leqslant k\,; s, m > k). \tag{15.8}$$

The potentials important in thermodynamics—including U for completeness—are set out in Table 2 (cf. eqns (10.11)).

TABLE 2

Thermodynamic potential	Definition	Differential form
Internal energy $U(S, \mathbf{V}, \mathbf{M})$		$\mathrm{d}U = \quad T\,\mathrm{d}S - \mathbf{p}\,.\,\mathrm{d}\mathbf{V} + \mathbf{g}\,.\,\mathrm{d}\mathbf{M}$
Free energy† $F(T, \mathbf{V}, \mathbf{M})$	$U - TS$	$\mathrm{d}F = -S\,\mathrm{d}T - \mathbf{p}\,.\,\mathrm{d}\mathbf{V} + \mathbf{g}\,.\,\mathrm{d}\mathbf{M}$
Enthalpy $H(S, \mathbf{p}, \mathbf{M})$	$U + \mathbf{p}\,.\,\mathbf{V}$	$\mathrm{d}H = \quad T\,\mathrm{d}S + \mathbf{V}\,.\,\mathrm{d}\mathbf{p} + \mathbf{g}\,.\,\mathrm{d}\mathbf{M}$
Gibbs function $G(T, \mathbf{p}, \mathbf{M})$	$U + \mathbf{p}\,.\,\mathbf{V} - TS$	$\mathrm{d}G = -S\,\mathrm{d}T + \mathbf{V}\,.\,\mathrm{d}\mathbf{p} + \mathbf{g}\,.\,\mathrm{d}\mathbf{M}$

† Also called the Helmholtz function.

From (15.6) and the differential form for $\mathrm{d}G$ we find

$$G = \mathbf{g}\,.\,\mathbf{M} = \sum_i (u_i + \mathbf{p}_i\,.\,\mathbf{v}_i - Ts_i)M_i. \tag{15.9}$$

15.3. Physical interpretation of the potentials

It is useful at this point to introduce two idealized thermodynamic systems known as *reversible heat* and *work sources.* A reversible heat (work) source is defined to be a simple system enclosed by a rigid (an adiabatic) impermeable wall and characterized by relaxation times so short that all processes of interest in it are equilibrium processes. When coupled to other systems, these idealized systems act as reversible sources or sinks of heat and work, and if they are very large they are known as *reservoirs.* More precisely, a heat reservoir R_T in thermal contact with a system P exchanges heat with P at constant temperature, and further maintains P at constant temperature regardless of any other interactions that may occur between P and the rest of its environment. Likewise, a work reservoir R_p in mechanical contact (for example, through a perfectly flexible wall or a frictionless piston) with P transfers work at constant pressure, and so maintains P at constant pressure. In practice, the atmosphere often plays the roles of both a heat and work reservoir. Clearly the work-reservoir concept can be extended to other forms of reversible work and $\mathrm{R}_\mathbf{p}$ will denote a reservoir for *all* the possible forms of deformation work available to the system. Thus such a reservoir will maintain $\mathbf{p}$ constant for a system.

One physical interpretation of the thermodynamic potentials can be given in terms of the reversible work $-\mathrm{d}'W_\mathrm{r}$ transmitted out of a system P in contact with one or two reservoirs. The useful reversible work is $\mathbf{p}\,.\,\mathrm{d}\mathbf{V}$ when no work reservoir is involved and $\mathbf{p}'\,.\,\mathrm{d}\mathbf{V}' = \mathbf{p}\,.\,\mathrm{d}\mathbf{V} - p\,\mathrm{d}V$ when a (pressure) work reservoir R_p is in contact with P; the work $p\,dV$ cannot appear as useful work in the latter case as it is done merely to keep the pressure constant. Let $H' = U+pV$ and $G' = U+pV-TS$, then, for **M** constant, Table 1 yields the following values for $d'W_r$ for the indicated combinations of system and reservoirs.

$$\begin{array}{lccccc} & P & P+R_T & P+R_p & P+R_p+R_T & \\ d'W_r \ : & (dU)_S & (dF)_T & (dH')_{p,S} & (dG')_{T,p} & \quad (15.10)\end{array}$$

Thus, for example, the isothermal reversible work done by a closed system equals the decrease in the free energy, and similar interpretations can be given for the other potentials.

Another set of characteristic functions, known as *Massieu functions*, can be obtained by performing Legendre transformations on the entropy representation, for example, with $\chi = S-\mathbf{p}\,.\,\mathbf{V}/T$ eqn (15.2) is replaced by

$$d\chi = T^{-1}\,dU - \mathbf{V}\,.\,d(\mathbf{p}T^{-1}) - T^{-1}\mathbf{g}\,.\,d\mathbf{M},$$

but such functions are not particularly useful.

16. Thermodynamic inequalities

16.1. The hierarchy of entropy functions

The inequality

$$T\,\delta S > dU + \mathbf{p}\,.\,d\mathbf{V}, \tag{16.1}$$

given in (13.7) applies to all natural processes in a one-component system. We shall use it in place of (14.4) for the *i*th component of a compound system and follow a similar analysis to that leading to eqn (14.8). Then we find in place of (14.8)

$$T\,\delta S > dU + \mathbf{p}\,.\,d\mathbf{V} - \mathbf{g}\,.\,d\mathbf{M} \tag{16.2}$$

for natural processes in a multi-component system P. In (16.2) δ is a differential operator in an enlarged state-space, the additional independent variables being unspecified internal coordinates $\{Y_\sigma\}$, which by (13.3) satisfy

$$T\,\delta S - (dU + \mathbf{p}\,.\,d\mathbf{V} - \mathbf{g}\,.\,d\mathbf{M}) = -\mathbf{z}\,.\,d\mathbf{Y} > 0.$$

Thus, replacing δ by d and prescribing the functional dependence of S fully, we have

$$T\,dS\,(U, \mathbf{V}, \mathbf{M}, \mathbf{Y}) > dU + \mathbf{p}\,.\,d\mathbf{V} - \mathbf{g}\,.\,d\mathbf{M} = T\,dS\,(U, \mathbf{V}, \mathbf{M}).$$

Of course, the set $\{M_i\}$ are also internal coordinates and play a role similar to $\{Y_\sigma\}$. In § 18.4 we shall show that, for natural processes, **M** satisfies the inequality

$$\left(\frac{\partial S}{\partial \mathbf{M}}\right)_{U,\mathbf{V}} . \mathrm{d}\mathbf{M} = -\mathbf{g} . \mathrm{d}\mathbf{M} > 0, \tag{16.3}$$

in which S is a function of U, $\mathbf{V}$, and $\mathbf{M}$. Hence

$$T \mathrm{d}S(U, \mathbf{V}, \mathbf{M}) > \mathrm{d}U + \mathbf{p} . \mathrm{d}\mathbf{V} = T \mathrm{d}S(U, \mathbf{V}).$$

For the hierarchy of entropy functions resulting from an increasing definition of the internal state, we have for natural processes

$$\mathrm{d}'Q/T < \mathrm{d}S(U, \mathbf{V}) < \mathrm{d}S(U, \mathbf{V}, \mathbf{M}) < \mathrm{d}S(U, \mathbf{V}, \mathbf{M}, \mathbf{Y}) < \dots , \tag{16.4}$$

where we have included the inequality in (7.8) for completeness.

16.2. Equilibrium conditions

So far we have been dealing with natural processes. Equilibrium will be secured by the absence of natural processes that could change the state of P, and by (16.2) and (16.3) there will be no such processes if for all conceivable displacements δU, $\delta \mathbf{V}$, $\delta \mathbf{M}$,

$$T \delta S \leqslant \delta U + \mathbf{p} . \delta \mathbf{V} - \mathbf{g} . \delta \mathbf{M} \leqslant \delta U + \mathbf{p} . \delta \mathbf{V}. \tag{16.5}$$

Of course these virtual displacements must conform with the general interaction conditions between P and its surroundings. From (16.5), corresponding to (16.4), we find that for virtual displacements from equilibrium

$$\mathrm{d}S(U, \mathbf{V}) \geqslant \mathrm{d}S(U, \mathbf{V}, \mathbf{M}) \geqslant \mathrm{d}S(U, \mathbf{V}, \mathbf{M}, \mathbf{Y}) \geqslant \dots , \tag{16.6}$$

so that the more detailed the description of the state, the smaller the associated entropy (see Fig. 5).

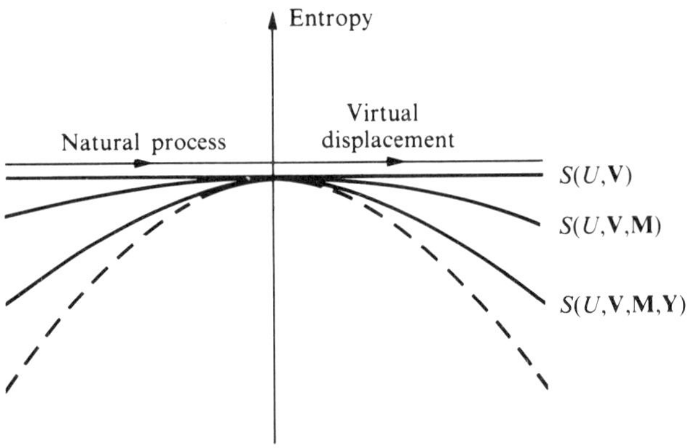

FIG. 5. Virtual entropy changes for an isolated system.

Various transformations of the first inequality in (16.5) are possible with the aid of the reservoirs introduced in § 15.3. In the notation of (15.10) we have the inequalities shown in Table 3.

TABLE 3

System and reservoir	Inequality for equilibrium	
P	$\delta U \geqslant T\delta S - \mathbf{p}.\delta\mathbf{V} + \mathbf{g}.\delta\mathbf{M}$	
$P + R_T$	$\delta F \geqslant -\mathbf{p}.\delta\mathbf{V} + \mathbf{g}.\delta\mathbf{M}$	(16.7)
$P + R_{\mathbf{p}}$	$\delta H \geqslant T\delta S + \mathbf{g}.\delta\mathbf{M}$	
$P + R_T + R_{\mathbf{p}}$	$\delta G \geqslant \mathbf{g}.\delta\mathbf{M}$	
$P + R_T + R_{\mathbf{p}} + R_{\mathbf{g}}$	$\delta\Omega \geqslant 0$	

For natural process replace $\geqslant$ by $<$.

where $R_{\mathbf{g}}$ maintains **g** constant and Ω is the so-called 'grand' potential function

$$\Omega \equiv U - TS + \mathbf{p}.\mathbf{V} - \mathbf{g}.\mathbf{M}. \tag{16.8}$$

From this table we may draw various conclusions such as 'F is a minimum for stability at constant values of T, **V**, and **M**', and so on.

17. Equilibrium and stability

17.1. Minimum internal energy

In (16.7) it is usual to refer to the *equations* as being *equilibrium conditions* and to the *inequalities* as *stability conditions*, the analogy with mechanical equilibrium and stability being obvious. For stable† equilibrium *both* conditions must be separately satisfied, which means that the size of the displacement δU cannot be the same in each case. There are three distinct problems to consider in stability (and equilibrium) theory. First, there is the *mutual* stability between contiguous phases in a heterogeneous system, secondly the *intrinsic* stability of a homogeneous system, and thirdly the *internal* or chemical stability between the components of a homogeneous system. Although the latter is really an aspect of intrinsic stability, it is convenient to consider it separately in the next section. Here it will be assumed that there are no chemical reactions between components, with the consequence that the component masses M_i can be changed only by material transport across phase boundaries.

The energy representation is most convenient for present purposes. On selecting S, **V**, **M** as independent variables and identifying the infinitesimals

† As truly unstable states are unrealizable in thermodynamics owing to ever-present fluctuations, the adjective 'stable' is apt to be confusing.

δS, $\delta \mathbf{V}$, $\delta \mathbf{M}$ as the independent differentials $\mathrm{d}S$, $\mathrm{d}\mathbf{V}$, $\mathrm{d}\mathbf{M}$, we may write the first of (16.7) as

$$\delta U \geqslant T\,\mathrm{d}S - \mathbf{p}\,.\,\mathrm{d}\mathbf{V} + \mathbf{g}\,.\,\mathrm{d}\mathbf{M},$$

with $=$ for equilibrium and $>$ for stability. Thus

$$\delta U \geqslant \mathrm{d}U\,(S, \mathbf{V}, \mathbf{M}), \tag{17.1}$$

and stable equilibrium follows if U is a *minimum* at constant $(S, \mathbf{V}, \mathbf{M})$ (cf. (16.6), which requires S to be a *maximum* at constant $(U, \mathbf{V}, \mathbf{M})$).

17.2. Mutual stability

Now suppose P consists of two phases P′ and P″. If P is heterogeneous, these sub-systems will be actual phases separated by a phase boundary C of known physical properties, whereas if P is homogeneous they will be merely conceptual phases arbitrarily chosen, with their interface possessing no properties. The internal energy U will now depend on the coordinates $S', \mathbf{V}', \mathbf{M}'$ of P′ and $S'', \mathbf{V}'', \mathbf{M}''$ of P″, or equivalently on

$$S = S'+S'', \qquad \mathbf{V} = \mathbf{V}'+\mathbf{V}'', \qquad \mathbf{M} = \mathbf{M}'+\mathbf{M}'', \tag{17.2}$$

plus the internal variables (internal for P) $S', \mathbf{V}', \mathbf{M}'$. By additivity

$$\delta U\,(S', \mathbf{V}', \mathbf{M}', S'', \mathbf{V}'', \mathbf{M}'') = \delta U'\,(S', \mathbf{V}', \mathbf{M}') + \delta U''\,(S'', \mathbf{V}'', \mathbf{M}'').$$

From Taylor's expansion $\delta U' = \mathrm{d}U' + \mathrm{d}^2 U' + \ldots$, $\delta U'' = \mathrm{d}U'' + \mathrm{d}^2 U'' + \ldots$, so that to second-order in differentials (17.1) reads

$$\delta U - \mathrm{d}U = (\mathrm{d}U' + \mathrm{d}U'' - \mathrm{d}U) + \mathrm{d}^2 U' + \mathrm{d}^2 U'' \geqslant 0. \tag{17.3}$$

The second-order differentials introduced here are important only when the displacement is large enough to make δU exceed $\mathrm{d}U$. Thus (17.3) splits into

$$\mathrm{d}U = \mathrm{d}U' + \mathrm{d}U'' \quad \text{(equilibrium)}, \tag{17.4}$$

$$\mathrm{d}^2 U' + \mathrm{d}^2 U'' > 0 \quad \text{(stability)}. \tag{17.5}$$

Alternatively, by considering the stability condition ($>$) in (17.3) first, we observe that as the first-order differentials may have either sign and would dominate the second-order differentials for small displacements, (17.4) is a necessary condition for stability (of course (17.4) is also true by additivity). Together, (17.4) and (17.5) are necessary and sufficient for *local* stable equilibrium; with finite displacements the equilibrium thus secured may not persist, in which event the equilibrium state is termed *metastable.*

17.3. Phase equilibrium

The equilibrium condition (17.4) can be expressed

$$T\,\mathrm{d}S-\mathbf{p}\,.\,\mathrm{d}\mathbf{V}+\mathbf{g}\,.\,\mathrm{d}\mathbf{M} = T'\,\mathrm{d}S'-\mathbf{p}'\,.\,\mathrm{d}\mathbf{V}'+\mathbf{g}'\,.\,\mathrm{d}\mathbf{M}'+ \\ +T''\,\mathrm{d}S''-\mathbf{p}''\,.\,\mathrm{d}\mathbf{V}''+\mathbf{g}''\,.\,\mathrm{d}\mathbf{M}''.$$

Choose $\mathrm{d}S$, $\mathrm{d}\mathbf{V}$, $\mathrm{d}\mathbf{M}$ to be zero, then by (17.2)

$$\mathrm{d}S'+\mathrm{d}S'' = 0, \qquad \mathrm{d}\mathbf{V}'+\mathrm{d}\mathbf{V}'' = 0, \qquad \mathrm{d}\mathbf{M}'+\mathrm{d}\mathbf{M}'' = 0,$$

and for equilibrium

$$(T''-T')\,\mathrm{d}S'-(\mathbf{p}''-\mathbf{p}')\,.\,\mathrm{d}\mathbf{V}'+(\mathbf{g}''-\mathbf{g}')\,.\,\mathrm{d}\mathbf{M}' = 0. \tag{17.6}$$

First suppose P is homogeneous, with the phase boundary imposing no constraint at all on the independent differentials in (17.6). Equilibrium reduces to

$$T'' = T', \qquad \mathbf{p}'' = \mathbf{p}', \qquad \mathbf{g}'' = \mathbf{g}'. \tag{17.7}$$

Secondly, let P be heterogeneous, then the phase boundary C may have one or more of the properties shown in Table 4.

TABLE 4

Diathermic	Semi-flexible	Semi-permeable
$\mathrm{d}S' \neq 0$	$\mathrm{d}V'_j \neq 0, \quad j = 1, 2, \ldots s$ $\mathrm{d}V'_j = 0, \quad j = s+1, \ldots n$	$\mathrm{d}M'_i \neq 0, \quad i = 1, 2, \ldots r$ $\mathrm{d}M'_i = 0, \quad i = r+1, \ldots m$
$T' = T''$	$p'_j = p''_j, \quad j = 1, 2, \ldots s$	$g'_i = g''_i, \quad i = 1, 2, \ldots r$

Thus C can respond only to s deformation changes and is permeable only to r of the components. The last row of Table 4 gives the corresponding equilibrium conditions, which are known as the *thermal*, *mechanical*, and *material* equilibrium conditions. The latter conditions must be modified for any of the r components that are electrically charged (Guggenheim 1967, Chapter VIII).

17.4. Intrinsic stability

Now consider the stability of a homogeneous system. The inequality (17.5) must hold for all choices of P′ and P″, the simplest being identical phases with $\mathrm{d}S' = \mathrm{d}S''$, $\mathrm{d}\mathbf{V}' = \mathrm{d}\mathbf{V}''$, $\mathrm{d}\mathbf{M}' = \mathrm{d}\mathbf{M}''$. Then $\mathrm{d}^2U' = \mathrm{d}^2U''$ and (17.5) reduces to $\mathrm{d}^2U' > 0$. Alternatively, choose P″ much larger than P′ but keep the independent differentials comparable in size; then as each of the second derivatives of U'' with respect to the extensive variables S'', $\mathbf{V}''$ and $\mathbf{M}''$ are inversely proportional to the size of P″, $\mathrm{d}^2U' \gg \mathrm{d}^2U''$, and (17.5) again reduces to $\mathrm{d}^2U' > 0$. Thus, omitting the prime, we arrive at the intrinsic stability condition

$$d^2U\,(S, \mathbf{V}, \mathbf{M}) > 0. \tag{17.8}$$

An alternative form follows from the differential of $dU = T\,dS - \mathbf{p}\,.\,d\mathbf{V} + \mathbf{g}\,.\,d\mathbf{M}$, namely,

$$dT\,dS - d\mathbf{p}\,.\,d\mathbf{V} + d\mathbf{g}\,.\,d\mathbf{M} > 0. \tag{17.9}$$

Note that if actual phases P′ and P″ each possess intrinsic stability, by (17.8) $d^2U' > 0$, $d^2U'' > 0$, (17.5) is automatically satisfied, and their mutual stability is assured.

To give a simple example of (17.8) that is easily generalized, we shall take P to be a simple closed system and use results given on p. 33 to find

$$d^2U = \frac{1}{2}\left[\frac{\partial^2 U}{\partial S^2}(dS)^2 + \frac{\partial^2 U}{\partial S\,\partial V}dS\,dV + \frac{\partial^2 U}{\partial V^2}(dV)^2\right]$$

$$= \tfrac{1}{2}[dS\,dV]\begin{bmatrix} \dfrac{T}{C_V} & \dfrac{-T\beta}{C_V\kappa_T} \\[2ex] \dfrac{-T\beta}{C_V\kappa_T} & \dfrac{1}{V\kappa_S} \end{bmatrix}\begin{bmatrix} dS \\ dV \end{bmatrix}.$$

The condition $d^2U > 0$ requires all the principal minors of the determinant of the matrix to be positive (choosing wording more suitable to the general case). By (10.6) and (10.7) this requirement leads back to the stability conditions given in (10.16).

Had we used the entropy representation (16.2), (17.5) and (17.9) would have been replaced by $d^2S' + d^2S'' < 0$, and

$$d^2S\,(U, \mathbf{V}, \mathbf{M}) = d\left(\frac{1}{T}\right)dU + d\left(\frac{\mathbf{p}}{T}\right).\,d\mathbf{V} - d\left(\frac{\mathbf{g}}{T}\right).\,d\mathbf{M} < 0, \tag{17.10}$$

but the final conclusions remain the same.

Stability conditions can be inferred directly from (17.9), each term of which can be non-zero when the others vanish. For (17.9) to be satisfied for all possible changes it is necessary—but not sufficient (see § 18)—that

$$\frac{C_V}{T} = \left(\frac{\partial S}{\partial T}\right)_{\mathbf{V},\mathbf{M}} > 0, \qquad -\left(\frac{\partial p_j}{\partial V_j}\right)_{T,\mathbf{M},\mathbf{V}(j)} > 0, \qquad \left(\frac{\partial g_i}{\partial M_i}\right)_{T,\mathbf{V},\mathbf{M}(i)} > 0, \tag{17.11}$$

thermal stability — mechanical stability — chemical stability

and clearly other combinations of independent variables are possible.†

† Of course it must be admitted that (17.11) have already been employed in § 13 and § 16 to deduce the law of entropy increase upon which we have built the theory of thermodynamic stability, and so we have merely recovered the initial input to the theory. It is usual to accept the law of entropy increase as a plausible hypothesis, then to derive (17.11) as apparently new results, but they are such fundamental requirements for the simple existence of thermodynamic systems, let alone the increase in entropy, that the usual axiomatic treatment of stability appears as a somewhat *a posteriori* approach, leading from the merely hypothetical to the inevitable. In other words, it is the law of entropy increase that requires explanation, not the stability conditions.

A final observation concerns intrinsic metastable equilibria, which can be studied with the help of the grand potential function Ω. By (16.7) Ω is a minimum when the intensive parameters T, $\mathbf{p}$, and $\mathbf{g}$ are held constant. Thus we have the stability condition

$$\Omega > \Omega_{eq},$$

in which the extensive variables may be given finite displacements.

18. Chemical equilibria and stability

18.1. Chemical reactions and equilibrium

The term $-\mathbf{g}.\mathrm{d}\mathbf{M}$ appearing in the Gibbs relation

$$T\,\mathrm{d}S(U,\mathbf{V},\mathbf{M}) = \mathrm{d}U+\mathbf{p}.\mathrm{d}\mathbf{V}-\mathbf{g}.\mathrm{d}\mathbf{M}, \tag{18.1}$$

will now be analysed for the case when the mass changes $\mathrm{d}M_1, \mathrm{d}M_2, \ldots \mathrm{d}M_l$ of the components are subject to the constraints imposed by chemical reactions.

For natural processes in an isolated system (16.2) shows that

$$T\,\delta S > T\,\mathrm{d}S(U,\mathbf{V},\mathbf{M}) = -\mathbf{g}.\mathrm{d}\mathbf{M},$$

and such processes continue until $\delta S = 0$, when

$$\mathbf{g}.\mathrm{d}\mathbf{M} = \sum_i g_i\,\mathrm{d}M_i = 0, \tag{18.2}$$

which is therefore the general condition for chemical equilibrium. Of course the $\mathrm{d}M_i$ in (18.2) are not independent, so we cannot infer that $g_i = 0$. For non-isolated systems changing on the long 'reversible' time scale $\mathscr{T}_R$ mentioned at the beginning of this chapter, (18.2) always applies.

The change in the mass M_i of component P_i may be split into d_eM_i due to external causes (mass exchange with neighbouring phases) and d_iM_i due to internal causes (chemical and physical reactions between components). Any chemical reaction may be expressed in the form $0 = \sum a_iP_i$, where $|a_i|$ is the stoichiometric number (number of interacting moles) of the ith species, and a_i is positive, zero, or negative according to whether P_i is produced, unchanged, or consumed in the reaction. For example, the reaction $2NH_3 \rightarrow N_2+3H_2$ may be written

$$\begin{array}{cccccccccc} 0 = & a_1 & P_1 & +a_2 & P_2 & +a_3 & P_3 & +a_4 & P_4\,. \\ & (-2) & (NH_3) & (1) & (N_2) & (3) & (H_2) & (0) & (O_2) \end{array}$$

Let m_i be the molecular mass of species P_i, then the amount of P_i produced in time $\mathrm{d}t$ may be written $a_im_i\,\mathrm{d}\xi$, where $\mathrm{d}\xi$ denotes a small number measuring the extent to which the reaction has taken place ($\mathrm{d}\xi/\mathrm{d}t$ is called the *reaction rate*). More generally, when P_i is involved in several reactions, $0 = \sum a_{ir}P_i$, $r = 1, 2, \ldots R$, formulated so as to be linearly independent, then

$$\mathrm{d}M_i = M \sum_r \nu_{ir}\,\mathrm{d}\xi_r + d_e M_i \qquad (\nu_{ir} \equiv a_{ir}m_i/M). \tag{18.3}$$

The new *independent* variable ξ_r introduced here is called the *progress variable* for the rth reaction. For the example reaction

$$\mathrm{d}\xi = -\mathrm{d}M_1/2m_1 = \mathrm{d}M_2/m_2 = \mathrm{d}M_3/3m_3,$$

so with the initial values $M_1 = M^0$, $M_2, M_3 = 0$, it follows that

$$M_1 = M_1^0 - 2m_1\xi, \qquad M_2 = m_2\xi, \qquad M_3 = 3m_3\xi \qquad (0 \leqslant \xi \leqslant M_1^0/2m_1).$$

The *affinity* of the rth reaction is a state variable defined by

$$A_r = -\sum_i \nu_{ir} g_i, \tag{18.4}$$

so that

$$-\mathbf{g}\,.\,\mathrm{d}\mathbf{M} = M \sum_r A_r\,\mathrm{d}\xi_r - \sum_i g_i\,\mathrm{d}_e M_i \tag{18.5}$$

and

$$T\,\mathrm{d}S = \mathrm{d}U + \mathbf{p}\,.\,\mathrm{d}\mathbf{V} + M \sum_r A_r\,\mathrm{d}\xi_r - \sum_i g_i\,\mathrm{d}_e M_i. \tag{18.6}$$

From (18.2), (18.5), and the independence of the progress variables, it follows that at chemical equilibrium

$$A_r = 0 \qquad (r = 1, 2, \ldots R), \qquad \sum_i g_i \mathrm{d}_e M_i = 0. \tag{18.7}$$

When the compound system is closed, exchanges with external phases are impossible and $\mathrm{d}_e M_i$ is zero; otherwise the second of (18.7) is satisfied as described in § 17.3.

18.2. Law of mass action

In § 21.1 it will be shown that the chemical potential of a perfect gas takes the form

$$m_i g_i = kT(\ln x_i + \psi_i(T) + \ln p),$$

where $\psi_i(T)$ depends only on the temperature and x_i is the fraction (termed the *mole fraction*) of molecular particles in the ith component. Thus for a given reaction the first of (18.7) may be written—omitting the subscript r—$\sum a_i m_i g_i = 0$, or

$$\prod x_i^{a_i} p^{\Sigma a_i} = K \qquad (K \equiv \exp(-\sum a_i \psi_i)).$$

This equation is known as the *law of mass action* and K, which depends only on T, is the *equilibrium constant*. From the definition of $\psi(T)$ given in § 21.1 we have

$$\frac{\mathrm{d}\psi}{\mathrm{d}T} = -\frac{1}{\mathscr{R}T^2}\left(h_0 + \int c_p\,\mathrm{d}T\right) = -\frac{h}{\mathscr{R}T^2},$$

therefore

$$\frac{\mathrm{d}(\ln K)}{\mathrm{d}T} = \frac{\sum a_i h_i}{\mathscr{R}T^2}.$$

Here $\sum a_i h_i$ is termed the *heat of reaction*, being in fact the heat transferred out of the system at constant pressure and temperature as a result of the reaction (cf. $T\,\mathrm{d}S = \mathrm{d}'Q = \mathrm{d}H - V\,\mathrm{d}p$).

18.3. Stability

On substituting the differential of (18.5) into (17.9) we obtain

$$\mathrm{d}T\,\mathrm{d}S - \mathrm{d}\mathbf{p}\,.\,\mathrm{d}\mathbf{V} - M\sum_{r=1}^{R}\mathrm{d}A_r\,\mathrm{d}\xi_r + \sum_i \mathrm{d}g_i\,\mathrm{d}_e M_i > 0. \qquad (18.8)$$

Thus at constant temperature and pressure in a closed system

$$\sum_{s,r}\frac{\partial A_r}{\partial \xi_s}\,\mathrm{d}\xi_s\,\mathrm{d}\xi_r < 0,$$

which requires all the principal minors of $|-\partial A_r/\partial \xi_s|$ to be positive:

$$\frac{\partial A_r}{\partial \xi_r} < 0, \qquad \frac{\partial A_r}{\partial \xi_r}\frac{\partial A_s}{\partial \xi_s} - \left(\frac{\partial A_r}{\partial \xi_s}\right)^2 > 0, \quad \text{etc.} \qquad (r, s = 1, 2, \ldots R, r \neq s), \qquad (18.9)$$

where the Maxwell relation (see (15.8)) $\partial A_r/\partial \xi_s = \partial A_s/\partial \xi_r$ has been used. It is clear from (18.8) that in (18.9) any one of the variable pairs $(T, \mathbf{p})$, $(T, \mathbf{V})$, $(S, \mathbf{p})$, $(S, \mathbf{V})$ may be chosen to be constant. Eqns (18.9) also apply with the variable pair $(U, \mathbf{V})$ constant, a conclusion that follows from (17.10) with the last term replaced by $M\sum \mathrm{d}(A_r/T)\,\mathrm{d}\xi_r$.

For open systems the requirement that the last term in (18.8) be independently positive yields inequalities similar to (18.9), of which the first one is the third inequality of (17.11).

18.4. Natural processes

A natural process as defined in § 13.3 is a spontaneous process that restores a system to an equilibrium state able to persist for a much longer time than the process time involved. Thus such processes are essentially stabilizing. It is worth remarking that some natural processes may initially appear to be destabilizing—for example, the so-called 'streaming instability' due to supersonic diffusion of one electrically charged gas through another—but, given time, such instabilities do lead to new equilibrium states.

Here **we** are interested in the direction of natural chemical processes. Consider a typical term $A_r \, d\xi_r$ of the sum over independent reactions in (18.5). By the first of (18.9), $d\xi_r > 0$ implies a decrease in A_r, and hence by Le Châletier's principle, which requires the change to bring A_r closer to its zero equilibrium value, $A_r > 0$. Similarly, $d\xi_r < 0$ implies $A_r < 0$, and in both cases $A_r \, d\xi_r > 0$. Thus for natural processes

$$A_r \, d\xi_r > 0 \qquad (r = 1, 2, \dots R). \tag{18.10}$$

It is important of course, that the reactions be independent. If an independent set is unknown, we may lose detailed information about these reactions by summing (18.10) to obtain the inequality

$$-\sum_i g_i(dM_i - d_e M_i) > 0, \tag{18.11}$$

by (18.5).

In order to determine the sign of $\sum_i g_i \, d_e M_i$ it is necessary to make assumptions about the environment P* from which the mass elements $d_e M_i$ are supplied. Let g_i^* be the value of g_i in P*, then assuming equilibrium in P* and that the mass change dM_i^* in P* is $-d_e M_i$, we have

$$\sum_i g_i^* \, d_e M_i = 0.$$

We shall show that

$$-\sum_i g_i \, d_e M_i = \sum_i (g_i^* - g_i) \, d_e M_i > 0. \tag{18.12}$$

If $d_e M_i > 0$, by eqn $(17.11)_3$ g_i increases, and as stability requires $g_i^* - g_i$ to be reduced, $g_i^* > g_i$. Similarly, $d_e M_i < 0$ implies $g_i^* < g_i$. In both cases $(g_i^* - g_i) \, d_e M_i > 0$, and (18.12) follows.

The inequality (18.11) may now be replaced by

$$-\mathbf{g} \, . \, d\mathbf{M} = -\sum_i g_i \, dM_i > 0. \tag{18.13}$$

19. The time variable in thermodynamics

19.1. Relaxation times

The progress variables $\xi_1, \xi_2, \dots \xi_R$ introduced in § 18.1 are internal coordinates defining the composition of a system during its transit between states of chemical equilibrium. Such evolution of a system is termed 'relaxation', a good non-chemical example being provided by the progressive equipartition of energy between the vibrational and translatory degrees of freedom of a diatomic molecule. A *relaxation time* τ_r may be associated with each coordinate ξ_r, this being the time it takes for the rth reaction to achieve chemical (or physical) equilibrium from some typical non-equilibrium state (cf. remarks on p. 4). Suppose that the times are arranged in order of decreasing size: $\tau_1 \geqslant \tau_2 \geqslant \ldots \geqslant \tau_R$, then on a macroscopic time scale $\mathscr{T} \gg \tau_{m+1}$, $\tau_{m+2}, \ldots$ the reactions characterized by $\xi_{m+1}, \xi_{m+2}, \ldots \xi_R$ may be neglected, or rather supposed to be in relaxed equilibrium with vanishing affinities, so that $\xi_{m+1}, \ldots \xi_R$ always have their equilibrium values. Internal coordinates often permit non-equilibrium situations to be elucidated by being described in equilibrium terms (cf. discussion in § 3.2) even in cases when the relaxation mechanisms belonging to the coordinates are unknown (see § 63).

To illuminate the concept of a relaxation time we shall consider a natural process in a closed homogeneous system, with the state coordinates including one internal variable ξ. Let the external variables U, $\mathbf{V}$ remain constant, then the affinity A will depend only on ξ, and by (18.6) and (18.10),

$$T\,\mathrm{d}S = MA(\xi)\,\mathrm{d}\xi > 0,$$

where $A(\xi) = 0$ at equilibrium. For a small departure $\mathrm{d}\xi = \xi_{\mathrm{eq}} - \xi$ from relaxed equilibrium

$$A = A_1\,\mathrm{d}\xi = -A_1(\xi - \xi_{\mathrm{eq}}) \qquad (A_1 > 0),$$

the sign of the constant A_1 following from the stability condition $(\partial A/\partial \xi)_{U,\mathbf{V}} < 0$ given in $(18.9)_1$. Thus

$$T\,\mathrm{d}S = MA_1 = MA_1(\mathrm{d}\xi)^2 > 0,$$

which is illustrated in Fig. 6 for the particular case when ξ_{eq} does not lie at an end of the range of ξ.

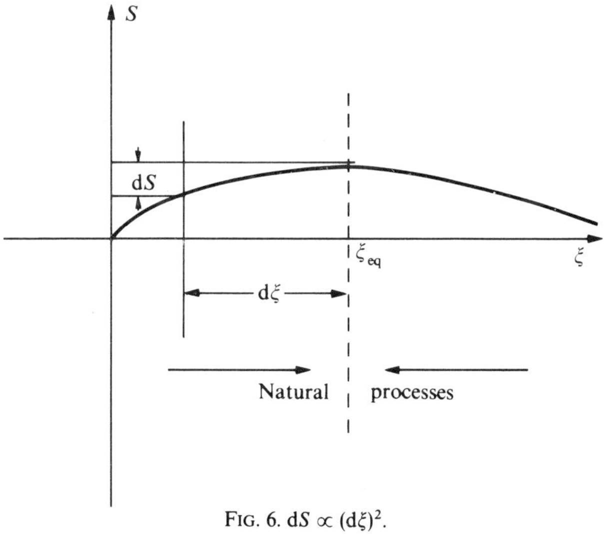

FIG. 6. $\mathrm{d}S \propto (\mathrm{d}\xi)^2$.

A plausible hypothesis for the rate at which ξ progresses towards ξ_{eq}—the reaction rate—is that it is proportional to the magnitude of the affinity A, which is itself a measure of the extent of the displacement $(\xi - \xi_{eq})$. Thus the affinity may be viewed as a 'thermodynamic force' tending to restore the system to equilibrium. With this hypothesis

$$\frac{\mathrm{d}\xi}{\mathrm{d}t} = lA = -lA_1(\xi - \xi_{eq}) = -(\xi - \xi_{eq})/\tau, \tag{19.1}$$

where l is a constant and $\tau = 1/lA_1$. Eqn (19.1) is called a *phenomenological law*; it yields the integral

$$\xi = \xi_{eq} + (\xi_1 - \xi_{eq})\,\mathrm{e}^{-t/\tau}, \tag{19.2}$$

where ξ_1 is the initial value of ξ and τ is called the *relaxation time* of the reaction. It follows from (19.2) that on a time scale $\mathscr{T} \gg \tau$, $\xi \approx \xi_{eq}$, $A \approx 0$, and the reaction proceeds in relaxed equilibrium with $S = S(U, \mathbf{V})$. On the other hand, on a time scale $\mathscr{T} \ll \tau$, $\xi \approx \xi_1$, the reaction appears 'frozen' and $S = S(U, \mathbf{V}, \xi)$. For the general case of several reactions let

$$\varepsilon_r \equiv \tau_r/\mathscr{T}, \qquad (r = 1, 2, \ldots R),$$

then the reactions for which $\varepsilon_r \gg 1$ are in *frozen equilibrium*, whereas those for which $\varepsilon_r \ll 1$ are in *relaxed equilibrium*.

19.2. Local thermodynamic equilibrium

The equilibrium process described in § 3.2 has a time scale $\mathrm{d}t$ of the same order of magnitude as the observer's time scale $\mathscr{T}$, and this latter scale determines the coordinates of the observer's state-space $X(\mathscr{T})$. As $\mathrm{d}t$ and $\mathscr{T}$ are comparable, we may divide the Gibbs relation for $X(\mathscr{T})$ by $\mathrm{d}t$ and convert it into a rate equation. Thus (18.6) may be written

$$T\frac{\mathrm{d}S}{\mathrm{d}t} = \frac{\mathrm{d}U}{\mathrm{d}t} + \mathbf{p}\cdot\frac{\mathrm{d}\mathbf{V}}{\mathrm{d}t} - \sum_i g_i \frac{\mathrm{d_e}M_i}{\mathrm{d}t} + M\sum_r A_r \frac{\mathrm{d}\xi_r}{\mathrm{d}t}, \tag{19.3}$$

where all relaxation processes that occur so rapidly that $\varepsilon_r \ll 1$ are omitted from the last summation.

As (19.3) is basic in the process thermodynamics of Part II, it is important to understand its status precisely. We have taken the Gibbs relation which *defines* the entropy appropriate to a given state-space $X(\mathscr{T})$, and divided by a time element $\mathrm{d}t$. But this element cannot be made arbitrarily small, in other words, as we want to represent the process entirely in $X(\mathscr{T})$ space there is some limit to the rates of change appearing in (19.3). Higher rates of change would require the addition of further terms to the right-hand side of (19.3).

As processes are frequently initiated in a system P by changes at its boundaries, the longest transit time of changes across P in the various thermodynamic variables imposes an obvious lower limit on dt for P to remain uniform. Clearly, large systems cannot be treated as being uniform on short time scales, and to obviate this we may divide P up into a large number of phases. And with P large and non-uniform, fluid velocities within P may be expected. The most convenient thermodynamic sub-system is the *fluid particle* $\mathscr{P}(\mathbf{r}, t)$, located at position vector **r** at time t. This is a convected fluid element of macroscopically infinitesimal volume dτ, having a typical dimension $|\mathrm{d}\mathbf{r}|$ much greater than a mean free path. As described in § 1.1, this latter constraint means that many physical effects are transmitted entirely locally, that is, between contiguous fluid particles.

For a fluid-particle system it is necessary to adopt specific variables in place of the total values appearing in (19.3). Thus on dividing (19.3) through by the constant mass M we obtain (cf. (14.10))

$$T\frac{\mathrm{d}s}{\mathrm{d}t}=\frac{\mathrm{d}u}{\mathrm{d}t}+\mathbf{p}\cdot\frac{\mathrm{d}\mathbf{v}}{\mathrm{d}t}-\sum_i g_i\frac{\mathrm{d_e}c_i}{\mathrm{d}t}+\sum_r A_r\frac{\mathrm{d}\xi_r}{\mathrm{d}t}. \tag{19.4}$$

In this equation the subscript e denotes causes external to $\mathscr{P}$, c_i is the concentration M_i/M, and d/dt is the time derivative in *any* convenient reference frame. The thermodynamically independent variables u, **v**, c_i, ξ_r are now functions of **r** and t.

There are two important constraints involved in (19.4), namely,

$$\mathrm{d}t \sim \mathscr{T} \gtrsim \tau_{\mathscr{P}}, \tag{19.5}$$

where $\tau_{\mathscr{P}}$ is the longest transit time for a signal involving thermodynamic information—for example, a sound wave—to cross $\mathscr{P}$. These restrictions ensure that processes are in temporal equilibrium and that during their progress the system $\mathscr{P}$ has nearly uniform properties. Their imposition is usually termed the hypothesis of *local thermodynamic equilibrium.*

A particularly important point to note from Fig. 6 is that the error committed in the Gibbs relation (19.4) by ignoring a relaxation process that is not quite fast enough to be relaxed, is only second-order in the displacement of the progress variable from its equilibrium value. This means that the $\sim$ relations in (19.5) may be liberally interpreted with little resulting error.

19.3. The central role of the time variable in thermodynamics

Although time does not appear explicitly in the various thermodynamic relations—unless formally introduced to convert differentials into time derivatives as in (19.4)—it is in fact the most important independent variable in thermodynamics. And its role is not merely that of a passive parameter on which the thermodynamic coordinates of state-space may depend; as described in § 1.2 the nature of state-space itself depends on this variable via the observer's time scale $\mathscr{T}$. Basic concepts like adiabaticity, closure, equilibrium (or quasi-static) process, relaxed and frozen equilibria, and non-equilibrium process all involve comparisons of time scales, and finally the actual value of the entropy depends on $\mathscr{T}$.

Unfortunately the absence of t as a variable has led to some confusion about the relevance of thermodynamics in situations, such as experienced in gas turbines or shock waves, where rapid time changes are inevitable. Sometimes the subject is even given the misleading name 'thermostatics', with the unfortunate implication that there is a 'thermo*dynamics*', possessing time as a natural independent variable, to follow in some further development. This misconception arises from supposing the concept of 'thermodynamic state' to be absolutely rather than relatively independent of time.

20. Irreversibility

A process $X_1 \to X_2$ in which an isolated system P evolves from state X_1 to state X_2 in a finite time is *reversible* if the process $X_2 \to X_1$ is possible and *irreversible* if it is not. If P is in some form of contact with its surroundings P* whose state is X^*, then reversibility requires the combined process $X_1 \to X_2$, $X_1^* \to X_2^*$ to be reversible, which may be regarded as a process in the isolated system (P + P*). It is therefore sufficient to consider isolated systems. When the evolution of a system is specified by an equation in which time appears as an independent variable, it is a simple matter to decide whether the process is reversible or not—if the equation is invariant to a change in the sign of t, the process is reversible, otherwise not. Although the transformation is termed 'time-reversal', the object of changing the sign of t is not really to reverse the direction of time but to reverse the *motions* (including spins) of all the microscopic particles and waves comprising the system. This motion-reversal requires a change of sign for all fluid velocities, for electric currents, and for magnetic inductions (cf. (11.8) and (11.9)). However, no sign change occurs in the purely thermodynamic variables like T, U, and S. Process reversibility requires *all* the relevant equations to be invariant under time-reversal; for example, (19.4) is clearly invariant under $t \to -t$, whereas the inequality (18.10), $A_r \, d\xi_r/dt > 0$, for an isolated system is not.

The irreversibility of thermodynamic processes, which appears directly in the form we have taken for the second law (p. 17) and indirectly as consequences of thermal (p. 26), mechanical (p. 34), and chemical stability (p. 62), is seemingly contradicted by two famous 'paradoxes'.

Loschmidt's microscopic reversibility paradox: Since the laws of mechanics (Newton's equations) governing molecular and other particle motions are invariant under time-reversal, to each process there is a motion-reversed process.†

† It is now believed that microscopic reversibility itself can be violated in rare circumstances (see Overseth 1969).

Zermelo's recurrence paradox: There is a theorem due to Poincaré to the effect that any given (microscopically defined) state of a system of interacting particles in a closed region of space will recur to within any assigned accuracy. Thus any non-equilibrium state once experienced by a system will occur again after long enough time.

There is, however, as we shall see, no conflict between macroscopic irreversibility and microscopic reversibility or recurrence.

The paradoxes may be illustrated by considering the free expansion of a gas from a small region in one corner of an otherwise empty container. The molecules experience collisions with each other and the walls, and after a very little time, t_1 say, will more or less uniformly fill the container. Now halt the particles and reverse their motions. Because of microscopic reversibility, in a time t_1 the particles will have returned to the corner region from which they started, provided of course that any microscopic motions of the container boundaries are also reversed. Macroscopic properties will revert to their original values, and in particular the entropy will have been reduced. Irreversibility *could* be introduced into this example through the boundary conditions by either reversing the motions of the gas molecules *alone*, or allowing the boundary conditions to be stochastic rather than determinate. For this reason it is often claimed that irreversibility has its origin in the interaction between a system and its environment. Indeed any uncertainties either in the boundary conditions, or for that matter in the initial conditions of the reversed motion, would prevent *our knowing* the outcome of the motion-reversal, but this fact is not to be equated with system irreversibility. Loschmidt's paradox is predicated on the determinate laws of microscopic reversibility and to introduce stochastic elements at the wall molecules is quite contrary to this basis. Again the failure to reverse the motion of determinate boundaries violates the condition for reversibility mentioned at the beginning of this article, namely, that the environment must be included in the reversal process. It is conceptually simpler to extend and thereby isolate the system by including its interacting environment as part of it.

It is in the quite exceptional nature of the *prepared* initial conditions of the gas molecules at the instant after motion-reversal that we resolve the paradox. While these conditions are possible, only an Act of God could establish them in the way described above, and they have almost zero probability of being found in an unprepared gas occupying the container. In short, in our example we have 'cooked' the initial conditions, without noting their extreme improbability among all possible initial conditions; there is no paradox—merely an extremely improbable event. The extremely small probability p of any specified initial condition is obviously a direct consequence of the enormous number N of molecules involved in any closed thermodynamic system. Zermelo's paradox is resolved in a similar way. The very large value of N means that the Poincaré cycle time $\mathfrak{T}$ (time for the recurrence of a given micro-state) is gigantic, for example, millions of times the present age of the Universe for just 1 cm^3 of gas at NTP. We conclude that microscopic reversibility has an extremely small chance of yielding macroscopic reversibility except after an almost infinite lapse of time.

Classical thermodynamics strictly applies only to a continuum and can take no cognizance of the $6N$ internal degrees of freedom ($3N$ coordinates, $3N$ momenta) involved in our example. Indeed the complete suppression of the particulate nature of a physical system, embodied in the definition of a continuum, requires $N \to \infty$, which is known as the 'thermodynamic limit'. This limit removes both paradoxes by sending the Poincaré time $\mathfrak{T}$ to infinity and by reducing the probability p of macroscopic reversibility to zero.

On the other hand, when N is not too large, macroscopic reversible changes, of which Brownian motion is a well-known example, become more probable. However, fluctuation theory (§ 29) is not a part of classical thermodynamics.

4

SYSTEMS POSSESSING INTERNAL DEGREES OF FREEDOM

IN this chapter we shall further reduce the observer's time scale to $\mathscr{T}_K$, where the subscript K denotes 'kinetic'. This time is quite short—of the order of the time a particle in a gas moves between successive collisions—and as chemical reaction times are much longer, we now have an equilibrium process time dt such that

$$\begin{array}{cccccc} dt \sim & \mathscr{T}_K & \ll & \mathscr{T}_I & \ll & \mathscr{T}_R \\ & \text{kinetic} & & \text{chemical} & & \text{engineering} \\ & \text{thermodynamics} & & \text{thermodynamics} & & \text{thermodynamics} \end{array}$$

Naturally such processes are irreversible. The subject may be termed 'kinetic thermodynamics', and is a branch of the kinetic theory of matter.

Here we shall be concerned with the thermodynamics of a system of particles—molecules, atoms, ions or electrons, etc.—sufficiently dilute to be in the gaseous state. The thermodynamic variables describing the state of such a system may be taken to be u the internal energy per unit mass, the specific volume v, the concentrations $c_1, c_2, \ldots c_l$ of the various species present, and the *velocity distribution functions* $f_1, f_2, \ldots f_l$, where $f_i(\mathbf{r}, \mathbf{w}, t)\, d\mathbf{w}\, d\mathbf{r}$ is the total number of particles of the ith species that at time t have positions lying in a volume element $d\mathbf{r}$ about $\mathbf{r}$ and velocities (of translation) lying within the velocity space element $d\mathbf{w}$ about $\mathbf{w}$.† These distribution functions are a set of internal coordinates prescribed over the 'internal' space $\mathbf{w}$, and they are thermodynamically important only on time scales small enough to be comparable with the relaxation times for f_i, $i = 1, 2, \ldots l$, to achieve their equilibrium values f_i^{eq}. It is necessary, of course, to choose the elements $d\mathbf{r}$ and $d\mathbf{w}$ so that $f_i(\mathbf{r}, \mathbf{w}, t)$ may be taken to be a smooth function of its arguments (see § 1.1).

In this model it is implicit that the particles are featureless, that is, lack an internal structure that needs specifying by internal degrees of freedom additional to $\mathbf{r}$ and $\mathbf{w}$. Another constraint on the model is that the gas is supposed to be sufficiently dilute to behave like a perfect gas. Thus in this chapter our aim is to pursue the short-time-scale thermodynamic behaviour of a rather special system, whereas in the earlier chapters the time scale is somewhat longer and the systems considered are more general in character.

† The notation for a volume element of physical space has been altered from $d\tau$ to $d\mathbf{r}$ to conform with the notation $d\mathbf{w}$ for a volume element of velocity space.

21. Internal degrees of freedom

21.1. Some relations for perfect gases

The system being considered is a mixture of perfect gases in thermal equilibrium, that is, at a single temperature T. Consider just one component. Let $\rho\,\mathrm{d}\mathbf{r}$ be the mass of this gas in a fluid particle system $\mathscr{P}(\mathbf{r}, t)$ (see § 19.2) of volume $\mathrm{d}\mathbf{r}$ and let n be the number of molecules or atomic particles of this gas per unit volume in $\mathscr{P}$. Denote by m, w the mass and molecular weight respectively of such a microscopic particle, then there are $\rho\,\mathrm{d}\mathbf{r}/w$ moles in $\mathscr{P}$ of density $\rho = mn$. The thermal equation of state (10.19) may be written $p\,\mathrm{d}\mathbf{r} = (\rho\,\mathrm{d}\mathbf{r}/w)RT$, or

$$p = \mathscr{R}\rho T = knT, \tag{21.1}$$

where

$$\mathscr{R} = R/w, \qquad k = \mathscr{R}m = Rm/w = R/N_{\mathrm{A}}.$$

The number $w/m = N_{\mathrm{A}}$, called the Avogadro number, is the number of microscopic particles in one mole and is the same for all gases; the universal constant k is Boltzmann's constant.†

On dividing eqns (10.21) by the mass $M = \rho\,\mathrm{d}\mathbf{r}$ of the thermodynamic system and introducing the specific heat capacities $c_V = C_V/M$, $c_p = C_p/M$, we find

$$\mathrm{d}u = c_V\,\mathrm{d}T, \qquad \mathrm{d}h = c_p\,\mathrm{d}T,$$
$$c_p - c_V = \mathscr{R}, \tag{21.2}$$
$$T\,\mathrm{d}s = c_V\,\mathrm{d}T + p\,\mathrm{d}\rho^{-1} = c_p\,\mathrm{d}T - \rho^{-1}\,\mathrm{d}p.$$

For a calorically perfect gas c_p, c_V are constants, and (21.2) immediately yield the integrals

$$u = c_V T = \frac{1}{\gamma-1}\frac{p}{\rho}, \qquad h = c_p T = \frac{\gamma}{\gamma-1}\frac{p}{\rho}, \qquad (\gamma = c_p/c_V), \tag{21.3}$$

$$s = s_0 + c_p \ln T - \mathscr{R}\ln p. \tag{21.4}$$

Hence the specific Gibbs function, $g = u + p/\rho - Ts$, may be written

$$g = \mathscr{R}T\left(\ln \rho - \frac{1}{\gamma-1}\ln T + \ln \mathscr{R} + \frac{c_p - s_0}{\mathscr{R}}\right). \tag{21.5}$$

For the more general case of a perfect gas with c_p a function of temperature, $h = h_0 + \int c_p\,\mathrm{d}T$, etc., and it is easily shown that (21.5) is generalized to

$$g = \mathscr{R}T[\ln p + \psi(T)], \tag{21.6}$$

† $N_{\mathrm{A}} = 6{\cdot}023\times 10^{23}$, $k = 1{\cdot}3804\times 10^{-23}\,\mathrm{JK^{-1}}$.

where

$$\psi(T) = \frac{h_0}{\mathscr{R}T} - \mathscr{R}\int T^{-2}\left(\int c_p\,\mathrm{d}T\right)\mathrm{d}T - \frac{s_0}{\mathscr{R}}.$$

For the ith component of a compound system the (partial) pressure $p_i = kn_iT$ is related to the total pressure $p = knT$ of the system by $p_i = x_ip$, where x_i is the number ratio n_i/n, termed the *mole fraction*. Hence by (21.6) the chemical potential for this component is given by

$$g_i = \frac{kT}{m_i}[\ln x_i + \ln p + \psi_i(T)]. \tag{21.7}$$

21.2. Generalized Gibbs relation

Here we shall start with the fundamental relation (14.10),

$$T\,\mathrm{d}s = \mathrm{d}u + \mathbf{p}\,.\,\mathrm{d}\mathbf{v} - \sum_i g_i\,\mathrm{d}c_i, \tag{21.8}$$

and, by assuming a more detailed description of each component than that provided by a single number c_i, extend it for a system of perfect gases.

Suppose that the ith component is subdivided into K distinct sub-fluids and that these are distinguished by a subscript k, $k = 1, 2, \ldots K$. If these sub-fluids are not in material equilibrium with each other, the last term in (21.8) must be replaced by a sum over all the component sub-fluids

$$-\sum_i\sum_k g_{ik}\,\mathrm{d}c_{ik} = -\sum_{i,k}\frac{m_i}{m}g_{ik}\,\mathrm{d}\left(\frac{n_{ik}}{n}\right). \tag{21.9}$$

Here g_{ik} is the specific Gibbs function for the kth sub-fluid within the ith component, c_{ik} is the mass concentration of this fluid, n_{ik} is its number density, and m is the average mass. Thus in terms of the mole fraction $x_{ik} = n_{ik}/n$,

$$m = \sum_{i,k} m_i x_{ik}, \qquad 1 = \sum_{i,k} x_{ik}. \tag{21.10}$$

When equilibrium prevails between the sub-fluids of the ith component, g_{ik} is the same for all values of k (see (17.7)), and as (2.19) must then reduce to the sum appearing in (21.8), it follows that $g_{i1}^{\mathrm{eq}} = g_{i2}^{\mathrm{eq}} = \ldots = g_{ik}^{\mathrm{eq}} = \ldots = g_i$. Hence (21.8) may be generalized to

$$T\,\mathrm{d}s = \mathrm{d}u + \mathbf{p}\,.\,\mathrm{d}\mathbf{v} - \sum_i g_i\,\mathrm{d}c_i + \sum_{i,k} A_{ik}\,\mathrm{d}x_{ik}, \tag{21.11}$$

where

$$A_{ik} = \frac{m_i}{m}(g_{ik}^{\mathrm{eq}} - g_{ik}) \tag{21.12}$$

are the affinities for the internal coordinates x_{ik}. The Euler relation for (21.11) (see rule in § 15.1) is

$$Ts = u+\mathbf{p}\,.\,\mathbf{v}-g+\sum_{i,k} A_{ik}x_{ik}, \tag{21.13}$$

where by (15.9) g is the specific Gibbs function for the system.†

21.3. A continuum of internal coordinates

Let the property that distinguishes the sub-fluids of the ith component be a continuous variable y_i, which may be a scalar such as the orientation of a molecular dipole in an electric field or a vector such as the velocity of a Brownian particle. Then as the number of sub-fluids K tends to infinity, the sums over k can be replaced by integrals over y_i. The sub-fluid mole fraction may be written in the form

$$x_{ik} = \xi_i(y_i)\,\mathrm{d}y_i, \tag{21.14}$$

where by $(21.10)_2$ the internal coordinate ξ_i satisfies the constraint

$$\sum_i \int \xi_i(y)\,\mathrm{d}y = 1, \tag{21.15}$$

the integration extending over the whole of y_i-space. The last term of (21.11) may now be replaced by

$$\sum_i \int A_i(y)\,\mathrm{d}\xi_i(y)\,\mathrm{d}y = \int \mathbf{A}(y)\,.\,\mathrm{d}\boldsymbol{\xi}(y)\,\mathrm{d}y,$$

where $A_i(y)$ is the affinity for the 'configuration' y and $\mathbf{A} = \{A_1, A_2, \ldots A_l\}$, $\boldsymbol{\xi} = \{\xi_1, \xi_2, \ldots \xi_l\}$. Hence the fundamental equation now assumes the vectorial form

$$T\,\mathrm{d}s = \mathrm{d}u+\mathbf{p}\,.\,\mathrm{d}\mathbf{v}-\sum_i g_i\,\mathrm{d}c_i+\int \mathbf{A}(y)\,.\,\mathrm{d}\boldsymbol{\xi}(y)\,\mathrm{d}y, \tag{21.16}$$

with the Euler integral

$$Ts = u+\mathbf{p}\,.\,\mathbf{v}-g+\int \mathbf{A}(y)\,.\,\boldsymbol{\xi}(y)\,\mathrm{d}y. \tag{21.17}$$

Notice that s is now a function of u, $\mathbf{v}$, and c_i, $i = 1, 2, \ldots$, and a functional of $\mathbf{A}(y)$, which must be prescribed over the 'internal' space y. Such a thermodynamic system is said to possess an *internal degree of freedom.*

21.4. The affinity of a perfect gas

From (21.7)

$$g_{ik} = \frac{kT}{m_i}[\ln x_{ik}+\ln p+\psi_{ik}(T)],$$

† It is important to notice that the entropy involved in the definition of g is the equilibrium ($A_{ik} = 0$) entropy and not that appearing on the left-hand side of (21.13).

so that (21.12) gives

$$A_{ik} = -\frac{kT}{m}\ln\left(\frac{x_{ik}}{x_{ik}^{\text{eq}}}\right), \tag{21.18}$$

x_{ik}^{eq} being the equilibrium mole fraction at the fixed temperature T. For a continuous distribution of mole fraction, (21.14) gives

$$A_{ik} = A_i(y) = -\frac{kT}{m}\ln\left(\frac{\xi_i(y)}{\xi_i^{\text{eq}}(y)}\right),$$

and eqn (21.16) becomes

$$T\,\mathrm{d}s = \mathrm{d}u + \mathbf{p}\,.\,\mathrm{d}\mathbf{v} - \sum_i g_i\,\mathrm{d}c_i - \frac{kT}{m}\sum_i \int \mathrm{d}\xi_i(y)\ln\left(\frac{\xi_i(y)}{\xi_i^{\text{eq}}(y)}\right)\mathrm{d}y. \tag{21.19}$$

In order that the final term here be really independent we require the change $\mathrm{d}\xi_i$ to occur at constant values of u, $\mathbf{v}$, and c_i. Eqn (21.19) is a general result for perfect or ideal systems, applicable to any type of distribution function ξ specified over an appropriate inner variable y.

21.5. *Velocity distribution functions*

In this chapter our main concern will be with 'velocity distribution functions' $f_i(\mathbf{r}, \mathbf{w}, t)$ in velocity space (w_x, w_y, w_z). Let $n_i(\mathbf{r}, t)$ be the number density of microscopic particles of the ith fluid at $\mathscr{P}(\mathbf{r}, t)$ then $f_i(\mathbf{r}, \mathbf{w}, t)$ is defined to be that fraction of n_i having velocities in the range $\mathbf{w}$, $\mathbf{w}+\mathbf{dw}$, where $\mathbf{dw}$ is a volume element in velocity space. Thus integrating over the whole of this space

$$\rho_i = m_i n_i = m_i \int f_i\,\mathbf{dw}, \qquad \sum_i \int f_i\,\mathbf{dw} = n. \tag{21.20}$$

Comparison with (21.15) shows that $\xi_i = f_i/n$ and $y_i = \mathbf{w}$, so that (21.19) takes the form

$$T\,\mathrm{d}s = \mathrm{d}u + p\,\mathrm{d}\rho^{-1} - \sum_i g_i\,\mathrm{d}c_i - \frac{kT}{\rho}\sum_i \int \mathrm{d}f_i \ln\left(\frac{f_i}{f_i^{\text{eq}}}\right)\mathbf{dw}, \tag{21.21}$$

on assuming that the only work term is due to density changes, and omitting such changes from the integral in accord with the remark following (21.19).

The Euler relation is

$$\rho T s = \rho u + p - \rho g - kT\sum_i \int f_i \ln\left(\frac{f_i}{f_i^{\text{eq}}}\right)\mathbf{dw}. \tag{21.22}$$

22. Kinetic Definitions for the independent variables

From eqn. (21.21) we observe that the state equations will be known when T, p, g_i, and the affinities

$$A_i \equiv -\frac{kT}{m}\ln\left(\frac{f_i}{f_i^{\text{eq}}}\right) \tag{22.1}$$

are specified as functions of the independent variables $u, \rho, c_1, c_2, \ldots$ and as functionals of $f_i(\mathbf{w})$, $i = 1, 2, \ldots$. In general such state equations are phenomenological in nature, to be determined by experiment or possibly by some theory ancillary to thermodynamics. For the compound system of featureless particles under present study, these equations can be deduced from kinetic theory without difficulty.

The first step is to express the independent variables u, ρ, c_i, $i = 1, 2, \ldots$, as functionals of the distribution $f_i(\mathbf{w})$; such relations provide *definitions* of u, ρ, and c_i, replacing their introduction as 'primitive concepts' as described in §§ 1.3 and 9.3.

Averages over velocity space of a function $\varphi_i(\mathbf{r}, \mathbf{w}_i, t)$ associated with the ith component will be denoted by $\bar{\varphi}_i(\mathbf{r}, t)$,

$$n_i\bar{\varphi}_i \equiv \int \varphi_i f_i \,\mathrm{d}\mathbf{w}_i, \tag{22.2}$$

the functional dependence being omitted for brevity. The average velocity of the ith component is $\mathbf{v}_i$, that is,

$$\mathbf{v}_i = \overline{\mathbf{w}}_i. \tag{22.3}$$

The total mass density ρ and the barycentric velocity $\mathbf{v}$ of the system are given by

$$\rho = \sum_i \rho_i = \sum_i m_i n_i, \qquad \rho\mathbf{v} = \sum_i \rho_i \mathbf{v}_i. \tag{22.4}$$

As the particles are featureless, the internal energy u_i is merely the kinetic energy of the particle motion relative to the centre of mass of the ith component,

$$\rho_i u_i = \int \tfrac{1}{2}m_i(\mathbf{w}_i - \mathbf{v}_i)^2 f_i \,\mathrm{d}\mathbf{w}_i = \tfrac{1}{2}\rho_i\overline{c_i^2} \qquad (\boldsymbol{c}_i \equiv \mathbf{w}_i - \mathbf{v}_i), \tag{22.5}$$

where $\boldsymbol{c}_i$ is sometimes termed the 'peculiar' velocity of the ith component.

The total internal energy density of a purely *thermodynamic* nature is

$$\rho u = \sum_i \rho_i u_i = \sum_i \tfrac{1}{2}\rho_i\overline{c_i^2}, \tag{22.6}$$

whereas the energy density defined relative to the centre of mass of the compound system is

$$\rho u_{\mathrm{m}} = \sum_i \int \tfrac{1}{2} m_i (\mathbf{w}_i - \mathbf{v})^2 f_i \,\mathrm{d}\mathbf{w}_i = \sum_i \tfrac{1}{2}\rho_i \overline{\mathscr{C}_i^2} \qquad (\mathscr{C}_i \equiv \mathbf{w}_i - \mathbf{v}),$$

or (22.7)

$$\rho u_{\mathrm{m}} = \rho u + \rho u' \qquad \left(\rho u' \equiv \sum_i \tfrac{1}{2}\rho_i (\mathbf{v}_i - \mathbf{v})^2\right),$$

where $\rho u'$ is the *diffusion* energy density. It is usual to take ρu_{m} as defining the energy density, but as will be explained later (§§ 23.5, 53.2), in process thermodynamics this mixing together of the energy due to purely random motions and of energy due to the often controllable relative fluid velocities is sometimes unsatisfactory.

The required definitions of $c_i = \rho_i/\rho$, ρ, and u are thus given by eqns (21.20), (22.4), and (22.6). The next step is to deduce state equations for p, T, g_i, and A_i—or equivalently f_i^{eq} for substitution in (22.1)—from physical considerations.

23. The state equations

23.1. *Pressure*†

Consider a surface element $\mathbf{n}\,\mathrm{d}\Sigma$ being convected with the barycentric velocity $\mathbf{v}$ through the system. The net flux of particles of all kinds across $\mathbf{n}\,\mathrm{d}\Sigma$ during a time $\mathrm{d}t$ is found by summing over all velocity groups,

$$\mathrm{d}t\,\mathrm{d}\Sigma \sum_i \int \mathbf{n} \,.\, \mathscr{C}_i f_i \,\mathrm{d}\mathbf{w}_i,$$

$\mathscr{C}_i \equiv \mathbf{w}_i - \mathbf{v}$ being the velocity relative to the convected surface. These particles transport properties like their energy, momentum, etc.; let $\varphi_i(\mathbf{r}, \mathbf{w}_i, t)$ denote such a property for the ith species, then the amount of φ_i conveyed across $\mathbf{n}\,\mathrm{d}\Sigma$ per second is

$$\mathrm{d}\Sigma \sum_i \int \mathbf{n} \,.\, \mathscr{C}_i \varphi_i f_i \,\mathrm{d}\mathbf{w}_i.$$

Now assume $\mathbf{n}\,\mathrm{d}\Sigma$ to be a solid, perfectly reflecting surface, then the difference between the momentum of the impinging particles and that of the

† To appreciate the material in this section the reader should first read the account of the pressure tensor given in § 53.1.

rebounding particles per second is

$$\mathrm{d}\Sigma \sum_i \left[\int_{\mathbf{n}.\mathscr{C}_i > 0} \mathbf{n} \,.\, \mathscr{C}_i m_i \mathscr{C}_i f_i \,\mathrm{d}\mathbf{w}_i - \int_{\mathbf{n}.\mathscr{C}_i < 0} (-\mathbf{n} \,.\, \mathscr{C}_i) m_i \mathscr{C}_i f_i \,\mathrm{d}\mathbf{w}_i \right]$$

$$= \mathrm{d}\Sigma\, \mathbf{n} \,.\, \sum_i \int m_i \mathscr{C}_i \mathscr{C}_i f_i \,\mathrm{d}\mathbf{w}_i,$$

on using the fact that at each collision the particle momentum normal to the surface is reversed in sign. Hence the pressure normal to the surface owing to

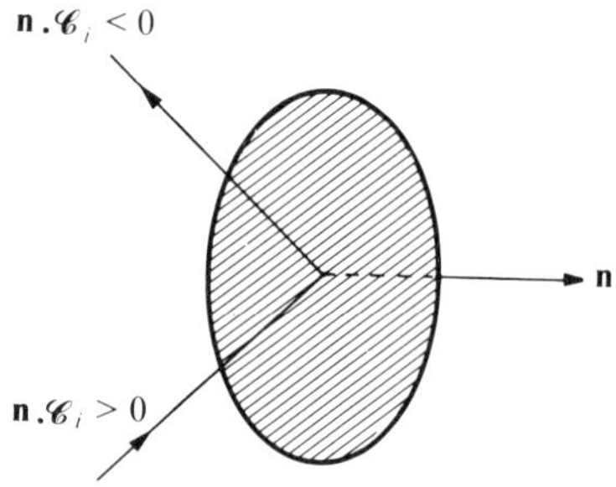

FIG. 7. Illustrating molecular pressure.

the change of momentum of the impinging particles is $\mathbf{n} \,.\, \mathbf{p}$, where $\mathbf{p}$ is the symmetric pressure tensor,

$$\mathbf{p} \equiv \sum_i \int m_i \mathscr{C}_i \mathscr{C}_i f_i \,\mathrm{d}\mathbf{w}_i. \tag{23.1}$$

Now $\mathscr{C}_i = \mathbf{w}_i - \mathbf{v} = \boldsymbol{c}_i + (\mathbf{v}_i - \mathbf{v})$, whence (23.1) yields

$$\mathbf{p} = \sum_i [\mathbf{p}_i + \rho_i(\mathbf{v}_i - \mathbf{v})(\mathbf{v}_i - \mathbf{v})],$$

where

$$\mathbf{p}_i \equiv \rho_i \overline{\boldsymbol{c}_i \boldsymbol{c}_i}, \tag{23.2}$$

is the pressure tensor for the ith component.

In the absence of internal structure of the particles, there is no bulk viscosity (see remarks on p. 193), which means that the trace of $\mathbf{p}_i$ is $3p_i$, where p_i is the (partial) pressure due to the ith component. Thus, taking the trace of eqn $(23.2)_2$ and adopting Dalton's law for the total pressure, we have

$$p_i = \tfrac{1}{3}\rho \overline{c_i^2}, \qquad p = \sum_i p_i. \tag{23.3}$$

The trace of $(23.2)_1$ yields

$$p_{\mathrm{m}} = p + p' = \tfrac{1}{3}\,\mathrm{trace}\,\mathbf{p} \qquad \left(p' \equiv \tfrac{1}{3} \sum_i \rho_i (\mathbf{v}_i - \mathbf{v})\right), \tag{23.4}$$

where p' is the 'diffusion' pressure. It now follows from (22.5), (22.6), and (22.7) that

$$p_i = \tfrac{2}{3}\rho_i u_i, \qquad p = \tfrac{2}{3}\rho u, \qquad p_m = \tfrac{2}{3}\rho u_m \qquad (p' = \tfrac{2}{3}\rho u'), \tag{23.5}$$

which are state equations of the same form as that for a perfect gas in equilibrium with a value of $\frac{5}{3}$ for γ (see (21.3)).†

23.2. Temperature

The fact that the caloric state equations in (23.5) hold for non-equilibrium states suggests that the equilibrium thermal state equations for a perfect gas, namely,

$$p_i = n_i kT, \qquad p = \sum_i p_i = nkT \qquad \left(n = \sum_i n_i\right), \tag{23.6}$$

may also apply to non-equilibrium states. To establish this we shall consider a single component, omitting the subscript i. For this component (23.5) and (23.6) yield for equilibrium states

$$\tfrac{1}{2}kT = \frac{\rho u}{3n} = \frac{\text{energy per unit volume}}{\text{degrees of freedom per unit volume}}, \tag{23.7}$$

where by 'degrees of freedom' is meant the total number of translatory degrees of freedom, there being three for each particle. This equation shows that in the equilibrium state the energy is $\frac{1}{2}kT$ per degree of freedom. Clearly for all states—equilibrium or not—of an *isolated* system the ratio $\rho u/3n$ must have the same value. Consequently, if non-equilibrium values of ρu and n are substituted into the right-hand side of (23.7), the equation remains valid and so it seems reasonable to interpret (23.7) as providing a *definition* of T for non-equilibrium states.

It remains to verify that such a definition conforms with the basic property of temperature, namely, that no heat flows between bodies at the same temperature. Let P_1 and P_2 be isolated systems in non-equilibrium states, with internal energies U_1, U_2 and total particle numbers N_1, N_2 respectively, such that

$$\rho_1 u_1/3n_1 = U_1/3N_1 = U_2/3N_2 = \rho_2 u_2/3n_2. \tag{23.8}$$

Now allow P_1 and P_2 to come into thermal contact; when complete equilibrium is finally attained, we may use (23.7) both for the combined system and for the separate systems to write

$$\tfrac{1}{2}kT = (U_1' + U_2')/3(N_1 + N_2) = U_1'/3N_1 = U_2'/3N_2,$$

† The terms 'equilibrium' and 'non-equilibrium' used in this chapter denote the conditions $f_i = f_i^{\text{eq}}$ and $f_i \neq f_i^{\text{eq}}$.

the dashes denoting final values. By energy conservation $U_1' + U_2' = U_1 + U_2$, and

$$\tfrac{1}{2}kT = (U_1 + U_2)/3(N_1 + N_2) = U_1/3N_1 = U_2/3N_2,$$

on using (23.8). It follows that $U_1' = U_1$, $U_2' = U_2$, so no heat has been transferred between P_1 and P_2 and (23.8) is therefore the condition for thermal equilibrium. This verifies (23.7) as a definition of T, and it follows that (23.6) also holds for non-equilibrium states.

Finally note that if in (23.7) we use the *total* energy density ρu_m, which is to assume that diffusion energy may be transferred as heat, then the appropriate temperature is T_m, where

$$nkT_m = p_m = \tfrac{2}{3}\rho u_m = \sum_i \tfrac{1}{3}\rho_i \overline{\mathscr{C}_i^2}, \tag{23.9}$$

which is larger than T. The Gibb's equation for this case will have u_m, p_m, and T_m in place of u, p, and T.

23.3. *Equipartition of energy*

At this point it is appropriate to introduce the very important statistical-mechanical theorem of *equipartition of energy* (see § 31.1) which is the following:

If the energy associated with any degree of freedom is a quadratic function of the variable specifying the degree of freedom, the equilibrium value of the corresponding energy is $\tfrac{1}{2}kT$.

Thus in the equilibrium state the energy is equally distributed over the various degrees of freedom, whereas eqn (23.7), which also holds for non-equilibrium states, merely states that the energy is $\tfrac{1}{2}kT$ *per* degree of freedom.

As a simple but important application, consider the case when the particles of the ith component are molecules, possibly with internal structure, each possessing f_i degrees of freedom of the type required by the theorem. Then by the theorem and (21.3)

$$\rho u = \left(\sum_i f_i n_i\right) \times \tfrac{1}{2}kT = \rho c_V T = p/(\gamma - 1). \tag{23.10}$$

Hence, with

$$\rho = mn = \sum_i m_i n_i, \qquad f \equiv \sum_i f_i n_i/n,$$

we have

$$c_V = \tfrac{1}{2}fk/m = \tfrac{1}{2}f\mathscr{R}, \qquad c_p = (\tfrac{1}{2}f + 1)\mathscr{R}, \tag{23.11}$$

and

$$\gamma = (2 + f)/f.$$

For monatomic particles $f = 3$ and

$$c_V = \frac{3}{2}\frac{k}{m}, \qquad c_p = \frac{5}{2}\frac{k}{m}, \qquad \gamma = \frac{5}{3}. \tag{23.12}$$

23.4. Gibbs function

An expression for the specific Gibbs function of a perfect gas in equilibrium was given in (21.5). On applying this to the ith component and using (23.12) we obtain

$$g_i = \frac{kT}{m_i}\left[\ln n_i - \tfrac{3}{2}\ln kT + \tfrac{5}{2}(1+\ln k) - \frac{m_i s_{0i}}{k}\right],$$

or

$$g_i = \frac{kT}{m_i}\ln\left[n_i\left(\frac{m_i}{2\pi kT}\right)^{\frac{3}{2}}\right] \qquad (i = 1, 2, \ldots l), \tag{23.13}$$

on choosing the disposable constant s_{0i} to have the value s'_{0i}, where

$$s'_{0i} \equiv \frac{k}{m_i}\left[\frac{5}{2} + \frac{5}{2}\ln k - \frac{3}{2}\ln\left(\frac{m_i}{2\pi}\right)\right]. \tag{23.14}$$

As will be seen below, this choice for s_{0i} simply amounts to a convenient allocation of the origin for entropy—absolute entropies have no significance in either classical or kinetic thermodynamics and can be introduced only via the concepts of statistical thermodynamics and quantum theory (§ 25.3). The effect of not setting $s_{0i} - s'_{0i}$ equal to zero will be noted later to assure the reader of this point.

The result for g_i in (23.13)—which could easily be expressed in terms of the independent variables u, ρ, and c_i—will now be taken as *defining* g_i in non-equilibrium states. It is readily verified that this extension of (23.13) preserves the role of g_i as a chemical potential (§ 17.3).

23.5. The equilibrium distribution function

Finally we require the state equation for the affinity A_i defined in (22.1); for this it is necessary to know the equilibrium distribution function f_i^{eq}. This distribution may be obtained in several ways; in § 75.3 we give the usual account based on Boltzmann's collision integral, and in the next section a rather simpler treatment that depends on equations already derived in § 21.5. Here we shall quote the result (due to Maxwell)

$$f_i^{\mathrm{eq}} = f_{ic}^{\mathrm{eq}} \equiv \exp\left[(m_i g_i - \tfrac{1}{2}m_i c_i^2)/kT\right] = n_i\left(\frac{m_i}{2\pi kT}\right)^{\frac{3}{2}}\exp\left(-\tfrac{1}{2}m_i c_i^2/kT\right)$$

or (23.15)

$$f_i^{\mathrm{eq}} = f_{im}^{\mathrm{eq}} \equiv \exp\left[(m_i g_{im} - \tfrac{1}{2}m_i \mathscr{C}_i^2)/kT_{\mathrm{m}}\right] = n_i\left(\frac{m_i}{2\pi kT_{\mathrm{m}}}\right)^{\frac{3}{2}}\exp\left(-\tfrac{1}{2}m_i \mathscr{C}_i^2/kT_{\mathrm{m}}\right).$$

The distribution function in the first of (23.15) is for *component* equilibrium, whereas that in the second of (23.15) is for *mixture* equilibrium. The theory may be developed in terms of either equilibrium—with component equilibrium we use the purely thermodynamic variables p, u, and T, whereas with mixture equilibrium it is necessary to replace these by p_m, u_m, and T_m. The difference is important in process thermodynamics, for component equilibrium is sometimes attained more rapidly in a relaxing process than is mixture equilibrium, so that on some time scales one equilibrium may be relaxed and the other frozen. Taking $(23.15)_1$, we have

$$mA_i = -kT\ln f_i + m_i g_i - \tfrac{1}{2}m_i c_i^2 \qquad (i = 1, 2, \ldots l).$$

24. The Maxwellian distribution function

24.1. Derivation of the distribution

The result just quoted for f_i^{eq} in the first of (23.15) may be deduced from the Euler relation (21.22), namely,

$$\rho Ts = \rho u + p - \rho g - kT\sum_i \int f_i \ln (f_i/f_i^{\text{eq}})\, \mathrm{d}\mathbf{w}_i, \tag{24.1}$$

plus the observation from (21.21) that s must be expressible as a function of the independent variables u, ρ^{-1}, and c_i and as functionals of the distributions f_i. The state equations enable us to eliminate T, p, and g_i in favour of u, ρ^{-1}, and c_i, and it follows that it must also be possible to remove the terms involving f_i^{eq} from (24.1) in a similar way. By using

$$p = kT\sum_i n_i = kT\sum_i \int f_i\, \mathrm{d}\mathbf{w}_i, \qquad \rho u = \tfrac{3}{2}nkT = \sum_i \int \tfrac{1}{2}m_i c_i^2 f_i\, \mathrm{d}\mathbf{w}_i, \tag{24.2}$$

we can write (24.1) in the form

$$\rho s = -k\sum_i \int f_i(\ln f_i - 1)\, \mathrm{d}\mathbf{w}_i - k\sum_i \int f_i\left[\frac{m_i g_i - \frac{1}{2}m_i c_i^2}{kT} - \ln f_i^{\text{eq}}\right] \mathrm{d}\mathbf{w}_i, \tag{24.3}$$

which immediately reveals an obvious means of removing f_i^{eq}.

Now f_i^{eq} is subject to the restriction that when it replaces f_i the values of the independent variables u, ρ, and ρ_i are unchanged (see remarks following (21.19)). Hence the values

$$\rho u = \tfrac{3}{2}nkT = \sum_i \int \tfrac{1}{2}m_i c_i^2 f_i^{\text{eq}}\, \mathrm{d}\mathbf{w}_i, \qquad n_i = \int f_i^{\text{eq}}\, \mathrm{d}\mathbf{w}_i, \tag{24.4}$$

are the same as those given in (24.2). We must now choose the distribution f_i^{eq} so that (1) it satisfies the constraints in (24.4) and (2) it can be removed from (24.3) in favour of u, ρ^{-1}, c_i, and f_i *alone*. The reader will soon convince himself that these conditions can be met only by the Maxwellian distribution (eqn $(23.15)_1$), which reduces (24.3) to the simple formula

$$\rho s = -k\sum_i \int f_i(\ln f_i - 1)\,\mathbf{dw}_i. \tag{24.5}$$

A similar discussion holds for f_{im}^{eq} with p, u, and T given their mixture values p_m, u_m, and T_m.

Eqn (24.5) may be replaced by

$$\rho s = -k\sum_i \int f_i \ln f_i\,\mathbf{dw}_i, \tag{24.6}$$

as the omitted term adds only a constant k/m to s. Then, provided s'_{0i} in (23.14) is likewise reduced by k/m_i and hence g_i modified to $g'_i = g_i + kT/m_i$ we find in place of $(23.15)_1$,

$$f_i^{eq} = \exp[-\beta m_i(\tfrac{1}{2}c_i^2 + p_i/\rho_i - g'_i)] \qquad (\beta = 1/kT).$$

Of course the term $-\beta m_i p_i/\rho_i$ appearing here is just -1. Had we used a more general work term $\mathbf{p}\,.\,\mathbf{l}$ (where $\mathbf{l}$ is the specific deformation vector denoted by $\mathbf{v}$ in (21.19)) in place of p/ρ, we would have found

$$f_i^{eq} = \exp[-\beta m_i(\tfrac{1}{2}c_i^2 + \mathbf{p}_i\,.\,\mathbf{l}_i - g'_i)], \tag{24.7}$$

on assuming $\mathbf{p} = \sum_i \mathbf{p}_i$, $\rho_i \mathbf{l}_i = \rho \mathbf{l}$ (§ 14.2). The interesting point about this formula is that it is in the standard form for equilibrium distribution functions, namely, exp [$-\beta$(energy − mechanical work − chemical work)], as we shall see in the next chapter.

24.2. Maximum entropy density

We have shown above that the Maxwellian distribution eliminates f_i^{eq} from the expression for the entropy density and leads to the kinetic equation given in (24.5). Now we shall prove that the converse is also true, that is, that starting with (24.5) it is possible to deduce the form of the Maxwellian distribution.

Consider an isolated system, which will have constant values of $p = nkT = \frac{2}{3}\rho u$ and of ρ. It follows from the theory given in § 16.2 that at equilibrium the variation $\delta(\rho s)$ in the entropy per unit volume due to changes δf_i in f_i—the only changes possible in an isolated system—must be zero. Hence from (24.5) and (24.2)

$$-\delta(\rho s)/k = \sum_i \int \delta[f_i(\ln f_i - 1)]\,\mathbf{dw} = 0,$$

subject to the constraints

$$\delta n_i = \int \delta f_i\,\mathbf{dw} = 0, \qquad \delta(\rho u) = \sum_i \int \tfrac{1}{2}m_i c_i^2\,\delta f_i\,\mathbf{dw} = 0.$$

Let λ, α_i, $i = 1, 2, \ldots l$, be Lagrangian multipliers to accommodate these constraints, then the required equilibrium distribution function satisfies

$$\sum_i \{\delta[f_i(\ln f_i - 1)] + (\alpha_i + \beta_i c_i^2)\,\delta f_i\} = 0 \qquad (\beta_i \equiv \tfrac{1}{2}m_i\lambda),$$

or

$$\sum_i (\ln f_i + \alpha_i + \beta_i c_i^2)\,\delta f_i = 0,$$

where the variations in f_i are now unconstrained. Thus each term in this sum vanishes:

$$\ln f_i = -\alpha_i - \beta_i c_i^2,$$

or

$$f_i^{\text{eq}} = \exp(-\alpha_i - \beta_i c_i^2). \tag{24.8}$$

The constants in (24.8) are determined by the second and third relations in (24.4).

Eqn $(24.4)_3$ gives†

$$\exp(-\alpha_i)\int_{-\infty}^{\infty}\int_{-\infty}^{\infty}\int_{-\infty}^{\infty} \exp[-\beta_i(c_x^2 + c_y^2 + c_z^2)]\,\mathrm{d}c_x\,\mathrm{d}c_y\,\mathrm{d}c_z = n_i,$$

where c_x, c_y, c_z are the Cartesian components of c_i. Carrying out these integrations we find that

$$\exp(-\alpha_i) = n_i\left(\frac{\beta_i}{\pi}\right)^{\frac{3}{2}}. \tag{24.9}$$

Further, by (22.2) †

$$\overline{c_x^2} = \left(\frac{\beta_i}{\pi}\right)^{\frac{3}{2}} \int_{-\infty}^{\infty}\left\{\int_{-\infty}^{\infty}\left[\int_{-\infty}^{\infty} c_x^2 \exp(-\beta_i c_x^2)\,\mathrm{d}c_x\right]\exp(-\beta_i c_y^2)\,\mathrm{d}c_y\right\}$$
$$.\exp(-\beta_i c_z^2)\,\mathrm{d}c_z = \frac{1}{2\beta_i};$$

similarly $\overline{c_y^2} = \overline{c_z^2} = 1/2\beta_i$, and therefore by $(24.4)_2$

$$nkT = \sum_i n_i\,\tfrac{1}{3}m_i\overline{c_i^2} = \sum_i n_i\,\tfrac{1}{2}\frac{m_i}{\beta_i} = \frac{n}{\lambda},$$

that is,

$$\beta_i = \frac{m_i}{2kT}. \tag{24.10}$$

† The following definite integrals are frequently required in kinetic theory:

$$\int_{-\infty}^{\infty}\begin{pmatrix}1\\x^2\end{pmatrix}\exp(-\beta x^2)\,\mathrm{d}x = \left(\frac{\pi}{\beta}\right)^{1/2}\begin{pmatrix}1\\1/2\beta\end{pmatrix}; \qquad \int_0^{\infty}\begin{pmatrix}x\\x^3\end{pmatrix}\exp(-\beta x^2)\,\mathrm{d}x = \frac{1}{2\beta}\begin{pmatrix}1\\1/\beta\end{pmatrix}.$$

Eqns (24.8), (24.9), and (24.10) now yield the Maxwellian distribution,

$$f_i^{\text{eq}} = n_i\left(\frac{m_i}{2\pi kT}\right)^{\frac{3}{2}} \exp\left(-\tfrac{1}{2}m_i c_i^2/kT\right).$$

25. Kinetic entropy

25.1. Entropy density

Eqns (21.22) and (24.5), namely,

$$\rho Ts = \rho u + p - \rho g - kT\sum_i \int f_i \ln\left(f_i/f_i^{\text{eq}}\right) \mathrm{d}\mathbf{w}_i \tag{25.1}$$

and

$$\rho s = -k\sum_i \int f_i(\ln f_i - 1)\,\mathrm{d}\mathbf{w}_i, \tag{25.2}$$

provide equivalent definitions of the entropy in kinetic thermodynamics. They are based on the presumption that, amongst all the possible relaxation processes within a monatomic perfect gas, that taking f_i towards its (component) equilibrium value is the longest lived. In other words, correlations between pairs or triplets of particles, for example, are negligible on the kinetic time scale $\mathscr{T}_K$.

There are two points to make about (25.1). First had we not set s_{0i} equal to the constant s'_{0i} defined in (23.14), but left it arbitrary, g_i would have been increased by $T(s_{0i} - s'_{0i})$, with the effect of adding $\sum_i c_i(s_{0i} - s'_{0i})$ to s on the left-hand side of (25.1). As remarked in § 23.4 this is merely a change in origin for entropy. The second point is that the definition of g_i is

$$g_i = u_i + p_i/\rho_i - Ts_i^{\text{eq}}, \tag{25.3}$$

where s_i^{eq} is the value when f_i equals f_i^{eq}; as stated in § 23.4 g_i has the same value whether kinetic equilibrium holds or not.

From (25.1) and (25.3)

$$\rho(s - s^{\text{eq}}) = -k\sum_i \int f_i \ln\left(f_i/f_i^{\text{eq}}\right) \mathrm{d}\mathbf{w}_i, \tag{25.4}$$

where

$$\rho s^{\text{eq}} = \sum_i \rho_i s_i^{\text{eq}} = \frac{1}{T}(\rho u + p - \rho g). \tag{25.5}$$

With the definition

$$\Omega \equiv \exp\left(ms^{\text{eq}}/k\right) \qquad \left(m \equiv \frac{1}{n}\sum_i m_i n_i\right), \tag{25.6}$$

(25.4) may be written

$$\rho s = -k \sum_i \int f_i \ln\left(\frac{f_i}{f_i^{\text{eq}}\Omega}\right) \mathrm{d}\mathbf{w}_i, \tag{25.7}$$

which is sometimes useful.

25.2. Entropy as a statistical average

For simplicity of presentation we shall now consider a single component, the extension of the following discussion to several components being a trifling matter. Let N particles of mass m lie within a volume V, then integrating the number density $n(\mathbf{r})$ over V we have

$$\int F_1(\mathbf{x})\,\mathrm{d}\mathbf{x} = 1, \tag{25.8}$$

where

$$F_1 \equiv f/Nm^3 \tag{25.9}$$

and

$$\mathrm{d}\mathbf{x} = \mathrm{d}(m^3\mathbf{w})\,\mathrm{d}\mathbf{r}.$$

Clearly $F_1(\mathbf{x})\,\mathrm{d}\mathbf{x}$ is the probability that a particular particle selected at random lies in the volume element $\mathrm{d}\mathbf{x}$ of six-dimensional phase-space. This single particle phase-space is usually called 'μ-space'.

With spatially uniform systems (25.1) may be integrated over the volume V to give

$$S = \frac{1}{T}(U + pV - \mu N) - kN \int F_1 \ln\left(\frac{F_1}{F_1^{\text{eq}}}\right) \mathrm{d}\mathbf{x}, \tag{25.10}$$

where $\mu = mg$ is the chemical potential per particle. The corresponding integral of (25.2), namely,

$$S = -kN \int F_1 \ln F_1\,\mathrm{d}\mathbf{x} - kN(\ln N - 1) - 3kN \ln m, \tag{25.11}$$

has the advantage of not being restricted to uniform systems. This formula gives the entropy to within a constant, in terms of the *one-particle* distribution function $F_1(\mathbf{x})$, and it is valid on time scales equal to or greater than the kinetic time scale $\mathscr{T}_{\mathrm{K}}$.

The rth *particle* distribution function $F_r(\mathbf{x}_1, \mathbf{x}_2, \ldots \mathbf{x}_r)$ is defined by requiring $F_r\,\mathrm{d}\mathbf{x}_1\,\mathrm{d}\mathbf{x}_2 \ldots \mathrm{d}\mathbf{x}_r$ to be the probability that particle 1 lies in $\mathrm{d}\mathbf{x}_1$, particle 2 lies in $\mathrm{d}\mathbf{x}_2$, and so on, to particle r in $\mathrm{d}\mathbf{x}_r$. As the particles are identical, the function F_r must be symmetric in $\mathbf{x}_1, \ldots \mathbf{x}_r$ and independent of any particular choice for the r particles. When there are no correlations between the locations of the particles in phase-space, that is, if the probabilities $F_1(\mathbf{x}_1)\,\mathrm{d}\mathbf{x}_1, \ldots$ $F_1(\mathbf{x}_r)\,\mathrm{d}\mathbf{x}_r$, are independent, then F_r factorizes,

$$F_r(\mathbf{x}_1, \mathbf{x}_2, \ldots \mathbf{x}_r) = F_1(\mathbf{x}_1)F_1(\mathbf{x}_2) \ldots F_1(\mathbf{x}_r). \tag{25.12}$$

On the hypothesis (due to Bogoliubov, 1946) that correlations disappear in a much shorter time than $\mathscr{T}_\mathrm{K}$, (25.12) applies when (25.11) defines S.

Now assume N to be very large, so that Stirling's asymptotic expansion

$$\ln N! \sim N \ln N - N \tag{25.13}$$

can be used. Then by (25.8), (25.12), and the fact that $F_1(\mathbf{x}_1)$, $F_1(\mathbf{x}_2), \ldots$ are identical functions, we may write (25.11) as

$$\begin{aligned} S &= -k \int\int \ldots \int F_1(\mathbf{x}_1)F_1(\mathbf{x}_2) \ldots F_1(\mathbf{x}_N) \ln [F_1(\mathbf{x}_1)F_1(\mathbf{x}_2) \ldots F_1(\mathbf{x}_N)N!] \times \\ &\qquad \times \mathrm{d}\mathbf{x}_1\, \mathrm{d}\mathbf{x}_2 \ldots \mathrm{d}\mathbf{x}_N - 3kN \ln m \\ &= -k \int F_N \ln [F_N N!]\, \mathrm{d}\Gamma - 3kN \ln m, \end{aligned}$$

where

$$\mathrm{d}\Gamma = \mathrm{d}\mathbf{x}_1\, \mathrm{d}\mathbf{x}_2 \ldots \mathrm{d}\mathbf{x}_N$$

is a volume element in a phase-space of $6N$ dimensions. It is convenient here to include in the expression for S the volume integral of the constant $\rho(s_0 - s_0')$ described in the second paragraph of § 25.1. By (23.14) the volume integral of $\rho s_0'$ is

$$S_0' = \tfrac{1}{2}Nk[5 + 5 \ln k - 3 \ln (m/2\pi)], \tag{25.14}$$

hence

$$S = -k \int F_N \ln (F_N N!)\, \mathrm{d}\Gamma + S_0 - \tfrac{1}{2}Nk[5 + 5 \ln k + 3 \ln (2\pi m)] \tag{25.15}$$

which, apart from the additive constant, reveals $-S/k$ to be the average value of $\ln (F_N N!)$ over the probability distribution F_N.

25.3. *Quantum effects on entropy*

While $F_N\, \mathrm{d}\Gamma$ is dimensionless, $\mathrm{d}\Gamma$ has the dimensions of 'action' (momentum component $\times$ distance) raised to the $3N$ power. Let h be a constant having the dimensions of action, then F_N has dimensions h^{-3N}, and clearly any change in the unit of action will alter the division of S between the integral and the additive constant in (25.15). In fact, as

$$\int F_N\, \mathrm{d}\Gamma = 1, \tag{25.16}$$

we may write (23.15) in the form

$$S = -\frac{k}{N!h^{3N}}\int P_N \ln P_N \, d\Gamma + S_0 - \tfrac{1}{2}Nk[5+5\ln k + 3\ln(2\pi m/h^2)],$$

where P_N is a non-dimensional N-particle distribution function defined by

$$P_N \equiv N!h^{3N}F_N. \tag{25.17}$$

The meaning of P_N is that $P_N(d\Gamma/h^{3N})$ is the probability of finding *a* particle—not necessarily particle 1—in $d\mathbf{x}_1$, *a* particle in $d\mathbf{x}_2$, and so on, for the first particle (for $d\mathbf{x}_1$) may be chosen in N ways, the second in $(N-1)$ ways, and so on. P_N is sometimes termed a *generic* distribution in contrast to the *specific* distribution F_N.

At this stage we shall allocate values to the as-yet undefined constants h and S_0 and justify the choices subsequently. First we take h to be Planck's constant,† and then set

$$S_0 = \tfrac{1}{2}Nk[5+5\ln k + 3\ln(2\pi m/h^2)], \tag{25.18}$$

thereby reducing the formula for S to

$$S = -\frac{k}{N!h^{3N}}\int P_N \ln P_N \, d\Gamma. \tag{25.19}$$

Heisenberg's uncertainty relation‡ leads to the conclusion that the position of a point $\mathbf{X}$ in the $6N$-dimensional phase-space—hereafter termed 'Γ-space'—that completely represents the state of the system, cannot be prescribed more accurately than to within a cell of size h^{3N}. Thus with Γ-space completely divided into such cells, the state is defined as fully as possible by specifying the cell into which $\mathbf{X}$ falls. Such a state is termed a *micro-state*, and many such states correspond to a single *macro-state*, which requires just a few external variables like U and V to define it. The total number of micro-states in a volume element $d\Gamma$ is $d\Gamma/h^{3N}$, but corresponding to any given cell there are $N!$ others derived from it merely by exchanging the individual particles, so the number of *distinct* micro-states is just $d\Gamma/(N!h^{3N})$. 'Distinct' here refers to states with different *number* distributions in μ-space (six-dimensional phase-space), for it is clear from (25.11) that as far as entropy is concerned this is the essential feature.

If in (25.19) we integrate over elements $d\Gamma_1, d\Gamma_2, \ldots d\Gamma_j, \ldots$ each chosen to cover one distinct micro-state, the formula for S is reduced to the sum

$$S = -k\sum_j P_j \ln P_j, \tag{25.20}$$

† $h = 6{\cdot}624\times10^{-34}$ J s.
‡ See discussion on pp. 78–9 of ter Haar (1966).

where P_j is the probability that the system is in the micro-state $d\Gamma_j$. This simple expression for entropy is often taken as providing its definition (see discussions in §§ 26.4 and 36.5).

25.4. Absolute entropy

Imagine an experiment in which the system of particles comprising the perfect gas is cooled towards absolute zero temperature, while remaining in mechanical equilibrium with its surroundings. This constraint imposes a constant pressure on the system, so that U and V tend to zero as T tends to zero. This conclusion follows from the state equations (§ 23.2)

$$U = \tfrac{3}{2}NkT, \qquad p = nkT = \tfrac{2}{3}U/V, \tag{25.21}$$

where by (22.5) and (25.9)

$$U = N\int \tfrac{1}{2}mc^2 F_1\, d\mathbf{x}. \tag{25.22}$$

It is clear from these equations that as $T \to 0$ the representative point $\mathbf{X}$ tends towards a quantum cell, $d\Gamma_1$ say, in which $\mathbf{w}_1, \mathbf{w}_2, \ldots \mathbf{w}_N$ are zero ($U = 0$) and $\mathbf{r}_1, \mathbf{r}_2, \ldots \mathbf{r}_N$ have the same value ($V = 0$). This is the 'ground state' of the system. At $T = 0$, $\mathbf{X}$ is certainly in $d\Gamma_1$, that is, $P_1 = 1$, and $P_j = 0$ for $j = 2, 3, \ldots$. Thus (25.20) gives

$$\lim_{T\to 0} S = 0, \tag{25.23}$$

which justifies the particular choice made for S_0 in (15.18). Any other choice would merely result in an additive constant, S_1 say, being attached to the right-hand sides of (25.20) and (25.23), and such constant entropy would be undetectable, being unaffected by any change in the independent variables for the system. As the value of S_1 is never required, setting it equal to zero is the obvious choice.

Eqn (25.23) is one form of the *third law of thermodynamics*, which applies to all systems. For general systems (our perfect gas is very special) this law cannot be derived, but must be added to the other laws of thermodynamics as a basic postulate.

With the entropy scale adjusted to satisfy (25.23), S becomes 'absolute' entropy and acquires the physical interpretation via (25.20) of being a measure of the 'uncertainty' inherent in the system, a point that will be amplified later (§ 36.5).

25.5. The Sackur–Tetrode relation

The simplest formula for the absolute entropy of a system is that of a perfect gas. Eqns (21.4) and (23.12) give

$$\rho s = \rho s_0 + \tfrac{5}{2}nk \ln T - nk \ln p$$

or

$$S = S_0 + \tfrac{5}{2}Nk \ln T - Nk \ln\left(\frac{N}{V}kT\right),$$

on integrating over the volume of a spatially uniform system. By (25.18) this may be expressed

$$S = Nk \ln\left[\left(\frac{2\pi mkT}{h^2}\right)^{\frac{3}{2}}\frac{V}{N}\right] + \tfrac{5}{2}Nk, \tag{25.24}$$

or by the first of (25.21)

$$S = kN \ln\left[\left(\frac{U}{N}\right)^{\frac{3}{2}}\frac{V}{N}\right] + \tfrac{1}{2}kN\left[5 + 3\ln\left(\frac{4\pi m}{3h^2}\right)\right], \tag{25.25}$$

which is known as the Sackur–Tetrode relation.

This result may also be obtained by employing the Maxwellian distribution

$$F_1^{\text{eq}} = \frac{n}{N}(2\pi mkT)^{-\frac{3}{2}} \exp(-\tfrac{1}{2}mc^2/kT), \tag{25.26}$$

in

$$S = -kN \int F_1 \ln F_1 \, d\mathbf{x} - kN(\ln N - 1) - 3kN \ln h, \tag{25.27}$$

which follows from adding the constant $(S_0 - S_0')$ (initially omitted) to (25.11).

25.6. Entropy on the shortest possible time scale

To this point the expression in (25.19) for S, which by (25.13) and (25.17) may be written

$$S = -k \int F_N \ln F_N \, d\Gamma - kN(\ln N - 1) - 3kN \ln h, \tag{25.28}$$

is simply an alternative to (25.27), for on the kinetic time scale $\mathscr{T}_K$, F_N factorizes as in (25.12). In the next chapter we shall make the ultimate refinement of the time scale and show that (25.28) is valid on time scales $\mathscr{T}_c$ very much shorter than $\mathscr{T}_K$ when F_N cannot be factorized into independent one-particle distributions.

For small values of r like 2, 3, or 4 it is of interest to define *correlations* G_{1r} and G_{rN} by

$$F_N = \underbrace{F_r F_r \dots F_r}_{\leftarrow N/r \rightarrow} + G_{rN}, \qquad F_r = \underbrace{F_1 F_1 \dots F_1}_{\longleftarrow r \longrightarrow} + G_{1r}, \tag{25.29}$$

and on the hypothesis that G_{rN} vanishes much more rapidly than G_{1r} in any non-equilibrium state permitted to relax, it is useful to introduce an entropy valid on the same time scale as G_{1r} persists. This, as shown later, is (cf. (25.27) and (25.28))

$$S_r = -k\frac{N}{r}\int F_r \ln F_r \, d\mathbf{x}_1 \dots d\mathbf{x}_r - kN(\ln N - 1) - 3kN \ln h; \tag{25.30}$$

the cases $r = 1, 2$ will be useful in our treatment of kinetic theory (§ 74.3).

Thus in the 'initial stage' of any very disequilibrated system the general expression in (25.28) reduces, as G_{2N} tends to zero, to the case $r = 2$ of (25.30). After a few particle collisions G_{12} disappears and we enter the 'kinetic stage' of the approach to equilibrium. Finally, when most particles in the system have each experienced several collisions, we enter the 'hydrodynamic stage' in which F_1 has its equilibrium form.

$$\begin{array}{ccc} \mathscr{T}_c & \mathscr{T}_K & \mathscr{T}_R \\ \leftarrow \text{initial stage} \rightarrow & \leftarrow \text{kinetic stage} \rightarrow & \leftarrow \text{hydrodynamic stage} \rightarrow \end{array}$$

$$\frac{2}{N}F_N(\mathbf{w}_1, \mathbf{w}_2, \dots \mathbf{w}_N) \rightarrow F_2(\mathbf{w}_1, \mathbf{w}_2) \rightarrow F_1(\mathbf{w}_1)\, F_1(\mathbf{w}_2) \rightarrow F_1^{\mathrm{eq}}(\mathbf{w}_1) F_1^{\mathrm{eq}}(\mathbf{w}_2)$$

We shall return to this process of successive relaxation when discussing kinetic processes in § 74.

It is clear from the theory given in § 16 (see Fig. 5), that the greater the detailed knowledge of the internal structure of the system, the lower the entropy. Hence in the present context this means that

$$S_1 \geqslant S_2 \geqslant \dots \geqslant S_r \geqslant S_{r+1} \geqslant \dots \geqslant S_N, \tag{25.31}$$

which may be proved as follows.†

As $f \ln f$ is a convex function, $f \ln f \geqslant f - 1$ for f non-negative. Thus, replacing f by f/g, we have

$$f \ln f - f \ln g \geqslant f - g \qquad (f \geqslant 0, g > 0). \tag{25.32}$$

† See Grad (1961), in which a correction to (25.30) is described that incorporates the symmetry of F_r in its arguments.

Take f and g to be probability distributions defined over some domain μ, then integrating (25.32) over μ, we obtain

$$\int f \ln f \, \mathrm{d}\mu \geqslant \int f \ln g \, \mathrm{d}\mu, \tag{25.33}$$

in which equality holds only when $f = g$.

Let $g = F_1(\mathbf{x}_1)F_1(\mathbf{x}_2) \ldots F_1(\mathbf{x}_N)$, $f = F_N(\mathbf{x}_1, \mathbf{x}_2, \ldots \mathbf{x}_N)$, and $\mathrm{d}\mu = \mathrm{d}\Gamma$, whence

$$\int F_N \ln F_N \, \mathrm{d}\Gamma \geqslant \sum_{r=1}^{N} \int F_N \ln F_1(\mathbf{x}_r) \, \mathrm{d}\Gamma = \sum_{r=1}^{N} \int F_1(\mathbf{x}_r) \ln F_1(\mathbf{x}_r) \, \mathrm{d}\mathbf{x}_r$$

$$= N \int F_1(\mathbf{x}_1) \ln F_1(\mathbf{x}_1) \, \mathrm{d}\mathbf{x}_1 .$$

By (25.30) this yields $S_1 \geqslant S_N$. Similarly the choice

$$g = F_2(\mathbf{x}_1, \mathbf{x}_2)F_2(\mathbf{x}_3, \mathbf{x}_4) \ldots \begin{matrix} F_2(\mathbf{x}_{N-1}, \mathbf{x}_N) & (N \text{ even}) \\ F_1(\mathbf{x}_N) & (N \text{ odd}) \end{matrix}$$

plus use of the inequality $S_1 \geqslant S_N$ for N odd enables us to show that $S_2 \geqslant S_N$ $(N \geqslant 2)$, and so on.

5

STATISTICAL THERMODYNAMICS

In this final chapter on 'equilibrium' thermodynamics, we shall further refine the observer's time scale, so that even the particle collision time τ_0, where

$$\tau_0 = \frac{\text{range of interparticle forces}}{\text{mean thermal speed of the particles}} = \frac{r_0}{\bar{c}}$$

is no longer 'instantaneous'. On this time scale, to be denoted by $\mathscr{T}_S$, particles are essentially stationary and the adjective 'equilibrium' has no relevance so it will be omitted. We now have the hierarchy of scales and associated branches of thermodynamics,

$$\mathscr{T}_S \quad \ll \quad \mathscr{T}_K \quad \ll \quad \mathscr{T}_I \quad \ll \quad \mathscr{T}_R$$

← engineering —

← chemical ——

←—— kinetic ——

←—— statistical ——

Each refinement of the time scale produces a theory of apparently wider applicability as just indicated, but it should be remembered that such refinement requires an increasing specificity for the system. Statistical thermodynamics thus has the widest range of validity as far as the time variable is concerned, but does require some detailed model of the particle behaviour to be provided. The return for these details is a direct expression for entropy in the form

$$S = -k \sum_j P_j \ln P_j,$$

where P_j is the probability of the jth micro-state and the summation is over distinct micro-states (cf. (25.20)).

Thus, whereas the entropy for systems of indeterminate particulate behaviour must be deduced from the Gibbs relation and phenomenological state equations (see discussion in § 9.3), the entropy in statistical thermodynamics is determined by the above expression for S and some fundamental hypotheses about the behaviour of the particles in the given system. As shown below, this formula follows from a generalized form of the Gibbs relation (cf. deduction of (25.20) from (21.21)), although it is usually introduced as a new definition of S and termed the 'Gibbs entropy postulate'.

26. Entropy as a statistical average

26.1. The Gibbs relation for an ensemble of systems

In § 21.2 we obtained a generalization of the Gibbs relation that we shall make the starting point of our treatment of statistical thermodynamics. In terms of the chemical potential vector $\boldsymbol{\mu} = (\mu_1, \mu_2, \dots \mu_i, \dots)$ and the number density vector $\mathbf{n} = (n_1, n_2, \dots n_i, \dots)$, where $\mu_i = m_i g_i$ and $n_i = \rho c_i/m_i$, this relation may be written

$$T\,\mathrm{d}s = \mathrm{d}u + \mathbf{p}\,.\,\mathrm{d}\mathbf{v} - \boldsymbol{\mu}\,.\,\mathrm{d}(\mathbf{n}/\rho) + \sum_y A_y\,\mathrm{d}x_y. \tag{26.1}$$

In (26.1) y is an integer replacing the double index ik used earlier and specifying a particular sub-fluid of mole fraction x_y and affinity A_y. Thus y denotes a state that can be possessed by single particles, for example, it may specify the particle species and the region of phase-space to which the particle belongs; x_y is the probability that a particle selected at random is in state y.

For a perfect gas it follows from (21.18) that

$$A_y = -\frac{kT}{m}\ln(x_y/x_y^{\mathrm{eq}}), \tag{26.2}$$

where the mass m is an average value for all the particles of the system.

In what follows we shall generalize (26.1) by upgrading the particles—whose precise nature is still unspecified—into complete 'representative' systems, and the original compound system will then become an *ensemble* of such systems. Our interest will be in the representative systems, with the ensemble purely a means of obtaining a statistical spread in the properties of these systems. This means that the ensemble—thinking of its component systems as 'particles' for the moment—need not retain the same physical behaviour as its component systems. Thus the nature of the interactions between the systems in the ensemble may be freely prescribed, and in fact we shall suppose that they interact weakly, just like the particles of a perfect gas. During these weak interactions the systems may, in general, transfer energy, material particles, and deformations, but we are free to restrict these exchanges between systems in a variety of ways, for example, to energy alone, or even to no transfer of properties.

An average representative system $\bar{\mathrm{P}}$ will have a mass m (originally the average particle mass) and as n_i/ρ is the number of particles of the ith species per unit mass, $N_i \equiv mn_i/\rho$ will be the number of these particles in $\bar{\mathrm{P}}$. Similarly, $S = ms$, $U = mu$, $\mathbf{V} = m\mathbf{v}$ will be values applying to $\bar{\mathrm{P}}$. We shall write ξ_y for the probability that any system P chosen at random is in state y. Now multiply (26.1) by the constant m and assume weak interactions between the systems of the ensemble. The result is the Gibbs relation for an ensemble, namely,

$$T\,dS = dU + \mathbf{p}\,.\,d\mathbf{V} - \boldsymbol{\mu}\,.\,d\mathbf{N} - kT\sum_y d\xi_y \ln(\xi_y/\xi_y^{eq}). \tag{26.3}$$

In (26.3) the properties U, $\mathbf{V}$, $\mathbf{N}$ are *average values* for the systems of the ensemble and they define the *macro-state* of a typical system. On the other hand, the intensive variable T, $\mathbf{p}$, and $\boldsymbol{\mu}$ are ensemble values, *not* averages over the corresponding system values. The property y defines the particular state of a system P and, for example, could be a very detailed description of the position and velocities of all the particles in P; it determines the *micro-state* of P (§ 25.3).

26.2. *Ergodicity*

In classical thermodynamics we are interested in the macroscopic properties of a single system over relatively long time intervals, but owing to fluctuations in the system, these properties are time-averages of instantaneous values. The difficulty of calculating these time-averages was overcome by Gibbs as follows. Instead of time-averages over a single system we take averages over an ensemble of similar systems, suitably randomized, at a single instant of time. The problem of proving the equivalence of the two kinds of averages is called ergodic theory, and the ensemble is said to be ergodic if these averages are equal. The equivalence has been shown for systems whose representative points in Γ-space (see § 25.3) trace orbits that are not confined to a sub-domain of accessible Γ-space. However, it is very difficult to establish this property for systems with a given Hamiltonian, and we shall accept as a hypothesis that the systems in which we are interested are ergodic.

The purpose of the ensemble model then is not to represent a real physical system by the complete ensemble itself, but simply to afford a means of generating a stochastic process from which the expected properties of a typical system may be obtained via ensemble averages. It is therefore no real limitation to suppose, as we have done above, that the systems interact with each other like the particles of a perfect gas. In the special case when the systems are *themselves* perfect gases, the ensemble can be given the physical significance of comprising a system P plus its environment P*, which contains all the other systems interacting with P.

26.3. *Relation between the entropy of a micro-state and its probability*

Consider a single system P in a micro-state y. Denote by $U_y, \mathbf{V}_y, \mathbf{N}_y$ the extensive variables defining the state of P, then the average value over the ensemble of these variables are

$$U = \sum_y \xi_y U_y, \qquad \mathbf{V} = \sum_y \xi_y \mathbf{V}_y, \qquad \mathbf{N} = \sum_y \xi_y \mathbf{N}_y. \tag{26.4}$$

As $U_y, \mathbf{V}_y, \mathbf{N}_y$ are determined entirely by the micro-state y and hence are constants for fixed y, U, $\mathbf{V}$, and $\mathbf{N}$ are functions of the probability distribution

$\{\xi_y\}$. To change the macro-state $(U, \mathbf{V}, \mathbf{N})$ we must alter the distribution $\{\xi_y\}$, but because of the enormous number of micro-states y, there is a vast number of sets $\{d\xi_y\}$ that leave the macro-state unchanged. It is in this sense that dU, $d\mathbf{V}$, $d\mathbf{N}$ are independent of $\{d\xi_y\}$ in (26.3), permitting the inference that $S = S(U, \mathbf{V}, \mathbf{N}, \{\xi_y\})$. By (26.4) it must be possible to express S as a function of the set $\{\xi_y\}$ alone, albeit with the constants U_y, $\mathbf{V}_y$, $\mathbf{N}_y$ possibly appearing as parameters,

$$S = S(\xi_y; U_y, \mathbf{V}_y, \mathbf{N}_y) \quad (\text{all } y). \tag{26.5}$$

We now need to give some substance to the equilibrium distribution ξ_y^{eq}. This we shall do by requiring it to yield the same ensemble averages as ξ_y, that is,

$$U = \sum_y \xi_y^{\text{eq}} U_y, \qquad \mathbf{V} = \sum_y \xi_y^{\text{eq}} \mathbf{V}_y, \qquad \mathbf{N} = \sum_y \xi_y^{\text{eq}} \mathbf{N}_y \tag{26.6}$$

(cf. (24.4)). Then remembering that T, $\mathbf{p}$, and $\boldsymbol{\mu}$ have fixed values determined by the ensemble, we may set $\xi_y = \xi_y^{\text{eq}}$ in (26.3) to find

$$T\, dS^{\text{eq}} = dU + \mathbf{p} \,.\, d\mathbf{V} - \boldsymbol{\mu} \,.\, d\mathbf{N},$$

which has the Euler relation

$$S^{\text{eq}} = T^{-1}(U + \mathbf{p} \,.\, \mathbf{V} - \boldsymbol{\mu} \,.\, \mathbf{N}) = T^{-1} \sum_y \xi_y^{\text{eq}}(U_y + \mathbf{p} \,.\, \mathbf{V}_y - \boldsymbol{\mu} \,.\, \mathbf{N}_y) = \sum_y \xi_y^{\text{eq}} S_y,$$

where

$$S_y = k\beta(U_y + \mathbf{p} \,.\, \mathbf{V}_y - \boldsymbol{\mu} \,.\, \mathbf{N}_y) \qquad (\beta = (kT)^{-1}) \tag{26.7}$$

is the entropy of P in state y. Eqn (26.7) is merely the Euler relation (15.5) for a single system in equilibrium. By (26.4) and (26.6)

$$S^{\text{eq}} = \sum_y \xi_y^{\text{eq}} S_y = \sum_y \xi_y S_y, \tag{26.8}$$

consequently the Euler relation that follows from (26.3), namely

$$S = S^{\text{eq}} - k \sum_y \xi_y \ln (\xi_y / \xi_y^{\text{eq}}),$$

may be written

$$S = -k \sum_y \xi_y \ln \xi_y + \sum_y \xi_y (S_y + k \ln \xi_y^{\text{eq}}). \tag{26.9}$$

To reduce (26.9) to the form given in (26.5) we require

$$k \ln \xi_y^{\text{eq}} = F_y(U_y, \mathbf{V}_y, \mathbf{N}_y),$$

where by the extensivity of S the unknown function F_y must be linear in its arguments. By (26.4) and (26.6) this linearity means that $\sum_y \xi_y F_y = \sum_y \xi_y^{\text{eq}} F_y$, so that (26.9) may be written

$$S = -k\sum_y \xi_y \ln \xi_y + \sum_y \xi_y^{\mathrm{eq}}(S_y + k \ln \xi_y^{\mathrm{eq}}). \tag{26.10}$$

Eqns (26.8), (26.9), and (26.10) are Euler relations, which means that they are identities (§ 15.1), that is, true for all values of the variables involved. In particular, corresponding to (26.5) we have

$$S^{\mathrm{eq}} = S^{\mathrm{eq}}(\xi_y^{\mathrm{eq}}; U_y, \mathbf{V}_y, \mathbf{N}_y) \quad (\text{all } y)$$

the functional form being the same as in (26.5). Now (26.8), (26.9), and (26.10) can have the same functional form only if

$$S_y = -k \ln \xi_y^{\mathrm{eq}} + \text{constant}, \tag{26.11}$$

which reduces (26.10) to

$$S = -k\sum_y \xi_y \ln \xi_y + \text{constant}, \tag{26.12}$$

and (26.8) to a similar formula.

From (26.7) and (26.11) we have

$$\xi_y^{\mathrm{eq}} = K \exp\left[-\beta(U_y + \mathbf{p}\cdot\mathbf{V}_y - \boldsymbol{\mu}\cdot\mathbf{N}_y)\right], \tag{26.13}$$

where the constant K must be chosen to satisfy $\sum_y \xi_y^{\mathrm{eq}} = 1$.

An important property of the equilibrium distribution is that, for a given macro-state $(U, \mathbf{V}, \mathbf{N})$, it *maximizes the entropy*, which is of course a consequence of the stability of thermodynamic systems (see § 13.3 and § 16.2). To see this we note that with U, $\mathbf{V}$, and $\mathbf{N}$ constant in (26.3), at $\xi_y = \xi_y^{\mathrm{eq}}$ (26.3) yields

$$\mathrm{d}S = 0, \qquad \mathrm{d}^2S = -k\sum_y (\mathrm{d}\xi_y)^2/\xi_y < 0.$$

Eqn (26.11) relating the entropy of a micro-state y to its probability was introduced as a hypothesis by Boltzmann, and is consequently termed the *Boltzmann entropy postulate*. Eqn (26.12) for the average entropy of a system is due to Gibbs, and when it is introduced axiomatically it is called the *Gibbs entropy postulate*. We shall discuss this axiomatic approach shortly.

Let y be chosen so that to each value in (26.12) there corresponds just one distinguishable micro-state of the systems; to emphasize this limiting subdivision of state-space we shall replace y by j. Now we assume that there exists a ground state $j = 1$ such that at $T = 0$ the system certainly occupies this state (see discussion in § 25), then $\xi_j = 1$ for $j = 1$ and $\xi_j = 0$ otherwise. The constant in (26.12) may be set equal to zero, and we arrive at the absolute entropy

$$S = -k \sum_j P_j \ln P_j, \tag{26.14}$$

where P_j is the probability that the system is in micro-state j.

26.4. Axiomatic statistical thermodynamics

Almost all treatments of statistical mechanics start from the Gibbs entropy postulate and then show that the statistical entropy so defined may be identified with the (equilibrium) entropy of classical thermodynamics. To do this it is necessary to add another postulate, which may take several forms. For a completely isolated system one form of the additional postulate may be stated:

A *all conceivable distinct micro-states of a system have* a priori *equal probabilities of occurrence.*

It is easy to see that in the absence of any knowledge about these probabilities, any other assumption would be arbitrary. For an ensemble of isolated systems then

$$\xi_y^{\text{eq}} = \Omega^{-1} = \text{constant}, \tag{26.15}$$

where Ω is the total number of possible micro-states, and (26.12) becomes (dropping the constant as in (26.14))

$$S^{\text{eq}} = k \ln \Omega. \tag{26.16}$$

It is easily shown that such a choice for ξ_y^{eq} maximizes S in the more general form of (26.12), so that the hypothesis of *a priori* equal probabilities is equivalent to requiring the statistical entropy to have the same maximal property as possessed by the classical entropy (§ 13). In fact a better additional hypothesis, which is valid also for a system interacting with the rest of the ensemble would be contained in the assertion that:

B *in equilibrium states the statistical entropy defined by* (26.12) *behaves like the classical entropy, in that it is a maximum subject to the constraints imposed by these interactions.*

This hypothesis can be given another interpretation via information theory (§ 36.3).

There would appear to be a contradiction between the assertion that entropy is undefined in non-equilibrium states (§ 7.5) and the apparently unambiguous expressions in (26.12) and (26.14) regardless of the state of the system. However, the precision of these formulae in defining S is illusionary, for without the additional postulates just described—or something equivalent—the value of S cannot be determined from them. Postulate B refers to 'constraints' on the system, and as these are not absolute, but depend on the observer's time scale, we are returned to the situation described in § 7.5. The statistical formula for entropy has possibly given false comfort to those for whom entropy is a primitive concept with the same physical status as mass or energy.

Now returning to the general case, to find ξ_y^{eq} we have to maximize (26.12) subject to the constraints in (26.4), in which the macro-state $(U, \mathbf{V}, \mathbf{N})$ is assumed to be known. We must also satisfy the constraint $\sum \xi_y = 1$. The calculation is straightforward (cf. § 24.2) and yields $\xi_y^{\text{eq}} \propto \exp(aU_y + \mathbf{b}\,.\,\mathbf{V}_y + + \mathbf{c}\,.\,\mathbf{N}_y)$, where a, $\mathbf{b}$, $\mathbf{c}$ are Lagrange multipliers. To complete the identification of the statistical entropy with the classical entropy, it remains only to substitute the expression for ξ_y^{eq} into (26.12) to find

$$S^{\text{eq}} = -k(aU + \mathbf{b}\,.\,\mathbf{V} + \mathbf{c}\,.\,\mathbf{N}) + \text{constant},$$

and comparison with the Euler relation of classical thermodynamics, namely,

$$S^{\text{eq}} = T^{-1}(U + \mathbf{p}\,.\,\mathbf{V} + \boldsymbol{\mu}\,.\,\mathbf{N}),$$

gives the appropriate values for a, $\mathbf{b}$, and $\mathbf{c}$. The expression for ξ_y^{eq} is now the same as that given in eqn (26.13).

For a comparison of the axiomatic approach to statistical thermodynamics just outlined and the rather different treatment presented in this text the reader is referred to § 36.5.

27. The principal ensembles of statistical thermodynamics

27.1. The generalized ensemble

Special cases of the equilibrium distribution $\xi_y^{\text{eq}} = \xi_y^{\text{eq}}(U_y, \mathbf{V}_y, \mathbf{N}_y)$ associated with a given macro-state $(U, \mathbf{V}, \mathbf{N})$ will occupy our attention for most of the rest of this chapter, which means we are now restricting ourselves to the 'chemical' and 'engineering' time scales described on p. 95. Therefore we may simplify our notation by omitting the superscript eq. It is usual to specify the micro-states y by an index, say $j = 0, 1, 2, \ldots$, for the energy, by the integers $N_1, N_2, \ldots$, for the particle numbers, and either by continuous variables $V_1, V_2, \ldots$, or by integers $r_1, r_2, \ldots$ in $V_1 = \Delta V_1 r_1$, $V_2 = \Delta V_2 r_2, \ldots$ for the deformations. Quantum theory imposes lower bounds on the energy increments $\Delta U = U_{j+1} - U_j$, and on the volume increments $\Delta V_1, \Delta V_2, \ldots$, and when these bounds are attained an absolute entropy may be introduced. (It is not our object in this text to describe quantum theory beyond a point required to give meaning to zero entropy.)

Let the inflexions ^ identify the variables that are able to change in value from one representative system to another, then we may omit the subscript

y and write $\hat{U}, \hat{\mathbf{V}}, \hat{\mathbf{N}}$ in place of $U_y, \mathbf{V}_y, \mathbf{N}_y$. With these conventions eqn (26.13) may be written

$$\xi = \exp\left[-\beta(\hat{U}+\mathbf{p}\,.\,\hat{\mathbf{V}}-\boldsymbol{\mu}\,.\,\hat{\mathbf{N}})\right]/\Xi, \tag{27.1}$$

where

$$\Xi = \sum_{\hat{}} \exp\left[-\beta(\hat{U}+\mathbf{p}\,.\,\hat{\mathbf{V}}-\boldsymbol{\mu}\,.\,\hat{\mathbf{N}})\right]$$

and (27.2)

$$\sum_{\hat{}} = \underbrace{\sum_{N_1=0}^{\infty}\sum_{N_2=0}^{\infty}\cdots}_{\text{(numbers)}}\ \underbrace{\sum_{r_1=0}^{\infty}\sum_{r_2=0}^{\infty}\cdots}_{\text{(deformations)}}\ \underbrace{\sum_{j}}_{\text{(energy)}}.$$

Sometimes it is satisfactory to approximate these sums by integrals. Eqns (27.1) and (27.2) refer to the *generalized ensemble*, and we shall derive a number of important special cases in the sequel. The number Ξ introduced to normalize ξ is known as the *partition function* for the ensemble.

For a given system, the rest of the ensemble is an idealized environment, and the statement that the systems of the ensemble are in some form of contact with each other may be rephrased by saying that a representative system P is in similar contact with its environment P*. We may set up an ensemble of systems plus environments by taking all possible choices of P from the original ensemble. Hence in the generalized ensemble T, $\mathbf{p}$, $\boldsymbol{\mu}$ may be regarded as being the values applying to every member of an ensemble of reservoirs associated with the ensemble of systems and with which the systems experience thermal, mechanical and material contact.†

By (26.12)

$$S = -k\sum_{\hat{}} \xi \ln \xi, \tag{27.3}$$

the constant being set equal to zero on the assumption that the increments in $\hat{\mathbf{V}}$ and $\hat{U}$ have their lower bounds as set by quantum theory. By (27.1)

$$S = k\sum_{\hat{}} \xi[\beta(\hat{U}+\mathbf{p}\,.\,\hat{\mathbf{V}}-\boldsymbol{\mu}\,.\,\hat{\mathbf{N}})+\ln \Xi],$$

or

$$\Omega \equiv -kT\ln \Xi = U - TS + \mathbf{p}\,.\,\mathbf{V} - \boldsymbol{\mu}\,.\,\mathbf{N}. \tag{27.4}$$

As

$$T\,\mathrm{d}S = \mathrm{d}U + \mathbf{p}\,.\,\mathrm{d}\mathbf{V} - \boldsymbol{\mu}\,.\,\mathrm{d}\mathbf{N}, \tag{27.5}$$

then

$$\mathrm{d}\Omega = -S\,\mathrm{d}T + \mathbf{V}\,.\,\mathrm{d}\mathbf{p} - \mathbf{N}\,.\,\mathrm{d}\boldsymbol{\mu}, \tag{27.6}$$

† Cf. the reservoir $R_T + R_{\mathbf{p}} + R_{\boldsymbol{\mu}}$ in (16.7)

showing $\Omega = \Omega(T, \mathbf{p}, \boldsymbol{\mu})$ to be a thermodynamic potential.† Once the partition function has been obtained, S, $\mathbf{V}$, and $\mathbf{N}$ may be calculated directly. For example, by (27.4) and (27.6)

$$S = -\left(\frac{\partial \Omega}{\partial T}\right)_{\mathbf{p},\boldsymbol{\mu}} = kT\left(\frac{\partial}{\partial T}\ln \Xi\right)_{\mathbf{p},\boldsymbol{\mu}},$$

and similarly for $\mathbf{V}$ and $\mathbf{N}$.

By the Gibbs–Duhem relation (15.6) the right-hand side of (27.6) is zero, so that $\Omega = 0$, a result that also follows from (27.4) and the Euler relation. Hence $\Xi = 1$, and the distribution (27.1) reduces to

$$\xi = \exp\left[-\beta(\hat{U} + \mathbf{p}\,.\,\hat{\mathbf{V}} - \boldsymbol{\mu}\,.\,\hat{\mathbf{N}})\right]. \tag{27.7}$$

For normal thermodynamic purposes eqn (27.3) is quite accurate even if $\Delta V_1, \Delta V_2, \ldots \Delta U$ are not at their lower bounds. For example, suppose the small volume increment ΔV_1 is changed to $\Delta V'_1$, then

$$\Xi'/\Xi = \sum_{r=0}^{\infty} \exp(-\beta p_1 \Delta V'_1 r) \Big/ \sum_{r=0}^{\infty} \exp(-\beta p_1 \Delta V_1 r) \approx \Delta V_1/\Delta V'_1,$$

which adds a constant $k \ln(\Delta V'_1/\Delta V_1)$ to S. This change in S is negligible compared with a typical contribution to S like $kN \ln(V_1/V_2)$ due to a volume change in a perfect gas. A similar discussion holds for changes in ΔU.

27.2. The grand canonical ensemble

The generalized ensemble is little used in statistical thermodynamics, although it is quite important for the general theory. More convenient is the ensemble in which the systems are in thermal and material contact with each other, but have no 'fluctuations' in their deformations $\mathbf{V}$. This means that $\hat{\mathbf{V}} = \mathbf{V}$ for all systems, when (27.7) reduces to

$$\xi(\hat{U}, \hat{\mathbf{N}}) = \mathrm{e}^{-\beta \mathbf{p}.\mathbf{V}}\,\mathrm{e}^{-\beta(\hat{U} - \boldsymbol{\mu}.\hat{\mathbf{N}})} = \mathrm{e}^{-\beta(\hat{U} - \boldsymbol{\mu}.\hat{\mathbf{N}})}/\mathscr{Z}, \tag{27.8}$$

where

$$\mathscr{Z} \equiv \sum_{\hat{}} \mathrm{e}^{-\beta(\hat{U} - \boldsymbol{\mu}.\hat{\mathbf{N}})} = \mathrm{e}^{\beta \mathbf{p}.\mathbf{V}} \tag{27.9}$$

is the new partition function.

The thermodynamic potential is

$$\mathbf{p}\,.\,\mathbf{V} = kT \ln \mathscr{Z} = -U + TS + \boldsymbol{\mu}\,.\,\mathbf{N}, \tag{27.10}$$

and as (27.5) gives

$$\mathrm{d}(\mathbf{p}\,.\,\mathbf{V}) = S\,\mathrm{d}T + \mathbf{p}\,.\,\mathrm{d}\mathbf{V} + \mathbf{N}\,.\,\mathrm{d}\boldsymbol{\mu}, \tag{27.11}$$

† Cf. the potential Ω defined in (16.8).

the independent macroscopic variables are $(T, \mathbf{V}, \boldsymbol{\mu})$. As $\mathscr{Z} = \mathscr{Z}(T, \mathbf{V}, \boldsymbol{\mu})$, it follows from (27.9) that $\hat{U}$ and $\hat{\mathbf{N}}$ must now be functions of $\mathbf{V}$, that is, the constraint on $\hat{\mathbf{V}}$ imposes on the remaining fluctuations a functional dependence on $\mathbf{V}$. As will be shown in § 29 the fluctuations $(\hat{\mathbf{V}} - \mathbf{V})$ in $\mathbf{V}$ suppressed in this ensemble, known as the *grand canonical ensemble*, are not important in classical thermodynamics.

Notice from (27.9) and (27.11) that

$$\mathbf{N} = kT\left(\frac{\partial}{\partial \boldsymbol{\mu}} \ln \mathscr{Z}\right)_{T,\mathbf{V}}; \tag{27.12}$$

similarly S and $\mathbf{p}$ may be derived from the grand canonical partition function.

27.3. *The canonical ensemble*

Because of its mathematical convenience, the ensemble most commonly used in calculations is the *canonical ensemble*, which has fluctuations only in the internal energy. The systems of this ensemble thus have thermal contact only with their surroundings. For all systems $\hat{\mathbf{N}} = \mathbf{N}$, $\hat{\mathbf{V}} = \mathbf{V}$, and (27.7) reduces to

$$\xi(\hat{U}) = \mathrm{e}^{-\beta(\mathbf{p}.\mathbf{V} - \boldsymbol{\mu}.\mathbf{N})}\,\mathrm{e}^{-\beta\hat{U}}$$

or

$$\xi(\hat{U}) = \mathrm{e}^{\beta(F - \hat{U})}, \tag{27.13}$$

where

$$F = -\mathbf{p}\,.\,\mathbf{V} + \boldsymbol{\mu}\,.\,\mathbf{N} = U - TS$$

is the free energy (§ 15.2). As F is the appropriate thermodynamic potential and (27.5) gives

$$\mathrm{d}F = -S\,\mathrm{d}T - \mathbf{p}\,.\,\mathrm{d}\mathbf{V} + \boldsymbol{\mu}\,.\,\mathrm{d}\mathbf{N},$$

the independent variables are $(T, \mathbf{V}, \mathbf{N})$.

We may write (27.13) as

$$\xi \doteq \mathrm{e}^{-\beta\hat{U}}/Z, \tag{27.14}$$

where

$$Z(T, \mathbf{V}, \mathbf{N}) = \sum_{\wedge} \mathrm{e}^{-\beta\hat{U}} = \mathrm{e}^{-\beta F} \tag{27.15}$$

is the partition function for the ensemble. Notice from (27.15) that $\hat{U} = \hat{U}(\mathbf{V}, \mathbf{N})$.

The fluctuations $\hat{\mathbf{N}}$ suppressed in this ensemble will be considered below; like $\hat{\mathbf{V}}$ they are insignificant for classical thermodynamics.

27.4. The micro-canonical ensemble

The final reduction is to completely isolate the systems of the ensemble, that is, to add the constraint $\hat{U} = U$ to $\hat{\mathbf{N}} = \mathbf{N}$, $\hat{\mathbf{V}} = \mathbf{V}$ and reduce (27.7) to

$$\xi = \mathrm{e}^{-\beta(U+\mathbf{p}.\mathbf{V}-\boldsymbol{\mu}.\mathbf{N})} = \mathrm{e}^{-S/k} = 1/\Gamma, \tag{27.16}$$

where Γ is a constant for given values of U, $\mathbf{V}$, $\mathbf{N}$. The thermodynamic potential is now

$$S(U, \mathbf{V}, \mathbf{N}) = k \ln \Gamma(U, \mathbf{V}, \mathbf{N}), \tag{27.17}$$

so that when Γ is known the intensive thermodynamic variables may be calculated from its partial derivations with the aid of (27.5). These equations define the *micro-canonical ensemble.* Notice from the discussion given in § 26.4 that Γ is the total number of distinguishable micro-states. The ensemble has the disadvantage that Γ is often difficult to evaluate.

27.5. Table of ensembles

TABLE 5

Name	Probability ξ	Partition function	Potential	Differential form
Generalized	$\mathrm{e}^{-\beta(\hat{U}+\mathbf{p}.\hat{\mathbf{V}}-\boldsymbol{\mu}.\hat{\mathbf{N}})}$	$\Xi(T, \mathbf{p}, \boldsymbol{\mu})$	$\Omega = -kT \ln \Xi$	$\mathrm{d}\Omega = -S\,\mathrm{d}T + \mathbf{V}\,.\,\mathrm{d}\mathbf{p} - \mathbf{N}\,.\,\mathrm{d}$
Grand canonical	$\mathrm{e}^{-\beta(\hat{U}-\boldsymbol{\mu}.\hat{\mathbf{N}})}/\mathscr{Z}$	$\mathscr{Z}(T, \mathbf{V}, \boldsymbol{\mu})$	$\mathbf{p}\,.\,\mathbf{V} = kT \ln \mathscr{Z}$	$\mathrm{d}(\mathbf{p}\,.\,\mathbf{V}) = S\,\mathrm{d}T + \mathbf{p}\,.\,\mathrm{d}\mathbf{V} + \mathbf{N}\,.\,\mathrm{d}$
Canonical	$\mathrm{e}^{-\beta\hat{U}}/Z$	$Z(T, \mathbf{V}, \mathbf{N})$	$F = -kT \ln Z$	$\mathrm{d}F = -S\,\mathrm{d}T - \mathbf{p}\,.\,\mathrm{d}\mathbf{V} + \boldsymbol{\mu}\,.\,\mathrm{d}$
Micro-canonical	Γ^{-1}	$\Gamma(U, \mathbf{V}, \mathbf{N})$	$S = k \ln \Gamma$	$T\,\mathrm{d}S = \mathrm{d}U + \mathbf{p}\,.\,\mathrm{d}\mathbf{V} - \boldsymbol{\mu}\,.\,\mathrm{d}$

Table 5 gives the four most widely known ensembles. Other ensembles exist, in fact four others are immediately obvious from the table, namely, two similar to the grand canonical ensemble and two similar to the canonical ensemble. The eight ensembles based on

$$\xi = \mathrm{e}^{-\hat{S}/k} = \mathrm{e}^{-\beta(\hat{U}+\mathbf{p}.\hat{\mathbf{V}}-\boldsymbol{\mu}.\hat{\mathbf{N}})}$$

do not complete the list of possible ensembles, for we are free to choose any one of $\hat{S}$, $\hat{U}$, $\hat{\mathbf{V}}$, $\hat{\mathbf{N}}$ to be the dependent fluctuation, which suggests a total of 32 ensembles. However, few of these are of interest—one that is is discussed in the next section.

28. Other ensembles

25.1. Independent entropy fluctuations

To this point in ensemble theory entropy has been given the role of a dependent variable, which is a natural choice as entropy is a *mathematically* defined quantity rather than one that can be directly measured (see § 9.3).

However, from a theoretical stand-point it is sometimes useful to put S into the role of an independent variable. One method of doing this is to interchange $T\,\mathrm{d}S$ and $\boldsymbol{\mu}\,.\,\mathrm{d}\mathbf{N}$ in (26.3). Then as $\boldsymbol{\mu}\,.\,\mathbf{N} = G$ (cf. (15.9)) we find by an argument similar to that given in § 26.4, that

$$G = G^{\mathrm{eq}} - kT\sum_y \xi_y \ln(\xi_y/\xi_y^{\mathrm{eq}}), \qquad G^{\mathrm{eq}} = U + \mathbf{p}\,.\,\mathbf{V} - TS.$$

Continuing to modify the theory of § 26.4 in this way, we arrive at

$$G_y^{\mathrm{eq}} = U_y + \mathbf{p}\,.\,\mathbf{V}_y - TS_y, \qquad \xi_y^{\mathrm{eq}} = K\exp(-\beta G_y^{\mathrm{eq}}),$$

and finally (dropping the superscript eq) instead of (27.7) we have

$$\xi(\hat{U}, \hat{\mathbf{V}}, \hat{S}) = \exp[-\beta(\hat{U} + \mathbf{p}\,.\,\hat{\mathbf{V}} - T\hat{S})]. \tag{28.1}$$

This ensemble is particularly useful in fluctuation theory to be discussed shortly. It may be rewritten in terms of the fluctuations from the mean

$$\Delta U \equiv \hat{U} - U, \qquad \Delta\mathbf{V} \equiv \hat{\mathbf{V}} - \mathbf{V}, \qquad \Delta S \equiv \hat{S} - S, \tag{28.2}$$

as

$$\xi = \mathrm{e}^{-\beta\hat{G}} = \mathrm{e}^{-\beta G}\,\mathrm{e}^{-\beta(\Delta U + \mathbf{p}.\Delta\mathbf{V} - T\Delta S)}. \tag{28.3}$$

By restricting the fluctuations in $\hat{G}$ to be small, that is, $\hat{G} \simeq G$, we impose a constraint on the fluctuations ΔU, $\Delta\mathbf{V}$ and ΔS, which in the limit $\Delta G = 0$ is clearly $\Delta U - T\Delta S + \mathbf{p}\,.\,\Delta\mathbf{V} = 0$. Short of this limit we may expand ΔU as a Taylor series and obtain

$$\Delta U = \frac{\partial U}{\partial S}\Delta S + \frac{\partial U}{\partial \mathbf{V}}\,.\,\Delta\mathbf{V} + \frac{1}{2}\frac{\partial^2 U}{\partial S^2}(\Delta S)^2 + \frac{\partial^2 U}{\partial S\,\partial\mathbf{V}}\,.\,\Delta\mathbf{V}\,\Delta S + \frac{1}{2}\frac{\partial^2 U}{\partial \mathbf{V}^2}:(\Delta\mathbf{V})^2 + \ldots$$

that is,

$$\begin{aligned}\Delta U - T\,\Delta S + \mathbf{p}\,.\,\Delta\mathbf{V} &= \tfrac{1}{2}\Delta S\left\{\frac{\partial}{\partial S}\left(\frac{\partial U}{\partial S}\right)\Delta S + \frac{\partial}{\partial\mathbf{V}}\left(\frac{\partial U}{\partial S}\right)\,.\,\Delta\mathbf{V}\right\}\\ &\quad + \tfrac{1}{2}\Delta\mathbf{V}\,.\left\{\frac{\partial}{\partial S}\left(\frac{\partial U}{\partial\mathbf{V}}\right)\Delta S + \frac{\partial}{\partial\mathbf{V}}\left(\frac{\partial U}{\partial\mathbf{V}}\right)\,.\,\Delta\mathbf{V}\right\} + \ldots\\ &= \tfrac{1}{2}\Delta S\,\Delta\left(\frac{\partial U}{\partial S}\right)_{\mathbf{V}} + \tfrac{1}{2}\Delta\mathbf{V}\,.\,\Delta\left(\frac{\partial U}{\partial\mathbf{V}}\right)_S + \ldots\end{aligned}$$

Now

$$\Delta\left(\frac{\partial U}{\partial S}\right)_{\mathbf{V}} = \left(\frac{\partial\hat{U}}{\partial S}\right)_{\mathbf{V}} - \left(\frac{\partial U}{\partial S}\right)_{\mathbf{V}} = \hat{T} - T = \Delta T$$

and

$$\Delta\left(\frac{\partial U}{\partial \mathbf{V}}\right)_S = \left(\frac{\partial \hat{U}}{\partial \mathbf{V}}\right)_S - \left(\frac{\partial U}{\partial \mathbf{V}}\right)_S = -\hat{\mathbf{p}}+\mathbf{p} = -\Delta\mathbf{p}.$$

Eqn (28.3) now assumes the approximate form

$$\xi \propto e^{(\Delta\mathbf{p}.\Delta\mathbf{V} - \Delta T\,\Delta S)/2kT}. \qquad (28.4)$$

28.2. A non-equilibrium ensemble

Suppose that additional but unspecified internal coordinates **Y** contribute to the entropy (see § 16.1) then (26.3) is replaced by

$$T\,\mathrm{d}S = \mathrm{d}U + \mathbf{p}\,.\,\mathrm{d}\mathbf{V} - \boldsymbol{\mu}\,.\,\mathrm{d}\mathbf{N} - \mathbf{x}\,.\,\mathrm{d}\mathbf{Y} - kT\sum_y \ln(\xi_y/\xi_y^{\mathrm{eq}})\,\mathrm{d}\xi_y, \qquad (28.5)$$

where the components of **Y** are extensive variables and **x** is the vector of conjugate intensive variables. For natural processes

$$\mathbf{x}\,.\,\mathrm{d}\mathbf{Y} < 0. \qquad (28.6)$$

If the theory is now developed as in § 28.1, the ensemble defined by (28.1) is replaced by

$$\xi(\hat{U}, \hat{\mathbf{V}}, \hat{S}, \hat{\mathbf{Y}}) = \exp[-\beta(\hat{U} + \mathbf{p}\,.\,\hat{\mathbf{V}} - \mathbf{x}\,.\,\hat{\mathbf{Y}} - T\hat{S})], \qquad (28.7)$$

and as the components of **Y** are unspecified, we may regard this as being the distribution function of a non-equilibrium ensemble. When the fluctuations are in thermodynamic equilibrium, **x** is zero and (28.7) reduces to (28.1).

Suppose now that all the systems have the same energy and deformation vector, then we may reduce (28.7) to

$$\xi = e^{-\beta\hat{G}} = e^{-\beta G}\, e^{(\Delta S + \mathbf{x}.\Delta\mathbf{X})/k},$$

where $\hat{\mathbf{X}} = \hat{\mathbf{Y}}/T$, $\Delta S = \hat{S} - S$, and $\Delta\mathbf{X} = \hat{\mathbf{X}} - \mathbf{X}$. The next step is to take ΔS to be dependent on **X** and to expand it in the Taylor series

$$\Delta S = -\mathbf{x}\,.\,\Delta\mathbf{X} - \tfrac{1}{2}\Delta\mathbf{x}\,.\,\Delta\mathbf{X} + \ldots \qquad (28.8)$$

and thus arrive at the approximate distribution (cf. (28.4))

$$\xi \propto e^{-\Delta\mathbf{x}.\Delta\mathbf{X}/2k}. \qquad (28.9)$$

We are free to choose the origin for the coordinates $\hat{X}_1, \hat{X}_2, \ldots$ at their ensemble average, that is, $\mathbf{X} = 0$ and $\Delta\mathbf{X} = \hat{\mathbf{X}}$. Now consider fluctuations about the thermodynamic equilibrium $\mathbf{x} = 0$. Eqns (28.8) and (28.9) reduce to

$$\Delta S = \hat{S} - S = -\tfrac{1}{2}\hat{\mathbf{x}}\,.\,\hat{\mathbf{X}} \qquad (28.10)$$

and

$$\xi \propto e^{-\hat{\mathbf{x}}.\mathbf{X}/2k}. \qquad (28.11)$$

28.3. Gaussian distribution of fluctuations about equilibrium

As an important example consider the case when $\hat{X}_1, \hat{X}_2, ... \hat{X}_n$ are continuously distributed, then from (28.10) and (28.11) the joint probability that $\hat{X}_1$ falls in $d\hat{X}_1$, $\hat{X}_2$ falls in $d\hat{X}_2$, etc. is given by

$$W(\hat{X}_1, \hat{X}_2, ... \hat{X}_n)\, d\hat{X}_1 ... d\hat{X}_n = A\, e^{\Delta S/k}\, d\hat{\mathbf{X}} = A\, e^{-\hat{\mathbf{x}}.\hat{\mathbf{X}}/2k}\, d\hat{\mathbf{X}}, \qquad (28.12)$$

where the constant A is to be chosen to satisfy $\int W\, d\hat{\mathbf{X}} = 1$ in which the integral covers the whole range of $\hat{\mathbf{X}}$.

For small departures from equilibrium the state equations relating the intensive variables $\hat{x}_i$, $i = 1, 2, ... n$, to the independent extensive variables $\hat{X}_i$ will be linear; thus

$$\hat{x}_i = \sum_j \beta_{ij} \hat{X}_j \quad \text{or} \quad \hat{\mathbf{x}} = \boldsymbol{\beta} \,.\, \hat{\mathbf{X}}, \qquad (28.13)$$

where β_{ij} are constants that may depend on other constants like U and $\mathbf{V}$. That $\boldsymbol{\beta}$ is a symmetric tensor follows from

$$\hat{S} - S = -\tfrac{1}{2}\hat{\mathbf{X}} \,.\, \boldsymbol{\beta} \,.\, \hat{\mathbf{X}} \qquad (28.14)$$

and

$$\beta_{ij} = -\left(\frac{\partial^2 \hat{S}}{\partial \hat{X}_i\, \partial \hat{X}_j}\right)_{\text{eq}}.$$

Furthermore, as thermodynamic stability requires the equilibrium entropy S to be a maximum for the fluctuations, $S > \hat{S}$, and so $\boldsymbol{\beta}$ is positive definite.

With the help of the first integral in the footnote of p. 86, when (28.13) is introduced into (28.12) we can evaluate the constant A, with the result that the distribution may be written

$$W\, d\hat{\mathbf{X}} = \sqrt{\left[\frac{|\boldsymbol{\beta}|}{(2\pi k)^n}\right]} \exp(-\boldsymbol{\beta} : \hat{\mathbf{X}}\hat{\mathbf{X}}/2k)\, d\hat{\mathbf{X}}, \qquad (28.15)$$

where $|\boldsymbol{\beta}|$ is the determinant of the matrix $[\beta_{ij}]$. Note from (28.12), (28.13), and (28.14) that

$$\hat{\mathbf{x}} = -\partial(\hat{S} - S)/\partial \hat{\mathbf{X}} = -k\, \partial(\ln W)/\partial \hat{\mathbf{X}}. \qquad (28.16)$$

Let us denote mean values over the ensemble thus

$$\langle \varphi \rangle = \int \varphi W\, d\hat{\mathbf{X}},$$

then by (28.16)

$$\langle \hat{x}_i \rangle = -k \int \frac{\partial(\ln W)}{\partial \hat{X}_i} W\, d\hat{\mathbf{X}} = -k \int \frac{\partial W}{\partial \hat{X}_i}\, d\hat{\mathbf{X}} = 0,$$

as W vanishes at the integration boundaries. This result has already been assumed. Also

$$\langle \hat{x}_i \hat{X}_j \rangle = -k \int \hat{X}_j \frac{\partial W}{\partial \hat{X}_i} \mathrm{d}\hat{\mathbf{X}} = k \int W \frac{\partial \hat{X}_j}{\partial \hat{X}_i} \mathrm{d}\hat{\mathbf{X}} = k\delta_{ij} \int W \mathrm{d}\hat{\mathbf{X}},$$

where δ_{ij} is the Kronecker delta, equal to unity if $i = j$, otherwise zero. Hence

$$\langle \hat{\mathbf{x}}\hat{\mathbf{X}} \rangle = k\mathbf{1}, \tag{28.17}$$

and if (28.13) is introduced into this result we obtain

$$\langle \hat{\mathbf{X}}\hat{\mathbf{X}} \rangle = k\boldsymbol{\beta}^{-1}, \tag{28.18}$$

where $\boldsymbol{\beta}^{-1}$ is the reciprocal tensor of $\boldsymbol{\beta}$.

In particular,

$$\langle \Delta S \rangle = \langle \hat{S} \rangle - S = -\tfrac{1}{2}\langle \hat{\mathbf{x}} \cdot \hat{\mathbf{X}} \rangle,$$

that is,

$$\langle \hat{S} \rangle = S - \tfrac{1}{2}nk. \tag{28.19}$$

As S is of order Nk, where N is enormously larger than the number of degrees of freedom n, there is a negligible difference between the classical entropy viewed as being a maximum and this entropy considered as an average over the distribution function. In other words, the distribution in (28.15) is very sharply peaked about $\hat{\mathbf{X}} = 0$.

29. Fluctuation theory

29.1. Mean square deviation

By the assumption of ergodicity the ensemble fluctuations with which we have been concerned may be considered as being system fluctuations occurring with the passage of time as a result of the interactions of the system with its environment. We shall now study the magnitude of these fluctuations.

Save near certain critical points, that will not be considered in this text, fluctuations are not important in classical thermodynamics because those quantities allowed to fluctuate in particular ensembles are sharply peaked about their mean values. We shall prove this fact shortly, in order to establish that for the classical theory it is a matter of indifference which ensemble is adopted. Fluctuation theory is important for small systems, but these will not be considered here.

Ensemble averages of a typical variable $\hat{A}$ are denoted by $\langle \hat{A} \rangle = \Sigma \hat{A}\xi$, and a useful measure of the spread of fluctuations about the mean $A = \langle \hat{A} \rangle$ is the mean square deviation, or variance

$$\sigma_A^2 = \langle (\Delta A)^2 \rangle \qquad (\Delta A \equiv \hat{A} - A).$$

For a normal distribution the probability of $\hat{A}$ falling in $\mathrm{d}\hat{A}$ is

$$W(\hat{A})\,\mathrm{d}\hat{A} = \frac{1}{\sigma_A\sqrt{(2\pi)}}\exp\left[-(\Delta A)^2/2\sigma_A^2\right]\mathrm{d}\hat{A}. \tag{29.1}$$

It will be shown below that for large values of the particle number N in a system, all the thermodynamic variables have the same asymptotic form for their relative deviation, namely,

$$\sigma_A/A \sim KN^{-\frac{1}{2}}, \tag{29.2}$$

where K is a constant of order unity. This simple result shows that all fluctuations vanish in the so-called *thermodynamic limit*

$$N \to \infty, \qquad V \to \infty, \qquad N/V \text{ finite}, \tag{29.3}$$

the condition on V being necessary to maintain finite pressure and density. Classical thermodynamics, which is not concerned with fluctuations, is thus an asymptotic theory when derived from statistical thermodynamics as described in § 26.4.

29.2. Fluctuations in number density and energy

Consider a one-component system having the volume V as its only deformation variable. By (27.9)

$$\mathscr{L}(T, V, \mu) = \sum_{\wedge} \mathrm{e}^{-\beta[\hat{U}(V) - \mu\hat{N}(V)]},$$

whence

$$\left(\frac{\partial \mathscr{L}}{\partial \mu}\right)_{T,V} = \beta \sum_{\wedge} \hat{N}\,\mathrm{e}^{-\beta(\hat{U} - \mu\hat{N})} = \beta\langle\hat{N}\rangle\mathscr{L} = \beta N\mathscr{L}.$$

Thus when

$$N\mathscr{L} = \sum_{\wedge} \hat{N}\,\mathrm{e}^{-\beta(\hat{U} - \mu\hat{N})}$$

is differentiated with respect to μ and then divided by $\mathscr{L}$ the result is

$$\left(\frac{\partial N}{\partial \mu}\right)_{T,V} + \beta N^2 = \beta\langle\hat{N}^2\rangle,$$

that is,

$$\frac{\sigma_N^2}{N^2} = \frac{\langle(\hat{N} - N)^2\rangle}{N^2} = \frac{kT}{N^2}\left(\frac{\partial N}{\partial \mu}\right)_{T,V}. \tag{29.4}$$

An estimate of the usual magnitude of the partial derivative in (29.4) may be obtained from the special case of an ideal gas. Eqn (21.5) gives $\mu = kT[\ln(N/V) - (\gamma - 1)^{-1}\ln T + \text{constant}]$, whence $(\partial N/\partial \mu)_{T,V} = N/(kT)$.

It follows from (29.3) that the relative deviation σ_N/N tends to zero like $N^{-\frac{1}{2}}$, and with typical values of N this deviation is very small indeed. This result is generally true save near critical points.

To calculate the mean square deviation for the energy we shall apply a similar method to the canonical ensemble

$$UZ(T, V, N) = U \sum_{\hat{}} e^{-\beta\hat{U}} = \sum_{\hat{}} \hat{U}\, e^{-\beta\hat{U}};$$

thus

$$\left(\frac{\partial U}{\partial \beta}\right)_{V,N} Z - U\langle\hat{U}\rangle Z = -\langle\hat{U}^2\rangle Z,$$

and by $-(\partial U/\partial\beta)_{V,N} = kT^2C_V$ it follows that

$$\sigma_{\hat{U}}^2 = kT^2C_V. \tag{29.5}$$

Now $C_V = O(Nk)$, $U = O(NkT)$ (cf. a perfect gas), whence $\sigma_U/U = O(N^{-\frac{1}{2}})$, and the deviation from the mean is again negligibly small.

An alternative treatment is based on (28.4) or better on

$$\xi \propto \exp\{[\Delta T^{-1}\Delta U + \Delta(pT^{-1})\Delta V]/2k\}, \tag{29.6}$$

which follows from (28.3) on choosing U and V to be the independent variables. Now

$$\Delta T^{-1} = -T^{-1}\left[\left(\frac{\partial T}{\partial U}\right)_V \Delta U + \left(\frac{\partial T}{\partial V}\right)_U \Delta V\right] = -\frac{\Delta U}{T^2C_V} + O(\Delta V),$$

hence at constant volume (29.6) may be written $\xi \propto \exp[-(\Delta U)^2/(2kT^2C_V)]$, and comparing this with (29.1) we recover the result in (29.5).

29.3. Pressure, temperature, and volume fluctuations

Fluctuations in an intensive variable may be derived either approximately by choosing it to be an independent variable in (28.4), or exactly by a method modelled on the following. The canonical ensemble has the partition function

$$Z(T, V, N) = e^{-\beta F} = \sum_{\hat{}} e^{-\beta\hat{U}(V,N)},$$

and as

$$p = -\left(\frac{\partial F}{\partial V}\right)_{T,N} = kT\frac{\partial}{\partial V}\ln\sum_{\hat{}} e^{-\beta\hat{U}} = -\sum_{\hat{}}\left(\frac{\partial\hat{U}}{\partial V}\right)_N e^{-\beta\hat{U}}\Big/Z = \left\langle-\left(\frac{\partial\hat{U}}{\partial V}\right)_N\right\rangle,$$

it follows that $p = \langle\hat{p}\rangle$, where $\hat{p} \equiv -(\partial\hat{U}/\partial V)_N$. The derivative of

$$pZ = \sum_{\hat{}} \hat{p}\, e^{-\beta\hat{U}(V,N)}$$

with respect to V yields

$$\langle(\hat{p}-p)^2\rangle = kT\left[\left(\frac{\partial p}{\partial V}\right) - \left\langle\frac{\partial \hat{p}}{\partial V}\right\rangle\right]. \qquad (29.7)$$

On the assumption that the derivatives on the right-hand side are of the same order, we find $\sigma_p/p \sim N^{-\frac{1}{2}}$ as in (29.2).

We shall illustrate the approximate method based on (28.4) by using it to calculate the temperature fluctuations. Choose T and V as the independent variables, then as (see § 10)

$$\Delta S = \frac{C_V}{T}\Delta T + \left(\frac{\partial p}{\partial T}\right)_V \Delta V, \qquad \Delta p = \left(\frac{\partial p}{\partial T}\right)_V \Delta T + \left(\frac{\partial p}{\partial V}\right)_T \Delta V,$$

$$\xi \propto \exp\left[-\frac{C_V}{2kT^2}(\Delta T)^2 + \frac{1}{2kT}\left(\frac{\partial p}{\partial V}\right)_T (\Delta V)^2\right].$$

It follows that temperature and volume fluctuate independently, that is, $\langle \Delta T\, \Delta V\rangle = 0$ (cf. (28.17)) and $\sigma_T^2 = kT^2/C_V$, $\sigma_V^2 = -kT(\partial V/\partial p)_T$. The thermodynamic stability conditions (17.11) ensure that the right-hand sides of these expressions are positive. It is easily shown that both σ_T/T and σ_V/V have the same asymptotic form as for all the other variables, namely that given in (29.2).

30. Ensembles in phase space

30.1. Hamiltonians and Γ-space

In classical mechanics the state of an isolated system possessing n degrees of freedom is prescribed by giving the values of its n generalized coordinates q_i, $i = 1, 2, \ldots n$, and its n conjugate momenta p_i. The representative point $(\mathbf{p}, \mathbf{q}) = (p_1, p_2 \ldots p_n, q_1, q_2, \ldots q_n)$ moves in a phase-space of $2n$ dimensions known as Γ-space. The trajectory in this space is governed by the $2n$ linear equations

$$\dot{q}_i = \frac{\partial H}{\partial p_i}, \qquad \dot{p}_i = -\frac{\partial H}{\partial q_i}, \qquad (30.1)$$

where $H(\mathbf{p}, \mathbf{q}, t)$ is the total energy of the system expressed in terms of the coordinates and the momenta. These are Hamilton's canonical equations, and H is the system Hamiltonian.

The representative points for the large number of systems comprising an ensemble are distributed over Γ-space, and if they are sprinkled thickly enough we can define a continuous function of $\mathbf{p}$, $\mathbf{q}$, and t, known as the phase-density $D(\mathbf{p}, \mathbf{q}, t)$, such that $D\,\mathbf{dp}\,\mathbf{dq}$ is the number of phase-points to be found at time t in the element of phase-space $\mathbf{dp}\,\mathbf{dq} = \mathrm{d}\Gamma$, say, around the point $(\mathbf{p}, \mathbf{q})$. The probability that a given system lies in $\mathrm{d}\Gamma$ is

$$\xi(\mathbf{p}, \mathbf{q}, t)\,\mathrm{d}\Gamma = D(\mathbf{p}, \mathbf{q}, t)\,\mathrm{d}\Gamma \bigg/ \int D(\mathbf{p}, \mathbf{q}, t)\,\mathrm{d}\Gamma, \tag{30.2}$$

where the integral is over the whole of Γ-space. The ensemble average of a property $\varphi(\mathbf{p}, \mathbf{q}, t)$ of the system is

$$\langle\varphi\rangle = \int \varphi D\,\mathrm{d}\Gamma \bigg/ \int D\,\mathrm{d}\Gamma. \tag{30.3}$$

30.2. Liouville's theorem

As the number of systems in an ensemble is conserved, the density D must satisfy the continuity equation (see § 38.1) in Γ-space, namely,

$$\frac{\partial D}{\partial t} + \nabla\,.\,(D\mathbf{V}) = 0,$$

where ∇ is the gradient operator in Γ-space and $\mathbf{V}$ is the Γ-space 'velocity', $\mathbf{V} = (\dot{\mathbf{p}}, \dot{\mathbf{q}})$. Now

$$\nabla\,.\,\mathbf{V} = \left(\frac{\partial}{\partial \mathbf{p}}, \frac{\partial}{\partial \mathbf{q}}\right)\,.\,(\dot{\mathbf{p}}, \dot{\mathbf{q}}) = \frac{\partial}{\partial \mathbf{p}}\,.\,\dot{\mathbf{p}} + \frac{\partial}{\partial \mathbf{q}}\,.\,\dot{\mathbf{q}},$$

which is zero by (30.1). Thus D satisfies

$$\frac{\partial D}{\partial t} + \mathbf{V}\,.\,\nabla D = \frac{\mathrm{d}D}{\mathrm{d}t} = 0, \tag{30.4}$$

where $\mathrm{d}/\mathrm{d}t$ denotes the rate of change following an element of ensemble 'fluid' through Γ-space.

Thus we have established *Liouville's theorem*, namely, that phase-points move through Γ-space as if they were embedded in an incompressible fluid. By (30.2) this important result may be written

$$\frac{\mathrm{d}\xi}{\mathrm{d}t} = 0, \tag{30.5}$$

which will play some role in our account of process thermodynamics (§ 74).

As the number of phase-points $D\,\mathrm{d}\Gamma$ in the element $\mathrm{d}\Gamma$ must remain fixed as $\mathrm{d}\Gamma$ is convected with the phase-fluid, that is, $\mathrm{d}(D\,\mathrm{d}\Gamma)/\mathrm{d}t = 0$, Liouville's theorem may be written alternatively

$$\frac{\mathrm{d}}{\mathrm{d}t}(\mathrm{d}\Gamma) = 0,$$

or, in words, the volume element associated with a given set of phase-points does not change in magnitude as it moves through Γ-space.

The N-particle distribution function F_N introduced in § 25.2 applies to a single system of identical particles and is defined by requiring $F_N \, d\Gamma$ to be the probability of finding this system in the element $d\Gamma$ of Γ-space. The assumption of ergodicity (§ 26.2) means that this probability is proportional to the number of systems $D \, d\Gamma$ of an ensemble that occupy $d\Gamma$. It follows from (30.2) that

$$\xi = F_N = D \bigg/ \int D \, d\Gamma.$$

From (26.12) we conclude that the entropy for a system of identical and featureless particles is given by

$$S = -k \int F_N \ln F_N \, d\Gamma + \text{constant},$$

confirming the result given in § 25.6.

30.3. The canonical ensemble

By $(27.2)_2$ the sum in (27.15) may be written

$$Z = \sum_j \exp(-\beta U_j) = \exp(-\beta F), \tag{30.6}$$

and is taken over all distinct micro-states of the system. When the state is defined by the continuous variables $(\mathbf{p}, \mathbf{q})$ of Γ-space, in order to transform the sum to an integral over Γ-space, we need to know (1) the relation between unit increment in j and an element $d\Gamma$ and (2) how many distinct elements $d\Gamma_1, d\Gamma_2, \ldots$ correspond to a single value of j. The inverse of this problem has already been discussed for the special case of a perfect gas in § 25.3. Quantum theory answers (1) by giving $d\Gamma/h^n$ as the total number of micro-states in $d\Gamma$. Thus

$$\sum_j \ldots \to \frac{1}{h^n} \int' \ldots d\Gamma,$$

where the accent on the integral sign means that the integration is restricted to those regions of phase-space corresponding to physically distinct states of the system. As described in § 25.3, when the system is composed of N identical particles, $n = 3N$ and the rule becomes

$$\sum_j \ldots \to \frac{1}{h^{3N}} \int' \ldots d\Gamma = \frac{1}{N! h^{3N}} \int \ldots d\Gamma, \tag{30.7}$$

the undashed integral being over the whole of phase-space. In this transformation the internal energy U_j is replaced by the Hamiltonian $H(\mathbf{p}, \mathbf{q}, t)$.

It follows from (30.6) that the partition function for a system of N identical particles is given by

$$Z(T, V, N) = e^{-\beta F} = \frac{1}{N!h^{3N}} \int e^{-\beta H(\mathbf{p},\mathbf{q})} \, d\Gamma, \tag{30.8}$$

and the phase-density is proportional to $\exp(-\beta H)$. When this function is known all the other thermoydnamic functions may be derived from the potential (§ 10) $F(T, V, N)$; for example,

$$p = -\left(\frac{\partial F}{\partial V}\right)_{T,N}, \qquad S = -\left(\frac{\partial F}{\partial T}\right)_{V,N}, \qquad U = F + TS. \tag{30.9}$$

30.4. The grand canonical ensemble

Suppose for simplicity that the system consists of one component and has the volume V as its only deformation coordinate. Then from the theory given in § 27.2 and (30.6) above we find that the partition function of the grand canonical ensemble is

$$\begin{aligned}\mathscr{Z}(T, V, z) = e^{\beta pV} &= \sum_{N=0}^{\infty} z^N Z(T, V, N) \\ &= \sum_{N=0}^{\infty} \frac{z^N}{N!h^{3N}} \int e^{-\beta H(\mathbf{p},\mathbf{q})} \, d\Gamma,\end{aligned} \tag{30.10}$$

where

$$z \equiv e^{\beta\mu} \tag{30.11}$$

is a number called the 'fugacity'.

The average number of particles N in the volume V is given by

$$N = \sum_{N'=0}^{\infty} N' z^{N'} Z(T, V, N') \Big/ \sum_{N'=0}^{\infty} z^{N'} Z(T, V, N'),$$

that is,

$$N = z\frac{\partial}{\partial z} \ln \mathscr{Z}(T, V, z), \tag{30.12}$$

and the state equation $p = p(T, V, N)$ is obtained by eliminating z between (30.12) and

$$pV = kT \ln \mathscr{Z}(T, V, z). \tag{30.13}$$

In a similar way we find that the internal energy $U = \langle H \rangle$ is given by

$$U = -\frac{\partial}{\partial \beta} \ln \mathscr{Z}(T, V, z), \tag{30.14}$$

and when z has been eliminated with the aid of (30.12), the other thermodynamic functions are readily calculated. Alternatively, we may derive these from F, which is given by

$$F = NkT \ln z - kT \ln \mathscr{Z}(T, V, z). \tag{30.15}$$

31. Some application of Hamiltonians

31.1. The equipartition theorem

The canonical ensemble will now be used to prove an important result, known as the *equipartition theorem*, concerning the average distribution of energy over the classical degrees of freedom of a system. The theorem is stated in § 23.3 and may be rephrased:

If the contributions to the total energy H are separable, the average energy associated with each single coordinate (position or momentum) that contributes a quadratic term to H is $\frac{1}{2}kT$.

Let $H(\mathbf{x})$ be the Hamiltonian and $\mathbf{x} = (\mathbf{p}, \mathbf{q})$ the position vector in Γ-space, then

$$\int \mathbf{x}\frac{\partial H}{\partial \mathbf{x}} \mathrm{e}^{-\beta H}\,\mathrm{d}\mathbf{x} = kT \int \left[\mathbf{1}\,\mathrm{e}^{-\beta H} - \frac{\partial}{\partial \mathbf{x}}(\mathbf{x}\,\mathrm{e}^{-\beta H})\right] \mathrm{d}\mathbf{x},$$

where $\mathbf{1} = \partial\mathbf{x}/\partial\mathbf{x}$ is unit tensor in Γ-space. The second right-hand integral is zero as $\exp(-\beta H)$ vanishes more rapidly than $|\mathbf{x}|$ as the integration boundaries are approached. It therefore follows from (30.3) and (30.8) that

$$\left\langle \mathbf{x}\frac{\partial H}{\partial \mathbf{x}} \right\rangle = kT\mathbf{1}. \tag{31.1}$$

This result may be decomposed into components; in particular

$$\left\langle q_i \frac{\partial H}{\partial q_i} \right\rangle = \langle -q_i \dot{p}_i \rangle = kT,$$

whence

$$\left\langle \sum_{i=1}^{3N} \dot{p}_i q_i \right\rangle = -3NkT, \tag{31.2}$$

which is known as the 'virial' theorem.

Most physical systems have Hamiltonians that can be put in the form

$$H = \sum_i A_i P_i^2 + \sum_i B_i Q_i^2,$$

where P_i and Q_i are canonically conjugate variables (see (30.1)) and A_i, B_i

are constants. For such systems

$$\sum_i \left(P_i \frac{\partial H}{\partial P_i} + Q_i \frac{\partial H}{\partial Q_i} \right) = 2H.$$

By (31.1)

$$\left\langle P_i \frac{\partial H}{\partial P_i} \right\rangle = kT, \qquad \left\langle Q_i \frac{\partial H}{\partial Q_i} \right\rangle = kT,$$

and so if f of the constants A_i, B_i are non-vanishing

$$\langle H \rangle = \tfrac{1}{2} f kT. \tag{31.3}$$

Thus each harmonic term in the Hamiltonian contributes $\frac{1}{2}kT$ to the average energy of the system, which establishes the equipartition theorem.

31.2. The classical perfect gas

We substitute the Hamiltonian for the perfect gas, namely,

$$H = \frac{1}{2m} \sum_{i=1}^{3N} p_i^2,$$

into (30.8) and integrate over the space coordinate to find

$$Z = \frac{V^N}{N!h^{3N}} \left[\int_{-\infty}^{\infty} \exp\left(-\frac{\beta}{2m} p_i^2 \right) \mathrm{d}p_i \right]^{3N}.$$

By

$$\sqrt{\pi} = \int_{-\infty}^{\infty} \exp(-t^2)\, \mathrm{d}t$$

and the definition

$$\lambda \equiv \frac{h}{(2\pi mkT)^{\frac{1}{2}}} \tag{31.4}$$

(known as the 'de Broglie wavelength') we obtain

$$\mathrm{e}^{-\beta F} = Z = \frac{1}{N!} \left(\frac{V}{\lambda^3} \right)^N. \tag{31.5}$$

Thus by Stirling's formula (25.13)

$$F = -NkT \ln \left(\frac{V}{N\lambda^3} \right) - NkT,$$

and (30.9) yield

$$S = kN \ln\left(\frac{V}{N\lambda^3}\right) + \tfrac{5}{2}kN,$$
$$U = \tfrac{3}{2}NkT, \qquad p = \frac{NkT}{V}. \tag{31.6}$$

The formula for S was derived by a different method in § 25.5.

32. The distributions of quantum statistical mechanics

32.1. Fermions and bosons

At low temperatures it is a consequence of quantum theory that the energy states of a system are not continuously distributed but possess a discrete spectrum of values. In the next few pages we shall give a brief account of the consequences of allowing for this quantum effect for systems in which the particles interact weakly enough for the system energy to be the sum of the one-particle energies.

Let $\varepsilon_1, \varepsilon_2, \dots \varepsilon_r, \dots$ be the one-particle energy spectrum, and suppose $n_1, n_2, \dots n_r, \dots$ are the corresponding 'occupation numbers', that is, the numbers of particles in these energy states. Then the state of a system is specified by the set $\{n_r\}$. Let U_j denote the energy eigenvalues for the system and let N be the total number of particles, then

$$U_j = \sum_r n_r\varepsilon_r, \qquad N = \sum_r n_r. \tag{32.1}$$

In nature two distinct types of particles exist, namely, those described by antisymmetric wave-functions and called *fermions* and those described by symmetric wave-functions and called *bosons*. The fundamental material particles—electrons, protons, and neutrons—belong to the first category, while photons belong to the second. Nuclei containing an odd number of elementary particles are fermions and those with an even number are bosons. Quantum theory shows that while any number of bosons can occupy a particular energy level, only one fermion can do this, that is,

$$n_r = \begin{cases} 0, 1 & \text{for fermions} \\ 0, 1, 2, \dots & \text{for bosons.} \end{cases} \tag{32.2}$$

The first constraint on n_r is known as 'Pauli's exclusion principle'.

Bosons are indistinguishable in that a given set of occupation numbers uniquely defines a single state of the system, and the same is true of fermions.

Hence an interchange of the momenta of two particles of a Bose gas or a Fermi gas does not yield a new state.

32.2. Fermi–Dirac and Bose–Einstein distributions

Consider a system of N particles of the same kind. The partition function of the grand canonical ensemble given in § 27.2 may be written

$$\mathscr{Z}(T, \mathbf{V}, z) = \sum_{N=0}^{\infty} \sum_{j} z^N \exp(-\beta U_j) = \sum_{N=0}^{\infty} \sum_{\{n_r\}} z^N \exp(-\beta \sum n_r \varepsilon_r),$$

where $z = \exp(\beta\mu)$ is the fugacity and the symbol $\{n_r\}$ under the final summation means summing over all occupation numbers such that $\sum n_r = N$. The final double summation is equivalent to summing over each n_r independently, as may be verified by noting that every term in each form of the sum appears once and only once in the other form. Hence

$$\begin{aligned}\mathscr{Z} &= \sum_{n_1} \sum_{n_2} \dots \{[z \exp(-\beta\varepsilon_1)]^{n_1} [z \exp(-\beta\varepsilon_2)]^{n_2} \dots \} \\ &= \prod_r \left\{ \sum_{n_r} [z \exp(-\beta\varepsilon_r)]^{n_r} \right\},\end{aligned}$$

where the sum $\sum_{n_r}$ extends over the values $n = 0, 1, 2, \dots$ for a Bose gas and the values $n = 0, 1$ for a Fermi gas. Thus for natural gases

$$\mathscr{Z}(T, \mathbf{V}, z) = \begin{cases} \prod_r \dfrac{1}{1 - z\exp(-\beta\varepsilon_r)} & \text{(Bose gas)} \\ \prod_r [1 + z\exp(-\beta\varepsilon_r)] & \text{(Fermi gas).} \end{cases} \tag{32.3}$$

Notice that convergence of the geometric series for the Bose gas requires that $\exp[\beta(\mu - \varepsilon_r)] < 1$, and as this must hold for all ε_r, including $\varepsilon_r = 0$, it is necessary that $\mu < 0$ for this case.

The equations of state are

$$pV = kT \ln \mathscr{Z} = \pm kT \sum_r \ln[1 \pm z \exp(-\beta\varepsilon_r)], \tag{32.4}$$

where the minus sign refers to the Bose gas and the plus sign to the Fermi gas. The fugacity may be eliminated from (32.4) by using (30.12), which gives

$$N = z\frac{\partial}{\partial z} \ln \mathscr{Z} = \sum_r \left[\frac{1}{\exp[\beta(\varepsilon_r - \mu)] \pm 1} \right] \quad \begin{matrix} (-\text{ bosons}) \\ (+\text{ fermions}) \end{matrix} \tag{32.5}$$

We shall return to this calculation in § 33.3.

The average occupation numbers are given by

$$\langle n_r \rangle = \frac{1}{\mathscr{L}} \sum_{N=0} z^N \sum_{\{n_r\}} n_r \exp(-\beta \sum n_r \varepsilon_r) = -\frac{1}{\beta} \frac{\partial}{\partial \varepsilon_r} \ln \mathscr{L}$$

or

$$\langle n_r \rangle = \frac{1}{\exp[\beta(\varepsilon_r - \mu)] \pm 1} \quad \begin{array}{l} (-\text{ Bose–Einstein}) \\ (+\text{ Fermi–Dirac}). \end{array} \tag{32.6}$$

These results are referred to as *Bose–Einstein statistics* and *Fermi–Dirac statistics* according to the sign indicated. Notice from (32.5) and (32.6)

$$N = \sum_r \langle n_r \rangle, \tag{32.7}$$

which is otherwise obvious.

32.3. The Maxwell–Boltzmann distribution

Now consider a model gas whose particles—unlike the particles of the Bose and Fermi gases—are distinguishable in the limited sense that an interchange of *momenta* between particles does yield a new state, but an interchange of momenta *and* position does not (see Fig. 8). For a given set $\{n_1, n_2, \ldots\}$ of occupation numbers and particle locations there are a total of $(N!/\prod_r n_r!)$ ways in which identifiable particles can be arranged in the single-particle states, and therefore summing over all sets $\{n_r\}$ that satisfy the constraint $N = \sum_r n_r$ we have a total of

$$\frac{M!}{(M-N)!} \sum_{\{n_r\}} N! \Big/ \prod_r (n_r!)$$

ways of arranging identifiable particles over M positions in space. However, $N!$ of these arrangements merely involve an interchange of both momentum and position. Hence the total number of *distinguishable* states in the gas is

$$\frac{M!}{(M-N)!} \sum_{\{n_r\}} 1 \Big/ \prod_r (n_r!) \quad \text{with} \quad N = \sum_r n_r.$$

The division by $N!$ to exclude mere labels from possessing physical significance is called 'the rule of correct Boltzmann counting'. Notice that with each set $\{n_r\}$ there is an associated weighting factor $[\prod_r (n_r!)]^{-1}$, so that upon summation over all sets the correct number of distinguishable states is obtained.

For the canonical function (27.15) we may now write

$$Z(T, V, N) = \sum_{\substack{\{n_r\} \\ \Sigma n_r = N}} \exp\left(-\beta \sum_r n_r \varepsilon_r\right)$$

$$= \sum_{\substack{n_1, n_2, \ldots \\ \Sigma n_r = N}} \left[\frac{\exp(-\beta n_1 \varepsilon_1)}{n_1!} \frac{\exp(-\beta n_2 \varepsilon_2)}{n_2!} \cdots \right]$$

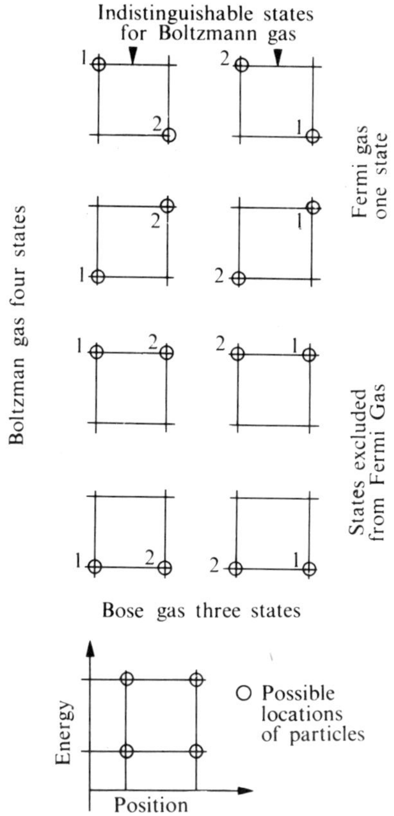

FIG. 8. Arrangements of two particles over two positions and two energies.

on summing over all distinguishable energetic states. By the multinomial theorem it follows that

$$Z = \frac{1}{N!}[\exp(-\beta\varepsilon_1) + \exp(-\beta\varepsilon_2) + \dots]^N. \tag{32.8}$$

By (30.10) the grand partition function is

$$\mathscr{Z}(T, \mathbf{V}, z) = \sum_{N=0}^{\infty} \frac{1}{N!}\left[z\sum_r \exp(-\beta\varepsilon_r)\right]^N = \exp\left[z\sum_r \exp(-\beta\varepsilon_r)\right],$$

hence

$$pV = kTz\sum_r \exp(-\beta\varepsilon_r), \tag{32.9}$$

from which z may be eliminated by using

$$N = z\frac{\partial}{\partial z}\ln \mathscr{Z} = z\sum_r \exp(-\beta\varepsilon_r). \tag{32.10}$$

Thus

$$pV = NkT. \tag{32.11}$$

As $N = \sum_r \langle n_r \rangle$, it follows from (32.10) that

$$\langle n_r \rangle = z \exp(-\beta\varepsilon_r) = \exp[\beta(\mu - \varepsilon_r)], \tag{32.12}$$

which is the Maxwell–Boltzmann distribution.

The entropy follows from $TS = U + pV - \mu N$. As $\beta = 1/kT$, we find

$$S = k\beta U - k\beta\mu N + kN. \tag{32.13}$$

Returning now to the corresponding equations for the natural gases we note that (32.4), (32.5), and (32.6) reduce to (32.9), (32.10), and (32.12) when

$$\exp[\beta(\mu - \varepsilon_r)] \ll 1, \quad \text{that is,} \quad \varepsilon_r - \mu \gg kT, \quad \text{or} \quad \langle n_r \rangle \ll 1. \tag{32.14}$$

In other words, dilute high-temperature natural gases behave like Boltzmann gases, a result that has important practical consequences.

33. State equations for perfect gases

33.1. Dense energy levels

The energy levels ε_r, $r = 1, 2, \ldots$, of a spinless particle of mass m in a perfect gas can be obtained by solving the stationary Schrödinger equation,

$$-\frac{h^2}{8\pi^2 m}\nabla^2\varphi(x, y, z) = \varepsilon\varphi(x, y, z), \tag{33.1}$$

in a cube of side L with φ zero on the cube boundaries. We readily find that (33.1) has non-zero solutions only when ε is one of the eigenvalues,

$$\varepsilon_r = \varepsilon_{n_1, n_2, n_3} = \frac{h^2}{2mL^2}(n_1^2 + n_2^2 + n_3^2) = \frac{1}{2m}\left(\frac{hn}{L}\right)^2$$
$$(n_1, n_2, n_3 = 0, \pm 1, \pm 2, \ldots),$$

n being the magnitude of the vector $\mathbf{n} = (n_1, n_2, n_3)$. Let $\mathbf{p}$ be the particle momentum, then this result yields

$$\mathbf{p} = \frac{h}{V^{\frac{1}{3}}}\mathbf{n} \qquad \left(\varepsilon_r = \frac{p^2}{2m}\right), \tag{33.2}$$

where $V = L^3$ is the volume of the system.

When $n \gg 1$, that is, $L \gg h/p$, the integers n_1, n_2, n_3 are sufficiently dense to permit the approximation

$$\sum_r \cdots = \sum_{n_1}\sum_{n_2}\sum_{n_3} \cdots \sim \int_{-\infty}^{\infty}\int_{-\infty}^{\infty}\int_{-\infty}^{\infty} \cdots \, dn_1\, dn_2\, dn_3$$

or

$$\sum_r \cdots \sim \frac{V}{h^3}\int \cdots \, d\mathbf{p} = \frac{V}{h^3}\int_0^{\infty} \cdots \, 4\pi p^2 \, dp \tag{33.3}$$

by (33.2) and the assumed isotropy. This approximation will be applied to the theory of the three gases described above.

33.2. The Boltzmann gas

By (32.10) and (33.3)

$$N = \frac{zV}{h^3}\int_0^\infty \exp(-\beta p^2/2m)\ 4\pi p^2\,\mathrm{d}p = \frac{zV}{\lambda^3} \qquad \left(\lambda \equiv \frac{h}{(2\pi mkT)^{\frac{1}{2}}}\right).$$

Hence

$$z = \lambda^3 N/V. \tag{33.4}$$

From (32.8) and (32.10) $Z = (N/z)^N/N! = (V/\lambda^3)^N/N!$ in accord with (31.5), and the Sackur–Tetrode relation (the first of (31.6)) now follows. Finally, by (32.12) and $U = \sum_r \varepsilon_r \langle n_r \rangle$, we have

$$U = z\sum_r \varepsilon_r \exp(-\beta\varepsilon_r) = \frac{zV}{h^3}\int_0^\infty \frac{p^2}{2m}\exp(-\beta p^2/2m)\ 4\pi p^2\,\mathrm{d}p = \frac{3}{2}\,\frac{zV}{\lambda^3}kT,$$

or by (33.4)

$$U = \tfrac{3}{2}NkT. \tag{33.5}$$

The third law of thermodynamics (§ 25.4) states that the entropy at absolute zero temperature is a universal constant that can be taken equal to zero. From eqn $(31.6)_1$, it is clear that the Boltzman gas does not satisfy this law. However, this gas is not a natural physical system but simply a model gas towards which the Fermi and Bose gases converge at high temperatures.

33.3. The natural gases

Let V be the only deformation coordinate, then for the Fermi gas (32.4), (32.5), and (33.3) give

$$\frac{p}{kT} = \frac{4\pi}{h^3}\int_0^\infty p^2 \ln[1 + z\exp(-\beta p^2/2m)]\,\mathrm{d}p,$$

$$\frac{N}{V} = \frac{4\pi}{h^3}\int_0^\infty \frac{p^2 z\,\mathrm{d}p}{\exp(\beta p^2/2m) + z},$$

or

$$\frac{p}{kT} = \frac{1}{\lambda^3}f_{\frac{5}{2}}(z), \qquad \frac{N}{V} = \frac{1}{\lambda^3}f_{\frac{3}{2}}(z), \tag{33.6}$$

where

$$f_{\frac{5}{2}}(z) \equiv \frac{4}{\sqrt{\pi}}\int_0^\infty x^2 \ln[1 + z\exp(-x^2)]\,\mathrm{d}x = \sum_{l=1}^\infty \frac{(-1)^{l+1}z^l}{l^{\frac{5}{2}}}$$

and

$$f_{\frac{3}{2}}(z) \equiv z\frac{\mathrm{d}f_{\frac{5}{2}}(z)}{\mathrm{d}z} = \sum_{l=1}^\infty \frac{(-1)^{l+1}z^l}{l^{\frac{3}{2}}}. \tag{33.7}$$

For the Bose gas the summands in (32.4) and (32.5) diverge as $z \to 1$ because of the term corresponding to $p = 0$ (see (33.2)). By first removing this term then replacing the remaining sums by integrals we find

$$\frac{p}{kT} = \frac{1}{\lambda^3} g_{\frac{5}{2}}(z) - \frac{1}{V}\ln(1-z), \qquad \frac{N}{V} = \frac{1}{\lambda^3} g_{\frac{3}{2}}(z) + \frac{1}{V}\left(\frac{z}{1-z}\right), \tag{33.8}$$

where

$$g_{\frac{5}{2}}(z) = -\frac{4}{\sqrt{\pi}}\int_0^\infty x^2 \ln[1 - z\exp(-x^2)]\,\mathrm{d}x = \sum_{l=1}^{\infty} \frac{z^l}{l^{\frac{5}{2}}}$$

and (33.9)

$$g_{\frac{3}{2}}(z) = z\frac{\mathrm{d}}{\mathrm{d}z} g_{\frac{5}{2}}(z) = \sum_{l=1}^{\infty} \frac{z^l}{l^{\frac{3}{2}}}.$$

The internal energy follows from

$$U = -\frac{\partial}{\partial \beta}(\ln \mathscr{Z}) = kT^2 \frac{\partial}{\partial T}\ln\left(\frac{pV}{kT}\right).$$

Thus

$$U = \frac{3}{2}\frac{kTV}{\lambda^3} f_{\frac{5}{2}}(z) \quad \text{(Fermi)}; \qquad U = \frac{3}{2}\frac{kTV}{\lambda^3} g_{\frac{5}{2}}(z) \quad \text{(Bose)}; \tag{33.10}$$

and it is not possible to eliminate z without obtaining a very complicated function. When the final term in eqn $(33.8)_2$ can be neglected (see § 33.6) both the natural gases have

$$U = \tfrac{3}{2}pV, \tag{33.11}$$

as for the Boltzmann gas.

Finally, the entropy may be obtained from $S = T^{-1}(U + pV - \mu N)$,

$$S = \frac{kV}{\lambda^3}[\tfrac{5}{2} f_{\frac{5}{2}}(z) - f_{\frac{3}{2}}(z)\ln z], \qquad \text{(Fermi)}$$

(33.12)

$$S = \frac{kV}{\lambda^3}[\tfrac{5}{2} g_{\frac{5}{2}}(z) - g_{\frac{3}{2}}(z)\ln z] - \frac{k}{1-z}[z\ln z + (1-z)\ln(1-z)] \quad \text{(Bose)}.$$

33.4. Expansions at high temperatures

For $z_0 \equiv \lambda^3 N/V \ll 1$, that is, when the mean interparticle distance $(V/N)^{\frac{1}{3}}$ greatly exceeds the de Broglie wavelength, quantum effects are very small. From (33.4) we see that z_0 is the Boltzmann fugacity.

For the Fermi gas $(33.6)_2$ yields $z_0 = z - z^2/2^{\frac{3}{2}} + \ldots$ or $z = z_0 + z_0^2/2^{\frac{3}{2}} + \ldots$, then $(33.6)_1$ may be expanded

$$pV = NkT\left[1 + \frac{\lambda^3}{2^{\frac{5}{2}}}\frac{N}{V} + O(z_0^3)\right]. \tag{33.13}$$

The Bose gas requires further consideration because of the terms involving $(1-z)$ in (33.8). Notice from (33.2) and (32.6) that at $p = 0$

$$\langle n_0 \rangle = \frac{z}{1-z}, \tag{33.14}$$

whence at small z, z/N is the fraction of all particles in the system occupying the zero energy or ground state. When this fraction is appreciable, we have the phenomenon known as *Bose–Einstein condensation.* If such condensation is negligible (see § 33.5), eqns (33.8) give an expansion similar to (33.13), save that the sign of the 'correction' term is changed.

These results may be interpreted by saying that in a Fermi (Bose) gas the quantum-mechanical effects lead to the appearance of an additional effective 'repulsion' ('attraction') between the particles.

Similar expansions may be found for the other thermodynamic variables.

33.5. *The Fermi gas at low temperature*

At low temperatures $z = \exp(\mu/kT)$ is a large number, for which case $(33.6)_2$ yields the asymptotic expansion†

$$\frac{N}{V} = \frac{1}{\lambda^3}\frac{4}{3\sqrt{\pi}}(\ln z)^{\frac{3}{2}}\left[1 + \frac{\pi^2}{8}(\ln z)^{-2} + \ldots\right].$$

The so-called *Fermi energy* is defined by

$$\varepsilon_{\mathrm{F}} = \frac{h^2}{2m}\left(\frac{3}{4\pi}\frac{N}{V}\right)^{\frac{2}{3}}, \tag{33.15}$$

and the expansion may be inverted to

$$kT \ln z = \varepsilon_{\mathrm{F}}\left[1 - \frac{\pi^2}{12}\left(\frac{kT}{\varepsilon_{\mathrm{F}}}\right)^2 + \ldots\right]. \tag{33.16}$$

The significance of ε_{F} is that it is the energy level below which there are exactly N states, for, suppose this is true, then as no two particles can be at the same level, each level is filled once so it follows from (33.3) that

$$N = \sum_{\varepsilon_r < \varepsilon_{\mathrm{F}}} \langle n_r \rangle = \frac{V}{h^3}\int_0^{p_{\mathrm{F}}} 4\pi p^2\, \mathrm{d}p = \frac{V}{h^3}\frac{4\pi}{3}p_{\mathrm{F}}^3 = \frac{V}{h^3}\frac{4\pi}{3}(2m\varepsilon_{\mathrm{F}})^{\frac{3}{2}},$$

† The reader requiring more details will find them in Huang (1963).

in agreement with (33.15). When $T \ll \varepsilon_F/k$ the gas is said to be *degenerate* as the particles tend to go into the lowest possible energy levels.

A similar treatment of $(33.6)_1$ yields

$$U = \tfrac{3}{2}pV = \tfrac{3}{5}N\varepsilon_F\left[1+\frac{5\pi^2}{12}\left(\frac{kT}{\varepsilon_F}\right)^2+\dots\right], \tag{33.17}$$

of which the derivative with respect to T yields

$$C_V = \frac{\pi^2}{2\varepsilon_F}Nk^2T. \tag{33.18}$$

As $C_V = T(\partial S/\partial T)_V$, (33.18) shows that near $T = 0$, $S \propto T$, in conformity with the third law of thermodynamics (§ 25.4).

33.6. The Bose gas at low temperatures

As μ is negative for a Bose gas (§ 32.2), the fugacity $z = \exp(\mu/kT)$ must lie in the range (0, 1). At the upper limit $g_{\frac{3}{2}}(z)$ achieves its maximum value of $g_{\frac{3}{2}}(1) = \zeta(\frac{3}{2}) = 2{\cdot}612\dots$, where $\zeta(x)$ is the Riemann zeta function. Hence by $(33.8)_2$ and (33.14)

$$\frac{\langle n_0\rangle}{V} = \frac{N}{V}-\lambda^{-3}g_{\frac{3}{2}}(z) \geqslant \frac{1}{v}-\lambda^{-3}g_{\frac{3}{2}}(1), \tag{33.19}$$

where $v \equiv V/N$ is the volume per particle. From this we conclude that a finite fraction of the particles will occupy the ground state $\mathbf{p} = 0$ when $\lambda^3/v > g_{\frac{3}{2}}(1)$.

We may use either $\lambda_c^3/v \equiv g_{\frac{3}{2}}(1)$ or $\lambda^3/v_c \equiv g_{\frac{3}{2}}(1)$ to define a critical temperature T_c for given v or a critical specific volume v_c for a given T,

$$kT_c = \frac{h^2/(2\pi m)}{[vg_{\frac{3}{2}}(1)]^{\frac{2}{3}}}, \qquad v_c = \frac{\lambda^3}{g_{\frac{3}{2}}(1)}, \tag{33.20}$$

and then state that Bose–Einstein condensation is important if either

$$T < T_c \quad \text{or} \quad v < v_c. \tag{33.21}$$

Notice that at criticality the de Broglie wavelength λ is comparable with the mean interparticle separation $v^{\frac{1}{3}}$.

The general appearance of the solution of $(33.8)_2$, namely,

$$\frac{z}{N(1-z)} = 1-\frac{v}{\lambda^3}g_{\frac{3}{2}}(z)$$

is shown in Fig. 9. In the limit $N \to \infty$, $V \to \infty$, and v finite, we have

$$z = \begin{cases} 1 & (\lambda^3/v \geqslant g_{\frac{3}{2}}(1)) \\ \text{root of } g_{\frac{3}{2}}(z) = \lambda^3/v & (\lambda^3/v \leqslant g_{\frac{3}{2}}(1)) \end{cases} \tag{33.22}$$

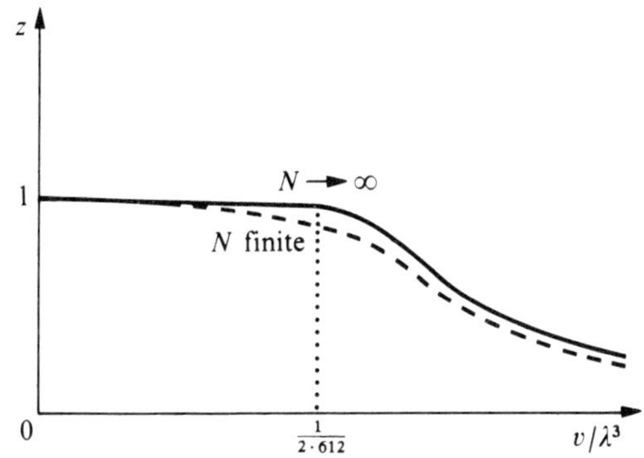

FIG. 9. Fugacity of a Bose gas.

Then from

$$\frac{\langle n_0\rangle}{N} = \frac{z}{N(1-z)} = 1 - \frac{(v/\lambda^3)g_{\frac{3}{2}}(z)}{(v/\lambda^3)_c g_{\frac{3}{2}}(1)}$$

we can write

$$\frac{\langle n_0\rangle}{N} = \begin{cases} 1-\left(\dfrac{T}{T_c}\right)^{\frac{3}{2}} = 1-\dfrac{v}{v_c} & (\lambda^3/v \geqslant g_{\frac{3}{2}}(1)) \\ 0 & (\lambda^3/v \leqslant g_{\frac{3}{2}}(1)) \end{cases} \tag{33.23}$$

When (33.22) is substituted into (33.8) and (33.12) we find in the 'thermodynamic limit' ($V, N \to \infty$, V/N finite) that

$$\frac{p}{kT} = \begin{cases} \lambda^{-3} g_{\frac{5}{2}}(z) & (v > v_c) \\ \lambda^{-3} g_{\frac{5}{2}}(1) \simeq 1{\cdot}342\lambda^{-3} & (v < v_c) \end{cases} \tag{33.24}$$

and

$$\frac{S}{kN} = \begin{cases} \frac{5}{2}v\lambda^{-3} g_{\frac{5}{2}}(z) - \ln z & (v > v_c) \\ \frac{5}{2}v\lambda^{-3} g_{\frac{5}{2}}(1) & (v < v_c) \end{cases} \tag{33.25}$$

Thus in the condensed phase p is independent of v and proportional to $T^{\frac{5}{2}}$. The entropy satisfies the third law, being proportional to $T^{\frac{3}{2}}$ below the critical condition.

34. Crystalline solids

34.1. Two thermodynamic models

Although a discussion of crystalline solids seems out of place in a text on fluid systems, a brief account will serve as a useful basis for our treatment of the thermodynamics of radiation. And of course the basic results given here have intrinsic importance in thermodynamic theory.

Consider a lattice of N identical, regularly spaced atoms, free to vibrate in three independent directions without mutual interference. This system can be regarded as being an assembly of $3N$ independent localized harmonic oscillators.

There are two distinct treatments of the problem. The first, due to Einstein, treats the oscillators as all having the same frequency ν and distinguishable by their permanent spatial location or orientation. Linear harmonic oscillators behave like particles in a potential field $V = kx^2$. The time-free Schrödinger equation can be written

$$\frac{\partial^2 \varphi}{\partial x^2} + \frac{8\pi^2 m}{h^2}(\varepsilon - kx^2)\varphi = 0,$$

and the eigenvalues ε_r of ε that will yield solutions $\varphi(x)$ bounded at $x = \infty$ are found to be

$$\varepsilon_r = (r+\tfrac{1}{2})h\nu \qquad (r = 0, 1, 2, \ldots), \tag{34.1}$$

where ν is the classical frequency $(k/2\pi^2 m)^{\frac{1}{2}}$. Maxwell–Boltzmann statistics (§ 32.3) are relevant.

The second approach, due to Debye, uses the range of frequencies determined by the periodic boundary conditions for waves in an elastic continuum. Classically, each normal mode is a wave of distortion of the lattice planes—a sound wave. In quantum theory these normal modes give rise to quanta called *phonons*, and at a very low temperature a solid can be regarded as a volume containing a gas of non-interacting phonons. As a phonon is a quantum of a certain harmonic oscillator, it has a characteristic frequency ν_i and an energy $h\nu_i$. Since the excited state of an oscillator may contain any number of quanta—which are indistinguishable—the phonons obey Bose statistics, with no conservation of their total number.

34.2. Localized oscillators (Einstein)

The number of independent oscillators is fixed at $3N$, so the canonical ensemble is appropriate. By (27.15) and (34.1) the partition function is (cf. § 31.2)

$$Z(V, T, N) = \mathrm{e}^{-\beta F} = \left[\sum_{r=0}^{\infty} \mathrm{e}^{-\beta(r+\frac{1}{2})h\nu}\right]^{3N} = \left(\frac{\mathrm{e}^{-\frac{1}{2}\beta h\nu}}{1-\mathrm{e}^{-\beta h\nu}}\right)^{3N}.$$

Thus

$$F = 3NkT\ln(1-e^{-\beta h\nu})+\tfrac{3}{2}Nh\nu$$

and

$$S = -\left(\frac{\partial F}{\partial T}\right)_V = \frac{3Nh\nu}{T(e^{\beta h\nu}-1)} - 3Nk\ln(1-e^{-\beta h\nu}).$$

Therefore

$$U = F+ST = U_0+\frac{3Nh\nu}{e^{\beta h\nu}-1} \qquad (U_0 \equiv \tfrac{3}{2}Nh\nu) \tag{34.2}$$

and

$$C_V = \left(\frac{\partial U}{\partial T}\right)_V = 3Nk\left(\frac{h\nu}{kT}\right)^2\frac{e^{h\nu/kT}}{(e^{h\nu/kT}-1)^2}. \tag{34.3}$$

When $kT \gg h\nu$, $U \sim 3NkT$ and $C_V \sim 3Nk$, which is the classical Dulong and Petit law (Fig. 10).

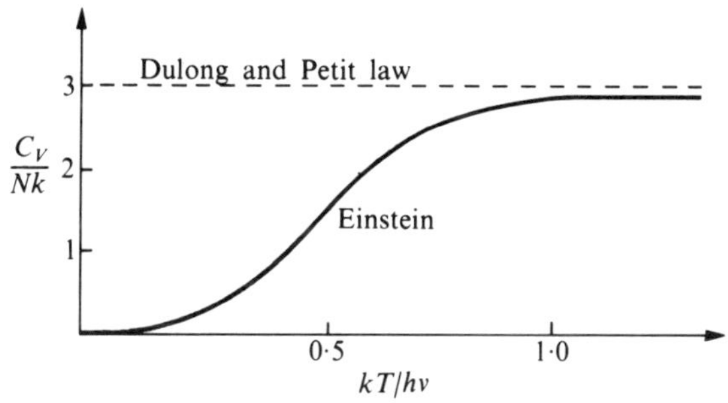

FIG. 10. Heat capacity of a crystalline solid.

34.3. Bose system of phonons (Debye)

Each phonon corresponds to a sound wave in which the displacement vector **u** in the elastic continuum may be represented by

$$\mathbf{u} = \boldsymbol{\varepsilon}\, e^{i(\mathbf{k}.\mathbf{r}-\omega t)} \qquad (\omega = 2\pi\nu = ck), \tag{34.4}$$

where **k** is the propagation vector, $\omega/2\pi$ is the frequency, and c is the sound speed for the particular wave under consideration. The polarization vector $\boldsymbol{\varepsilon}$ possesses three independent coordinates. For a given value of **k** there are three distinct wave states (or phonons), namely, one longitudinal wave that propagates with velocity $c_l = \omega_l/k$ and two transverse waves with propagation velocities $c_t = \omega_t/k$.†

† In terms of the Lamé constants λ and μ of elasticity theory and the density ρ, $c_l = [(\lambda+2\mu)/\rho]^{\frac{1}{2}}$, $c_t = (\mu/\rho)^{\frac{1}{2}}$.

A solid with N atoms present has $3N$ normal modes and hence $3N$ different phonons with characteristic frequencies $\omega_1, \omega_2, \ldots \omega_{3N}$ which depend on the lattice. For the purpose of finding these frequencies—and only this purpose—Debye assumed the solid to be an elastic continuum, and for $\omega_1, \omega_2, \ldots \omega_{3N}$ took the lowest $3N$ normal frequencies of the solid. The calculation of the energy spectrum for large systems is similar to the calculation of the momentum spectrum for perfect gases given in § 33.1. We consider the standing waves in $\mathbf{u}$ in an elastic cube, with $\mathbf{u}$ zero on the boundaries, and find from the imaginary part of (34.4) that $\mathbf{k} = (2\pi/L)\mathbf{n}$, where $\mathbf{n}$ is a vector with integer components (including zero) and $L = V^{\frac{1}{3}}$ is the size of the cube. Comparison of this result with (33.2) shows that provided $L \gg 2\pi/k$, that is, many wavelengths fit into the cube, the possible values of $\mathbf{k}$ are sufficiently continuous to permit the approximation (cf. (33.3))

$$\sum_{\mathbf{k},\varepsilon} \ldots \sim 3\sum_{\mathbf{k}} \ldots \sim \frac{3V}{(2\pi)^3}\int \ldots \mathrm{d}\mathbf{k} = \frac{3V}{2\pi^2}\int_0^\infty \ldots k^2\,\mathrm{d}k. \tag{34.5}$$

The additional factor 3 arises from the three distinct wave states associated with each $\mathbf{k}$.

We may introduce frequencies by noting that for each of the three wave types included in (34.5)

$$\sum_{\substack{\mathbf{k}\\(k=\omega/c)}} \ldots \sim \frac{V}{2\pi^2}\int_0^\infty \ldots k^2\,\mathrm{d}k = \frac{V}{2\pi^2}\frac{1}{c^3}\int_0^\infty \ldots \omega^2\,\mathrm{d}\omega = \frac{4\pi V}{c^3}\int_0^\infty \ldots \nu^2\,\mathrm{d}\nu.$$

Thus if $f(\nu)\,\mathrm{d}\nu$ is the number of wave states in the frequency range ν, $\nu+\mathrm{d}\nu$,

$$f(\nu)\,\mathrm{d}\nu = 4\pi V\left(\frac{1}{c_l^3}+\frac{2}{c_t^3}\right)\nu^2\,\mathrm{d}\nu = \frac{12\pi V}{c^3}\nu^2\,\mathrm{d}\nu, \tag{34.6}$$

where c is an average sound speed defined by

$$\frac{3}{c^3} = \frac{1}{c_l^3}+\frac{2}{c_t^3}.$$

The maximum or 'cut-off' frequency ν_m follows from the condition that

$$\int_0^{\nu_\mathrm{m}} f(\nu)\,\mathrm{d}\nu = 3N,$$

which gives

$$\nu_\mathrm{m}^3 = \frac{3}{4\pi}\frac{N}{V}c^3. \tag{34.7}$$

The corresponding wavelength is $\lambda_m = (4\pi V/3N)^{\frac{1}{3}} \sim$ mean interparticle distance, which is physically reasonable as for smaller wavelengths a wave displacement of atoms becomes meaningless. Eqn (34.6) may now be rewritten

$$f(\nu) = \begin{cases} \dfrac{9N}{\nu_{\mathrm{m}}^3}\nu^2 & \text{for } \nu \leqslant \nu_{\mathrm{m}} \\ 0 & \text{for } \nu > \nu_{\mathrm{m}} \end{cases}. \tag{34.8}$$

34.4. The thermodynamics of Debye's model

The energy of the state in which there are n_i phonons of frequency ν_i is

$$\varepsilon\{n_i\} = \sum_{i=1}^{3N} n_i h\nu_i,$$

and so the partition function of the canonical ensemble is

$$Z = \sum_{\{n_i\}} \exp(-\beta\varepsilon\{n_i\}) = \prod_{i=1}^{3N} \sum_{\{n_i\}} \exp(-\beta h\nu_i n_i) = \prod_{i=1}^{3N} \left[\frac{1}{1-\exp(-\beta h\nu_i)}\right],$$

as $\{n_i\}$ is simply the set $\{0, 1, 2, \ldots\}$. Hence

$$\ln Z = -\sum_{i=1}^{3N} \ln[1-\exp(-\beta h\nu_i)].$$

In the limit as $V \to \infty$ we may use (34.8) to replace this sum by

$$\ln Z = -\frac{9N}{\nu_{\mathrm{m}}^3}\int_0^{\nu_{\mathrm{m}}} \nu^2 \ln(1-\mathrm{e}^{-h\nu/kT})\,\mathrm{d}\nu, \tag{34.9}$$

from which all the thermodynamic properties may be obtained by differentiation.

On introducing a characteristic temperature (the Debye temperature) T_{D}, by $T_{\mathrm{D}} \equiv h\nu_{\mathrm{m}}/k$, we have

$$F(T, V, N) = -kT\ln Z = \frac{9NkT^4}{T_{\mathrm{D}}^3}\int_0^{T_{\mathrm{D}}/T} \xi^2 \ln(1-\mathrm{e}^{-\xi})\,\mathrm{d}\xi$$

$$= 3NkT\ln[1-\exp(-T_{\mathrm{D}}/T)] - \frac{3NkT^4}{T_{\mathrm{D}}^3}\int_0^{T_{\mathrm{D}}/T} \frac{\xi^3}{\mathrm{e}^{\xi}-1}\,\mathrm{d}\xi, \tag{34.10}$$

the second form following an integration by parts. Comparing this with the corresponding formula of the Einstein theory, we see that the last term of (34.10) can be considered a correction to the Einstein model.

The internal energy follows from $U = F - T(\partial F/\partial T)_V$; we find

$$U = 9Nk\frac{T^4}{T_{\mathrm{D}}^3}\int_0^{T_{\mathrm{D}}/T} \frac{\xi^3}{\mathrm{e}^{\xi}-1}\,\mathrm{d}\xi, \tag{34.11}$$

and a further differentiation with respect to T yields

$$C_V = 3NkD(T_D/T), \qquad D(x) \equiv \frac{3}{x^3}\int_0^x \frac{e^\xi \xi^4}{(e^\xi - 1)^2}\,d\xi.$$

Limiting forms of the *Debye function* $D(x)$ give

$$C_V = \begin{cases} 3Nk\left\{1 - \dfrac{1}{20}\left(\dfrac{T_D}{T}\right)^2 + O\left[\dfrac{T_D}{T}\right]^4\right\} & (T \gg T_D) \\ 3Nk\left\{\dfrac{4\pi^4}{5T_D^3}T^3 + O\left[\dfrac{T}{T_D}\exp(-T_D/T)\right]\right\} & (T \ll T_D) \end{cases}, \tag{34.12}$$

the second of which shows that the model satisfies the third law of thermodynamics (cf. final sentence of § 33.5).

A more fundamental approach to the problem was initiated by the work of Born and von Kármán, an account of which is given by Grossman (1969).

35. The thermodynamics of radiation

35.1. *The photon gas*

We shall consider the radiation or photons within a vacuum enclosure of volume V whose walls are maintained at a temperature T. The electric field $\mathbf{E}$ of the electromagnetic waves within the vacuum satisfies the equations

$$\nabla^2\mathbf{E} = \frac{1}{c^2}\frac{\partial^2\mathbf{E}}{\partial t^2}, \qquad \nabla \cdot \mathbf{E} = 0, \tag{35.1}$$

where c is the velocity of light. The wave solution of (35.1) is

$$\mathbf{E}(\mathbf{r}, t) = \boldsymbol{\varepsilon}\, e^{i(\mathbf{k}\cdot\mathbf{r} - \omega t)}, \qquad \mathbf{k} \cdot \boldsymbol{\varepsilon} = 0, \qquad (\omega = 2\pi\nu = ck), \tag{35.2}$$

where, unlike the corresponding expression for the elastic displacement vector given in (34.4), the polarization vector has only two independent coordinates. Thus to each value of $\mathbf{k}$ there corresponds two distinct wave solutions (or photons). By considering the standing waves in the enclosure we find, as in § 34.3, that

$$\mathbf{k} = 2\pi V^{-\frac{1}{3}}\mathbf{n}, \tag{35.3}$$

and provided a typical dimension $V^{\frac{1}{3}}$ of the enclosure greatly exceeds the wavelength $2\pi/k$, we may adopt the approximation

$$\sum_{\mathbf{k},\boldsymbol{\varepsilon}} \ldots = 2\sum_{\mathbf{k}} \ldots \sim \frac{2V}{(2\pi)^3}\int \ldots\, d\mathbf{k} = \frac{V}{\pi^2}\int_0^\infty \ldots\, k^2\, dk = \frac{8\pi V}{c^3}\int_0^\infty \ldots\, \nu^2\, d\nu. \tag{35.4}$$

Let $f(\nu)\,d\nu$ be the number of photons in the frequency range ν, $\nu+d\nu$, then (35.4) yields

$$f(\nu)\,d\nu = \frac{8\pi V}{c^3}\nu^2\,d\nu \qquad (0 \leqslant \nu < \infty). \tag{35.5}$$

Each photon of frequency ν possesses an energy $h\nu$, and being indistinguishable from other photons with the same value of $\boldsymbol{\varepsilon}$ and $\mathbf{k}$, of which there can be any number, it obeys Bose statistics. Also, as atoms can emit and absorb photons, the number in a black-body cavity is not definite. This last point suggests that only the grand canonical ensemble would provide an appropriate representation for radiation, but phonons are likewise not conserved, despite which we used the canonical ensemble for them (in § 34.4). The difficulty is removed by setting the chemical potential μ equal to zero for systems in which even average particle numbers have no assignable value. In this case it follows from (27.9) and (27.15) that $\mathscr{Z} = Z$ and the ensembles are identical.

35.2. The partition function

Let $n_{\mathbf{k},\boldsymbol{\varepsilon}}$ denote the number of photons with given values of $\mathbf{k}$ and $\boldsymbol{\varepsilon}$, then the energy in this state is

$$\varepsilon\{n_{\mathbf{k},\boldsymbol{\varepsilon}}\} = \sum_{\mathbf{k},\boldsymbol{\varepsilon}} h\nu n_{\mathbf{k},\boldsymbol{\varepsilon}}.$$

The partition function is given by

$$\mathscr{Z} = Z = \sum_{\{n_{\mathbf{k},\boldsymbol{\varepsilon}}\}} \exp(-\beta\varepsilon\{n_{\mathbf{k},\boldsymbol{\varepsilon}}\}) = \sum_{\{n_{\mathbf{k},\boldsymbol{\varepsilon}}\}} \exp\left(-\beta\sum_{\mathbf{k},\boldsymbol{\varepsilon}} h\nu n_{\mathbf{k},\boldsymbol{\varepsilon}}\right),$$

and as for each value of $(\mathbf{k}, \boldsymbol{\varepsilon})$, $\{n_{\mathbf{k},\boldsymbol{\varepsilon}}\}$ is the set of positive integers, $r = 0, 1, 2, \ldots$,

$$Z = \prod_{\mathbf{k},\boldsymbol{\varepsilon}} \sum_{r=0}^{\infty} e^{-\beta h\nu r} = \prod_{\mathbf{k},\boldsymbol{\varepsilon}} (1-e^{-\beta h\nu})^{-1},$$

that is,

$$\ln Z = -\sum_{\mathbf{k},\boldsymbol{\varepsilon}} \ln(1-e^{-\beta h\nu}) = -2\sum_{\mathbf{k}} \ln(1-e^{-\beta h\nu}). \tag{35.6}$$

The average occupation number for photons of given wave-vector $\mathbf{k}$, regardless of polarization, is

$$\langle n_{\mathbf{k}}\rangle = -\frac{1}{\beta}\frac{\partial}{\partial(h\nu)}\ln Z = \frac{2}{e^{\beta h\nu}-1}, \tag{35.7}$$

a result that could have been inferred directly from the case $\mu = 0$ of the Bose–Einstein statistics given in (32.6). The internal energy may be expressed

$$U = -\frac{\partial}{\partial\beta}\ln Z = \sum_{\mathbf{k}} h\nu\langle n_{\mathbf{k}}\rangle. \tag{35.8}$$

From (35.3) $\nu = c|\mathbf{n}|V^{-\frac{1}{3}}$, and with this expression for ν substituted in (35.6) we may determine the pressure by a differentiation,

$$p = \frac{1}{\beta}\frac{\partial}{\partial V}\ln Z = \frac{1}{3V}\sum_{\mathbf{k}} h\nu\langle n_{\mathbf{k}}\rangle,$$

which by (35.8) yields the equation of state

$$pV = \tfrac{1}{3}U. \tag{35.9}$$

35.3. The laws of Planck and others

In the limit $V \to \infty$ we may use (35.4) and (35.7) to write (35.8) as

$$U = \frac{8\pi V}{c^3}\int_0^\infty \frac{h\nu^3\,\mathrm{d}\nu}{\mathrm{e}^{h\nu/kT}-1}, \tag{35.10}$$

from which we infer that if $E_\nu\,\mathrm{d}\nu$ is the radiation energy in ν, $\nu+\mathrm{d}\nu$, then

$$E_\nu = \frac{8\pi V}{c^3}\,\frac{h\nu^3}{\mathrm{e}^{h\nu/kT}-1}. \tag{35.11}$$

This is the famous spectral-energy-distribution law of Planck. It has the two limiting forms,

$$E_\nu \approx \frac{8\pi V}{c^3}\begin{cases}\nu^2 kT & (h\nu/kT \ll 1)\\ h\nu^3\,\mathrm{e}^{-h\nu/kT} & (h\nu/kT \gg 1)\end{cases},$$

the first of which is the classical *Rayleigh–Jeans law*, obtained simply by assigning the energy kT to each degree of freedom, that is, by multiplying (35.5) by kT. The second form is known as *Wien's formula*.

The integral in (35.10) can be evaluated in terms of the gamma and Riemann zeta functions. The result is

$$U = aVT^4, \qquad a \equiv \frac{8\pi^5 k^4}{15h^3c^3},$$

and it follows

$$p = \tfrac{1}{3}aT^4, \qquad C_V = 4aVT^3, \qquad S = \left(\frac{\partial(pV)}{\partial T}\right)_V = \tfrac{4}{3}aVT^3.$$

If a black-body cavity is opened to the external world through a small hole, the amount of energy radiated per second per unit area of the hole is

$$I(T) = cU/V = caT^4,$$

as the radiation escapes from the cavity with a velocity c. This is *Stefan's law*, and the universal constant $\sigma = ca$ is known as Stefan's constant.

36. Information theory and thermodynamics

36.1. The uncertainty function

Information theory is the study of the information content of such entities as transmitted messages, experimental observations, and computer output; to make this study precise a quantitative definition of information is required.

One 'bit' of information is that required to make a binary choice between (two) equally probable items. Thus if one out of say 16 equally probable events is identified as being the actual outcome, the information required for this is $4 = \log_2 16$ bits, as 4 binary choices in sequence are involved. If P different events or final states are possible, with *a priori* equal probabilities, the appearance of one of them has an information content of $I = \log_2 P = -\log_2 p$, $p = 1/P$ being the probability of a particular event occurring. More generally if a system X has states $x_1, x_2, \dots x_n$ with *a priori* probabilities $p_1, p_2, \dots p_n$ of occurrence, the average information necessary to identify the actual state that occurs is

$$I(X) = -\sum_{i=1}^{n} p_i \log_2 p_i. \qquad (36.1)$$

Complete information conveyed to an observer about the state of X means that for him the initial uncertainty in this state has been removed, consequently (36.1) suggests the following definition:

Let a random variable X take on a finite number of possible states $x_1, x_2, \dots x_n$ with probabilities $p_1, p_2, \dots p_n$ ($p_i > 0, \sum p_i = 1$), then the uncertainty $H(X)$ in the initial state of X is defined by

$$H(X) \equiv -\sum_{i=1}^{n} p_i \log_2 p_i. \qquad (36.2)$$

It remains to be seen whether this somewhat artificially defined quantity is useful in dealing with real problems. The close analogy between the uncertainty function and the statistical entropy is apparent from eqn (26.14), namely,

$$S = -k \sum_{j} P_j \ln P_j. \qquad (36.3)$$

36.2. Some desirable properties for the uncertainty function

Suppose for the moment we ignore the definition in (36.2) and list some of the properties that a measure H of the uncertainty in a system X should have. These properties are arrived at intuitively and taken as axioms should enable us to deduce the actual form of $H(p_1, p_2, \dots p_n)$.

First for fixed n it is reasonable to require H to be a maximum when all probabilities are equal, that is, when $p_i = 1/n$, $i = 1, 2, \dots n$. Clearly if one state, say x_1, has a much larger probability p_1 of occurrence than any other state, the uncertainty in X will be small, and in the limit as $p_1 \to 1, p_2, p_3, \dots \to 0$ the uncertainty must vanish. Thus

$$H(p_1, p_2, \dots p_n) \text{ is maximal for } p_i = 1/n. \tag{36.4}$$

Now let Y denote a second random variable with states $y_1, y_2, \dots y_m$ and probabilities $p'_1, p'_2, \dots p'_m$ of occurrence. Denote by $H(X, Y)$ the *joint uncertainty* in the states of X and Y, that is, $H(X, Y)$ is the average value of the information conveyed by specifying joint states like $\{X = x_i, Y = y_j\}$. In another situation suppose the state of X is specified whereas that of Y remains uncertain and denote by $H(Y \mid X)$ the average value of the information conveyed by specifying Y in this circumstance, then $H(Y \mid X)$ is termed a *conditional uncertainty*. For any measure of uncertainty it is intuitively reasonable to expect that the average value of the information conveyed by the joint states $\{X, Y\}$ equals the information conveyed on average by knowing X alone, plus the additional information conveyed by knowing Y when X is already specified, that is,

$$H(X, Y) = H(X) + H(Y \mid X). \tag{36.5}$$

Of course if there is no connection at all between the states of X and Y, $H(Y \mid X)$ is the same as $H(Y)$.

If an impossible state $x_{n+1} (p_{n+1} = 0)$ is added to the system X, the uncertainty in X should not be altered, that is,

$$H(p_1, p_2, \dots p_n, 0) = H(p_1, p_2, \dots p_n). \tag{36.6}$$

Another obviously desirable property is that

$$H(p_1, p_2, \dots p_n) \text{ is continuous in all its arguments.} \tag{36.7}$$

By starting with the above four properties as axioms, Khinchine (1957) shows that, save for a positive multiplicative constant, H must have the form given in eqn (36.2).

36.3. The principle of minimum prejudice

Suppose that the probabilities $p_1, p_2, \dots p_n$ to be assigned to the states $x_1, x_2, \dots x_n$ are unknown and further that there is no information at all available about the system X. In this case no choice of values for the probabilities other than the completely unbiased one of giving them all equal values can be justified and such a choice we note from (36.4) is the one that maximizes the uncertainty. More generally we have the principle of minimum prejudice due to Jaynes:

The least prejudiced assignment of the probabilities is that which maximizes H subject to the given information.

If a choice is preferred that does not maximize H in this way, then clearly some states will receive a preference over others without justification.

Suppose that all that is known about X is contained in the set of l constraints

$$\bar{g}_s = \sum_{i=1}^{n} p_i g_{si} \qquad (s = 1, 2, \dots l), \tag{36.8}$$

where both the functions g_{si} and their expectations $\bar{g}_s$ are known. Besides (36.8) the p_i also satisfy

$$1 = \sum_{i=1}^{n} p_i. \tag{36.9}$$

To find the most unbiased set of probabilities we must maximize

$$H = -K \sum_i p_i \ln p_i \qquad (K = 1/\ln 2) \tag{36.10}$$

(converting to natural logarithms), subject to (36.8) and (36.9). Now

$$\delta H = 0 = -K \sum_i (\ln p_i + 1)\delta p_i, \qquad \sum_i \delta p_i = 0, \qquad \sum_i \delta p_i g_{si} = 0,$$

whence

$$\sum_i \left[\ln p_i + 1 + (\lambda_0 - 1) + \sum_s \lambda_s g_{si} \right] \delta p_i = 0,$$

where $(\lambda_0 - 1)$, λ_s, $s = 1, 2, \dots l$, are Lagrange multipliers. As the δp_i may be supposed independent in this sum, it follows that each term is zero, that is,

$$p_i = \exp\left(-\lambda_0 - \sum_s \lambda_s g_{si} \right), \tag{36.11}$$

where the numbers λ_0, λ_s must be chosen to satisfy the constraints (36.8) and (36.9). To show that (36.11) actually corresponds to a maximum it is necessary to verify that $\delta^2[H + (\lambda_0 - 1)\sum_i p_i + \sum_i \sum_s \lambda_s p_i g_{si}] < 0$ for p_i given by (36.11); this is easily established.

36.4. *The uncertainty in a system of energetic particles*

As a simple example of the principle of minimum prejudice we shall use (36.11) to estimate the number distribution $n_1, n_2, \dots$ of identical particles over energy levels $\varepsilon_1, \varepsilon_2, \dots$, subject only to the information that their total number N and total energy U have known values,

$$N = \sum_i n_i, \qquad U = \sum_i n_i \varepsilon_i. \tag{36.12}$$

With $p_i = n_i/N$ the probability that a particle chosen at random has the energy ε_i, then (36.11) and (36.12) yield the estimate (cf. (32.12))

$$n_i = p_i N = N \exp(-\alpha - \beta\varepsilon_i), \tag{36.13}$$

where α and β satisfy

$$\alpha = \ln \sum_i \exp(-\beta\varepsilon_i) \tag{36.14}$$

and

$$U = N \sum_i \varepsilon_i \exp[-(\alpha + \beta\varepsilon_i)] = \sum_i \varepsilon_i \exp(-\beta\varepsilon_i) \Big/ \sum_i \exp(-\beta\varepsilon_i). \tag{36.15}$$

Now p_i is the probability that a given *particle*, say X_i, is in state ε_i, hence by (36.10) and (36.13) the uncertainty for the particle is

$$H(X_1) = -K \sum_i p_i \ln p_i \qquad (p_i = \exp[-(\alpha + \beta\varepsilon_i)]).$$

However, the system of interest is not just one particle but a collection $X_1, X_2, \ldots X_N$ of identical particles in which the labels $1, 2, \ldots N$ have no physical significance. Denote this system by X and the system of labels by Y, then by the independence of X and Y eqn (36.5) gives

$$H(X) = H(X, Y) - H(Y \mid X) = H(X_1, X_2, \ldots X_N) - H(Y).$$

As $X_1, X_2, \ldots$ are independent, repeated application of (36.5) gives $H(X_1, X_2, \ldots X_N) = NH(X_1)$. Also, since there are $N!$ equally probable sequences of labelled particles, to attach the labels requires an increase in conveyed information of amount $H(Y) = K \ln N! \sim KN(\ln N - 1)$ for large N.

Collecting these results we have

$$H(X) = -KN \sum_i p_i \ln p_i - KN(\ln N - 1) = K\beta U - KN \ln(N\,\mathrm{e}^{-\alpha}) + KN.$$

Define μ by $\beta\mu \equiv \ln[N \exp(-\alpha)]$, so that (36.13) may be written

$$n_i = \exp[\beta(\mu - \varepsilon_i)], \tag{36.16}$$

and the uncertainty in the system reads

$$H(X) = K\beta U - K\beta\mu N + KN. \tag{36.17}$$

36.5. *The analogy between uncertainty and entropy*

If it is now *assumed* that the principle of minimum prejudice applied to the system X yields a number distribution n_i, $i = 1, 2, \ldots$, the same as the averages $\langle n_i \rangle$ taken over an ensemble of Xs, then the results given in § 32.3 may be used to identify the constants β and μ appearing in (36.16) and (36.17). Comparison of (32.12) and (36.16) leads to the conclusion that the β and μ used above have the same significance as in § 32.3, in particular that $\beta^{-1} = kT$. Also (36.17) and (32.13) reveal that

$$H = KS/k.$$

For the chosen example then, uncertainty and entropy are proportional once the hypothesis is adopted that the notion of minimum prejudice has real physical significance.

Despite the success of this example in linking H and S, it is certainly not correct to associate the uncertainty in *any* system with entropy. They are analogous but distinct concepts. In general, H is a non-dimensional measure of the uncertainty in any system capable of being in a variety of states—it is *statistical* rather than a physical concept. We can talk of the information content of a transmitted message, of a computer output, or of the DNA code, but not of their entropy. 'Entropy' applies only to those systems in which the concepts of heat and temperature are also relevant; in such cases Boltzmann's dimensional constant (heat content per degree) supplies the essential link between mere statistics and physics.

Incidentally it is sometimes stated that 'entropy is a measure of the disorder in a system', an empty and misleading statement unless of course the term 'disorder' can be given a scientific meaning. It is rather the *capacity* of a system to be in distinct micro-states that is important, its state of apparent disorder or otherwise at a particular instant being irrelevant. 'Disorder' would have to be interpreted to be 'uncertainty' to be a useful (and then redundant) concept.

Turning back now to the usual axiomatic approach to statistical mechanics, we can see that postulates A and B in § 26.4 are immediate consequences of the principle of minimum prejudice, and to this extent information theory appears to underpin the subject. But its contribution to statistical mechanics is quite small and easily avoidable, although much exaggerated by some authors. The single idea involved is that of 'minimum prejudice', which is equivalent to the physical principle that stability requires the entropy to be a maximum at equilibrium. Thermodynamic systems are much more complicated than simple information systems, having many distinct properties for which a physical rather than a statistical description is essential. The approach via information theory may produce a verbiage evocative of the observer's role (which we have emphasized in this book), but it adds nothing to the subject that cannot be clearer stated in physical terms.

The axiomatic approach described in § 26.4 is quite different from that adopted in this book. Our method has been to start with classical thermodynamics and to extend its range to shorter time scales by enlarging its state-space to include internal coordinates. The generalized Gibbs relation for this enlarged state-space then provides the basis for our introduction of statistical thermodynamics, and quite general considerations applied to this relation enables us to arrive at the statistical formula for the entropy. No additional postulates—save the usual ergodicity requirement—are necessary, and in particular information theory is irrelevant. However, this latter is merely a pedagogical point; any approach that adds at all to our understanding of the rather elusive nature of entropy is worth having.

36.6. Life and entropy fluctuations

Finally consider the frequently encountered remark that life is anomalous *vis à vis* the law of entropy increase (§ 13). The complexity of biological molecules appears as the very antithesis of disorder, and their growth must therefore correspond to a *reduction* in entropy, at least in some regions. There are various misleading resolutions of this paradox. First, on the longest time scale, life cannot survive in the closed system required by the law; secondly, on a shorter time scale life is a 'fluctuation' and entropy decreases are therefore possible (§ 28)—equivalently we could regard life as a kind of 'instability', to which the law, based on a postulated trend towards stable equilibrium, cannot apply. Neither explanation pays sufficient attention to the way in which entropy depends on the observer's knowledge of the system.

As a simple illustration of this point, consider a closed system consisting of three types of molecule uniformly dispersed throughout its volume. Suppose that below some threshold temperature T_c a combination of these molecules can fuse into a complex molecule, which occurs whenever certain chance fluctuations in the concentrations c_1, c_2, c_3 of the elementary molecules makes fusion energetically favourable. The entropies above and below T_c are defined in value by

$$T\,\mathrm{d}s = -\sum_{i=1}^{3} g_i\,\mathrm{d}c_i, \qquad T\,\mathrm{d}s_c = -\sum_{i=1}^{4} g_i\,\mathrm{d}c_i = T\,\mathrm{d}s - g_4\,\mathrm{d}c_4,$$

c_4 being the concentration of the new molecule. To an observer ignorant of the existence of T_c all changes in c_1, c_2, c_3 away from what he believes to be their equilibrium values, based on experiments above T_c, are chance fluctuations that reduce s. But as soon as he recognizes that certain types of fluctuation persist for long times, he will postulate the presence of a new component and hence abandon s in favour of the lower entropy s_c. Now the random fluctuations that persistently reduced s are seen as being chemical reactions that increase s_c. An unstable departure from equilibrium in (c_1, c_2, c_3)-space is now an approach towards stable equilibrium in (c_1, c_2, c_3, c_4)-space. We can always preserve the law of entropy increase by a suitable enlargement of state-space.

PART II
PROCESS THERMODYNAMICS

6

GENERAL PRINCIPLES

PROCESS thermodynamics provides a general framework for the description of the irreversible processes of macroscopic physics, such as the molecular transport of matter (diffusion), momentum (fluid stress), energy (heat), and electric charge (current). It therefore has important applications to gas dynamics, magnetohydrodynamics, plasma dynamics, superfluids, and other systems in which gradients and time dependence are usual. In this text it is the applications of the ideas of the subject to fluid systems that will be the main concern, but process thermodynamics also plays a role in elasticity, plasticity, and other branches of continuum mechanics.

The subject is usually described as being 'non-equilibrium thermodynamics', and in the sense that the variables change with position or time the name is appropriate. However, in view of the various meanings the word 'equilibrium' may have in thermodynamics (see § 1.2) and also of the particular meaning already given in § 3.2 to 'non-equilibrium', the name is unfortunate and confusing. In fact the processes we shall consider below are what we have termed 'irreversible equilibrium processes' in § 3.2, which means that a thermodynamic state may be assigned to each element of the system at any given time. A better name for the subject, sometimes used, is 'irreversible thermodynamics'; certainly all real processes are irreversible, but it is convenient to include the limiting case of ideal reversible processes in the discussion, so the term 'process thermodynamics' seems more appropriate. The central problem in process thermodynamics is that of separating the processes into reversible and irreversible components, which requires some intuitive grasp of the physical behaviour of the medium under thermodynamic forces. It proves simplest to postulate which of the component processes may be considered reversible on the observers time scale, then by knowing all the relevant processes, to identify the irreversible components by subtraction. Time scales play an important role in this identification.

37. The basic equilibrium thermodynamics

37.1. The Clausius inequality

Consider a compound system P (Fig. 11) extending over a volume V and comprised of (infinitesimal) fluid-particle systems $\mathscr{P}$. The surface Σ of P is assumed to be convected along at the local fluid velocity $\mathbf{v}(\mathbf{r}, t)$, where $\mathbf{r}$ is the position vector and t is the time, and therefore the mass of P is constant.

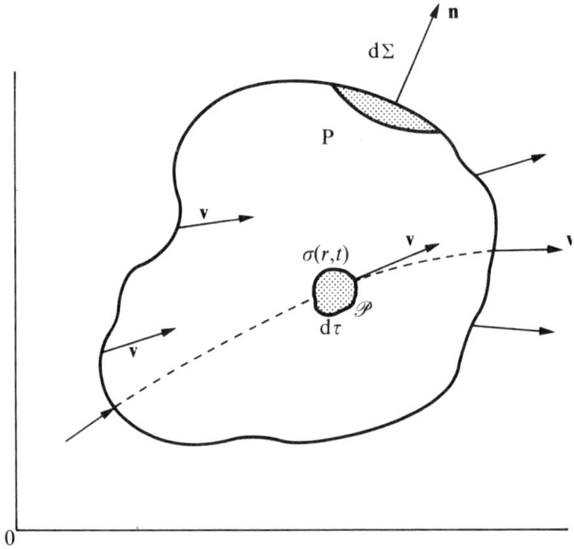

FIG. 11. Convected thermodynamic systems P and $\mathscr{P}$.

The mass of $\mathscr{P}$ will also be taken to be constant, but in general the component systems P_i and $\mathscr{P}_i$ will have variable masses. Also in the general case the component systems P_i and $\mathscr{P}_i$ will move relative to P and $\mathscr{P}$, a process known as *diffusion*.

In § 12.3 it was established that the entropy S of a non-diffusing compound system P, whose components P_i are of constant mass, increases as a result of heat transfers alone at a rate such that

$$\frac{\mathrm{d}S}{\mathrm{d}t} \geqslant -\int_{\Sigma} \sum_i \frac{\mathbf{n} \cdot \mathbf{q}_i}{T_i} \mathrm{d}\Sigma + \int_V \sum_i \left(\sum_j \frac{\varphi_{ij}}{T_j} + \frac{\psi_i}{T_i} \right) \mathrm{d}\tau, \tag{37.1}$$

where φ_{ij} and ψ_i are the heats received by P_i per second per unit volume from P_j and an external radiative source $\mathscr{S}$ respectively, and $\mathbf{q}_i$ is the heat flux vector for P_i. This generalization of Clausius's heat transfer inequality holds *a fortiori* when work and mass transfers are also permitted between the components (see §§ 13.1 and 18.4), and it remains only to modify (37.1) to allow for diffusion. This will be done in § 51.3 and has no direct bearing on the general principles to be described in this chapter.

The introduction of $\sigma(\mathbf{r}, t)$, a non-negative entropy production rate per unit volume, permits us to rewrite (37.1) as an equation,

$$\frac{\mathrm{d}S}{\mathrm{d}t} = -\int_{\Sigma} \sum_i \frac{\mathbf{n} \cdot \mathbf{q}_i}{T_i} \mathrm{d}\Sigma + \int_V \left[\sum_i \left(\sum_j \frac{\varphi_{ij}}{T_j} + \frac{\psi_i}{T_i} \right) + \sigma \right] \mathrm{d}\tau, \tag{37.2}$$

with

$$\sigma \geqslant 0. \tag{37.3}$$

The important point to remember here is that the inequality in (37.3), which is basic to the whole of process thermodynamics, is a consequence of assuming mechanical, thermal, and material stability for the system P.

37.2. Local thermodynamic equilibrium

Our immediate objective is to find an expression for σ, from which we shall be able to infer appropriate forms for the *constitutive relations* of process thermodynamics. These relations, of which Fourier's law for heat conduction is a typical example, play the same role in process thermodynamics as state equations in equilibrium thermodynamics. Before (37.2) can be used to evaluate σ, it is necessary to give the derivative $\mathrm{d}S/\mathrm{d}t$ a precise meaning, for in non-equilibrium states, as has been pointed out in § 7.5, entropy is undefined. This problem has been discussed in § 19.2, where it is shown that provided the time element $\mathrm{d}t$ is everywhere comparable with the time scale $\mathscr{T}$ on which $\mathrm{d}S$ is defined, we may write for a fluid particle $\mathscr{P}$

$$T\frac{\mathrm{d}s}{\mathrm{d}t} = \frac{\mathrm{d}u}{\mathrm{d}t} + \mathbf{p}\cdot\frac{\mathrm{d}\mathbf{l}}{\mathrm{d}t} - \sum_i g_i \frac{\mathrm{d_e}c_i}{\mathrm{d}t} + \sum_r A_r \frac{\mathrm{d}\xi_r}{\mathrm{d}t}, \tag{37.4}$$

where $\mathbf{l}$ is the specific deformation vector, so written to avoid confusion with the fluid velocity vector $\mathbf{v}$ to be introduced shortly. The variables s, u, $\mathbf{l}$, ... in (37.4) are functions of $\mathbf{r}$ and t.

It is particularly important for later considerations to note that by *definition* $\mathbf{p}\,.\,\mathrm{d}\mathbf{l}/\mathrm{d}t$ is the rate of doing *reversible* work (§ 3.4), whereas the final term of (37.4) is an irreversible contribution to $\mathrm{d}s/\mathrm{d}t$ (§ 20). The diffusion represented by the term $(-\sum_i g_i\, \mathrm{d_e}c_i/\mathrm{d}t)$ is partly reversible and partly irreversible, but as we shall defer the theory of mixtures until Chapter 9, we need not elaborate this point here.

For (37.4) to apply it is also necessary that the fluid particle $\mathscr{P}$ be small enough to always possess nearly uniform properties, but on the other hand for a valid macroscopic description $\mathscr{P}$ cannot be made smaller than a mean free path. The validity of (37.4) therefore depends on a nice balance between time and length scales, and when it may be assumed that this is achieved, we have a state of 'local thermodynamic equilibrium'.

To (37.4) we must add the state equations expressing the intensive variables T, $\mathbf{p}$, g_i, and A_r as function of the thermodynamic coordinates u, $\mathbf{l}$, c_i, and ξ_r, for example,

$$T = T[u(\mathbf{r}, t), \mathbf{l}(\mathbf{r}, t), c_1(\mathbf{r}, t), c_2(\mathbf{r}, t) \ldots, \xi_1(\mathbf{r}, t), \xi_2(\mathbf{r}, t), \ldots],$$

and similarly for the other functions of state.

When the component fluid particles $\mathscr{P}_i$, $i = 1, 2, \ldots$, are at different temperatures T_i, it is necessary to regard each one as being a distinct thermodynamic system, in which case instead of (37.4) we need (see (14.6))

$$T_i \frac{\mathrm{d}s_i}{\mathrm{d}t} = \frac{\mathrm{d}u_i}{\mathrm{d}t} + \mathbf{p}_i \cdot \frac{\mathrm{d}\mathbf{l}_i}{\mathrm{d}t} \qquad (i = 1, 2, \ldots). \tag{37.5}$$

38. Entropy production rate in a fluid

38.1. Conservation of mass

In the next few sections, to simplify the account, we shall deal with a fluid comprised of a single component, deferring the much more complicated case of a mixture of fluids to Chapter 9. Our first task is to briefly outline the derivation of the conservation laws for a fluid-particle system $\mathscr{P}$, which is moving with a velocity $\mathbf{v}$ relative to a fixed coordinate frame in which $\partial/\partial t$ denotes the time derivative. In a frame convected with $\mathscr{P}$ the time derivative will be denoted by

$$\mathrm{D} = \frac{\partial}{\partial t} + \mathbf{v} \cdot \nabla, \tag{38.1}$$

where ∇ is the gradient vector operator; D is known as the *material derivative*. Let $\mathrm{d}\tau$ denote the volume of $\mathscr{P}$, then it is easily shown that

$$\mathrm{D}(\mathrm{d}\tau) = \nabla \cdot \mathbf{v}\, \mathrm{d}\tau.$$

Suppose that $X(\mathbf{r}, t)\, \mathrm{d}\tau$ denotes some physical property of the fluid particle $\mathscr{P}(\mathbf{r}, t)$, then for the rate of change of this property in a frame moving with $\mathscr{P}$ we may write

$$\frac{1}{\mathrm{d}\tau}\mathrm{D}(X\, \mathrm{d}\tau) = \mathrm{D}X + X\, \nabla \cdot \mathbf{v} = \frac{\partial X}{\partial t} + \nabla \cdot (X\mathbf{v}) = \mathscr{S}(X), \tag{38.2}$$

where $\mathscr{S}(X)$ is the source term for the supply of X due to external causes.

First let X be the density ρ, then as the mass $\rho\, \mathrm{d}\tau$ of $\mathscr{P}$ is constant, $\mathscr{S}(\rho) = 0$, and (38.2) yields the various forms

$$\begin{aligned} \mathrm{D}(\rho\, \mathrm{d}\tau) &= 0, \qquad & \rho\mathrm{D}(1/\rho) &= \nabla \cdot \mathbf{v}, \\ \mathrm{D}\rho + \rho\, \nabla \cdot \mathbf{v} &= 0, \qquad & \partial\rho/\partial t + \nabla \cdot \rho\mathbf{v} &= 0, \end{aligned} \tag{38.3}$$

of the law of mass conservation.

38.2. Conservation of momentum

Secondly let X be the momentum $\rho\mathbf{v}$ per unit volume, a property that is changed by body forces acting on $\mathscr{P}$ and by stresses applied over the surface of $\mathscr{P}$ by neighbouring fluid elements. The body force acting on $\mathscr{P}$ will be written $\rho\mathbf{F}\,\mathrm{d}\tau$, so that $\mathbf{F}$ is the body force per unit mass.

The *pressure tensor* $\mathbf{p}$ may be defined by requiring $\mathbf{n}\,.\,\mathbf{p}\,\mathrm{d}\Sigma$ to be the force transmitted across a surface element $\mathbf{n}\,\mathrm{d}\Sigma$ in the positive direction along $\mathbf{n}$. Thus the surface integral of $-\mathbf{n}\,.\,\mathbf{p}\,\mathrm{d}\Sigma$ over a system P (see Fig. 11, p. 144) is the force acting on the fluid within P due to the fluid outside P. With the usual assumption of continuity in all properties throughout P, we infer from the divergence theorem that $-\nabla\,.\,\mathbf{p}\,\mathrm{d}\tau$ is the corresponding force on the fluid particle $\mathscr{P}$. Thus

$$\mathscr{S}(\rho\mathbf{v}) = -\nabla\,.\,\mathbf{p}+\rho\mathbf{F}.$$

By (38.2) and $(38.3)_3$ we now have two forms of the momentum balance equation

$$\frac{\partial}{\partial t}(\rho\mathbf{v})+\nabla\,.\,(\rho\mathbf{v}\mathbf{v}+\mathbf{p}) = \rho\mathbf{F},$$
$$\rho\mathrm{D}\mathbf{v}+\nabla\,.\,\mathbf{p} = \rho\mathbf{F}. \tag{38.4}$$

38.3. Conservation of energy

Now let X be the total energy possessed by $\mathscr{P}$ per unit volume, then $X = \rho(u+\frac{1}{2}v^2)$, u being the specific internal energy and $\frac{1}{2}v^2$ the corresponding kinetic energy.† This energy is changed by the work that the body force does on $\mathscr{P}$ and also by work and energy transfers to $\mathscr{P}$ from contiguous fluid particles. The body force does work at the rate $\rho\mathbf{F}\,.\,\mathbf{v}\,\mathrm{d}\tau$ on $\mathscr{P}$, while the flow of heat from neighbouring particles adds energy to $\mathscr{P}$ at the rate $-\nabla\,.\,\mathbf{q}\,\mathrm{d}\tau$ (cf. (12.7)), where $\mathbf{q}$ is the heat flux vector.

The mechanical force transmitted across a convected surface element $\mathbf{n}\,\mathrm{d}\Sigma$ does work at the rate $-(\mathbf{n}\,.\,\mathbf{p}\,\mathrm{d}\Sigma)\,.\,\mathbf{v}$ on the fluid lying on the negative side of the element, and thus the surface integral of $-\mathbf{n}\,.\,\mathbf{p}\,.\,\mathbf{v}\,\mathrm{d}\Sigma$ over the system P of Fig. 11 is the rate at which the surface stress does work on the fluid within P. By the divergence theorem $-\nabla\,.\,(\mathbf{p}\,.\,\mathbf{v})\,\mathrm{d}\tau$ is therefore the rate surface stresses do work on a fluid particle $\mathscr{P}$.

Finally $\mathscr{P}$ may receive heat at the rate $\psi\,\mathrm{d}\tau$, say, by radiation from sources external to P. Collecting the various source terms together, we have

$$\mathscr{S}[\rho(u+\tfrac{1}{2}v^2)] = -\nabla\,.\,(\mathbf{p}\,.\,\mathbf{v}+\mathbf{q})+\rho\mathbf{F}\,.\,\mathbf{v}+\psi.$$

It now follows from (38.2) that

$$\frac{\partial}{\partial t}[\rho(u+\tfrac{1}{2}v^2)]+\nabla\,.\,[\rho\mathbf{v}(u+\tfrac{1}{2}v^2)+\mathbf{p}\,.\,\mathbf{v}+\mathbf{q}] = \rho\mathbf{F}\,.\,\mathbf{v}+\psi, \tag{38.5}$$

† Intrinsic rotational energy due to any angular velocity possessed by $\mathscr{P}$ is ignored here but will be discussed in § 49.2.

which with the help of (38.1) and $(38.3)_4$ can be written as

$$\rho \mathrm{D}(u+\tfrac{1}{2}v^2)+\nabla\,.\,(\mathbf{p}\,.\,\mathbf{v}+\mathbf{q}) = \rho\mathbf{F}\,.\,\mathbf{v}+\psi. \tag{38.6}$$

The vector product of $(38.4)_2$ and $\mathbf{v}$, namely,

$$\rho \mathrm{D}(\tfrac{1}{2}v^2)+\mathbf{v}\,.\,\nabla\,.\,\mathbf{p} = \rho\mathbf{F}\,.\,\mathbf{v},$$

and the vector relation $\nabla\,.\,(\mathbf{p}\,.\,\mathbf{v})-\mathbf{v}\,.\,\nabla\,.\,\mathbf{p} = \tilde{\mathbf{p}}:\nabla\mathbf{v}$, where $\tilde{\mathbf{p}}$ denotes the *transpose* of $\mathbf{p}$, enable us to reduce (38.6) to the form

$$\rho \mathrm{D}u+\tilde{\mathbf{p}}:\nabla\mathbf{v} = -\nabla\,.\,\mathbf{q}+\psi. \tag{38.7}$$

Comparing this equation with the first law of thermodynamics, $\mathrm{d}U-\mathrm{d}'W = \mathrm{d}'Q$, we infer that $-\tilde{\mathbf{p}}:\nabla\mathbf{v}$ is the rate at which work is done on $\mathscr{P}$ due to the fluid stresses in a frame moving with $\mathscr{P}$. Some of this work will be reversible and the rest irreversible.

38.4. The entropy balance equation

Let $s = s(\mathbf{r}, t)$ denote the specific entropy within a volume V, then the total entropy within this volume is

$$S = \int_V \rho s\,\mathrm{d}\tau.$$

And as $\rho\,\mathrm{d}\tau$ is constant for a given fluid particle,

$$\frac{\mathrm{d}S}{\mathrm{d}t} = \int_V \mathrm{D}(\rho s\,\mathrm{d}\tau) = \int_V \rho\mathrm{D}s\,\mathrm{d}\tau.$$

If the divergence theorem is applied to the surface integral in eqn (37.2), and continuity assumed for the variables within the volume, we may conclude that

$$\sigma = \rho\mathrm{D}s+\nabla\,.\sum_i(\mathbf{q}_i/T_i)-\sum_i\left(\sum_j\varphi_{ij}/T_j+\psi_i/T_i\right), \tag{38.8}$$

an equation that will be corrected at a later stage (§ 51.2) to allow for the effects of diffusion.

An expression for Ds follows from (37.4), in which the time derivative may be taken to be the rate of change in a frame moving with $\mathscr{P}$,

$$T\mathrm{D}s = \mathrm{D}u+\mathbf{p}\,.\,\mathrm{D}\mathbf{l}-\sum_i g_i\mathrm{D_e}c_i+\sum_r A_r\dot{\xi}_r, \tag{38.9}$$

where $\dot{\xi}_r \equiv \mathrm{D}\xi_r$ is the *reaction rate* of the rth reaction.

In the present application there is only one component, when (38.8) and (38.9) reduce to

$$\left.\begin{aligned}\sigma &= \rho\mathrm{D}s+\nabla\,.\,(\mathbf{q}/T)-\psi/T,\\ T\mathrm{D}s &= \mathrm{D}u+\mathbf{p}\,.\,\mathrm{D}\mathbf{l},\end{aligned}\right\} \tag{38.10}$$

whence on eliminating Du by (38.7),

$$T\sigma = -\frac{\mathbf{q}}{T}\cdot\nabla T+(\rho\mathbf{p}\,.\,\mathrm{D}\mathbf{l}-\tilde{\mathbf{p}}:\nabla\mathbf{v}) \geqslant 0. \tag{38.11}$$

The first right-hand term is irreversible heat transfer to $\mathscr{P}$, while the second, being the difference between total and reversible work, must be the irreversible work done on $\mathscr{P}$.

To simplify the work term in (38.11) an hypothesis is now necessary about the nature of a fluid, namely, that if an incremental stress is applied to $\mathscr{P}$ and then removed, the only deformation coordinate able to recover its original value is the specific volume, $1/\rho$. In other words, unlike elastic bodies, shear strains in fluids are not reversible. Hence the reversible term in (38.11) is $\rho p\mathrm{D}(1/\rho)$, where p is the thermodynamic pressure. By mass conservation (eqn $(38.3)_2$), the irreversible work rate may be written

$$\rho p\mathrm{D}(1/\rho)-\tilde{\mathbf{p}}:\nabla\mathbf{v} = p\nabla\,.\,\mathbf{v}-\tilde{\mathbf{p}}:\nabla\mathbf{v} = -(\tilde{\mathbf{p}}-p\mathbf{1}):\nabla\mathbf{v},$$

where $\mathbf{1}$ is unit tensor. It is convenient to introduce the *viscous stress* tensor $\boldsymbol{\pi}$ defined by

$$\boldsymbol{\pi} \equiv \mathbf{p}-p\mathbf{1}, \tag{38.12}$$

so that the irreversible work rate Φ, also known as the viscous dissipation function, is written

$$\Phi = -\tilde{\boldsymbol{\pi}}:\nabla\mathbf{v},$$

and (38.11) reads

$$T\sigma = -\frac{\mathbf{q}}{T}\cdot\nabla T-\tilde{\boldsymbol{\pi}}:\nabla\mathbf{v} \geqslant 0. \tag{38.13}$$

It is particularly important to recognize the key role that the concept of 'irreversibility' has played in this derivation; σ is not merely non-negative but also due *only* to irreversible processes. To identify these processes it is first necessary to determine the reversible processes involved by reference to the properties of the medium under consideration. For example, had we been dealing with a perfect elastic body in which all strains are supposed reversible, we would have found the reversible work rate in (38.11) exactly equal to the total work rate and the resulting contribution to σ would be zero.

38.5. Analysis of the dissipation function

Mathematical Note. In general a tensor $\mathbf{A}$ possesses a symmetric part $\mathbf{A}^{\mathrm{s}}$, an antisymmetric part $\mathbf{A}^{\mathrm{a}}$, a scalar trace $\check{A}$ and a deviator $\mathring{\mathbf{A}}$, which has zero trace,

$$\mathbf{A}^{\mathrm{s}} \equiv \tfrac{1}{2}(\mathbf{A}+\tilde{\mathbf{A}}), \qquad \mathbf{A}^{\mathrm{a}} \equiv \tfrac{1}{2}(\mathbf{A}-\tilde{\mathbf{A}}), \qquad \check{A} \equiv \mathbf{1}:\mathbf{A} = \mathbf{1}:\tilde{\mathbf{A}},$$
$$\mathring{\mathbf{A}} \equiv \mathbf{A}^{\mathrm{s}}-\tfrac{1}{3}\check{A}\mathbf{1}, \qquad \mathbf{1}:\mathring{\mathbf{A}} = 0.$$

Any tensor $\mathbf{A}$ may be expanded thus,

$$\mathbf{A} = \tfrac{1}{3}\check{A}\mathbf{1} + \mathbf{A}^{\mathrm{a}} + \mathring{\mathbf{A}}. \tag{38.14}$$

As $\mathbf{A}:\mathbf{B} = \tilde{\mathbf{A}}:\tilde{\mathbf{B}}$, it follows that $\mathbf{A}^{\mathrm{s}}:\mathbf{B}^{\mathrm{a}} = \mathbf{A}^{\mathrm{s}}:(-\mathbf{B}^{\mathrm{a}}) = 0$. Hence

$$\mathbf{A}:\mathbf{B} = \mathbf{A}^{\mathrm{s}}:\mathbf{B}^{\mathrm{s}} + \mathbf{A}^{\mathrm{a}}:\mathbf{B}^{\mathrm{a}} = \tfrac{1}{3}\check{A}\check{B} + \mathbf{A}^{\mathrm{a}}:\mathbf{B}^{\mathrm{a}} + \mathring{\mathbf{A}}:\mathring{\mathbf{B}}, \tag{38.15}$$

and

$$\mathbf{A}:\mathbf{B}^{\mathrm{s}} = \mathbf{A}^{\mathrm{s}}:\mathbf{B}^{\mathrm{s}}, \qquad \mathbf{A}:\mathring{\mathbf{B}} = \mathring{\mathbf{A}}:\mathring{\mathbf{B}}. \tag{38.16}$$

Returning to (38.13) we note from (38.15) that as the trace of $\nabla\mathbf{v}$ is $\nabla \cdot \mathbf{v}$, $T\sigma$ may be written as the sum

$$T\sigma = -\tfrac{1}{3}\check{\pi}\nabla \cdot \mathbf{v} + \boldsymbol{\pi}^{\mathrm{a}}:(\nabla\mathbf{v})^{\mathrm{a}} - \frac{\mathbf{q}}{T} \cdot \nabla T - \mathring{\boldsymbol{\pi}}:\mathring{\nabla}\mathbf{v} \geqslant 0, \tag{38.17}$$

scalars   axial vectors   polar vectors   deviators

that is, a bilinear form involving scalars, both kinds of vectors, and second-order tensors. While no further types of tensor will appear in the bilinear forms to be obtained later for other fluid systems, (38.17) does lack one typical feature, namely, there is only one term of each type in the form. The study of process thermodynamics began in fact with the thermoelectric effect, which involves dissipation from two independent processes both involving polar vectors.

38.6. Dissipation in a conducting fluid

If the fluid is electrically conducting, then additional terms are required in the momentum and energy balance equations. By summing (11.8) over all current carrying components, we find that the additional force is $(Q\mathbf{E}+\mathbf{j}\times\mathbf{B})$ per unit volume, where Q is the electric charge density. Also, as follows from the discussion leading to eqn (11.10), the additional work term for the right-hand side of the energy equation is $\mathbf{j} \cdot \mathbf{E}$. Hence eqns (38.4) and (38.6) are modified to read

$$\rho \mathrm{D}\mathbf{v} + \nabla \cdot \mathbf{p} = \rho\mathbf{F} + Q\mathbf{E} + \mathbf{j}\times\mathbf{B} \tag{38.18}$$

and

$$\rho \mathrm{D}(u + \tfrac{1}{2}v^2) + \nabla \cdot (\mathbf{p} \cdot \mathbf{v} + \mathbf{q}) = \rho\mathbf{F} \cdot \mathbf{v} + \mathbf{j} \cdot \mathbf{E} + \psi, \tag{38.19}$$

and the resulting modification to (38.7) is

$$\rho \mathrm{D}u + \tilde{\mathbf{p}}:\nabla\mathbf{v} = -\nabla \cdot \mathbf{q} + \psi + (\mathbf{j} - Q\mathbf{v}) \cdot (\mathbf{E} + \mathbf{v}\times\mathbf{B}). \tag{38.20}$$

Notice that the additional term is simply $\mathbf{j}' \cdot \mathbf{E}'$, where $\mathbf{j}'$ and $\mathbf{E}'$ are the current density and electric field measured in a frame moving with the fluid particle.

Eqn (38.11) will now read

$$T\sigma = -\frac{\mathbf{q}}{T} \cdot \nabla T + \mathbf{j}' \cdot \mathbf{E}' + \rho \mathbf{p} \cdot \mathrm{D}\mathbf{l} - \tilde{\mathbf{p}} : \nabla \mathbf{v} \geqslant 0, \tag{38.21}$$

and it remains to decide the form of the reversible work term $\rho\mathbf{p} \cdot \mathrm{D}\mathbf{l}$. If we assert, as previously, that only the specific volume is a reversible coordinate, then the expression reduces to†

$$T\sigma = -\frac{\mathbf{q}}{T} \cdot \nabla T + (\mathbf{j} - Q\mathbf{v}) \cdot (\mathbf{E} + \mathbf{v} \times \mathbf{B}) - \tilde{\boldsymbol{\pi}} : \nabla \mathbf{v} \geqslant 0, \tag{38.22}$$

which has two distinct processes with polar vectors. On the other hand, if the electromagnetic work is also reversible, as assumed in § 11.2, there would be a term equivalent to $-\mathbf{j}' \cdot \mathbf{E}'$ incorporated in $\rho\mathbf{p} \cdot \mathrm{D}\mathbf{l}$ (cf. eqn (11.26)) that would cancel with $\mathbf{j}' \cdot \mathbf{E}'$ explicitly occurring in (38.21). In this case (38.21) would reduce to the original form (38.13).

38.7. Dissipation from internal coordinates

As a final example of a bilinear form for σ, consider an isolated system consisting of several interacting components at the same temperature T. For this case (38.9) and $(38.10)_1$ reduce to

$$\sigma = \rho \mathrm{D}s = \frac{\rho}{T} \sum_r A_r \dot{\xi}_r \geqslant 0, \tag{38.23}$$

and these internal processes are certainly irreversible. In fact this property of internal processes is a direct consequence of the stability of natural processes (cf. (38.23) and (18.10)).

The various bilinear forms for σ given above may be expressed in terms of scalar components in the general expression

$$\sigma = \sum_\alpha J_\alpha X_\alpha \geqslant 0 \qquad (\alpha = 1, 2, \ldots). \tag{38.24}$$

Each J_α is a scalar or a scalar component and the associated X_α is the 'conjugate' scalar; they may be chosen in any way from the pairs of scalars comprising the separate terms of σ. In the next few sections we shall base our discussions on the general expression (38.24), so our results will be independent of the particular choices available in (38.22) and (38.23) for J_α and X_α.

† A more accurate expression for σ is given in § 56.2.

39. Linear constitutive relations

39.1. Processes and affinities

To eqns (38.3), (38.4), and (38.7) we must add two thermodynamic state equations, say

$$u = u(\rho, T), \qquad p = p(\rho, T),$$

giving a total of one vector and four scalar equations for the dependent variables $\mathbf{v}$, u, p, ρ, T, $\mathbf{q}$, and $\boldsymbol{\pi}$. With $\mathbf{F}$ and ψ given, equations that relate $\mathbf{q}$ and $\boldsymbol{\pi}$ to the other dependent variables are required to complete the set. Such equations, known as *constitutive relations*, cannot be derived by the methods of continuum mechanics and usually appear as phenomenological additions to the subject, supposedly based on experiment. For dilute gases kinetic theory (Chapter 13) enables good approximations to be derived for these relations, provided fluid turbulence is absent.

A typical constitutive relation is Fourier's law for heat conductivity in an isotropic medium,

$$\mathbf{q} = -\kappa \nabla T, \tag{39.1}$$

which is a linear relation between the heat flux vector and the vector conjugate to it in the bilinear form (38.13) for σ. When the viscous stress tensor $\boldsymbol{\pi}$ can be set equal to zero

$$\sigma = -\frac{\mathbf{q}}{T^2} \cdot \nabla T = \kappa(\nabla T/T)^2 \geqslant 0, \tag{39.2}$$

that is, the inequality is equivalent to requiring a non-negative thermal conductivity,

$$\kappa \geqslant 0. \tag{39.3}$$

In (39.1) we may think of ∇T as an independent vector, with $\mathbf{q}$ dependent on it.

More generally in place of (39.2) we have (38.24),

$$\sigma = \sum_\alpha J_\alpha X_\alpha \geqslant 0, \tag{39.4}$$

from which expression we shall infer a general form for linear constitutive relations. The values of the scalars $X_1, X_2, \ldots$ (which may be chosen in any way, one from each pair of the bilinear form) will be supposed to be *independently* assignable, and shall be termed 'affinities',† while the conjugate

† Not to be confused with the chemical affinities of § 18.1, although in some cases the affinities introduced here may happen to be chemical affinities. The clumsy term 'thermodynamic force' is also used, but as affinities and processes may change roles it seems inappropriate.

scalars $J_1, J_2, \ldots$, termed 'processes',† will be assumed to be dependent on the affinities. In support of the dependence of the Js, note that when an affinity, X_1 say, vanishes, so must its conjugate scalar J_1, otherwise the inequality could not be satisfied for *all* values of the affinities. Indeed J_1 must be dominantly an odd function of X_1.

Let β denote a particular value of α, then J_β depends on X_β. It is reasonable to expect it to depend also on the remaining (independent) affinities X_α, $\alpha = 1, 2, \ldots \beta-1, \beta+1, \ldots$. Suppose further that J_β could also depend explicitly on another, yet unspecified variable W. Then as σ does not depend explicitly on W, the independent variables X_α and W satisfy

$$\frac{\partial \sigma}{\partial W} = \sum_\alpha \frac{\partial J_\alpha}{\partial W} X_\alpha = 0,$$

from which it is clear that J_β cannot depend directly on W. Of course J_β may depend implicitly on variables other than $X_\alpha, \alpha = 1, 2, \ldots$.

Thus, as J_β must be dominantly an odd function of X_β, for small deviations from the equilibrium state $X_\alpha = 0$, $\alpha = 1, 2, \ldots$, the processes will depend linearly on the X_α:

$$J_\beta = \sum_\alpha L_{\beta\alpha} X_\alpha \qquad (\alpha, \beta = 1, 2, \ldots). \tag{39.5}$$

The phenomenological or 'transport' coefficients $L_{\beta\alpha}$ introduced here depend on such local physical properties of the medium as the thermodynamic state, the strength of any magnetic field, the presence or not of turbulence on a scale length somewhat smaller than the fluid element $\mathscr{P}$, and so on. Their actual values must be determined by experiment or if possible by kinetic theory.

39.2. The viscous stress tensor

As an example of a constitutive relation consider the linear law relating the viscous stress tensor to the velocity tensor. We start from the special case of (38.17),

$$T\sigma = -\tfrac{1}{3}\overset{\times}{\pi}\nabla \cdot \mathbf{v} - \mathring{\boldsymbol{\pi}} : \overset{\circ}{\nabla}\mathbf{v}, \tag{39.6}$$

which holds in the absence of heat conductivity and when there are physical grounds (§ 49.2) for believing the pressure tensor $\mathbf{p} = p\mathbf{1} + \boldsymbol{\pi}$ to be symmetric, that is,

$$\boldsymbol{\pi}^{\mathrm{a}} = 0. \tag{39.7}$$

† Processes are usually termed 'fluxes', but as genuine fluxes have negative parity under particle motion-reversal (§ 40.2) and we do not wish to impose this restriction on our processes, the term is appropriate only in special cases. The distinction made here between processes and affinities is merely the mathematical distinction between dependent and independent variables.

The most general linear constitutive relation will relate each component of the viscous stress tensor to all the velocity gradient components, which fact may be expressed

$$\tfrac{1}{3}\overset{\times}{\pi} = -\rho v'\nabla \cdot \mathbf{v} - \rho\mathring{\boldsymbol{\beta}} : \nabla\mathring{\mathbf{v}},$$
$$\mathring{\boldsymbol{\pi}} = -\rho\mathring{\boldsymbol{\alpha}}\nabla \cdot \mathbf{v} - \rho\mathring{\boldsymbol{\nu}} : \nabla\mathring{\mathbf{v}}, \qquad (39.8)$$

where v' is a scalar, $\mathring{\boldsymbol{\alpha}}$, $\mathring{\boldsymbol{\beta}}$ are deviators, and $\mathring{\boldsymbol{\nu}}$ is a fourth-order tensor with component dyads that are deviators (§ 47.4). That (39.8) are the most general relations of the type required follows from (38.16) and the fact that $\mathbf{1}:\mathring{\boldsymbol{\pi}}$ must vanish.

In (39.8) $\mathring{\boldsymbol{\beta}}$ couples the scalar process $\overset{\times}{\pi}$ with the tensor affinity $\nabla\mathring{\mathbf{v}}$, while $\mathring{\boldsymbol{\alpha}}$ couples the tensor process $\mathring{\boldsymbol{\pi}}$ with the scalar affinity $\nabla \cdot \mathbf{v}$. It will be shown later (§ 48.3) that in an isotropic medium these coupling tensors vanish and further that $\mathring{\boldsymbol{\nu}}$ reduces to $2\nu\mathring{\boldsymbol{K}}$, where $\mathring{\boldsymbol{K}}$ is a fourth-order tensor with the property $\mathring{\boldsymbol{K}}:\mathbf{A} = \mathring{\mathbf{A}}$ for any tensor $\mathbf{A}$. Thus it follows from (38.12) that in an isotropic medium the pressure tensor has the form

$$\mathbf{p} = p\mathbf{1} + \boldsymbol{\pi} = (p - \rho v'\nabla \cdot \mathbf{v})\mathbf{1} - 2\rho\nu\nabla\mathring{\mathbf{v}}. \qquad (39.9)$$

The scalars v', ν are known as the bulk and shear kinematic viscosities respectively.

39.3. Thermoelectric coupling

For another important example, consider the vectorial processes of (38.22):

$$T\sigma = -\frac{\mathbf{q}}{T} \cdot \nabla T + \mathbf{j}' \cdot \mathbf{E}' \geqslant 0. \qquad (39.10)$$

The most general linear constitutive relations between ∇T and $\mathbf{E}'$, chosen arbitrarily to be the independent affinities, and the processes $\mathbf{q}$, $\mathbf{j}'$ are

$$\mathbf{j}' = \mathbf{L}_{11} \cdot \mathbf{E}' + \mathbf{L}_{12} \cdot \nabla T,$$
$$-\mathbf{q}/T = \mathbf{L}_{21} \cdot \mathbf{E}' + \mathbf{L}_{22} \cdot \nabla T, \qquad (39.11)$$

where $\mathbf{L}_{ij}$, $i, j = 1, 2$, are phenomenological second-order tensors. The coupling terms involving $\mathbf{L}_{12}$ and $\mathbf{L}_{21}$ are known as the thermoelectric effects, which for solid conductors are associated with the names Seebeck ($\mathbf{L}_{12}$) and Peltier ($\mathbf{L}_{21}$).

Had $\mathbf{q}$ and $\mathbf{E}$ been chosen to be the affinities, in place of (39.11) there would be

$$\mathbf{j}' = \mathbf{L}'_{11} \cdot \mathbf{E}' - \mathbf{L}'_{12} \cdot \mathbf{q}/T,$$
$$\nabla T = \mathbf{L}'_{21} \cdot \mathbf{E}' - \mathbf{L}'_{22} \cdot \mathbf{q}/T, \qquad (39.12)$$

where $\mathbf{L}'_{ij}$ are another set of phenomenological tensors, but expressible in terms of $\mathbf{L}_{ij}$ because (39.11) and (39.12) must represent the same physical phenomenon.

39.4. Range of linear constitutive relations

The adoption of linear constitutive relations may appear to somewhat restrict the range of validity of process thermodynamics. However, well-known linear laws such as those of Fourier (heat conduction), Ohm (electrical conduction), and Fick (mass diffusion) have been found experimentally to apply to systems well removed from equilibrium. The success of the Navier–Stokes equations based on (39.9) (§ 49.3) in describing shock-wave structure is a striking example of the way linear laws appear to operate well beyond their expected range. Chemical reactions are exceptional, as most reactions that actually occur are so far from equilibrium that a linear dependence of the reaction rates on the affinities does not arise. It may be shown (see, for example, Haase (1968, p. 115)) that in place of the linear law of (19.1) we have

$$\dot{\xi} = lRT(1-\mathrm{e}^{-A/RT}), \tag{39.13}$$

valid for any displacement from equilibrium. Thus the linear law is subject to the constraint $|A| \ll RT$, and when $|A| \gg RT$, $\dot{\xi}$ 'saturates' at a value of lRT.

For processes that are known to saturate, (39.13) suggests that a suitable empirical generalization of (39.5) would be

$$J_\alpha = \sum L_{\alpha\beta} X_\beta^* [1-\exp(-|X_\beta/X_\beta^*|)], \tag{39.14}$$

where X_β^* is the critical value of X_β. For $|X_\beta| \ll |X_\beta^*|$, (39.14) reduces to (39.5), whereas for $|X_\beta| \gg |X_\beta^*|$ it gives the saturated process

$$J_\alpha^* = \sum L_{\alpha\beta} X_\beta^*.$$

The influence of the *history* of the particle $\mathscr{P}$ appears to be neglected in (39.5), where the X_α are functions of $\mathbf{r}$ and the local time t. But this need not be so. It is necessary only to include among the affinities the time derivatives of some internal variables, $\xi_1, \xi_2, \ldots$, the present values of which are determined entirely by recent processes over a time comparable with the relaxation times involved. This aspect of the constitutive relations will be pursued in later sections (§§ 62–5).

Another tacit assumption involved in (39.5) is that J_β is determined entirely *locally*. More generally we could write

$$J_\beta(\mathbf{r}) = \sum_\alpha \int L_{\beta\alpha}(\mathbf{r}, \mathbf{r}') X_\alpha(r')\,\mathrm{d}\tau', \tag{39.15}$$

a move that is really necessary only if the fluid particle $\mathscr{P}$ is smaller than the range of the intermolecular effects responsible for the relation between J_β and X_α.

Finally, recent attempts (Coleman 1964; Eringen 1967; Truesdell 1969) to set up somewhat more general constitutive relations should be noted. The theory is a 'thermodynamics of processes' in which equilibrium thermodynamics is not an essential part as in the treatment described here, but is merely a limiting case. A non-equilibrium entropy η and a non-equilibrium temperature θ are introduced axiomatically as primitive quantities, which means of course that η and θ cannot replace s and T in the fundamental Gibbs equation. However, they can do this in the Clausius–Duhem inequality, although as long as these variables remain undefined, the inequality is of little practical value. The theory is not widely accepted, principally because it is well known that a unique non-equilibrium entropy function does not exist (cf. eqn (16.4)).

40. Constraints on the phenomenological coefficients

40.1. The entropy production rate is non-negative

The constitutive relations

$$J_\alpha = \sum_\alpha L_{\alpha\beta} X_\beta \qquad (\alpha, \beta = 1, 2, \ldots) \tag{40.1}$$

substituted into

$$\sigma = \sum_\alpha J_\alpha X_\alpha \tag{40.2}$$

yield the quadratic form

$$\sigma = \sum_{\alpha,\beta} X_\alpha L_{\alpha\beta} X_\beta = \tfrac{1}{2} \sum_{\alpha,\beta} X_\alpha (L_{\alpha\beta} + L_{\beta\alpha}) X_\beta . \tag{40.3}$$

And a standard theorem about non-negative quadratic forms applied to (40.3) provides the necessary and sufficient condition for $\sigma \geqslant 0$ to be that the determinant $|L_{\alpha\beta} + L_{\beta\alpha}|$ and all its principal minors are non-negative. Thus

$$L_{\alpha\alpha} \geqslant 0, \qquad L_{\alpha\alpha} L_{\beta\beta} \geqslant \tfrac{1}{4}(L_{\alpha\beta} + L_{\beta\alpha})^2, \ldots \qquad (\alpha, \beta = 1, 2, \ldots). \tag{40.4}$$

For example, for the viscous stress tensor of § 39.2 we find

$$\sigma = \frac{\rho}{T}[\nu'(\nabla \cdot \mathbf{v})^2 + \nabla \cdot \mathbf{v}\, (\mathring{\boldsymbol{\alpha}} + \mathring{\boldsymbol{\beta}}) : \mathring{\nabla \mathbf{v}} + \mathring{\nabla \mathbf{v}} : \mathring{\boldsymbol{\nu}} : \mathring{\nabla \mathbf{v}}] \geqslant 0 \quad \text{(in general)}$$

$$\sigma = \frac{\rho}{T}[\nu'(\nabla \cdot \mathbf{v})^2 + 2\nu \mathring{\nabla \mathbf{v}} : \mathring{\nabla \mathbf{v}}] \geqslant 0 \quad \text{(isotropic medium)}$$

and from the second of these the constraints

$$\nu' \geqslant 0, \qquad \nu \geqslant 0$$

are a consequence of the independence of the affinities $\nabla \cdot \mathbf{v}$ and $\mathring{\nabla \mathbf{v}}$.

40.2. The entropy is produced irreversibly

Not only must the entropy production rate σ be non-negative, but also it must be due to *irreversible* processes occurring in the fluid-particle system $\mathscr{P}$. As all natural processes are irreversible, it would appear that this constraint is both obvious and empty of content, but it should be remembered that constitutive relations are merely algebraic expressions in which positive entropy production need not imply that all the new entropy is created irreversibly. The problem is to identify these processes, if any, represented in the constitutive relations (40.1), that are macroscopically reversible. Such processes cannot be allowed to contribute to σ.

It is necessary to distinguish clearly between 'microscopic reversibility' and 'macroscopic reversibility'. Suppose that at $t = 0$ the motions of all microscopic particles in the thermodynamic system $\mathscr{P}$ are reversed, then in an infinitesimal time interval τ, so small that there is insufficient time for the surroundings of $\mathscr{P}$, either via force fields or particle collisions, to upset the reversed motions (see discussion in third paragraph of § 20), any affinity X_α that is a basic physical property of $\mathscr{P}$ will either remain constant or will change its sign but not its magnitude. Any X_α not possessing this character may be decomposed into affinities that do. We should make it quite clear that τ is very much smaller than the time between particle collisions that are termed 'microscopic' in § 1.1, although in the sequel we shall describe τ as being the 'microscopic time scale'.

The property just described is the 'principle of microscopic reversibility' and may be represented by the parity equation

$$X_\alpha(\tau) = \eta_\alpha X_\alpha(-\tau), \tag{40.5}$$

in which η_α is either 1 or -1 and is termed the *parity* of X_α under microscopic motion-reversal. This equation could equally be thought of as being the result of a microscopic time-reversal, but motion-reversal, with the observer's sense of time unchanged, gives a clearer physical picture.

To achieve exact motion-reversal for charged particles moving in magnetic fields, it is necessary to reverse the magnetic fields at the same instant; symbolically

$$\tau \to -\tau \Rightarrow \mathbf{B} \to -\mathbf{B}. \tag{40.6}$$

Examples of (40.5) are $\nabla T(\tau) = \nabla T(-\tau)$, and $\mathring{\nabla}\mathbf{v}(\tau) = -\mathring{\nabla}\mathbf{v}(-\tau)$. It is easily seen that all of the thermodynamic state variables must have even parity ($\eta_\alpha = 1$), whereas their time derivatives will have odd parity ($\eta_\alpha = -1$). In particular,

$$\sigma(\tau) = -\sigma(-\tau), \tag{40.7}$$

and it follows from (40.2) that J_α has the opposite parity from its conjugate variable X_α, that is

$$J_\alpha(\tau) = -\eta_\alpha J_\alpha(-\tau). \tag{40.8}$$

Now suppose that the motion-reversal initiated at $t = 0$ is continued over an infinitesimal *macroscopic* time interval $\mathrm{d}t$, then because σ is non-negative on such a time scale, eqn (40.7) cannot apply when τ is replaced by the much larger time $\mathrm{d}t$. What in fact happens after $t = 0$ is that irreversible effects caused by particle collisions and interactions rapidly suppress the motion-reversal fluctuation and restores σ to its original value, that is,

$$\sigma(\mathrm{d}t) = \sigma(-\mathrm{d}t). \tag{40.9}$$

While we shall call $\mathrm{d}t$ the 'macroscopic time scale', it is really an *infinitesimal* macroscopic time interval, comparable to or perhaps a little greater than the average collision time termed 'microscopic' in § 1.1.

Consider, for example, the formula given in (38.23) for σ. The affinities A_r are thermodynamic state variables with even parities on both time scales, whereas this is not true of the reaction rates $\dot{\xi}_r$. These rates will have instantaneous negative parity, but as chemical reactions are independent of the directions of the particle motions, they will certainly proceed in their original positive direction on the macroscopic time scale. Accordingly all the terms in (38.23) will satisfy (40.9).

Consider again the reasoning that leads from (38.11) to (38.13). The first step is the identification of the rate of doing reversible work, say $\dot{W}_\mathrm{r}(t)$. By definition such a rate has a negative parity on *both* the microscopic and macroscopic time scales,

$$\dot{W}_\mathrm{r}(\tau) = -\dot{W}_\mathrm{r}(-\tau), \qquad \dot{W}_\mathrm{r}(\mathrm{d}t) = -\dot{W}_\mathrm{r}(-\mathrm{d}t). \tag{40.10}$$

And we distinguish the irreversible work rate $\dot{W}_\mathrm{i}$ from $\dot{W}_\mathrm{r}$ by requiring the former to satisfy

$$\dot{W}_\mathrm{i}(\tau) = -\dot{W}_\mathrm{i}(-\tau), \qquad \dot{W}_\mathrm{i}(\mathrm{d}t) = \dot{W}_\mathrm{i}(-\mathrm{d}t); \tag{40.11}$$

indeed there is no other way of distinguishing the two types of work. It follows then that the bracketed quantity in (38.11) is

$$-\dot{W}_\mathrm{r} + \dot{W} = -\dot{W}_\mathrm{r} + (\dot{W}_\mathrm{r} + \dot{W}_\mathrm{i}) = \dot{W}_\mathrm{i},$$

and the net contribution to σ clearly satisfies (40.9).

Similar considerations apply to all other types of entropy production and we conclude that, in contrast to the microscopic reversibility specified in (40.7), eqn (40.9) is the essential signature of macroscopic irreversibility. If we discover that a phenomenon contributes a reversible term σ' to σ, that is, a term that satisfies $\sigma'(\mathrm{d}t) = -\sigma'(-\mathrm{d}t)$, we may conclude either that $\sigma' = 0$ or that there is another reversible term in σ that will cancel with σ' (see remarks following (38.22)).

It will be shown in the next chapter that (40.9) leads directly to constraints on the coefficients $L_{\alpha\beta}$ appearing in the constitutive relations of the reciprocal type,

$$L_{\alpha\beta}(\mathbf{B}) = \eta_\alpha \eta_\beta L_{\beta\alpha}(-\mathbf{B}) \qquad (\alpha, \beta = 1, 2, \dots). \tag{40.12}$$

A special case of these relations was obtained by W. Thomson in 1854. The case $\eta_\alpha = 1$, $\eta_\beta = 1$ was inferred from fluctuation theory by L. Onsager in 1931, and the final generalization to (40.12) was provided by H. B. G. Casimir, also using fluctuation theory, in 1945. Only Thomson considered genuine process thermodynamics in which fluctuations (§ 29) are negligibly small.

40.3. Invariance to choice of reference frame

There are two other important constraints on the constitutive equations still to be described, namely, 'reference frame indifference' and 'material indifference'. These concepts are not of course peculiar to process thermodynamics, being general physical principles of wide applicability. First consider the reference frame.

The Cartesian reference frame in which the constitutive equations (40.1) are specified may be replaced by another Cartesian frame derived from the first by a rigid body rotation or an inversion of the axes, termed an *orthogonal transformation*, and clearly such a change cannot influence the physical nature of the processes involved. Hence the *form* of the constitutive equations must be invariant under orthogonal transformations, a property that can be automatically secured by writing the equations in a three-dimensional tensorial form.

Suppose that X_α, J_α are partitioned into m_0 scalar (subscript s), m_1 vector (subscript v), and m_2 second-order tensor (subscript t) terms as indicated in the following scheme,

$$\begin{array}{lllll} n & m_0 & m_1 & m_2 & \\ (X_\alpha) = & (X_s & \mathbf{X}_v & \mathbf{X}_t) & s = 1, 2, \dots m_0 \\ & & & & v = 1, 2, \dots m_1 \\ (J_\alpha) = & (J_s & \mathbf{J}_v & \mathbf{J}_t) & t = 1, 2, \dots m_2. \end{array}$$

It is important to note that this partitioning must be based on our prior knowledge of the physical nature (that is, scalar, vector, or second-order tensor) of the processes and affinities. The entropy may now be written

$$\sigma = \sigma_0 + \sigma_1 + \sigma_2, \tag{40.13}$$

where

$$\sigma_0 \equiv \sum_{s=1}^{m_0} J_s X_s, \qquad \sigma_1 \equiv \sum_{v=1}^{m_1} \mathbf{J}_v \,.\, \mathbf{X}_v, \qquad \sigma_2 \equiv \sum_{t=1}^{m_2} \mathbf{J}_t : \mathbf{X}_t.$$

It is clear from the tensor product expansion (38.15) that this partitioning is not unique.

The constitutive relation may now be written in the covariant form†

$$\left.\begin{aligned} J_s &= \sum_\sigma L_{s\sigma} X_\sigma + \sum_\nu \mathbf{L}_{s\nu} \cdot \mathbf{X}_\nu + \sum_\tau \mathbf{L}_{s\tau} : \mathbf{X}_\tau \qquad (s, \sigma = 1, 2, \ldots m_0), \\ \mathbf{J}_v &= \sum_\sigma \mathbf{L}_{v\sigma} X_\sigma + \sum_\nu \mathbf{L}_{v\nu} \cdot \mathbf{X}_\nu + \sum_\tau L_{v\tau} : \mathbf{X}_\tau \qquad (v, \nu = 1, 2, \ldots m_1), \\ \mathbf{J}_t &= \sum_\sigma \mathbf{L}_{t\sigma} X_\sigma + \sum_\nu L_{t\nu} \cdot \mathbf{X}_\nu + \sum_\tau L_{t\tau} : \mathbf{X}_\tau \qquad (t, \tau = 1, 2, \ldots m_2), \end{aligned}\right\} \quad (40.14)$$

where the various Ls are transport coefficients of tensional order equal to the sum of the values $(s, v = 0;\ v, \nu = 1;\ t, \tau = 2)$ of the two subscripts. The important point is that (40.14) is not merely (40.1) rewritten in a more complicated form, but incorporates the knowledge provided by physics about the transformation properties of the processes and affinities.

40.4. Rotating frames

To this point we have been considering transformations between coordinate frames not in relative motion; but clearly the notions of reversible and irreversible behaviour must be invariant under both Galilean transformations and transformations to rotating frames. Of course in the latter case additional force fields appear, but these are reversible and cannot contribute to σ. The conclusion follows that when the constitutive relations (40.14) are properly formulated, they can contain only *relative* velocities of translation and rotation.

Now eqn (38.13), for the entropy production rate in a simple fluid, contains an absolute velocity $\mathbf{v}$ in the gradient $\nabla\mathbf{v}$, and while such a gradient is invariant to Galilean transformation—which merely add a uniform velocity to $\mathbf{v}$—it is not invariant to transformations to rotating frames.

Let $\mathscr{P}'$ be a frame whose origin is coincident with $\mathscr{P}$, and which has an angular velocity $\mathbf{\Omega}$ with respect to $\mathscr{P}$ then a typical velocity in $\mathscr{P}'$ is $\mathbf{v}' = \mathbf{v} + \mathbf{\Omega} \times \mathbf{r}$. As $\nabla(\mathbf{\Omega} \times \mathbf{r}) = -\nabla\mathbf{r} \times \mathbf{\Omega} = -\mathbf{1} \times \mathbf{\Omega}$, the dissipation function in $\mathscr{P}'$ is

$$\Phi' = -\tilde{\boldsymbol{\pi}} : \nabla\mathbf{v}' = \Phi + \tilde{\boldsymbol{\pi}} : \mathbf{1} \times \mathbf{\Omega} = \Phi + \tilde{\boldsymbol{\pi}} : [\Omega_x(\mathbf{kj} - \mathbf{jk}) + \Omega_y(\mathbf{ik} - \mathbf{ki}) + \Omega_z(\mathbf{ji} - \mathbf{ij})].$$

The last term vanishes, that is, Φ is invariant, only when the physical properties of the medium permit the inference that $\boldsymbol{\pi}$ is a symmetric tensor. If this is so, it follows from (38.16) that only the symmetric part of $\nabla\mathbf{v}$ need be retained in Φ and the constitutive relations are then $\boldsymbol{\pi}^{\mathrm{a}} = 0$ plus a linear relation between $\boldsymbol{\pi}^{\mathrm{s}}$ and $(\nabla\mathbf{v})^{\mathrm{s}}$.

† The notation here is a trifle deficient for $\mathbf{L}_{s\tau}$ is meant to be quite different from $\mathbf{L}_{v\nu}$ and $\mathbf{L}_{t\sigma}$, even with identical subscripts; similarly for other tensors of the same order.

The problem just described also disappears if, in the formula for Φ, $\mathbf{v}$ can be replaced by the relative velocity $(\mathbf{v}-\boldsymbol{\omega}\times\mathbf{r})$, where $\boldsymbol{\omega}$ is the angular velocity of some reference frame embedded in the fluid. For in this case $\boldsymbol{\omega}' = \boldsymbol{\omega}+\boldsymbol{\Omega}$ and

$$\Phi = -\tilde{\boldsymbol{\pi}}:\nabla(\mathbf{v}-\boldsymbol{\omega}\times\mathbf{r}) = -\tilde{\boldsymbol{\pi}}:\nabla(\mathbf{v}+\boldsymbol{\Omega}\times\mathbf{r}-\boldsymbol{\omega}'\times\mathbf{r}) = -\tilde{\boldsymbol{\pi}}:\nabla(\mathbf{v}'-\boldsymbol{\omega}\times\mathbf{r}) = \Phi'.$$

This possibility is pursued further in § 49.2.

Constitutive relations that possess the essential invariance properties described in this and the previous article are said to be 'frame-indifferent', which should not be confused with the special property of material invariance to be discussed next.

40.5. *Material invariance*

The material properties of the system $\mathscr{P}$ are introduced via the phenomenological tensors appearing in (40.14). In important practical cases there are a great number of scalar coefficients involved in these tensors, so it is important to exploit such material symmetries as may exist to reduce this number.†

If, to take the simplest example, the properties of $\mathscr{P}$ are the same in all directions, $\mathscr{P}$ is said to be *isotropic*, and however $\mathscr{P}$ is rotated in its coordinate frame, the numbers descriptive of its transport properties contained in the phenomenological tensors $L_{s\sigma}$, $\mathbf{L}_{sv}$, $\mathbf{L}_{v\sigma}$, $\mathbf{L}_{s\tau}$, $\mathbf{L}_{vv}$, ... will be unaffected. Equivalently we may hold $\mathscr{P}$ fixed and make orthogonal transformations of the coordinate system. Tensors that have the same components in all coordinate frames arrived at in this way are said to be *isotropic*, and if $\mathscr{P}$ is isotropic only isotropic phenomenological tensors can occur in the constitutive relations. Now these relations are not merely frame-indifferent in *form*, but absolutely invariant under orthogonal transformations.

Fluid systems lack the complicated symmetry structure possible in crystal lattices, and for the purposes of this text, besides isotropic systems, it is sufficient to consider systems with *transverse isotropy*, for which the constitutive relations are invariant only to rotations about one specified direction. The details will be deferred to Chapter 8.

† There are $\mathscr{N} = (m_0+3m_1+9m_2)^2$ scalars in (40.14). For a system of m components, without mass transfer between components, eqns (52.11) and (52.12) give $m_0 = \frac{1}{2}m(m-1)$, $m_1 = \frac{1}{2}m(m+1)$, $m_2 = m$, whence $\mathscr{N} = 4m^2(m+5)^2$. Thus for $m = 1, 2, 3, \ldots, \mathscr{N} = 144, 784, 2304, \ldots$!

7

THE RECIPROCAL RELATIONS

41. Historical survey

41.1. Thomson's hypothesis

THE history of the reciprocal relations of process thermodynamics began in 1854 with W. Thomson's (Lord Kelvin) remarkable work on the thermoelectric phenomenon. The result he obtained for the isothermal flow of heat due to an electric current flowing in the wires of a thermocouple can be deduced from eqns (39.11).

In the absence of changing magnetic fields we may introduce the electric potential φ and write (39.11) as

$$\mathbf{j} = -L_{11}\nabla\varphi + L_{12}\nabla T, \qquad -\mathbf{q}/T = -L_{21}\nabla\varphi + L_{22}\nabla T, \tag{41.1}$$

as scalar phenomenological coefficients suffice in thin wires. The 'absolute thermoelectric power' is defined by $\epsilon = -(\partial\varphi/\partial T)$ at zero electric current, whence by the first of (41.1) $\epsilon = L_{12}/L_{11}$. Now consider isothermal conduction in the thermocouple. Eqns (41.1) yield $\mathbf{q} = -T\mathbf{j}L_{21}/L_{11}$, and provided it is true that

$$L_{12} = L_{21}, \tag{41.2}$$

we arrive at Thomson's (1882, p. 232) famous 'second relation', namely,

$$\mathbf{q} = -T\epsilon\,\mathbf{j}. \tag{41.3}$$

Thomson justified his derivation with the aid of the hypothesis:

A *The electromotive forces produced by inequalities of temperature in a circuit of different metals, and the thermal effects of electric currents circulating in it, are subject to the laws which would follow from the general principles of the dynamical theory of heat if there were no conduction of heat from one part of the circuit to another.*

Just how this oracular assumption leads to (41.2) will be shown shortly. Regarding it Thomson cautiously remarked: 'In adopting this hypothesis, it must be distinctly understood that it is only a hypothesis, and that, however probable it may appear, experimental evidence in the special phenomena of thermo-electricity is quite necessary to prove it.'

41.2. Onsager's hypothesis

In 1931 L. Onsager (1931*a*, *b*) established the reciprocal relations that now bear his name, namely,

$$L_{\alpha\beta} = L_{\beta\alpha}, \tag{41.4}$$

for the phenomenological coefficients occurring in constitutive relations of the special type

$$\dot{x}_\alpha = \sum_\beta L_{\alpha\beta} X_\beta, \tag{41.5}$$

where x_α and X_α are conjugate internal state variables, whose values are subject to the small fluctuations consequent on *not* proceeding to the thermodynamic limit (29.3). The proof of (41.4) (see § 43.2) introduces the concept of microscopic reversibility and also requires that the ensemble variables $x_\alpha(t)$ and $X_\alpha(t)$ be stationary stochastic processes. Concerning Thomson's treatment of thermoelectrical processes, Onsager remarked that ' it is easy to show that Thomson's relation [(41.3)] is no thermodynamic necessity'. However to be able to apply his own theory to such processes, he in effect introduced in place of hypothesis A an even more remarkable hypothesis, which may be stated:

B *The laws for the average regression of small fluctuations are also satisfied by the large and persisting deviations from temporal and spatial uniformity that characterize thermodynamic processes in the thermodynamic limit.*

In other words, the two distinct types of phenomenon are assumed to behave analogously as far as their constitutive relations are concerned. Note that even if hypothesis B is accepted, in order to infer Thomson's relations it is also necessary to show that the fluxes **q** and **j** appearing in (41.1) can be expressed as the time derivatives of thermodynamic state variables. This can often be contrived but requires some ingenuity (de Groot and Mazur 1962).

In his second (1931*b*) paper Onsager generalized his reciprocal relations to allow for the presence of magnetic fields; then in 1941 J. Meixner (1941, 1942, 1943) started to develop the general approach of calculating the entropy production rate σ as a first step.

In (41.5) the affinities X_β are all assumed to have the same parity $\eta_\beta = 1$ under microscopic motion-reversal, and of course the processes $\dot{x}_\alpha$ are all fluxes (negative parity). H. B. G. Casimir (1945) added a final touch of some importance to Onsager's theory by generalizing (41.4) to apply to affinities X_α of arbitrary parity. The reciprocal relations now achieved their final form,

$$L_{\alpha\beta}(\mathbf{B}) = \eta_\alpha \eta_\beta L_{\beta\alpha}(-\mathbf{B}). \tag{41.6}$$

The proof of (41.6) to be described below was devised to overcome the shortcomings of the Onsager–Casimir theory; it is closest in spirit to that of Thomson but makes essential use of the concept of microscopic reversibility, which is also central in the Onsager–Casimir approach.

42. Proof of the reciprocal relations

42.1. Proof from Thomson's hypothesis

In order to appreciate the meaning of Thomson's hypothesis given in § 41, we need to know that the 'laws' referred to are simply the reversible processes of classical thermodynamics in which entropy is conserved. So Thomson is saying that the relation between $\nabla\varphi$ and ∇T in the absence of either $\mathbf{q}$ or $\mathbf{j}$, and the relation between $\mathbf{j}$ and $\mathbf{q}$ in the absence of either $\nabla\varphi$ or ∇T, are reversible in the macroscopic sense. To make this point clear we rewrite (41.1) in the equivalent form

$$\begin{aligned} \mathbf{j} &= -L_{22}^{-1}(L_{11}L_{22}-L_{12}L_{21})\nabla\varphi - \overset{\downarrow}{L_{22}^{-1}}L_{12}\mathbf{q}/T, \\ \nabla T &= \overset{\downarrow}{L_{22}^{-1}}L_{21}\nabla\varphi \qquad\qquad\qquad\qquad - L_{22}^{-1}\mathbf{q}/T, \end{aligned} \tag{42.1}$$

and observe that if $\nabla\varphi$ is absent from the first of (42.1) the remaining relation between the two fluxes is invariant to a reversal of the flow directions, and similarly if $\mathbf{q}$ is absent from the second of (42.1) the remaining relation is likewise invariant. Thomson's conclusion is that the terms indicated by arrows are reversible effects and hence will not contribute to the creation of entropy.

The entropy production rate is given by (see (39.10))

$$\begin{aligned} T\sigma &= -\frac{\mathbf{q}}{T}\cdot\nabla T - \mathbf{j}\cdot\nabla\varphi \\ &= L_{22}^{-1}\left[\left(\frac{\mathbf{q}}{T}\right)^2 + (L_{11}L_{22}-L_{12}L_{21})(\nabla\varphi)^2 + \frac{\mathbf{q}}{T}\cdot\nabla\varphi(\overset{\downarrow}{L}_{12}-\overset{\downarrow}{L}_{21})\right], \end{aligned} \tag{42.2}$$

and as the reversible effects must cancel, we have the reciprocal relations $L_{12} = L_{21}$. This method is not precisely that used by Thomson, but it is equivalent to his procedure. It has been adapted to many other phenomena, but always with that element of uncertainty, which Thomson himself had recognized. We shall justify Thomson's heuristic approach later.

42.2. Irreversibility via the constitutive relations

We return to the discussion initiated in § 40.2. The equations expressing the principle of microscopic reversibility are

$$X_\alpha(\tau) = \eta_\alpha X_\alpha(-\tau), \tag{42.3}$$

$$J_\alpha(\tau) = -\eta_\alpha J_\alpha(-\tau), \qquad \sigma(\tau) = -\sigma(-\tau), \tag{42.4}$$

where

$$\sigma = \sum_\alpha J_\alpha X_\alpha \tag{42.5}$$

is the entropy production rate in a fluid particle system $\mathscr{P}$. Here τ is really an *infinitesimal* microscopic time interval, although we shall omit the adjective 'infinitesimal'; also the reversal $\tau \to -\tau$ implies a change of sign for any magnetic field **B** present.

Now arrange conditions on the boundary of $\mathscr{P}$—or on more distant boundaries if necessary—so that the independent affinities X_α continue to satisfy (42.3) over a macroscopic time interval dt, that is,

$$X_\alpha(\mathrm{d}t) = \eta_\alpha X_\alpha(-\mathrm{d}t). \tag{42.6}$$

Thus, the affinities are controlled so that their microscopic behaviour after motion-reversal is maintained over an infinitesimal macroscopic time interval. In general, the process will not have a definite parity over the same time interval, for

$$J_\alpha(\mathrm{d}t) = \sum_\beta L_{\alpha\beta}(\mathbf{B}) X_\beta(\mathrm{d}t) \tag{42.7}$$

becomes

$$J_\alpha(-\mathrm{d}t) = \sum_\beta L_{\alpha\beta}(-\mathbf{B}) \eta_\beta X_\beta(\mathrm{d}t)$$

after the reversal. Here we have used the fact that $L_{\alpha\beta}$ depends on the thermodynamic state and so must have even parity provided **B** changes sign. As X_α and J_α have opposite parities on the microscopic time scale τ, and as the X_α may be chosen freely from each term in the bilinear form for σ, we can make a choice in which all $\eta_\beta = 1$ or all $\eta_\beta = -1$. With either choice $J_\alpha(\mathrm{d}t)$ does have a definite parity, which is in fact the same as that chosen for the affinities. It follows from (42.5) that for these two special cases $\sigma(\mathrm{d}t) = \sigma(-\mathrm{d}t)$. As explained in § 40.2 this result is true generally, regardless of the parities of the individual affinities.

It is clear that the constitutive relations (42.7) introduce irreversibility into thermodynamics by preventing the microscopic parities of the processes persisting for small macroscopic times. See, for example, eqn (39.11) in which the affinities have even parity and the processes are both fluxes with odd parity, or eqn (39.8) in which the affinities have odd parity and the processes even parity.

However, not all constitutive relations need be irreversible. For example, the thermoelectric phenomenon in a plasma can be described approximately by the equations (cf. (41.1) and (41.2))

$$\mathbf{E} = \eta\mathbf{j} - \delta\nabla T, \qquad \mathbf{q} = -T\delta\mathbf{j} - \kappa\nabla T,$$

and from kinetic theory (see, for example, Braginskii (1965)) it can be shown that in the absence of plasma turbulence $\delta \simeq 0{\cdot}7\,k/e$, $\eta \simeq m_e/(ne^2\tau_e)$, $\kappa \simeq 1{\cdot}6 p_e\tau_e k/m_e$, where the subscript e denotes electron properties, n is the electron number density, and τ_e is the appropriate electron–ion collision time. Clearly by choosing plasma conditions such that τ_e is either very large or very small we may obtain one or other of the reversible constitutive relations,

$$\mathbf{E} = -\delta\nabla T, \qquad \mathbf{q} = -T\delta\mathbf{j} \qquad (\delta \sim 0{\cdot}7k/e).$$

Notice that δ does not depend on particle collisions.

42.3. *Reciprocal relations derived from the condition of irreversibility*

In § 40.2 we showed that because of its very definition, σ must satisfy the important *irreversibility condition*

$$\sigma(\mathrm{d}t) = \sigma(-\mathrm{d}t), \tag{42.8}$$

which we shall now use to derive the reciprocal relations. We start with (42.6) written as $X_\alpha(-\mathrm{d}t) = \eta_\alpha X_\alpha(\mathrm{d}t)$, and the quadratic forms for σ before and after motion-reversal,

$$\sigma(\mathrm{d}t) = \tfrac{1}{2}\sum_{\alpha,\beta} X_\alpha(\mathrm{d}t)[L_{\alpha\beta}(\mathbf{B}) + L_{\beta\alpha}(\mathbf{B})]X_\beta(\mathrm{d}t),$$

$$\sigma(-\mathrm{d}t) = \tfrac{1}{2}\sum_{\alpha,\beta} X_\alpha(-\mathrm{d}t)[L_{\alpha\beta}(-\mathbf{B}) + L_{\beta\alpha}(-\mathbf{B})]X_\beta(-\mathrm{d}t).$$

As the affinities are independent, it follows from (42.8) that

$$L_{\alpha\beta}(\mathbf{B}) + L_{\beta\alpha}(\mathbf{B}) = \eta_\alpha\eta_\beta[L_{\alpha\beta}(-\mathbf{B}) + L_{\beta\alpha}(-\mathbf{B})]. \tag{42.9}$$

With $\alpha = \beta$ this shows that

$$L_{\alpha\alpha}(\mathbf{B}) = L_{\alpha\alpha}(-\mathbf{B}). \tag{42.10}$$

Now choose *specific* values for α and β in (42.7) and (42.9), and set all the affinities equal to zero save X_α, X_β, so that

$$J_\alpha = L_{\alpha\alpha}X_\alpha + L_{\alpha\beta}X_\beta, \qquad J_\beta = L_{\beta\alpha}X_\alpha + L_{\beta\beta}X_\beta. \tag{42.11}$$

It is the independence of the affinities that permits us to make this choice, and had we chosen J_α and X_α to be the non-zero independent affinities then in the place of (42.11) we would have

$$X_\alpha = L'_{\alpha\alpha}J_\alpha + L'_{\alpha\beta}X_\beta, \qquad J_\beta = L'_{\beta\alpha}J_\alpha + L'_{\beta\beta}X_\beta, \tag{42.12}$$

where, since (42.11) and (42.12) refer to the same physical phenomena,

$$\begin{gathered} L'_{\alpha\alpha} = L^{-1}_{\alpha\alpha}, \qquad L'_{\alpha\beta} = -L^{-1}_{\alpha\alpha}L_{\alpha\beta}, \qquad L'_{\beta\alpha} = L^{-1}_{\alpha\alpha}L_{\beta\alpha}, \\ L'_{\beta\beta} = L^{-1}_{\alpha\alpha}(L_{\beta\beta}L_{\alpha\alpha} - L_{\beta\alpha}L_{\alpha\beta}). \end{gathered} \tag{42.13}$$

The general relation in (42.9) does not depend on the particular choice of independent variables, so that with the choice in (42.12)

$$L'_{\alpha\beta}(\mathbf{B}) + L'_{\beta\alpha}(\mathbf{B}) = \eta'_\alpha\eta_\beta[L'_{\alpha\beta}(-\mathbf{B}) + L'_{\beta\alpha}(-\mathbf{B})],$$

where η'_α is the parity of the 'affinity' J_α, that is, by eqn $(42.4)_1$ $\eta'_\alpha = -\eta_\alpha$. Substitution of the values given in (42.13) into this result, plus use of (42.10) to remove the common factor $L^{-1}_{\alpha\alpha}$, which has been tacitly assumed not to be zero, leads to

$$L_{\beta\alpha}(\mathbf{B}) - L_{\alpha\beta}(\mathbf{B}) = \eta_\alpha\eta_\beta[L_{\alpha\beta}(-\mathbf{B}) - L_{\beta\alpha}(-\mathbf{B})]. \tag{42.14}$$

The reciprocal relations, namely,

$$L_{\alpha\beta}(\mathbf{B}) = \eta_\alpha\eta_\beta L_{\beta\alpha}(-\mathbf{B}) \qquad (\alpha, \beta, = 1, 2, ...) \tag{42.15}$$

now follow by subtracting (42.14) from (42.9), and noting that the particular choice made for α and β is not restrictive.

An important corollary of the reciprocal relations is that coupling between affinities is a *reversible* process, that is, makes no contribution to σ provided either

$$\eta_\alpha\eta_\beta = 1 \qquad \text{and } L_{\alpha\beta}(\mathbf{B}) = -L_{\alpha\beta}(-\mathbf{B})$$

or (42.16)

$$\eta_\alpha\eta_\beta = -1 \quad \text{and } L_{\alpha\beta}(\mathbf{B}) = L_{\alpha\beta}(-\mathbf{B}).$$

This follows from (42.8) and (42.15) which lead to

$$\sigma = \tfrac{1}{2}\sum_{\alpha,\beta} X_\alpha[L_{\alpha\beta}(\mathbf{B}) + \eta_\alpha\eta_\beta L_{\alpha\beta}(-\mathbf{B})]X_\beta$$

and

$$L_{\alpha\beta}(\mathbf{B}) + \eta_\alpha\eta_\beta L_{\alpha\beta}(-\mathbf{B}) = L_{\beta\alpha}(\mathbf{B}) + \eta_\alpha\eta_\beta L_{\beta\alpha}(-\mathbf{B}).$$

42.4. Discussion

Recapitulating the basis of the reciprocal relations, we note that there are four essential ingredients: (1) the independence of the affinities; (2) their possession of *definite* motion-reversal parities, which, owing to (1), may be maintained over infinitesimal macroscopic time intervals; (3) linear relations between processes and affinities; and (4) entropy production due only to irreversible effects, which are identified as those that destroy the microscopic motion-reversal parities of the processes after sufficient time. The constraints imposed by (1) and (2) will receive attention in § 44.

In the light of the derivation of the reciprocal relations given in § 42.3, it is easy to justify Thomson's procedure. Hypothesis A in § 41.1 is in fact equivalent to (4), and the final term in (42.2) must be zero as it has odd parity (that is, is reversible) over both microscopic and macroscopic time scales. What Thomson lacked to give his procedure precision were the ideas of microscopic reversibility and of two distinct time scales. This is not surprising for after all the atomic theory was scarcely established when Thomson made his great contribution.

From a purely continuum stand-point the theory has the defect of making reference to microscopic properties and particles, and it might be worthwhile indicating how such references may be omitted.† We only require a means of assigning the parities η_α that makes no reference to the particulate nature of the fluid. This is easily done by considering the conditions under which the continuum conservation equations for mass, momentum, and energy are reversible in time. The parities that must be assigned to the dependent variables in these equations in order to achieve this invariance are identical with those arrived at by considering the dependence of these variables on the microscopic velocities. They may be thought of as defining a reference condition, enabling the irreversible effects introduced via the constitutive relations to be identified. The thermodynamic affinities, being independent, retain their parities, and irreversibility manifests itself in the destruction of the parities of the processes; the theory is then developed as above.

A supposedly more general form of these relations for systems rotating with an angular velocity $\mathbf{\Omega}$ is often quoted in the literature. This is

$$L_{\alpha\beta}(\mathbf{B}, \mathbf{\Omega}) = \eta_\alpha \eta_\beta L_{\beta\alpha}(-\mathbf{B}, -\mathbf{\Omega}), \tag{42.17}$$

the changes in the sign of $\mathbf{\Omega}$ being required to ensure that the particles retrace their former paths after motion-reversal. But according to the principle of frame-indifference (see § 40.4) $L_{\alpha\beta}$ cannot depend on $\mathbf{\Omega}$, which unlike $\mathbf{B}$ is not characteristic of the *intrinsic state* of a fluid particle. Thus the appearance of $\mathbf{\Omega}$ in (42.17) is redundant.

† Axiomatic treatments of continuum mechanics, sometimes misnamed 'rational' treatments, have the virtue of setting the subject in a logical deductive frame and of eliminating relevant, if not vital, physical concepts drawn from wider fields. The net effect is to stifle intuitive reasoning—the frame becomes a straightjacket—and mathematical clarity is won at the expense of physical understanding.

43. The reciprocal relations of fluctuation theory

43.1. Onsager coordinates

As described in § 29, thermodynamic systems containing a finite number N of particles are subject to very small fluctuations with a relative mean square deviation of order $N^{-\frac{1}{2}}$. Process thermodynamics is mainly concerned with irreversible phenomena in systems sufficiently large for these fluctuations to be negligible, and unless we find Onsager's hypothesis (§ 41.2) compelling, the reciprocal relations that can be derived for the regression of fluctuations towards equilibrium must be considered as being merely analogous to those derived in § 42. Nevertheless, Onsager's contribution has great importance† and a brief account will be given. Certainly what *is* relevant to process thermodynamics is Onsager's use of microscopic reversibility in his derivation of the reciprocal relations.

We shall consider an isolated system, whose internal state is characterized by a set of n extensive variables $\hat{X}_1, \hat{X}_2, \ldots \hat{X}_n$ as described in §28.3. Just what these so-called Onsager coordinates represent is not important in the theory, but we may think of them to be perturbations in the energy, volume, and particle numbers of a large number of quite small sub-systems into which the system has been partitioned. The mean value of $\hat{X}$ taken either over an ensemble of systems or for one system over a long time interval will be taken to be zero, consequently the inflexions ^ can be omitted without ambiguity, both for the fluctuating coordinates and for their conjugate intensive variables $x_1, x_2, \ldots x_n$. By eqn (28.10)

$$\hat{S}(t) = S - \tfrac{1}{2}\mathbf{x}(t)\,.\,\mathbf{X}(t), \tag{43.1}$$

where, by (28.19), $\hat{S}$ fluctuates about a mean $\frac{1}{2}nk$ below the maximum possible value S. The probability that a given system lies in $\mathrm{d}\mathbf{X}$ is

$$W\,\mathrm{d}\mathbf{X} = A\,\mathrm{e}^{(\hat{S}-S)/k}\,\mathrm{d}\mathbf{X} = A\,\mathrm{e}^{-\mathbf{x}.\mathbf{X}/2k}\,\mathrm{d}\mathbf{X} \tag{43.2}$$

Eqn (43.1) may be written

$$\hat{S} = S - \tfrac{1}{2}\boldsymbol{\beta}:\mathbf{X}\mathbf{X}, \tag{43.3}$$

where

$$\mathbf{x} = \boldsymbol{\beta}\,.\,\mathbf{X}, \tag{43.4}$$

are the linearized state equations (28.13). Hence

$$\dot{\hat{S}} = -\mathbf{x}\,.\,\dot{\mathbf{X}} = -\boldsymbol{\beta}:\mathbf{X}\dot{\mathbf{X}} = -\dot{\mathbf{x}}\,.\,\mathbf{X}. \tag{43.5}$$

As $\langle \hat{S} \rangle$ is constant, $\langle \dot{\hat{S}} \rangle$ is zero and, on the other hand, over a very long time a single system will experience almost all values of $\hat{S}$. Thus a special device is

† Particularly for the fluctuation dissipation theorem; see final paragraph of § 70.2.

necessary to introduce irreversibility and the associated phenomenological equations.

Any system well displaced from the equilibrium state $\mathbf{X} = 0$, that is, with $|\hat{S} - S| \gg nk$, will have a higher probability of moving towards this state than away from it, for according to the probability curve W in Fig. 12, the system must spend most of its time near $\mathbf{X} = 0$. Certainly the *average* change over a family of systems that have the same initial state $\mathbf{X}'(t)$ must be towards equilibrium. This irreversible regression to the mean can be represented by the phenomenological equation

$$\bar{\dot{x}}'_\alpha(t+\tau) = \sum_\beta L_{\alpha\beta}(\mathbf{B})\bar{X}'_\beta(t+\tau) \qquad (\alpha = 1, 2, \ldots), \tag{43.6}$$

where the averages $\bar{\dot{x}}'_\alpha$, $\bar{X}'_\alpha$ are taken over those members of the ensemble that passed through the initial state $\mathbf{X}'(t)$, τ seconds earlier.

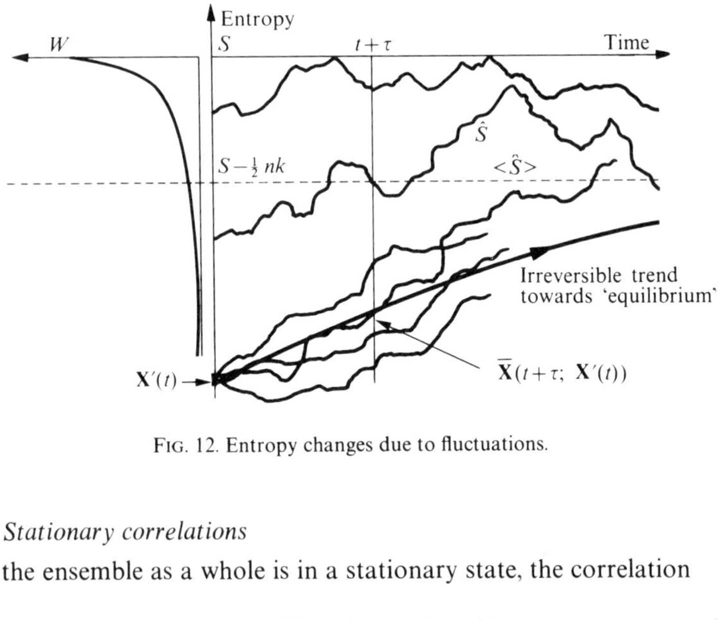

FIG. 12. Entropy changes due to fluctuations.

43.2. *Stationary correlations*

As the ensemble as a whole is in a stationary state, the correlation

$$\varphi_{\alpha\beta}(\tau\,; \mathbf{B}) \equiv \langle x_\alpha(t) x_\beta(t+\tau)\rangle \tag{43.7}$$

is independent of t. Let η_α, η_β denote the motion-reversal parities of x_α and x_β, which by (43.4) are the same as the parities of X_α and X_β, then combining stationarity with the principle of microscopic reversibility we have

$$\varphi_{\alpha\beta}(\tau\,; \mathbf{B}) = \eta_\alpha\eta_\beta\langle x_\alpha(-t)x_\beta(-t-\tau)\rangle = \eta_\alpha\eta_\beta\langle x_\alpha(t+\tau)x_\beta(t)\rangle = \eta_\alpha\eta_\beta\varphi_{\beta\alpha}(\tau\,; -\mathbf{B}).$$

Differentiating with respect to τ gives

$$\dot{\varphi}_{\alpha\beta}(\tau;\mathbf{B}) = \eta_\alpha\eta_\beta\dot{\varphi}_{\beta\alpha}(\tau;-\mathbf{B}) \tag{43.8}$$

or

$$\overline{\langle x'_\alpha(t)\dot{x}'_\beta(t+\tau)\rangle}' = \eta_\alpha\eta_\beta\overline{\langle x'_\beta(t)\dot{x}'_\alpha(t+\tau)\rangle}',$$

where we have first averaged over the subset of ensemble members that have the initial state $\mathbf{X}'(t)$ in common and then over all such initial states.

As $\overline{x'_\alpha(t)} = x'_\alpha(t)$, we can apply (43.6) to (43.8) and find

$$\overline{\langle x'_\alpha(t)\sum_\gamma L_{\beta\gamma}(\mathbf{B})X'_\gamma(t+\tau)\rangle}' = \eta_\alpha\eta_\beta\overline{\langle x'_\beta(t)\sum_\gamma L_{\alpha\gamma}(-\mathbf{B})X'_\gamma(t+\tau)\rangle}',$$

or

$$\sum_\gamma L_{\beta\gamma}(\mathbf{B})\langle x_\alpha(t)X_\gamma(t+\tau)\rangle = \eta_\alpha\eta_\beta\sum_\gamma L_{\alpha\gamma}(-\mathbf{B})\langle x_\beta(t)X_\gamma(t+\tau)\rangle. \tag{43.9}$$

From (28.17)

$$\langle x_\alpha X_\gamma\rangle = k\delta_{\alpha\gamma} \qquad \begin{pmatrix}\delta_{\alpha\gamma} = 1 \text{ if } \alpha = \gamma\\ \delta_{\alpha\gamma} = 0 \text{ if } \alpha \neq \gamma\end{pmatrix},$$

consequently the limit $\tau \to 0$ in (43.9) gives

$$kL_{\alpha\beta}(\mathbf{B}) = k\eta_\alpha\eta_\beta L_{\beta\alpha}(-\mathbf{B}), \tag{43.10}$$

and the reciprocal relations follow on division by k. The appearance of k, a mere $1{\cdot}38\times 10^{-23}\ \mathrm{J\,K^{-1}}$, in this final step is indicative of the fact that the magnitudes involved are microscopic rather than macroscopic, which is the main weakness of the theory.

Actually the averaging process adopted in (43.6) has contributed nothing to the final result, as may be seen by replacing (43.6) by

$$J_\alpha = \dot{x}_\alpha = \sum_\beta L_{\alpha\beta}X_\beta,$$

where (43.11)

$$X_\alpha = -\partial(\Delta S)/\partial x_\alpha \qquad (\Delta S = \hat{S}-S),$$

and using this directly in (43.8).

If t is replaced by $(t-\tau)$ in (43.7), $\varphi_{\alpha\beta}$ is unchanged, that is, $\varphi_{\alpha\beta}(\tau,\mathbf{B}) = \varphi_{\beta\alpha}(-\tau,\mathbf{B})$, so that

$$\dot{\varphi}_{\alpha\beta}(\tau,\mathbf{B}) = -\dot{\varphi}_{\beta\alpha}(-\tau,\mathbf{B}),$$

which appears to be in conflict with (43.8) in the limit $\tau \to 0$ unless

(a) $\begin{cases}\text{either} & \eta_\alpha\eta_\beta = 1 \quad \text{and} \quad \varphi_{\alpha\beta} \text{ is odd in } \mathbf{B}\\ \text{or} & \eta_\alpha\eta_\beta = -1 \quad \text{and} \quad \varphi_{\alpha\beta} \text{ is even in } \mathbf{B}\end{cases}$

or

(b) $\dot{\varphi}_{\alpha\beta}$ is either zero or discontinuous at $\tau = 0$.

Comparing (a) with (42.16) we see that it corresponds to a reversible process, whereas (b) is characteristic of an irreversible process. Physical continuity requires $\dot{\varphi}_{\alpha\beta}$ to vanish at $\tau = 0$, but to accept this would reduce (43.10), in which τ is zero, to $0 = 0$. However, both statements in (b) may be reconciled by taking $\dot{\varphi}_{\alpha\beta}(0)$ to be zero on a *micro*scopic time scale, but discontinuous on a *macro*scopic time scale. In the notation used to distinguish (42.3) and (42.6), we may write

$$\dot{\varphi}_{\alpha\beta}(\tau) = 0 \quad \text{at } \tau = 0,$$

$$\dot{\varphi}_{\alpha\beta}(\mathrm{d}t) \neq 0 \quad \text{at } \mathrm{d}t = 0,$$

and the limit $\tau \to 0$ of (43.9) should be replaced by $\mathrm{d}t \to 0$ and interpreted macroscopically as indicated in Fig. 13.

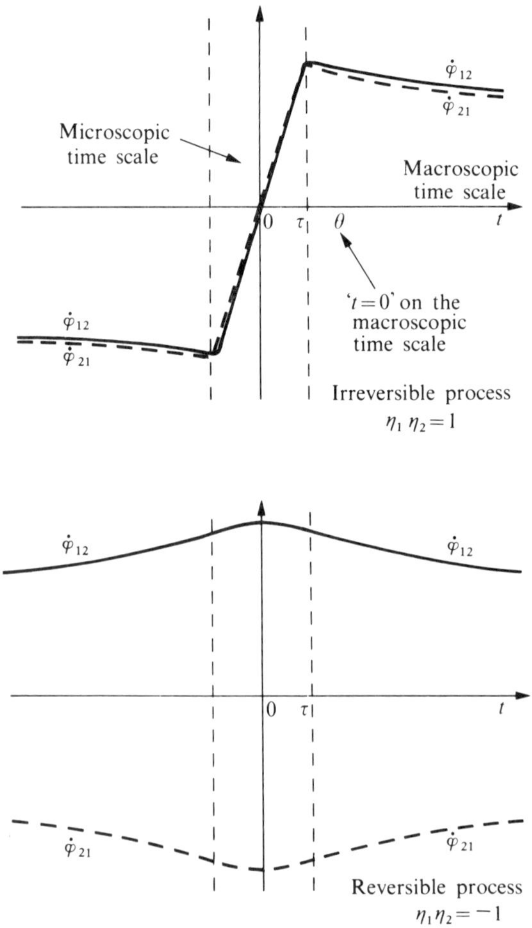

FIG. 13. The correlation function (no magnetic field).

43.3. The proper choice of affinities and processes

A practical drawback of Onsager's theory is that it requires the process variable J_α and the affinity X_α to be expressed in the particular form given in (43.11). In many important cases the affinities and fluxes either cannot be cast into this form or can be only with considerable effort and ingenuity. The upshot is that the rule contained in (43.11) is seldom used in practice, which often results in a vagueness or even a silence in the literature as to whether a given choice is valid or not.

The question of choice is important for it is in fact insufficient merely to ensure that the entropy production rate has the usual bilinear form,

$$\sigma = \sum_\alpha J_\alpha X_\alpha = \mathbf{J}\,.\,\mathbf{X}, \tag{43.12}$$

a point that may be established (Truesdell 1969, p. 127) as follows. Suppose that the invariance of the bilinear form is the only constraint on the choice of **J** and **X**, then we may adopt the linear transformation

$$\mathbf{J}' = \mathbf{J}+\mathsf{A}\,.\,\mathbf{X}, \qquad \mathbf{X}' = \mathbf{X} \qquad (\mathsf{A} = -\tilde{\mathsf{A}}), \tag{43.13}$$

for as $\mathbf{X}\,.\,\mathsf{A}\,.\,\mathbf{X}$ is zero,

$$\sigma = \mathbf{J}\,.\,\mathbf{X} = (\mathbf{J}+\mathbf{X}\,.\,\mathsf{A})\,.\,\mathbf{X} = \mathbf{J}'\,.\,\mathbf{X}'. \tag{43.14}$$

Then with constitutive equations $\mathbf{J} = \mathsf{L}\,.\,\mathbf{X}$, $\mathbf{J}' = \mathsf{L}'\,.\,\mathbf{X}'$, we find that $\mathsf{L}' = \mathsf{L}+\mathsf{A}$. In the absence of magnetic fields and with all components of **X** having even parity, the reciprocal relations are $\mathsf{L} = \tilde{\mathsf{L}}$, but clearly $\mathsf{L}' \neq \tilde{\mathsf{L}}'$. Thus while the new processes and affinities leave σ invariant, they are not a correct choice for the validity of the reciprocal relations.

The transformation in (43.13) is in fact excluded by Onsager's theory, for by (43.11) $J'_\alpha = \dot{x}_\alpha + \sum_\beta A_{\alpha\beta} X_\beta$, which is not a time derivative. This transformation is also excluded by the theory given in § 42.3, which requires the variables J_α, X_α to possess motion-reversal parities, whereas $\dot{x}_\alpha$ and X_α have opposite parities so that J'_α has none.† We shall return to this point in § 44.1.

However, the result given in (43.13) plus the difficulty of applying Onsager's theory to specific cases have led some authors writing on continuum mechanics to raise strong doubts as to the general validity of the theory itself, despite the overwhelming experimental evidence in support of reciprocal relations. Furthermore, the use of statistical arguments outside of the usual ambit of macroscopic thermodynamics, to establish the reciprocal relations as a general result of continuum mechanics, has been fairly criticized. To avoid the difficulty it is sometimes proposed that they should be given the status of a *fourth* law of thermodynamics, and so made an induction from experiment

† Indeed plain physical common sense, which requires our symbols to represent at least possibly real things, would exclude (43.13).

instead of a deduction from concepts of no standing in continuum theory (see footnote to p. 168). It was to circumvent these problems that the author developed the theory given in § 42.3.

44. Transformation properties of the reciprocal relations

44.1. Invariance of reciprocal relations

We return now to the theory given in § 42.3. The reciprocal relations (42.15) may be written in the matrix form

$$\eta L(\mathbf{B}) = \tilde{L}(-\mathbf{B})\eta, \tag{44.1}$$

where η is a diagonal matrix with elements $\eta_1, \eta_2, \ldots \eta_n$ along the leading diagonal. The constitutive equations are

$$J = L(\mathbf{B})X, \tag{44.2}$$

and it will be assumed that the components of X are independent. The most general linear transformation of the affinities and processes is of the form

$$J' = \mathscr{A}(\mathbf{B})J + \mathscr{B}(\mathbf{B})X, \qquad X' = \mathscr{C}(\mathbf{B})J + \mathscr{D}(\mathbf{B})X, \tag{44.3}$$

where $\mathscr{A}$, $\mathscr{B}$, $\mathscr{C}$, $\mathscr{D}$ are matrices. Our objective here is to show that L' in the transformed version of (44.2), namely $J' = L'X'$, also satisfies (44.1), that is,

$$\eta L'(\mathbf{B}) = \tilde{L}'(-B)\eta, \tag{44.4}$$

provided certain conditions are satisfied.

Now an essential feature in the proof of (44.1) is that the affinities and processes possess motion-reversal parities. Thus for (44.4) to hold the components of J' and X' must also have definite parities, a requirement that imposes restrictions on the components of $\mathscr{A}$, $\mathscr{B}$, $\mathscr{C}$, and $\mathscr{D}$. The situation is clearly illustrated by the case of two components, with parities $\eta_1 = 1$, $\eta_2 = -1$ for X_1 and X_2. For X'_1 and X'_2 to have definite parities, the second of eqn (44.3) can be no more general in form than

$$\begin{bmatrix} X'_1 \\ X'_2 \end{bmatrix} = \begin{bmatrix} c^-_{11} & c^+_{12} \\ c^+_{21} & c^-_{22} \end{bmatrix}\begin{bmatrix} J_1 \\ J_2 \end{bmatrix} + \begin{bmatrix} d^+_{11} & d^-_{12} \\ d^-_{21} & d^+_{22} \end{bmatrix}\begin{bmatrix} X_1 \\ X_2 \end{bmatrix} \quad \begin{matrix} (c^\pm_{\alpha\beta} = \frac{1}{2}[c_{\alpha\beta}(\mathbf{B}) \pm c_{\alpha\beta}(-\mathbf{B})]) \\ (d^\pm_{\alpha\beta} = \frac{1}{2}[d_{\alpha\beta}(\mathbf{B}) \pm d_{\alpha\beta}(-\mathbf{B})]). \end{matrix}$$

Under motion-reversal $(J_1, J_2) \to (-J_1, J_2)$, $(X_1, X_2) \to (X_1, -X_2)$, $c^\pm_{\alpha\beta} \to \pm c^\pm_{\alpha\beta}$, and $d^\pm_{\alpha\beta} \to \pm d^\pm_{\alpha\beta}$, so the above transformation yields parities $1, -1$ for X'_1 and X'_2. With $C = [c_{\alpha\beta}]$, $D = [d_{\alpha\beta}]$, the transformation may be expressed

$$X' = \tfrac{1}{2}[C(\mathbf{B}) - \eta C(-\mathbf{B})\eta]J + \tfrac{1}{2}[D(\mathbf{B}) + \eta D(-\mathbf{B})\eta]X, \tag{44.5}$$

in which form it holds for any number of components, as the operations indicated produce coefficients $c^-_{\alpha\beta}$, $d^+_{\alpha\beta}$ if $\eta_\alpha = \eta_\beta$ and $c^+_{\alpha\beta}$, $d^-_{\alpha\beta}$ if $\eta_\alpha \neq \eta_\beta$.

From eqns (44.5) and $(44.3)_2$, $\mathscr{C}(\mathbf{B}) = \frac{1}{2}[C(\mathbf{B})-\eta C(-\mathbf{B})\eta]$, $\mathscr{D}(\mathbf{B}) = \frac{1}{2}[D(\mathbf{B})+\eta D(-\mathbf{B})\eta]$, and as $\eta^2 = U$, where U is the unit matrix, it follows that $\mathscr{C}$ and $\mathscr{D}$ satisfy the relations

$$\eta\mathscr{C}(\mathbf{B}) = -\mathscr{C}(-\mathbf{B})\eta, \qquad \eta\mathscr{D}(\mathbf{B}) = \mathscr{D}(-\mathbf{B})\eta. \tag{44.6}$$

A similar discussion applied to eqn $(44.3)_1$ leads to

$$\eta\mathscr{A}(\mathbf{B}) = \mathscr{A}(-\mathbf{B})\eta, \qquad \eta\mathscr{B}(\mathbf{B}) = -\mathscr{B}(-\mathbf{B})\eta. \tag{44.7}$$

It is convenient to take the transformations (44.3) in two steps,

$$J' = J^1, \qquad J^1 = \mathscr{A}J+\mathscr{B}X,$$

$$X' = \mathscr{S}J^1+X^1, \qquad X^1 = \mathscr{T}X,$$

with $\mathscr{C} = \mathscr{S}\mathscr{A}$ and $\mathscr{D} = \mathscr{S}\mathscr{B}+\mathscr{T}$. Equivalently, our objective will be reached if we can show that (44.1) implies (44.4) for transformations (44.3) in which we put $\mathscr{B}$ and $\mathscr{C}$ equal to zero in turn.

First consider the case $\mathscr{C} = 0$, when

$$J'X' = (J\tilde{\mathscr{A}}+X\tilde{\mathscr{B}})\mathscr{D}X = J\tilde{\mathscr{A}}\mathscr{D}X+X\tilde{\mathscr{B}}\mathscr{D}X.$$

As $LX = J$, the bilinear form is preserved if

$$\tilde{\mathscr{A}}\mathscr{D} = U+aL, \qquad \tilde{\mathscr{B}}\mathscr{D} = b, \tag{44.8}$$

where a and b are antisymmetric matrices (cf. (43.14)). Now as $\mathscr{D} = \tilde{\mathscr{A}}^{-1}+$ $+\tilde{\mathscr{A}}^{-1}aL$, $X' = \mathscr{D}X = \tilde{\mathscr{A}}^{-1}aJ+\tilde{\mathscr{A}}^{-1}X$, so that $\mathscr{C}' \equiv \tilde{\mathscr{A}}^{-1}a$ must have the same properties as $\mathscr{C}$ in $(44.3)_2$. Hence by (44.6) and (44.7)

$$\begin{aligned}\eta a(\mathbf{B}) &= \eta\tilde{\mathscr{A}}(\mathbf{B})\mathscr{C}'(\mathbf{B}) = \tilde{\mathscr{A}}(-\mathbf{B})\eta\mathscr{C}'(\mathbf{B}) = -\tilde{\mathscr{A}}(-\mathbf{B})\mathscr{C}'(-\mathbf{B})\eta\\ &= -a(-\mathbf{B})\eta = \tilde{a}(-\mathbf{B})\eta\cdot\end{aligned}$$

that is,

$$\eta a(\mathbf{B}) = \tilde{a}(-\mathbf{B})\eta. \tag{44.9}$$

A similar calculation applied to $b = \tilde{\mathscr{B}}\mathscr{D}$ leads to

$$\eta b(\mathbf{B}) = \tilde{b}(-\mathbf{B})\eta. \tag{44.10}$$

Now $J' = \mathscr{A}J+\mathscr{B}X = (\mathscr{A}L+\mathscr{B})\mathscr{D}^{-1}X$, so $L' = \mathscr{A}L\mathscr{D}^{-1}+\mathscr{B}\mathscr{D}^{-1}$. By eqns (44.8)

$$\tilde{\mathscr{A}}\mathscr{D}L^{-1} = L^{-1}+a, \qquad \tilde{\mathscr{D}}\mathscr{B} = -b,$$

so

$$L\mathscr{D}^{-1}\tilde{\mathscr{A}}^{-1} = (L^{-1}+a)^{-1}, \qquad \mathscr{B} = -\tilde{\mathscr{D}}^{-1}b.$$

Therefore

$$L' = \mathscr{A}(L^{-1}+a)^{-1}\tilde{\mathscr{A}} - \tilde{\mathscr{D}}^{-1}b\mathscr{D}^{-1}. \tag{44.11}$$

It easily follows from (44.1), (44.6), (44.7), and $\eta^2 = U$ that

$$\eta L^{-1}(\mathbf{B})\eta = \tilde{L}^{-1}(-\mathbf{B}), \qquad \tilde{\mathscr{A}}(\mathbf{B})\eta = \eta\tilde{\mathscr{A}}(-\mathbf{B}), \qquad \mathscr{D}^{-1}(\mathbf{B})\eta = \eta\mathscr{D}^{-1}(-\mathbf{B});$$

hence from (44.11)

$$\begin{aligned}\eta L'(\mathbf{B}) = \{&\mathscr{A}(-\mathbf{B})\eta[L^{-1}(\mathbf{B})+a(\mathbf{B})]^{-1}\eta\tilde{\mathscr{A}}(-\mathbf{B}) \\ &-\tilde{\mathscr{D}}^{-1}(-\mathbf{B})\eta b(\mathbf{B})\eta\mathscr{D}^{-1}(-\mathbf{B})\}\eta.\end{aligned} \tag{44.12}$$

As

$$\eta[L^{-1}(\mathbf{B})+a(\mathbf{B})]^{-1}\eta = [\eta L^{-1}(\mathbf{B})\eta+\eta a(\mathbf{B})\eta]^{-1} = [\tilde{L}^{-1}(-\mathbf{B})+\tilde{a}(-\mathbf{B})]^{-1}$$

and

$$\eta b(\mathbf{B})\eta = \tilde{b}(-\mathbf{B}),$$

it follows from (44.11) that the right-hand side of (44.12) is $\tilde{L}'(-\mathbf{B})\eta$, and (44.4) is thus established for the special case $\mathscr{C} = 0$.

Secondly suppose that $\mathscr{B} = 0$ in (44.3). Then the equations of the special case just considered, namely,

$$X' = \mathscr{D}X, \qquad J' = \mathscr{A}J+\mathscr{B}X, \qquad J = LX$$

are replaced by

$$J' = \mathscr{A}J, \qquad X' = \mathscr{D}X+\mathscr{C}J, \qquad X = L^{-1}J,$$

and it is clear that with obvious interchanges of symbols, including the replacement of L by L^{-1}, we can duplicate the proof given and again deduce eqn (44.4) from (44.1).

We have now proved:

The reciprocal relations are invariant to any linear transformations of the affinities and processes that preserve the bilinear form for the entropy production, provided the new affinities and processes possess motion-reversal parities.

The question of the 'proper choice' of the affinities and processes is thus completely resolved by ensuring that in all choices motion-reversal parities exist. The reason why (43.13) does not offer a valid choice is now clear: for the special case there considered $\eta_\alpha = 1, \alpha = 1, 2, \ldots n$, so all J_α have opposite parities from all X_α, which means that some J'_α will have no parities at all unless $\mathbf{A} = 0$. Finally, eqns (43.11) of Onsager's theory play the role (in a round-about way) of ensuring that J_α and X_α always possess parities, this property being assumed tacitly in the theory for the Onsager coordinates x_α. Of course, the theory given in this article obviates the often difficult task of uncovering the appropriate coordinates x_α in any particular case.

44.2. Dependent processes and affinities

The processes $\mathbf{J}_i$ in (53.20) are not independent, but are subject to the linear relation $\sum \mathbf{J}_i = 0$, which raises the general question of whether the reciprocal relations still apply when there is a linear relation between either the affinities or the processes. In § 42.2 the 'independence' of the affinities meant that their microscopic parities could be extended over macroscopic times (cf. (42.3) and (42.6)), and clearly if the reciprocal relations are still to hold any additional constraints must not destroy the parities of the processes.

First suppose the n processes $J = \{J_1, J_2, \ldots J_n\}$ to be subject to the constraint $aJ = 0$, where $a = [a_1, a_2, \ldots a_{n-1}, 1]$ is a (row)† vector whose components are functions of the local thermodynamic state and of $\mathbf{B}$, and it has been assumed without real loss in generality that $a_n = 1$. We shall partition J, X, and the diagonal parity matrix η introduced in § 44.1 thus,

$$J = \{J' \vdots J_n\}, \qquad X = \{X' \vdots X_n\}, \qquad \eta = \begin{bmatrix} \eta' & \vdots & \cdot \\ \cdots & \vdots & \cdots \\ \cdot & \vdots & \eta_n \end{bmatrix}, \tag{44.13}$$

then with $a = [a' \vdots 1]$, the linear relation $aJ = 0$ may be written

$$J_n = -a'J'. \tag{44.14}$$

This may be used to eliminate J_n from $\sigma = JX$, with the result

$$\sigma = J'(X' - a'X_n). \tag{44.15}$$

To preserve the parities of the $(n-1)$ affinities $X' - a'X_n$ appearing in (44.15), we require that

$$\eta' a'(\mathbf{B}) = a'(-\mathbf{B})\eta_n \qquad (\eta'^2 = U), \tag{44.16}$$

for under motion-reversal $X' - a'(\mathbf{B})X_n \to \eta' X' - a'(-\mathbf{B})\eta_n X_n = \eta'[X' - -\eta' a'(-\mathbf{B})\eta_n X_n]$, so that if (44.16) holds, the affinities possess the parities of X', that is, $X' - a'(\mathbf{B})X_n \to \eta'[X' - a'(\mathbf{B})X_n]$. The $(n-1)$ affinities in (44.15) are independent, possess parities and there are no constraints on the associated processes. Hence by the general theory of § 42.3 we may write

$$J' = L'(X' - a'X_n), \tag{44.17}$$

where

$$\eta' L'(\mathbf{B}) = \tilde{L}(-\mathbf{B})\eta'. \tag{44.18}$$

By (44.14) and (44.17)

$$J_n = -a'L'X' + a'L'a'X_n, \tag{44.19}$$

which may be regarded as an ancillary equation for J_n.

† We shall not bother to distinguish between row and column vectors in the notation, as this omission causes no obscurity in the analysis.

When it is convenient to work with the original set of n affinities and processes, we may combine (44.17) and (44.19) into

$$J = \begin{bmatrix} J' \\ J_n \end{bmatrix} = \begin{bmatrix} L' & -L'a' \\ -a'L' & a'L'a' \end{bmatrix} \begin{bmatrix} X' \\ X_n \end{bmatrix} = LX. \qquad (44.20)$$

It follows from $a = [a' \vdots 1]$ and the definition of L implicit in (44.20) that

$$aL = 0, \qquad La = 0.$$

Also

$$\eta L(\mathbf{B})\eta = \begin{bmatrix} \eta' & \cdot \\ \cdot & \eta_n \end{bmatrix} \begin{bmatrix} L' & -L'a' \\ -a'L' & a'L'a' \end{bmatrix} \begin{bmatrix} \eta' & \cdot \\ \cdot & \eta_n \end{bmatrix} = \begin{bmatrix} \eta' L' \eta' & -\eta' L' a' \eta_n \\ -\eta_n a' L' \eta' & a'L'a' \end{bmatrix}$$

in which L' and a' are functions of $\mathbf{B}$, and it follows easily from (44.16) and (44.18) that the last matrix is $\tilde{L}(-\mathbf{B})$. Hence

$$\eta L(\mathbf{B}) = \tilde{L}(-\mathbf{B})\eta. \qquad (44.21)$$

The conclusion is that the reciprocal relations are unimpaired by a linear homogeneous dependency among the processes. If there are other linear relations between the processes, these may be used to eliminate J_{n-1}, $J_{n-2}, \ldots$ from $\sigma = JX$ and the treatment is similar to that just given for a single dependency; the same conclusion, namely (44.21), follows.

Secondly suppose that the n affinities are now subject to a dependency $bX = 0$, or $X_n = -b'X'$ (choosing $b_n = 1$) and that the processes are unconstrained. In place of (44.15) and (44.17) we now have

$$\sigma = (J' - b'J_n)X', \qquad J' - b'J_n = L^0 X',$$

where $b'(\mathbf{B})$ satisfies a relation like (44.16) and because the components of X' are independent, $L^0(\mathbf{B})$ satisfies (44.18). It is convenient to interchange the roles of X_n and J_n,

$$\begin{bmatrix} J' \\ X_n \end{bmatrix} = \begin{bmatrix} L^0 & b' \\ -b' & 0 \end{bmatrix} \begin{bmatrix} X' \\ J_n \end{bmatrix},$$

where the matrix coefficient here obviously satisfies the reciprocal relations (44.21).

Finally suppose that *both* the constraints $bX = 0$ and $aJ = 0$ are imposed. If $a_n \neq 0$, $b_n \neq 0$, on choosing $b_n = 1$, we have

$$X_n + b'X' = 0\cdot \qquad (44.22)$$

so that (44.17) is

$$J' = L'(U + a'b')X'.$$

Eqns (44.14) and (44.22) now merely define J_n and X_n; the components of X' are independent and there is no constraint on J'. Thus, as before, L' satisfies the reciprocal relations. However, by (44.22) eqn (44.20) can be written

$$J = \begin{bmatrix} J' \\ J_n \end{bmatrix} = \begin{bmatrix} L' & -L'a' \\ -a'L' + \lambda b' & a'L'a' + \lambda \end{bmatrix} \begin{bmatrix} X' \\ X_n \end{bmatrix} = L(\lambda)X, \qquad (44.23)$$

where λ is an arbitrary parameter. It follows that in this case L is neither uniquely defined nor satisfies the reciprocal relations, except at $\lambda = 0$. By omitting the term λX_n from (44.23), we see that the same conclusion holds for $b_n = 0$. The lesson to be drawn from (44.23) is that simultaneous constraints on the affinities and the processes should be avoided by reducing either the number of affinities or the number of processes appearing in the bilinear form for σ until an independent set of affinities or processes has been obtained.

45. Applications of the reciprocal relations

45.1. The symmetry of the thermal conductivity tensor

Suppose that heat flux is the only process occurring in a medium, then by (38.13) $\sigma = -\mathbf{q} \,.\, \nabla T/T^2$, and the most general constitutive relation is of the form

$$\mathbf{q} = -\boldsymbol{\kappa}\!:\!\nabla T, \qquad (45.1)$$

where $\boldsymbol{\kappa}$ is termed the thermal conductivity tensor.

To apply the reciprocal relations to $\boldsymbol{\kappa}$ we first write (45.1) in the scalar form

$$q_i = -\sum_j \kappa_{ij}(\nabla T)_i \qquad (i, j = 1, 2, 3).$$

Now each $(\nabla T)_i$ has even parity under motion-reversal, hence by the reciprocal relations (42.15) $\kappa_{ij}(\mathbf{B}) = \kappa_{ji}(-\mathbf{B})$, that is,

$$\boldsymbol{\kappa}(\mathbf{B}) = \tilde{\boldsymbol{\kappa}}(-\mathbf{B}). \qquad (45.2)$$

Thus in the absence of a magnetic field $\boldsymbol{\kappa}$ is a symmetric tensor.

45.2. Transport coefficients in general tensorial form

We shall now give forms for the reciprocal relations in terms of the tensor transport coefficients appearing in the general frame-indifferent expression (40.14) for the constitutive relations. For the scalars $L_{s\sigma}$ it is immediate that

$$L_{s\sigma}(\mathbf{B}) = \eta_\sigma \eta_s L_{\sigma s}(-\mathbf{B}), \qquad (45.3)$$

but to find results for the tensors $\mathbf{L}_{t\sigma}$, $\mathbf{L}_{vv}$, and $\mathbf{L}_{s\tau}$ we first need to write out (40.14) in scalar component form and then to exchange subscripts.

Consider, for example, the process

$$\mathbf{J}_{\upsilon} = \sum_{\nu} \mathsf{L}_{\upsilon\nu} \cdot \mathbf{X}_{\nu} \quad \text{or} \quad J_{\upsilon,p} = \sum_{\nu,q} L_{\upsilon\nu,pq} X_{\nu,q} \qquad (p, q = 1, 2, 3),$$

where p, q specify Cartesian coordinates. The appropriate values of α and β to substitute in the reciprocal relation (42.15) are $\alpha = (\upsilon, p)$, $\beta = (\nu, q)$; hence

$$L_{\upsilon\nu,pq}(\mathbf{B}) = \eta_{\upsilon,p}\eta_{\nu,q}L_{\nu\upsilon,qp}(-\mathbf{B}), \tag{45.4}$$

where $\eta_{\upsilon,p}$ and $\eta_{\nu,q}$ are the parities of $X_{\upsilon,p}$ and $X_{\nu,q}$. Of course, on physical grounds we cannot expect the components of the affinity $\mathbf{X}_{\nu}$ to possess different parities; accepting this we can now write (45.14) in the form

$$\mathsf{L}_{\upsilon\nu}(\mathbf{B}) = \eta_{\upsilon}\eta_{\nu}\tilde{\mathsf{L}}_{\nu\upsilon}(-\mathbf{B}); \tag{45.5}$$

similarly

$$\mathsf{L}_{s\tau}(\mathbf{B}) = \eta_{s}\eta_{\tau}\tilde{\mathsf{L}}_{\tau s}(-\mathbf{B}). \tag{45.6}$$

A like analysis for the fourth-order tensor in $(40.14)_3$ yields

$$L_{t\tau;ijkl}(\mathbf{B}) = \eta_{t}\eta_{\tau}L_{\tau t;lkji}(-\mathbf{B}) \qquad (i, j, k, l = 1, 2, 3),$$

or

$$\boldsymbol{L}_{t\tau}(\mathbf{B}) = \eta_{t}\eta_{\tau}\tilde{\tilde{\boldsymbol{L}}}_{\tau t}(-\mathbf{B}), \tag{45.7}$$

in which the meaning of the double transpose indicated by the double tilde is apparent from the scalar form.

We shall not bother to write down the reciprocal relation for the odd-order phenomenological tensors in (40.14), for as shown in the next chapter they are zero for all cases in which we shall be interested.

Examples of (45.6) and (45.7) may be obtained from the expressions derived in § 39.2 for the viscous stress tensor. In the notation of eqns (39.8) we have

$$\mathring{\boldsymbol{\beta}}(\mathbf{B}) = \mathring{\boldsymbol{\alpha}}(-\mathbf{B}), \qquad \mathring{\boldsymbol{\nu}}(\mathbf{B}) = \tilde{\tilde{\mathring{\boldsymbol{\nu}}}}(-\mathbf{B}), \tag{45.8}$$

as all the affinities involved have the same (odd) parity.

8

MATERIAL INVARIANCE AND VISCOSITY

WE shall now return to the discussion of material invariance, commenced in § 40.5. When the symmetry properties of the medium are known, the constitutive relations (40.14) can be specialized to account for these properties, with a resulting reduction in the number of independent scalars appearing in the phenomenological coefficients. Unlike some solids, fluids are structurally quite simple and consequently the complexity of the constitutive relations can be greatly reduced. The simplest and most commonly encountered fluid is *isotropic*, that is, there are no preferred directions within the body of the fluid. An important example of an anisotropic fluid is the magneto-plasma discussed in § 56. The magnetic induction **B** introduces a preferred direction, about which there is *transverse isotropy*, which means that the phenomenological coefficients are absolute invariants under rotations about the direction of **B**. Another very important symmetry, which may be traced back to the central character of the microscopic particle interaction forces, is known as *central symmetry*. This is invariance of material properties under an inversion of the coordinates, $\mathbf{i} \to -\mathbf{i}'$, $\mathbf{j} \to -\mathbf{j}'$, $\mathbf{k} \to -\mathbf{k}'$, where **i**, **j**, **k** are unit Cartesian vectors. This completes the list of symmetries of concern in what follows.

The reciprocal relations described in Chapter 7, which may be thought of as being a consequence of a general type of material symmetry in the (microscopic) time variable, reduce the possible number of numerically distinct phenomenological scalars by less than half. For example, the 36 distinct components of a general symmetric viscous stress tensor are reduced to 21 by the reciprocal relations (45.8). However, spatial symmetries are much more important, being able to make dramatic reductions in the number of distinct phenomenological scalars. For the example just cited, as will be shown in §50.1, transverse isotropy reduces the 36 scalar viscosities to just eight, only one of which is removed by a subsequent use of the reciprocal relations. Full spatial isotropy further reduces the remaining viscosities to only two (see eqn (39.9)).

46. Central symmetry

As all the fluids to be considered in later articles possess at least central symmetry,† we shall deal with invariance under coordinate inversion first.

† Due to the central forces acting between randomly oriented microscopic particles.

Under this transformation tensors like **k**, **ijk**, **ikk**, ... , that is, tensors of *odd* order change sign. Thus a polar vector, say $a_x\mathbf{i}+a_y\mathbf{j}+a_z\mathbf{k}$, changes sign, whereas an axial vector, which is the cross-product of two polar vectors, does not change sign. In fact axial vectors must be regarded as being second-order tensors in their transformation properties. We shall assume that all the vectors in the constitutive relations (40.14) are polar vectors, any axial vectors having been expressed in second-order tensor form and thus included in the tensors $\mathbf{J}_t$ and $\mathbf{X}_t$. Therefore under coordinate inversion the phenomenological tensors in (40.14) of odd order change sign, whereas those of even order do not. We conclude that material invariance requires all the odd-order phenomenological tensors to vanish, which reduces (40.14) to

$$\left.\begin{aligned} J_s &= \sum_\sigma L_{s\sigma}X_\sigma + \sum_\tau \mathbf{L}_{s\tau}:\mathbf{X}_\tau && (s,\sigma = 1,2,\dots m_0)\\ \mathbf{J}_v &= \sum_\nu \mathbf{L}_{v\nu}\cdot\mathbf{X}_\nu && (v,\nu = 1,2,\dots m_1)\\ \mathbf{J}_t &= \sum_\sigma \mathbf{L}_{t\sigma}X_\sigma + \sum_\tau \boldsymbol{L}_{t\tau}:\mathbf{X}_\tau && (t,\tau = 1,2,\dots m_2)\end{aligned}\right\} \qquad (46.1)$$

The entropy production may be written

$$\sigma = \sigma_1+(\sigma_0+\sigma_2), \qquad (46.2)$$

where

$$\left.\begin{aligned} \sigma_0 &\equiv \sum_s J_sX_s = \sum_{s,\sigma} X_sL_{s\sigma}X_\sigma + \sum_{s,\tau} X_s\mathbf{L}_{s\tau}:\mathbf{X}_\tau\\ \sigma_1 &\equiv \sum_v J_v\cdot\mathbf{X}_v = \sum_{v,\nu} \mathbf{X}_v\cdot\mathbf{L}_{v\nu}\cdot\mathbf{X}_\nu\\ \sigma_2 &\equiv \sum_t \mathbf{J}_t:\mathbf{X}_t = \sum_{t,\sigma} \mathbf{X}_t:\mathbf{L}_{t\sigma}X_\sigma + \sum_{t,\tau} \mathbf{X}_t:\boldsymbol{L}_{t\tau}:\mathbf{X}_\tau.\end{aligned}\right\} \qquad (46.3)$$

As the affinities are independent, the absence of affinities in common between σ_1 and $(\sigma_0+\sigma_2)$ means that the condition $\sigma \geqslant 0$ can be split into two distinct parts,

$$\sigma_1 \geqslant 0, \qquad \sigma_0+\sigma_2 \geqslant 0. \qquad (46.4)$$

This is as much as we can deduce for centrally symmetric fluids.

That the division of entropy production between σ_0 and σ_2 is not unique may be inferred by applying the expansion (38.15), namely,

$$\mathbf{A}:\mathbf{B} = \tfrac{1}{3}\check{A}\check{B}+\mathbf{A}^{\mathrm{a}}:\mathbf{B}^{\mathrm{a}}+\mathring{\mathbf{A}}:\mathring{\mathbf{B}} \qquad (46.5)$$

to the product $\mathbf{J}_t:\mathbf{X}_t$ in eqn $(46.3)_3$. Also $\mathbf{J}_t^{\mathrm{a}}:\mathbf{X}_t^{\mathrm{a}}$ could be expressed as a scalar product of axial vectors and included in σ_1, but for the fact that we have expressly restricted σ_1 to contain only polar vectors.

47. Transverse isotropy

47.1. Invariance of polyads under rotations

Transverse isotropy will be discussed in terms of unit Cartesian vectors **i**, **j**, **k** and various polyadic combinations of them like **ii**, **ikj**, **jiji**, ... , with **k** chosen to be the preferred direction. Tensors of second order may be represented as sums over 9 *dyads* like **ab**, where **a** = **i**, **j**, or **k** and similarly for **b**; likewise tensors of higher order are sums over *polyads* like **abc** (third order, 27 terms), **abcd** (fourth order, 81 terms), etc., where **a**, **b**, **c**, **d**, ... = **i**, **j**, **k**. First consider the transformation of a polar vector **a** into **a**′ by a rotation through a small angle θ. We have (see Fig. 14) $\mathbf{a}' = \mathbf{a}+\theta\mathbf{k}\times\mathbf{a}$. The dyad **ab** transforms according to

$$(\mathbf{ab})' = \mathbf{a}'\mathbf{b}' = (\mathbf{a}+\theta\mathbf{k}\times\mathbf{a})(\mathbf{b}+\theta\mathbf{k}\times\mathbf{b})$$
$$\simeq \mathbf{ab}+\theta(\mathbf{k}\times\mathbf{ab}+\mathbf{ak}\times\mathbf{b}),$$

and in general

$$(\mathbf{abc}\ldots)' = \mathbf{abc}\ldots+\theta(\mathbf{k}\times\mathbf{abc}\ldots+\mathbf{ak}\times\mathbf{bc}\ldots+\mathbf{abk}\times\mathbf{c}\ldots+\mathbf{abck}\times\ldots).$$

If the polyad is invariant under the transformation, $(\mathbf{abc}\ldots)' = \mathbf{abc}\ldots$, so that

$$\mathbf{k}\times\mathbf{abc}\ldots+\mathbf{ak}\times\mathbf{bc}\ldots+\mathbf{abk}\times\mathbf{c}\ldots+\mathbf{abck}\times\ldots = 0. \qquad (47.1)$$

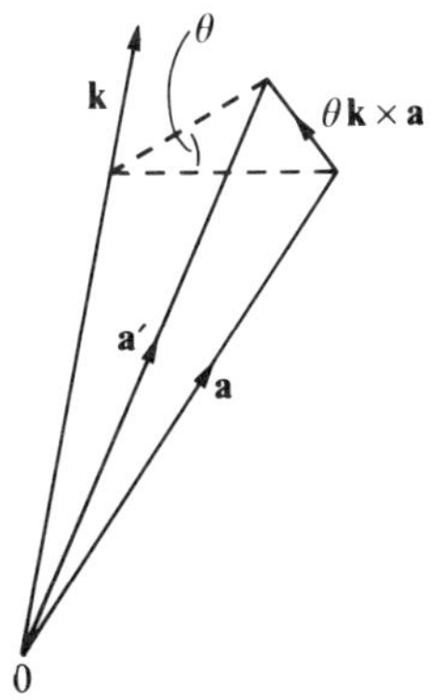

FIG. 14. Transformation of a vector by rotation.

Admittedly this condition for the invariance of **abc** ... is based on small angles of rotation, but as a large rotation can be divided into a large number of small rotations in each of which (47.1) holds, we must expect (47.1) to apply generally.

Suppose that we reverse the magneto-fluid system $\mathscr{P}$ by a 180° rotation about any axis normal to the preferred direction $0z$. Transverse isotropy ensures that changes in the orientation of $\mathscr{P}$ in the plane transverse to $0z$ cannot affect the constitutive relations, so apart from a change in the sign of **B**, the phenomenological tensors in (46.1) will be invariant under the proposed transformation. Thus choosing a rotation about $0y$, we have

$$\mathbf{i} \to -\mathbf{i}', \mathbf{j} \to \mathbf{j}', \mathbf{k} \to -\mathbf{k}'; \mathscr{L} \to \mathscr{L}': \qquad \mathscr{L}(\mathbf{B}) = \mathscr{L}'(-\mathbf{B}), \tag{47.2}$$

where $\mathscr{L}$ denotes a phenomenological tensor of any order.

47.2. Tensors possessing **k**-*symmetry*

Scalars are unaffected by coordinate transformations and so are isotropic; among *vectors*, **k** itself is the only one satisfying (47.1). The set $\{D_2\}$ of *dyads* possessing lateral isotropy about **k** satisfy

$$\mathbf{k} \times \mathbf{ab} - \mathbf{ab} \times \mathbf{k} = 0, \tag{47.3}$$

and it is readily verified that only the set

$$\{D_2\}: \qquad \mathbf{kk}, \qquad \mathbf{ii}+\mathbf{jj} = \mathbf{1}-\mathbf{kk}, \qquad \mathbf{ji}-\mathbf{ij} = \mathbf{k} \times \mathbf{1}, \tag{47.4}$$

meets this constraint. Thus the most general second-order tensor, possessing what shall be termed '**k**-symmetry', is the linear combination

$$\mathbf{a} = \alpha_{\parallel}\mathbf{kk} + \alpha_{\perp}(\mathbf{ii}+\mathbf{jj}) + \alpha_{\wedge}(\mathbf{ji}-\mathbf{ij}), \qquad \begin{pmatrix} \alpha_{\parallel}, \alpha_{\perp}, \text{ even in } \mathbf{B} \\ \alpha_{\wedge} \qquad \text{odd in } \mathbf{B} \end{pmatrix}, \tag{47.5}$$

where the dependence of the coefficients $\alpha_{\parallel}$, $\alpha_{\perp}$, and $\alpha_{\wedge}$ on the sign of **B** follows from (47.2). If, instead of **k**, an arbitrary unit vector **b** is the preferred direction, (47.5) is replaced by

$$\mathbf{a} = \alpha_{\parallel}\mathbf{bb} + \alpha_{\perp}(\mathbf{1}-\mathbf{bb}) + \alpha_{\wedge}\mathbf{b} \times \mathbf{1}.$$

Although we do not have **k**-symmetric third-order tensors in the constitutive relations, we shall include a discussion of these for completeness. By (47.1) the **k**-symmetric triads $\{D_3\}$ satisfy

$$\begin{aligned}(\mathbf{k} \times \mathbf{ab} - \mathbf{ab} \times \mathbf{k})\mathbf{c} + \mathbf{abk} \times \mathbf{c} &= \mathbf{a}(\mathbf{k} \times \mathbf{bc} - \mathbf{bc} \times \mathbf{k}) + \mathbf{k} \times \mathbf{abc} \\ &= \mathbf{k} \times \mathbf{abc} - \mathbf{abc} \times \mathbf{k} + \mathbf{ak} \times \mathbf{bc} = 0,\end{aligned}$$

and as the triads must either cancel in pairs or vanish individually, it follows from (47.3) and the first vanishing expression that **ab** may be any of $\{D_2\}$ with $\mathbf{c} = \mathbf{k}$, from the second expression that $\mathbf{bc} \in \{D_2\}$ with $\mathbf{a} = \mathbf{k}$, and from the third that $\mathbf{ac} \in \{D_2\}$ with $\mathbf{b} = \mathbf{k}$ being inserted between **a** and **c**. This yields the set

$$\{D_3\}: \qquad \begin{matrix} (\mathbf{ii}+\mathbf{jj})\mathbf{k}, \ \mathbf{k}(\mathbf{ii}+\mathbf{jj}), \ \mathbf{iki}+\mathbf{jkj}, \\ (\mathbf{ji}-\mathbf{ij})\mathbf{k}, \ \mathbf{k}(\mathbf{ji}-\mathbf{ij}), \ \mathbf{jki}-\mathbf{ikj}, \end{matrix} \qquad \mathbf{kkk}, \tag{47.6}$$

and every **k**-symmetric third-order tensor is a linear combination of $\{D_3\}$.

47.3. The general **k***-symmetric fourth-order tensor*

Finally the **k**-symmetric polyads of fourth order $\{D_4\}$ satisfy

$$(\mathbf{k}\times\mathbf{ab}-\mathbf{ab}\times\mathbf{k})\mathbf{cd}+\mathbf{ab}(\mathbf{k}\times\mathbf{cd}-\mathbf{cd}\times\mathbf{k}) = 0 \Rightarrow \mathbf{ab}\in\{D_2\}, \mathbf{cd}\in\{D_2\}, \quad \text{(a)}$$

$$\mathbf{k}\times\mathbf{abcd}-\mathbf{abcd}\times\mathbf{k}+\mathbf{a}(\mathbf{k}\times\mathbf{bc}-\mathbf{bc}\times\mathbf{k})\mathbf{d} = 0 \Rightarrow \mathbf{ad}\in\{D_2\}, \mathbf{bc}\in\{D_2\}, \quad \text{(b)}$$

$$\mathbf{a}(\mathbf{k}\times\mathbf{bcd}-\mathbf{bcd}\times\mathbf{k})+(\mathbf{k}\times\mathbf{abc}-\mathbf{abc}\times\mathbf{k})\mathbf{d} = 0 \Rightarrow \left\{\begin{matrix}\mathbf{bd}\in\{D_2\}, \mathbf{a}, \mathbf{c} = \mathbf{k} \\ \mathbf{ac}\in\{D_2\}, \mathbf{b}, \mathbf{d} = \mathbf{k}\end{matrix}\right\}, \quad \text{(c)}$$

$$(47.7)$$

where the equations are different arrangements of (47.1) and the inferences follow from (47.3). Thus the set $\{D_4\}$ may be determined from $\{D_2\}$ given in (47.4). The results are set out in Table 6, which also indicates where (47.7) (a), (b), and (c) have been employed.

TABLE 6

The set $\{D_4\}$ *of* **k***-symmetric fourth-order polyads*

r	$\boldsymbol{K}_r$		r	$\boldsymbol{K}_r$		r	$\boldsymbol{K}_r$	
1	$(\mathbf{ii}+\mathbf{jj})^2$	a	7	$\mathbf{kk}(\mathbf{ii}+\mathbf{jj})$	a	13	$\mathbf{kk}(\mathbf{ji}-\mathbf{ij})$	a
2	$(\mathbf{ji}-\mathbf{ij})^2$	a	8	$(\mathbf{ii}+\mathbf{jj})\mathbf{kk}$	a	14	$(\mathbf{ji}-\mathbf{ij})\mathbf{kk}$	a
3	$(\mathbf{ii}+\mathbf{jj})(\mathbf{ji}-\mathbf{ij})$	a	9	$\mathbf{ikki}+\mathbf{jkkj}$	b	15	$\mathbf{jkki}-\mathbf{ikkj}$	b
4	$(\mathbf{ji}-\mathbf{ij})(\mathbf{ii}+\mathbf{jj})$	a	10	$\mathbf{kiik}+\mathbf{kjjk}$	b	16	$\mathbf{kjik}-\mathbf{kijk}$	b
5	$\mathbf{i}(\mathbf{ii}+\mathbf{jj})\mathbf{i}+\mathbf{j}(\mathbf{ii}+\mathbf{jj})\mathbf{j}$	b	11	$\mathbf{ikik}+\mathbf{jkjk}$	c	17	$\mathbf{jkik}+\mathbf{ikjk}$	c
6	$\mathbf{j}(\mathbf{ii}+\mathbf{jj})\mathbf{i}-\mathbf{i}(\mathbf{ii}+\mathbf{jj})\mathbf{j}$	b	12	$\mathbf{kiki}+\mathbf{kjkj}$	c	18	$\mathbf{kjki}-\mathbf{kikj}$	c
			19	$\mathbf{kkkk}$	a, b or c			

Of course, while (47.7) covers all possibilities, it is necessary to check that the polyads derived from (47.7) are in fact independent. Such a check has eliminated two obvious possibilities for the first column of Table 6.

The most general **k**-symmetric fourth-order tensor is

$$\boldsymbol{\alpha} = \sum_{r=1}^{19} \alpha_r \boldsymbol{K}_r, \quad (47.8)$$

where $\boldsymbol{K}_r, r = 1, 2, \ldots 19$, are the polyads listed in Table 6 and $\alpha_1, \alpha_2, \ldots \alpha_{19}$ are scalar coefficients. The odd or even dependence of α_r on **B** is readily deduced from (47.2) and the table of $\boldsymbol{K}_r$.

With $0z$ aligned along the preferred direction in the fluid, material invariance under rotations about $0z$ requires the phenomenological tensors in (46.1) to have the form given in (47.5) if they are of second order, and the form given in (47.8) if they are of fourth order. Thus—neglecting for the moment any reductions due to the reciprocal relations—the tensors $\boldsymbol{L}_{t\tau}$ now have just 19 non-zero components instead of 81.

47.4. Symmetric processes

Further reductions in the number of non-vanishing components of $\boldsymbol{L}_{t\tau}$ can be made when $\mathbf{J}_t$ in eqn $(46.1)_3$ is known to be symmetric, that is, when the simple constitutive relation

$$\mathbf{J}_t^{\mathrm{a}} = 0, \quad \text{or} \quad \mathbf{J}_t = \tilde{\mathbf{J}}_t \tag{47.9}$$

applies. The viscous stress tensor $\boldsymbol{\pi}$ for ordinary fluids is the obvious example of a symmetric process; there may be some fluids comprised of long-chain molecules for which $\boldsymbol{\pi}^{\mathrm{a}} = 0$ is not a good approximation. The physical basis for this constitutive relation is discussed in §§ 40.4 and 49.2.

As $\mathbf{J}_t^{\mathrm{a}}:\mathbf{X}_t^{\mathrm{a}}$ is zero, the antisymmetric part of $\mathbf{X}_t$ disappears from the bilinear form (46.3) for σ, and so need not be included among the independent affinities. By (46.5) we may write $(46.1)_3$ as

$$\mathring{\mathbf{J}}_t + \tfrac{1}{3}\overset{\times}{J}_t\mathbf{1} = \sum_\sigma (\mathring{\mathbf{L}}_{t\sigma} + \tfrac{1}{3}\overset{\times}{L}_{t\sigma}\mathbf{1})X_\sigma + \sum_\tau \boldsymbol{\alpha}_{t\tau}\overset{\times}{X}_\tau + \sum_\tau \boldsymbol{L}_{t\tau}:\mathring{\mathbf{X}}_\tau, \tag{47.10}$$

where $\boldsymbol{\alpha}_{t\tau} \equiv \boldsymbol{L}_{t\tau}:\mathbf{1}$. It is readily verified by expanding $\boldsymbol{L}_{t\tau}$ in terms of $\boldsymbol{K}_r$ as in (47.8) that $\boldsymbol{\alpha}_{t\tau}$ is of the form given in (47.5), and a similar remark is true of $\boldsymbol{\beta}_{t\tau} \equiv \mathbf{1}:\boldsymbol{L}_{t\tau}$ to be introduced shortly. Now on taking the trace of (47.10) and subtracting $\frac{1}{3}\mathbf{1}$ times this from (47.10), we find the equations

$$\begin{aligned} \overset{\times}{J}_t &= \left(\sum_\sigma \overset{\times}{L}_{t\sigma}X_\sigma + \sum_\tau \overset{\times}{\alpha}_{t\tau}\overset{\times}{X}_\tau\right) + \sum_\tau \mathring{\boldsymbol{\beta}}_{t\tau}:\mathring{\mathbf{X}}_\tau, \\ \mathring{\mathbf{J}}_t &= \left(\sum_\sigma \mathring{\mathbf{L}}_{t\sigma}X_\sigma + \sum_\tau \mathring{\boldsymbol{\alpha}}_{t\tau}\overset{\times}{X}_\tau\right) + \sum_\tau \mathring{\boldsymbol{L}}_{t\tau}:\mathring{\mathbf{X}}_\tau, \end{aligned} \tag{47.11}$$

where (38.16) has permitted us to replace $\boldsymbol{\beta}_{t\tau}$ by $\mathring{\boldsymbol{\beta}}_{t\tau}$ and $\boldsymbol{L}_{t\tau} - \frac{1}{3}\mathbf{11}:\boldsymbol{L}_{t\tau}$ by

$$\mathring{\boldsymbol{L}}_{t\tau} = \boldsymbol{L}_{t\tau} - \tfrac{1}{3}\mathbf{11}:\boldsymbol{L}_{t\tau} - \tfrac{1}{3}\boldsymbol{L}_{t\tau}:\mathbf{11} + \tfrac{1}{9}\mathbf{11}:\boldsymbol{L}_{t\tau}:\mathbf{11}, \tag{47.12}$$

which is symmetric and has zero trace in its first and second component dyads:

$$\mathring{\boldsymbol{L}}_{t\tau} = \sum (\mathring{\mathbf{ab}})(\mathring{\mathbf{cd}}) \qquad (\mathbf{abcd} \in \{D_4\}) \qquad (\mathring{\mathbf{ab}} \equiv \tfrac{1}{2}(\mathbf{ab}+\mathbf{ba}) - \tfrac{1}{3}\mathbf{a}\,.\,\mathbf{b}\mathbf{1}),$$

with a similar definition for $(\mathring{\mathbf{cd}})$.

From the symmetric part of (47.5) we find

$$\mathring{\boldsymbol{\alpha}} = \alpha(\mathbf{kk} - \tfrac{1}{3}\mathbf{1}) \qquad (\alpha(\mathbf{B}) = \alpha(-\mathbf{B})), \tag{47.13}$$

where $\alpha = \alpha_{\parallel} - \alpha_{\perp}$. The tensors $\mathring{\mathbf{L}}_{t\sigma}$, $\mathring{\boldsymbol{\alpha}}_{t\tau}$, $\mathring{\boldsymbol{\beta}}_{t\tau}$ in (47.11) are all of this type.

To determine the form of $\mathring{\boldsymbol{L}}_{t\tau}$ the first step is to reduce $\{D_4\}$ to the set symmetrical in the component dyads. From the table of $\{D_4\}$ we easily obtain the eight polyads

$$(\mathbf{ii}+\mathbf{jj})^2,\ \mathbf{iiii}+\tfrac{1}{2}(\mathbf{ij}+\mathbf{ji})^2+\mathbf{jjjj},\ \boldsymbol{N}_3,\ \mathbf{kk}(\mathbf{ii}+\mathbf{jj}),\ (\mathbf{ii}+\mathbf{jj})\mathbf{kk},\ \boldsymbol{N}_4,\ \boldsymbol{N}_5,\ \mathbf{kkkk}, \qquad (47.14)$$

where $\boldsymbol{N}_3$, $\boldsymbol{N}_4$, and $\boldsymbol{N}_5$ are defined below in (47.15). If each component dyad is now rendered traceless, we arrive at just five independent polyads:

$$\left.\begin{aligned}
\boldsymbol{N}_1 &\equiv 2(\mathbf{ii}-\tfrac{1}{3}\mathbf{1})^2+2(\mathbf{jj}-\tfrac{1}{3}\mathbf{1})^2-(\mathbf{kk}-\tfrac{1}{3}\mathbf{1})^2+(\mathbf{ij}+\mathbf{ji})^2,\\
\boldsymbol{N}_2 &\equiv 3(\mathbf{kk}-\tfrac{1}{3}\mathbf{1})^2,\\
\boldsymbol{N}_3 &\equiv (\mathbf{ik}+\mathbf{ki})^2+(\mathbf{jk}+\mathbf{kj})^2,\\
\boldsymbol{N}_4 &\equiv (\mathbf{ij}+\mathbf{ji})(\mathbf{jj}-\mathbf{ii})+(\mathbf{ii}-\mathbf{jj})(\mathbf{ij}+\mathbf{ji}),\\
\boldsymbol{N}_5 &\equiv (\mathbf{ik}+\mathbf{ki})(\mathbf{jk}+\mathbf{kj})-(\mathbf{jk}+\mathbf{kj})(\mathbf{ik}+\mathbf{ki}).
\end{aligned}\right\} \qquad (47.15)$$

Hence by (47.2) the tensors $\mathring{\boldsymbol{L}}_{t\tau}$ are each of the form

$$\mathring{\boldsymbol{L}} = \sum_{i=1}^{5} \nu_i \boldsymbol{N}_i \qquad \begin{pmatrix} \nu_1, \nu_2, \nu_3 \text{ even in } \mathbf{B} \\ \nu_4, \nu_5 \quad \text{odd in } \mathbf{B} \end{pmatrix}, \qquad (47.16)$$

where ν_i, $i = 1, 2, \ldots 5$, are coefficients. If eqn $(46.1)_3$ is not expanded as in (47.11), the set (47.14) involving eight coefficients is required to define $\boldsymbol{L}_{t\tau}$. These coefficients correspond to the five appearing in (47.16), two coupling coefficients arising through $\mathring{\boldsymbol{\alpha}}_{t\tau}$ and $\mathring{\boldsymbol{\beta}}_{t\tau}$ as in (47.13), and $\check{\alpha}_{t\tau}$.

48. General isotropy

Let us now seek the sets of completely isotropic tensors up to the fourth order. First it is obvious that there are no isotropic vectors. Also as isotropic tensors must be symmetric about the directlon $0x$ as well as about $0z$, we require linear combinations of the sets $\{D_r\}$, $r = 2, 3, 4$, of § 47, that are also **i**-symmetric. Such tensors we shall find are also symmetric about any axis and so they are the tensors we want.

48.1. Isotropic tensors of second and third order

The tensor $\boldsymbol{\alpha}$ in (47.5) will be **i**-symmetric if it is invariant to the rotation $\mathbf{i} \to \mathbf{j}$, $\mathbf{j} \to \mathbf{k}$, $\mathbf{k} \to \mathbf{i}$, which places **i** along the preferred direction. This requirement leads to

$$\alpha_{\parallel} = \alpha_{\perp} = \alpha \quad \text{say,} \qquad \alpha_{\wedge} = 0, \qquad (48.1)$$

with

$$\boldsymbol{\alpha} = \alpha\mathbf{1} \qquad (48.2)$$

as a typical second-order isotropic tensor.

A similar argument applied to the set (47.6) shows that all third-order isotropic tensors are proportional to the so-called *alternating tensor* $\boldsymbol{\varepsilon}$ defined by

$$\boldsymbol{\varepsilon} \equiv \mathbf{1} \times \mathbf{1} = \mathbf{ikj} - \mathbf{ijk} + \mathbf{jik} - \mathbf{jki} + \mathbf{kji} - \mathbf{kij}. \tag{48.3}$$

Mathematical note. We pause here to note some of the interesting properties of $\boldsymbol{\varepsilon}$. From any vector $\mathbf{a}$ we can construct the second-order tensor,

$$\text{tens } \mathbf{a} \equiv \boldsymbol{\varepsilon} \,.\, \mathbf{a} = \mathbf{a} \,.\, \boldsymbol{\varepsilon} = \mathbf{1} \times \mathbf{a} = \mathbf{a} \times \mathbf{1}. \tag{48.4}$$

It also follows from (48.3) that

$$\mathbf{ab} : \boldsymbol{\varepsilon} = \mathbf{a} \,.\, \boldsymbol{\varepsilon} \,.\, \mathbf{b} = \boldsymbol{\varepsilon} : \mathbf{ab} = \mathbf{a} \times \mathbf{b} = -\boldsymbol{\varepsilon} : \mathbf{ba} = \boldsymbol{\varepsilon} : (\mathbf{ab})^{\mathrm{a}},$$

and more generally from any second-order tensor A with a dyadic expansion, $\mathbf{ab} + \mathbf{cd} + \mathbf{ef}$ say, we can construct the vector

$$2 \text{ vec } \mathsf{A} \equiv \boldsymbol{\varepsilon} : \mathsf{A} = \boldsymbol{\varepsilon} : \mathsf{A}^{\mathrm{a}} = \mathbf{a} \times \mathbf{b} + \mathbf{c} \times \mathbf{d} + \mathbf{e} \times \mathbf{f}. \tag{48.5}$$

Let $\mathbf{r}$ be any vector, then $\mathbf{r} \,.\, \boldsymbol{\varepsilon} \,.\, \boldsymbol{\varepsilon} : \mathbf{ab} = \mathbf{r} \times (\mathbf{a} \times \mathbf{b}) = -\mathbf{r} \,.\, (\mathbf{ab} - \tilde{\mathbf{ab}}) = -2\mathbf{r} \,.\, (\mathbf{ab})^{\mathrm{a}}$, whence, as $\mathbf{r}$ is arbitrary, $-\frac{1}{2}\boldsymbol{\varepsilon} \,.\, \boldsymbol{\varepsilon} : \mathbf{ab} = (\mathbf{ab})^{\mathrm{a}}$. From this result we conclude that

$$\boldsymbol{J} : \mathsf{A} = \boldsymbol{J} : \mathsf{A}^{\mathrm{a}} = \mathsf{A}^{\mathrm{a}}, \tag{48.6}$$

where

$$2\boldsymbol{J} \equiv -\boldsymbol{\varepsilon} \,.\, \boldsymbol{\varepsilon} = -[(\mathbf{ij} - \mathbf{ji})^2 + (\mathbf{jk} - \mathbf{kj})^2 + (\mathbf{ki} - \mathbf{ik})^2]. \tag{48.7}$$

It follows from (48.5), (48.6), and (48.7) that

$$\mathsf{A}^{\mathrm{a}} = -\boldsymbol{\varepsilon} \,.\, (\text{vec } \mathsf{A}). \tag{48.8}$$

Another useful result easily derived from (48.3) is

$$2\mathbf{1} = -\boldsymbol{\varepsilon} : \boldsymbol{\varepsilon}.$$

Finally we note that $\tilde{\mathsf{A}}^{\mathrm{a}} : \mathsf{B}^{\mathrm{a}} = \frac{1}{2}\mathsf{A}^{\mathrm{a}} : \boldsymbol{\varepsilon} \,.\, \boldsymbol{\varepsilon} : \mathsf{B}^{\mathrm{a}} = \frac{1}{2}(\boldsymbol{\varepsilon} : \mathsf{A}^{\mathrm{a}}) \,.\, (\boldsymbol{\varepsilon} : \mathsf{B}^{\mathrm{a}})$, that is,

$$\tilde{\mathsf{A}}^{\mathrm{a}} : \mathsf{B}^{\mathrm{a}} = 2(\text{vec } \mathsf{A}) \,.\, (\text{vec } \mathsf{B}). \tag{48.9}$$

48.2. Isotropic tensors of fourth order

The fourth-order isotropic tensors may be obtained by requiring $\boldsymbol{\alpha}$ in (47.8) to be invariant under the rotation $\mathbf{i} \to \mathbf{j}, \mathbf{j} \to \mathbf{k}, \mathbf{k} \to \mathbf{i}$. Thus, for example, $\mathbf{jjjj} \to \mathbf{kkkk}$ leaves $\boldsymbol{\alpha}$ invariant only if $\alpha_1 + \alpha_5 = \alpha_{19}$; similarly $\mathbf{jjii} \to \mathbf{kkjj}$ implies $\alpha_1 = \alpha_7 = \alpha_8$, $\mathbf{jkki} \to \mathbf{kiij}$ implies $\alpha_{15} = 0$, there being no term in $\mathbf{kiij}$, and so on. We soon find that under this constraint (47.8) reduces to the isotropic form

$$\boldsymbol{\alpha} = \alpha_1 \mathbf{1}\mathbf{1} + \alpha_2 \boldsymbol{J} + \alpha_5 \boldsymbol{U}, \tag{48.10}$$

where

$$\mathbf{11} = (\mathbf{ii}+\mathbf{jj}+\mathbf{kk})^2,$$

$$\boldsymbol{U} \equiv \mathbf{iiii}+\mathbf{jjjj}+\mathbf{kkkk}+\mathbf{ikki}+\mathbf{jkkj}+\mathbf{kiik}+\mathbf{kjjk}+\mathbf{jiij}+\mathbf{ijji},$$

and $\boldsymbol{J}$ is defined in (48.7). The tensor $\boldsymbol{U}$ is the fourth-order unit tensor, for

$$\boldsymbol{U}:\mathbf{A} = \mathbf{A}. \tag{48.11}$$

It is useful to replace $\boldsymbol{U}$ by $\boldsymbol{K}+\boldsymbol{J}$ or by $\mathring{\boldsymbol{K}}+\frac{1}{3}\mathbf{11}+\boldsymbol{J}$, where

$$\boldsymbol{K} \equiv \mathbf{iiii}+\mathbf{jjjj}+\mathbf{kkkk}+\tfrac{1}{2}(\mathbf{ij}+\mathbf{ji})^2+\tfrac{1}{2}(\mathbf{jk}+\mathbf{kj})^2+\tfrac{1}{2}(\mathbf{ki}+\mathbf{ik})^2,\ \mathring{\boldsymbol{K}} \equiv \boldsymbol{K}-\tfrac{1}{3}\mathbf{11}, \tag{48.12}$$

and by (48.6) and (48.11) have the properties

$$\boldsymbol{K}:\mathbf{A} = \mathbf{A}^{\mathrm{s}}, \qquad \mathring{\boldsymbol{K}}:\mathbf{A} = \mathring{\mathbf{A}}. \tag{48.13}$$

Thus the general fourth-order isotropic tensor may be expressed as

$$\boldsymbol{\alpha} = \tfrac{1}{3}\overset{\times}{\alpha}\mathbf{11}+\mathring{\alpha}\mathring{\boldsymbol{K}}+\alpha^{\mathrm{a}}\boldsymbol{J}, \tag{48.14}$$

where $\overset{\times}{\alpha}$, $\mathring{\alpha}$, α^{a} are scalar functions. Incidentally, the polyad $\mathring{\boldsymbol{K}}$ may also be found by constructing an isotropic polyad from the **k**-symmetric polyads of (47.15); we find $\boldsymbol{N}_1+\boldsymbol{N}_2+\boldsymbol{N}_3 = 2\mathring{\boldsymbol{K}}$.

48.3. Processes of different tensorial order uncoupled by isotropy

We have now established that, when the fluid is known to be isotropic, the second-order phenomenological tensors in (46.1) are proportional to the unit tensor, and the fourth-order tensors have the form given in (48.14). As $\mathring{\mathbf{1}} = 0$, the second-order tensors in (47.11) vanish, which eliminates the coupling between $\overset{\times}{J}_t$ and $\mathring{\mathbf{J}}_t$. Eqns (46.1) can now be reduced to†

$$\left.\begin{aligned}
J_s &= \sum_\sigma L_{s\sigma}X_\sigma+\tfrac{1}{3}\sum_\tau \overset{\times}{L}_{s\tau}\overset{\times}{X}_\tau, & \overset{\times}{J}_t &= \sum_\sigma \overset{\times}{L}_{t\sigma}X_\sigma+\sum_\tau \overset{\times}{\alpha}_{t\tau}\overset{\times}{X}_\tau & &\text{(scalars)},\\
\mathbf{J}_\upsilon &= \tfrac{1}{3}\sum_\nu \overset{\times}{L}_{\upsilon\nu}X_\nu, & \mathbf{J}^{\mathrm{a}}_t &= \sum_\tau \alpha^{\mathrm{a}}_{t\tau}\mathbf{X}^{\mathrm{a}}_\tau & &\text{(polar, axial vectors)},\\
& & \mathring{\mathbf{J}}_t &= \sum_\tau \mathring{\alpha}_{t\tau}\mathring{\mathbf{X}}_\tau & &\text{(deviators)},
\end{aligned}\right\} \tag{48.15}$$

making clear the absence of coupling between the four types of tensor involved.‡

† We must distinguish between $\overset{\times}{L}_{s\tau}$, $\overset{\times}{L}_{\tau s}$, and $\overset{\times}{L}_{\upsilon\nu}$ even when the subscripts are numerically identical; see footnote on p. 160.

‡ The independence of processes of different tensorial order in an isotropic system is often, wrongly it seems, attributed to Curie (1908) and termed 'Curie's principle'.

By (46.3) and (46.5) the entropy production may be written

$$\sigma = \left(\sum_s J_s X_s + \tfrac{1}{3}\sum_t \check{J}_t \check{X}_t\right) + \sum_v \mathbf{J}_v \cdot \mathbf{X}_v + \sum_t \mathbf{J}_t^{\mathrm{a}} : \mathbf{X}_t^{\mathrm{a}} + \sum_t \mathring{\mathbf{J}}_t : \mathring{\mathbf{X}}_t \geqslant 0,$$

or

$$\sigma = \sigma_0 + \sigma_1 + \sigma_2^{\mathrm{a}} + \mathring{\sigma}_2,$$

where by (48.15)

$$\begin{aligned}
\sigma_0 &= \sum_{s,\sigma} X_s L_{s\sigma} X_\sigma + \tfrac{1}{3}\sum_{s,t} X_s(\check{L}_{st} + \check{L}_{ts})\check{X}_t + \sum_{t,\tau} \check{X}_t \check{\sigma}_{t\tau} \check{X}_\tau \geqslant 0, \\
\sigma_1 &= \tfrac{1}{3}\sum_{v,\nu} \check{L}_{v\nu} \mathbf{X}_v \cdot \mathbf{X}_\nu \geqslant 0, \\
\sigma_2^{\mathrm{a}} &= \sum_{t,\tau} \alpha_{t\tau}^{\mathrm{a}} \mathbf{X}_t^{\mathrm{a}} : \mathbf{X}_\tau^{\mathrm{a}} \geqslant 0, \\
\mathring{\sigma}_2 &= \sum_{t,\tau} \mathring{\alpha}_{t\tau} \mathring{\mathbf{X}}_t : \mathring{\mathbf{X}}_\tau \geqslant 0,
\end{aligned} \tag{48.16}$$

the *individual* non-negative constraints following from the independence of the affinities involved and the lack of coupling. The inequalities in (48.16) are an extension of the result contained in (46.4) for centrally symmetric fluids.

The consequences of taking material invariance into account can be graphically illustrated by counting the number of scalar parameters required in the constitutive relations for the various symmetries considered above. This number is set out in Table 7 for the m-component fluid system to be described below in § 52 (see footnote on p. 161).

TABLE 7

m	Anisotropic	Central symmetry	Transverse isotropy	Isotropy
1	144	90	22 (11)	4 (3)
2	784	442	116 (72)	22 (18)
3	2304	1224	342 (243)	72 (63)

The numbers in brackets apply when the second-order tensorial processes are symmetric. The simplest case of three scalars is the thermal conductivity, the bulk viscosity, and the shear viscosity of a single isotropic fluid.

In the rest of this chapter we shall concentrate on the viscosity tensor, which, being of fourth-order, is the most complicated of all transport coefficients. The (second-order) heat and electric conductivity tensors will be discussed in the following chapter.

49. Viscosity in an isotropic fluid

49.1. Entropy production in a viscous fluid with internal coordinates

Our first step is to generalize slightly the formula for σ given in § 38.4. We start with (38.9) for a compound system written as

$$TDs = Du + pD(1/\rho) - \sum_i g_i D_e c_i + \sum_r A_r \dot{\xi}_r, \tag{49.1}$$

following a fluid particle $\mathscr{P}$. The total mass $\rho\, d\tau$ of $\mathscr{P}$ is constant, whereas the mass $\rho_i\, d\tau$ of its ith component may change, and as the concentration is $c_i = \rho_i/\rho$ it follows from (18.3) that

$$Dc_i = D_e c_i + \sum_r \nu_{ir} \dot{\xi}_r. \tag{49.2}$$

However, we are not here interested in the effects of diffusion, which introduces vector processes into the entropy production, so at present we may ignore $D_e c_i$ in (49.1). It now follows from (38.7), (38.8), (49.1), and (46.4) that the scalar and second-order tensorial contributions to σ for a centrally symmetric fluid satisfy

$$T(\sigma_0 + \sigma_2) = -\tilde{\boldsymbol{\pi}} : \nabla \mathbf{v} + \rho \sum_r A_r \dot{\xi}_r \geqslant 0.$$

By (46.5) and (48.9) we may expand the viscous dissipation term and arrive at

$$T(\sigma_0 + \sigma_2) = \rho \sum_r A_r \dot{\xi}_r - \tfrac{1}{3}\check{\pi} \nabla \cdot \mathbf{v} - \boldsymbol{\Pi} \cdot \nabla \times \mathbf{v} - \mathring{\boldsymbol{\pi}} : \mathring{\nabla \mathbf{v}} \geqslant 0, \tag{49.3}$$

where we have used the result $2 \operatorname{vec} \nabla \mathbf{v} = \nabla \times \mathbf{v}$ and defined a new axial vector $\boldsymbol{\Pi}$ by

$$\begin{aligned} \boldsymbol{\Pi} &= \operatorname{vec} \boldsymbol{\pi} = \tfrac{1}{2}\boldsymbol{\varepsilon} : \boldsymbol{\pi}^{a} = \tfrac{1}{2}\boldsymbol{\varepsilon} : \mathbf{p}, \\ \boldsymbol{\pi}^{a} &= -\boldsymbol{\varepsilon} \cdot \boldsymbol{\Pi}. \end{aligned} \tag{49.4}$$

When $\check{\pi}$, $\boldsymbol{\Pi}$, and $\boldsymbol{\pi}$ are known, the pressure tensor follows from

$$\mathbf{p} = (p + \tfrac{1}{3}\check{\pi})\mathbf{1} - \boldsymbol{\varepsilon} \cdot \boldsymbol{\Pi} + \boldsymbol{\pi}. \tag{49.5}$$

In § 40.4 the point was made that frame-indifference requires $\boldsymbol{\pi}^{a}$ (and hence $\boldsymbol{\Pi}$) to vanish, unless it is possible to modify (49.3) by replacing $\mathbf{v}$ by $\mathbf{v} - \boldsymbol{\omega} \times \mathbf{r}$, where $\boldsymbol{\omega}$ is an angular velocity intrinsic to the fluid. By results given on p. 188, $2 \operatorname{vec} \nabla(\mathbf{v} - \boldsymbol{\omega} \times \mathbf{r}) = \nabla \times \mathbf{v} + \boldsymbol{\varepsilon} : \boldsymbol{\varepsilon} \cdot \boldsymbol{\omega} = \nabla \times \mathbf{v} - 2\boldsymbol{\omega}$, so unless $\boldsymbol{\Pi}$ is zero we need to replace $\nabla \times \mathbf{v}$ in (49.3) by $(\nabla \times \mathbf{v} - 2\boldsymbol{\omega})$. With this modification, to be given a physical justification shortly, we arrive at three distinct types of viscosity: *bulk* (from $\nabla \cdot \mathbf{v}$ and relaxation processes $\dot{\xi}_r$), *rotational* (from $\nabla \times \mathbf{v} - 2\boldsymbol{\omega}$), and *shear* from $\mathring{\nabla \mathbf{v}}$.

49.2. Rotational viscosity

Rotational viscosity is usually neglected as it arises only in the exceptional circumstances when the microscopic particles are large 'rough' molecules and so capable of absorbing and emitting rotational energy. In this case (49.3) needs modifying to allow for the dissipation of such energy. If $\rho\mathbf{F}\,\mathrm{d}\tau$ is the body force on a fluid particle $\mathscr{P}$, the moment produced on $\mathscr{P}$ about the origin is $\rho\mathbf{F}\times\mathbf{r}\,\mathrm{d}\tau$, where $\mathbf{r}$ is the position coordinate of $\mathscr{P}$. Also the moment supplied by the fluid stress on a surface element $\mathbf{n}\,\mathrm{d}\Sigma$ of $\mathscr{P}$ is $-\mathbf{n}\,.\,\mathbf{p}\times\mathbf{r}\,\mathrm{d}\Sigma$, so that by the divergence theorem $-\nabla\,.\,(\mathbf{p}\times\mathbf{r})\,\mathrm{d}\tau$ is the moment of the fluid stress acting on $\mathscr{P}$. Let $\mathbf{J}$ denote the angular momentum per unit mass of $\mathscr{P}$ about the origin, then angular momentum balance requires that

$$\mathrm{D}(\rho\mathbf{J}\,\mathrm{d}\tau) = -\nabla\,.\,(\mathbf{p}\times\mathbf{r})\,\mathrm{d}\tau+\rho\mathbf{F}\times\mathbf{r}\,\mathrm{d}\tau.$$

Now (see p. 188)

$$\nabla\,.\,(\mathbf{p}\times\mathbf{r}) = \nabla\,.\,(\mathbf{pr}:\boldsymbol{\varepsilon}) = (\nabla\,.\,\mathbf{p})\mathbf{r}:\boldsymbol{\varepsilon}+(\tilde{\mathbf{p}}\,.\,\nabla\mathbf{r}):\boldsymbol{\varepsilon}$$
$$= (\nabla\,.\,\mathbf{p})\times\mathbf{r}+\tilde{\mathbf{p}}:\boldsymbol{\varepsilon} = (\nabla\,.\,\mathbf{p})\times\mathbf{r}+2\mathbf{\Pi},$$

by (49.4). Hence

$$\rho\mathrm{D}\mathbf{J} = (-\nabla\,.\,\mathbf{p}+\rho\mathbf{F})\times\mathbf{r}-2\mathbf{\Pi}.$$

Let $\boldsymbol{\omega}$ and θ denote the mean angular velocity and average moment of inertia per unit mass of the microscopic particles in $\mathscr{P}$, then $\theta\boldsymbol{\omega}$ is the *internal* contribution to $\mathbf{J}$. Thus removing the obviously external contribution to $\mathrm{D}\mathbf{J}$, and introducing the specific rotational energy $u_\mathrm{r} = \frac{1}{2}\theta\omega^2$, we obtain

$$\rho\theta\mathrm{D}\boldsymbol{\omega} = -2\mathbf{\Pi}, \qquad \rho\mathrm{D}u_\mathrm{r} = -\mathbf{\Pi}\,.\,\boldsymbol{\omega}. \tag{49.6}$$

Now in eqn (38.6), where u_r has been ignored, the first term should read $\rho\mathrm{D}(u+u_r+\frac{1}{2}v^2)$. Thus with eqn $(49.6)_2$ used to remove $\mathrm{D}u_r$, the result is to replace $\Phi = -\tilde{\boldsymbol{\pi}}:\nabla\mathbf{v}$ in (38.13) by $-\tilde{\boldsymbol{\pi}}:\nabla\mathbf{v}+2\mathbf{\Pi}\,.\,\boldsymbol{\omega}$, so that (49.3) now assumes the frame-indifferent form

$$T(\sigma_0+\sigma_2) = \rho\sum_r A_r\dot{\xi}_r-\tfrac{1}{3}\overset{\times}{\pi}\nabla\,.\,\mathbf{v}-\mathbf{\Pi}\,.\,(\nabla\times\mathbf{v}-2\boldsymbol{\omega})-\mathring{\boldsymbol{\pi}}:\overset{\circ}{\nabla\mathbf{v}} \geqslant 0. \tag{49.7}$$

Suppose now that the time scale is long enough to permit all the internal processes to relax, so that the first term of (49.7) is zero. Then, as shown in § 48.3, in an isotropic system there is no coupling between the terms in (48·7). We have a special case of the constitutive relations given in (49.15),

$$\left.\begin{aligned} \tfrac{1}{3}\overset{\times}{\pi} &= -\rho v'\,\nabla\,.\,\mathbf{v}, \\ \mathbf{\Pi} &= -\rho v_\mathrm{r}(\nabla\times\mathbf{v}-2\boldsymbol{\omega}), \\ \mathring{\boldsymbol{\pi}} &= -2\rho v\overset{\circ}{\nabla\mathbf{v}} = -\rho v(\nabla\mathbf{v}+\widetilde{\nabla\mathbf{v}}-\tfrac{2}{3}\mathbf{1}\nabla\,.\,\mathbf{v}), \end{aligned}\right\} \tag{49.8}$$

where the scalars ν', ν_r, ν are respectively the bulk (or volume), rotational, and shear kinematic viscosities. From (49.7)

$$T\sigma_0 = \rho\nu'(\nabla\,.\,\mathbf{v})^2 \geqslant 0, \qquad T\sigma_2 = \rho\nu_r(\nabla\times\mathbf{v}-2\boldsymbol{\omega})^2+2\rho\nu(\overset{\circ}{\nabla\mathbf{v}})^2 \geqslant 0, \quad (49.9)$$

whence, as the forces are independent,

$$\nu', \nu_r, \nu \geqslant 0. \qquad (49.10)$$

By eqn $(49.6)_1$ and the choice $\mathbf{A} = \frac{1}{2}\theta(\nabla\times\mathbf{v}-2\boldsymbol{\omega})$, $\boldsymbol{\xi} = \boldsymbol{\omega}$, we have $-\boldsymbol{\Pi}\,.\,(\nabla\times\mathbf{v}-2\boldsymbol{\omega}) = \rho\mathbf{A}\,.\,\dot{\boldsymbol{\xi}}$, which exhibits rotational viscosity as due to a relaxation process between the rotational and translatory molecular energies. If $\nabla\times\mathbf{v}$ is constant over the time scale of interest and if $\boldsymbol{\omega} = 0$ at $t = 0$, then by $(46.6)_1$ and $(49.8)_2$

$$\boldsymbol{\omega} = \tfrac{1}{2}\nabla\times\mathbf{v}(1-e^{-t/\tau}) \qquad (\tau \equiv \theta/4\nu_r).$$

For times $t \gg \tau$ the affinity $\mathbf{A}$ is sensibly zero and rotational viscosity may be neglected. The argument usually presented tacitly accepts θ to be zero, when $\boldsymbol{\Pi} = 0$ by eqn $(49.6)_1$, and then shows $\mathbf{p}$ to be a symmetric tensor.

49.3. The Navier–Stokes equation

When rotational viscosity is unimportant, the pressure tensor is given by

$$\mathbf{p} = (p-\rho\nu'\nabla\,.\,\mathbf{v})\mathbf{1}-2\rho\nu\overset{\circ}{\nabla\mathbf{v}},$$

which follows from $(49.8)_1$, $(49.8)_3$ and (49.5). Now

$$\nabla\,.\,\overset{\circ}{\nabla\mathbf{v}} = \tfrac{1}{2}\nabla\,.\,(\nabla\mathbf{v}+\widetilde{\nabla\mathbf{v}}-\tfrac{1}{3}\mathbf{1}\nabla\,.\,\mathbf{v}) = \tfrac{1}{2}\nabla^2\mathbf{v}+\tfrac{1}{6}\nabla\nabla\,.\,\mathbf{v},$$

so, ignoring gradients in $\rho\nu'$ and $\rho\nu$,

$$\nabla\,.\,\mathbf{p} = \nabla p-\rho(\nu'+\tfrac{1}{3}\nu)\nabla\nabla\,.\,\mathbf{v}-\rho\nu\nabla^2\mathbf{v}.$$

Eqn $(38.4)_2$ therefore may be written

$$\rho\frac{\partial\mathbf{v}}{\partial t}+\rho\mathbf{v}\,.\,\nabla\mathbf{v} = -\nabla p+\rho(\nu'+\tfrac{1}{3}\nu)\nabla\nabla\,.\,\mathbf{v}+\rho\nu\nabla^2\mathbf{v}+\rho\mathbf{F}, \quad (49.11)$$

which is the well-known Navier–Stokes equation.

49.4. Thermal relaxation

While bulk viscosity is significant in dense gases, bubbly liquids, and so on, its appearance in dilute gases is almost entirely through relaxation phenomena. We shall show that any type of delayed equilibrium—chemical or thermal—can produce bulk viscosity if the macroscopic time scale of the volume changes is sufficiently small. Thermal relaxation arises with polyatomic molecules, and in an expansion or contraction the work done by the pressure immediately alters the translatory energy of the molecules but affects their internal energies of rotation and vibration only after a time lag. In other words, the internal degrees of freedom of the molecules are not in thermal equilibrium with the translatory degrees of freedom.

It is the translatory modes that determine what is usually called the temperature T of the system, and provided it is legitimate to regard the jth internal mode of the ith species as being a thermodynamic system $\mathscr{P}_{ij}$ we may assign to it a temperature T_{ij}, a specific energy u_{ij}, and a specific entropy s_{ij}. The system $\mathscr{P}_{ij}$ will not possess deformation coordinates, and it will be assumed independent of the states of the other similar internal systems with different values of i or j. Hence u_{ij} depends only on T_{ij} and

$$T_{ij}\mathrm{D}s_{ij} = \mathrm{D}u_{ij}. \tag{49.12}$$

The entropy and internal energy in eqn (49.1) do not include the contribution from $\mathscr{P}_{ij}$, and so must be increased by

$$\sum_{i,j} c_i s_{ij} \quad \text{and} \quad \sum_{i,j} c_i u_{ij}.$$

To do this we use (49.2) to rewrite (49.12) in the form

$$T\mathrm{D}(c_i s_{ij}) = \mathrm{D}(c_i u_{ij}) + (Ts_{ij} - u_{ij})\mathrm{D_e}c_i + \sum_r (Ts_{ij} - u_{ij})\nu_{ir}\dot{\xi}_r + \frac{T - T_{ij}}{T_{ij}} c_i \mathrm{D}u_{ij},$$

sum over i and j, and then add to (49.1) to find

$$T\mathrm{D}s' = \mathrm{D}u' + p\mathrm{D}\left(\frac{1}{\rho}\right) - \sum_i g_i'\mathrm{D_e}c_i + \sum_r A_r'\dot{\xi}_r + \sum_{i,j} \left[\frac{T - T_{ij}}{T_{ij}} c_i \dot{u}_{ij}\right], \tag{49.13}$$

where s' and u' are the increased values of s and u, and

$$g_i' = g_i + \sum_j (u_{ij} - Ts_{ij}), \qquad A_r' = -\sum_i g_i'\nu_{ir}.$$

Thus the main effect of our modification is simply to introduce a new set of internal coordinates u_{ij} with affinities $c_i(T - T_{ij})/T_{ij}$. Thermal relaxation of $\mathscr{P}_{ij}$ proceeds until its temperature T_{ij} equals T.

49.5. Bulk viscosity

When the fluid is comprised of only one component, eqn (49.13) reduces to

$$T\mathrm{D}s = \mathrm{D}u + p\mathrm{D}(1/\rho) + \sum_r A_r\dot{\xi}_r,$$

where (49.14)

$$A_r \equiv (T - T_r)/T_r, \qquad \xi_r = u_r,$$

and T_r, u_r are the temperatures and specific energies of the rth internal mode of molecular vibration or rotation. The thermodynamic systems $\mathscr{P}_r$ are assumed to be independent of each other, so that selecting s, ρ, and ξ_r, $r = 1, 2, \ldots$, as the independent thermodynamic coordinates, we have the set of state equations $A_r = A_r(s, \rho, \xi_r)$. For small departures $\hat{s}$, $\hat{\rho}$, $\hat{\xi}_r$ from a time-independent equilibrium state, A_r may be expanded

$$A_r = \left(\frac{\partial A_r}{\partial \rho}\right)_{s,\xi_r} \hat{\rho} + \left(\frac{\partial A_r}{\partial \xi_r}\right)_{\rho,s} \hat{\xi}_r + \left(\frac{\partial A_r}{\partial s}\right)_{\rho,\xi_r} \hat{s}, \tag{49.15}$$

the partial derivatives being evaluated at equilibrium conditions, where $A_r = 0$. On setting $\mathring{\pi}$ equal to zero in (49.7), and using isotropy to separate σ_0 and σ_2, we obtain

$$T\sigma_0 = \rho \sum_r A_r \dot{\xi}_r \geqslant 0, \tag{49.16}$$

Also with the assumption that the systems $\mathscr{P}_r$ interact only with the translatory system $\mathscr{P}$, we may write

$$\dot{\xi}_r = \beta_r A_r \qquad (\beta_r \geqslant 0, \qquad r = 1, 2, \ldots) \tag{49.17}$$

for the constitutive relations. By (49.16) and the independence of the A_r, the coefficients β_r are non-negative. From (38.7), $(49.14)_1$, and the equation of continuity,

$$\rho T \dot{\hat{s}} = \Phi - \nabla \cdot \mathbf{q} + \rho \sum_r A_r \dot{\xi}_r \qquad (\Phi = -\tilde{\boldsymbol{\pi}} : \nabla \mathbf{v})$$

and

$$\dot{\hat{\rho}} = -\rho \nabla \cdot \mathbf{v},$$

consequently the time derivative of (49.15) may be written

$$-\tau_r \dot{A}_r = \tau_r \rho \left(\frac{\partial A_r}{\partial \rho}\right)_{s,\xi_r} \nabla \cdot \mathbf{v} + A_r - \frac{1}{\rho T}\left(\frac{\partial A_r}{\partial s}\right)_{\rho,\xi_r} \tau_r \left(\Phi - \nabla \cdot \mathbf{q} + \rho \sum_r A_r \dot{\xi}_r\right), \tag{49.18}$$

where

$$\tau_r \equiv -\frac{1}{\beta_r}\left(\frac{\partial \xi_r}{\partial A_r}\right)_{\rho,s} \geqslant 0 \tag{49.19}$$

is a relaxation time. The inequality here follows from $\beta_r \geqslant 0$ and the thermodynamic stability condition (see (18.9)) $(\partial \xi_r / \partial A_r) < 0$. Notice from (49.15) and (49.17) that, at constant ρ and s,

$$\dot{\xi}_r = \beta_r \left(\frac{\partial A_r}{\partial \xi_r}\right)_{\rho,s} \hat{\xi}_r = -\frac{1}{\tau_r} \hat{\xi}_r,$$

showing τ_r to be the time in which ξ_r relaxes to a fraction $1/e$ of its initial value at constant ρ and s.

We now agree to neglect products of the small relaxation times τ_r with the transport coefficients in the expansion of $\dot{s}$, and further to choose a macroscopic time scale somewhat larger than τ_r, $r = 1, 2, \ldots$, so that $\tau_r \dot{A}_r \ll A_r$.† These assumptions reduce (49.18) to

$$A_r = -\tau_r \rho \left(\frac{\partial A_r}{\partial \rho}\right)_{s,\xi_r} \nabla \cdot \mathbf{v} = -\frac{\rho}{\beta_r}\left(\frac{\partial \xi_r}{\partial \rho}\right)_{s,A_r} \nabla \cdot \mathbf{v}.$$

From this result, (49.16) and (49.17) it follows that

$$T\sigma_0 = \rho \sum_r \beta_r A_r^2 = \rho \sum_r \frac{\rho^2}{\beta_r}\left(\frac{\partial \xi_r}{\partial \rho}\right)_{s,A_r}^2 (\nabla \cdot \mathbf{v})^2,$$

whence (cf. $(49.10)_1$)

$$\nu' = \rho^2 \sum_r \frac{1}{\beta_r}\left(\frac{\partial \xi_r}{\partial \rho}\right)_{s,A_r}^2 \geqslant 0. \tag{49.20}$$

Thus for oscillatory motions like a sound wave, we can say that at low frequencies ω such that $\omega\tau_r \ll 1$, a relaxation phenomenon can be considered as having the same effect as a bulk viscosity. The theory leading to (49.20) is quite general, applying to any kind of delayed equilibrium, chemical or physical.

For the particular case of thermal relaxation eqns $(49.14)_2$ and (49.19) give

$$\beta_r = -\frac{1}{\tau_r}\left\{\frac{\partial u_r}{\partial[(T-T_r)/T_r]}\right\}_{\rho,s} = \frac{1}{\tau_r} c_{V_r} \frac{T_r^2}{T} \approx \frac{1}{\tau_r} c_{V_r} T,$$

where $c_{V_r} = \mathrm{d}u_r/\mathrm{d}T_r$ and we have used $T_r \approx T$. Also from the equilibrium relations $\mathrm{d}u = T\,\mathrm{d}s + (p/\rho^2)\,\mathrm{d}\rho$ and $\mathrm{d}T_r = \mathrm{d}u_r/c_{V_r} = \mathrm{d}u/c_V = \mathrm{d}T$, where c_V is the total specific heat, we find

$$\left(\frac{\partial \xi_r}{\partial \rho}\right)_{s,A_r} = \left(\frac{\partial u_r}{\partial \rho}\right)_{s,A_r} = \frac{c_{V_r}}{c_V}\left(\frac{\partial u}{\partial \rho}\right)_{s,A_r} = \frac{c_{V_r}}{c_V}\frac{p}{\rho^2}.$$

Therefore (49.20) becomes

$$\nu' = \frac{p^2}{T\rho^2 c_V^2} \sum_r c_{V_r} \tau_r. \tag{49.21}$$

Finally we remark that as the shear viscosity results directly from the random translatory motions of particles, it is not a relaxation phenomenon—at least on the macroscopic time scales for which a fluid description is really valid (see remarks in § 1.1).

† For a different treatment not involving this approximation see § 65.

50. Viscosity in a fluid with transverse isotropy

50.1. General theory

Let the fluid possess central symmetry, so that $(\sigma_0+\sigma_2)$ and σ_1 do not have any affinities in common, and suppose for simplicity that only viscosity contributes to $(\sigma_0+\sigma_2)$. With $\mathbf{\Pi}$ and A_r zero in (49.3), we have

$$T(\sigma_0+\sigma_2) = -\tfrac{1}{3}\overset{\times}{\pi}\nabla\,.\,\mathbf{v}-\overset{\circ}{\boldsymbol{\pi}}:\overset{\circ}{\nabla\mathbf{v}} \geqslant 0, \tag{50.1}$$

and the further separation of σ_0 and σ_2 into distinct non-negative terms is possible only in an isotropic system (§ 48.3). The constitutive relations have the general form (see (47.11))

$$\begin{aligned} \tfrac{1}{3}\overset{\times}{\pi} &= -\rho\nu'\nabla\,.\,\mathbf{v}-\overset{\circ}{\boldsymbol{\beta}}:\overset{\circ}{\nabla\mathbf{v}}, \\ \overset{\circ}{\boldsymbol{\pi}} &= -\overset{\circ}{\boldsymbol{\alpha}}\nabla\,.\,\mathbf{v}-\rho\overset{\circ}{\boldsymbol{\nu}}:\nabla\mathbf{v}, \end{aligned} \tag{50.2}$$

where the component dyads of the fourth-order viscosity tensor $\overset{\circ}{\boldsymbol{\nu}}$ are deviators. In an isotropic system $\overset{\circ}{\boldsymbol{\nu}} = 2\nu\overset{\circ}{\boldsymbol{K}}$, $\overset{\circ}{\boldsymbol{\alpha}} = \overset{\circ}{\boldsymbol{\beta}} = 0$ and we have just *two* coefficients of kinematic viscosity, ν' and ν (cf. (49.8)).

For a system with transverse isotropy, such as a magneto-plasma with $\mathbf{B}$ along $0z$, it follows from (47.13) and (47.16) that

$$\overset{\circ}{\boldsymbol{\alpha}} = \rho\nu''(\mathbf{kk}-\tfrac{1}{3}\mathbf{1}), \qquad \overset{\circ}{\boldsymbol{\beta}} = \rho\nu_0(\mathbf{kk}-\tfrac{1}{3}\mathbf{1}), \qquad \overset{\circ}{\boldsymbol{\nu}} = \sum_{i=1}^{5}\nu_i\boldsymbol{N}_i, \tag{50.3}$$

$$\text{where} \quad \begin{matrix} \nu', \nu'', \nu_0, \nu_1, \nu_2, \nu_3 \\ \nu_4, \nu_5 \end{matrix} \quad \text{are} \quad \begin{matrix} \textit{even} \\ \textit{odd} \end{matrix} \quad \text{functions of } \mathbf{B}. \tag{50.4}$$

(That ν' is even in $\mathbf{B}$ follows directly from (47.2) and the fact that ν' is a zero-order tensor.)

The reciprocal relations require that

$$\overset{\circ}{\boldsymbol{\alpha}}(\mathbf{B}) = \overset{\circ}{\boldsymbol{\beta}}(-\mathbf{B}), \qquad \overset{\circ}{\boldsymbol{\nu}}(\mathbf{B}) = \overset{\approx}{\overset{\circ}{\boldsymbol{\nu}}}(-\mathbf{B}),$$

which follow from (45.6), (45.7), and the fact that the processes in (50.2) both have odd parity. By (50.3) and (50.4) the first reciprocal relation gives

$$\nu_0(\mathbf{B}) = \nu''(\mathbf{B}); \tag{50.5}$$

and by (50.3) and (47.15) the second relation merely confirms the results for ν_i, $i = 1, 2, \ldots 5$, contained in (50.4) and arrived at from material invariance. Thus a system with transverse isotropy has *seven* distinct viscosity coefficients, namely, two bulk viscosities ν', ν'' and five shear viscosities ν_i, $i = 1, \ldots 5$.

50.2. Components of the pressure tensor

The results contained in (50.2) are more perspicuous in component form. With the definition

$$\mathbf{S} \equiv \overset{\circ}{\nabla}\mathbf{v} = \tfrac{1}{2}(\nabla\mathbf{v}+\widetilde{\nabla\mathbf{v}})-\tfrac{1}{3}\mathbf{1}\nabla\,.\,\mathbf{v}, \tag{50.6}$$

and some straightforward algebra we find for

$$\mathbf{p} = p\mathbf{1}-\rho[(v'-\tfrac{1}{3}v'')\mathbf{1}+v''\mathbf{kk}]\nabla\,.\,\mathbf{v}-\rho\left[\sum_{i=1}^{5} v_i\boldsymbol{N}_i+v''(\mathbf{kk}-\tfrac{1}{3}\mathbf{1})\right]:\mathbf{S}, \tag{50.7}$$

the components

$$\left.\begin{aligned}
p_{xx} &= p-\rho(v'-\tfrac{1}{3}v'')\nabla\,.\,\mathbf{v}-2\rho[v_1S_{xx}+\tfrac{1}{2}(v_2-v_1-v'')(S_{xx}+S_{yy})+v_4S_{xy}],\\
p_{yy} &= p-\rho(v'-\tfrac{1}{3}v'')\nabla\,.\,\mathbf{v}-2\rho[v_1S_{yy}+\tfrac{1}{2}(v_2-v_1-v'')(S_{xx}+S_{yy})-v_4S_{xy}],\\
p_{zz} &= p-\rho(v'+\tfrac{2}{3}v'')\nabla\,.\,\mathbf{v}-2\rho(v_2+\tfrac{1}{2}v'')S_{zz},\\
p_{xy} &= p_{yx} = -2\rho[v_1S_{xy}+\tfrac{1}{2}v_4(S_{yy}-S_{xx})],\\
p_{yz} &= p_{zy} = -2\rho[v_3S_{yz}-v_5S_{zx}],\\
p_{zx} &= p_{xz} = -2\rho[v_3S_{zx}+v_5S_{yz}].
\end{aligned}\right\} \tag{50.8}$$

By (50.2), (50.3), and (50.6) the constraint in (50.1) may be written

$$T(\sigma_0+\sigma_2) = \rho[v'(\nabla\,.\,\mathbf{v})^2+\tfrac{4}{3}v''S_{zz}\nabla\,.\,\mathbf{v}+\mathbf{S}:\overset{\circ}{\mathbf{v}}:\mathbf{S}],$$

or, on using the symmetry of $\mathbf{S}$ and the fact that $\overset{\times}{S} = 0$,

$$T(\sigma_0+\sigma_2) = \rho\{v'(\nabla\,.\,\mathbf{v})^2+\tfrac{4}{3}v''S_{zz}\nabla\,.\,\mathbf{v}+v_1[(S_{xx}-S_{yy})^2+4S_{xy}^2]+ \\ +3v_2S_{zz}^2+4v_3[S_{xz}^2+S_{yz}^2]\} \geqslant 0. \tag{50.9}$$

As $\nabla\,.\,\mathbf{v}$ and the components of $\mathbf{S}$ with S_{xx} replaced by $-(S_{yy}+S_{zz})$ may be independently assigned—these being independent affinities—it follows from (50.9) that

$$v',\,v_1,\,v_2,\,v_3 \geqslant 0, \qquad v'(v_1+3v_2) \geqslant \tfrac{4}{9}v''^2, \tag{50.10}$$

of which the last is the condition that $v'(\nabla\,.\,\mathbf{v})^2+\tfrac{4}{3}v''S_{zz}\nabla\,.\,\mathbf{v}+(v_1+3v_2)S_{zz}^2$ be non-negative. No restrictions are placed on v_4 and v_5.

50.3. Pressure tensor for a magneto-plasma

The fully ionized magneto-plasma provides a good example of the above theory. Except when the electrons are *very* much hotter than the ions, the viscous stresses are dominantly due to the much heavier particles, and it is the product of the ion-cyclotron frequency $\omega_i = eB/m_i$ with the ion–ion collision time τ_i that provides a measure of the importance of the magnetic field (see remarks in § 56.1). We shall write

$$\mathscr{s} \equiv \tfrac{2}{3}\omega_i \tau_i,$$

so that when $\mathscr{s} \to 0$, (50.8) tend to their isotropic form. First note that as ν_2 is the viscosity for motions parallel to **B**, it must be independent of **B** and so equal to the viscosity ν, say, of the plasma without magnetic field. Then from (50.4)

$$\lim_{\mathscr{s}\to 0} \nu_1, \nu_2, \nu_3 = \nu, \qquad \lim_{\mathscr{s}\to 0} \nu_4, \nu_5 = 0. \tag{50.11}$$

The bulk viscosities are zero as the ionized particles possess no internal structure. From kinetic theory Chapman and Cowling (1970, p. 378) obtain

$$\nu_1 = \nu/(1+4\mathscr{s}^2), \qquad \nu_2 = \nu, \qquad \nu_3 = \nu/(1+\mathscr{s}^2), \qquad \nu_4 = 2\mathscr{s}\nu/(1+4\mathscr{s}^2),$$
$$\nu_5 = \mathscr{s}\nu/(1+\mathscr{s}^2),$$

which are in accordance with the requirements of the continuum theory just described.

9

FLUID MIXTURES

WHILE a fluid system composed of several distinct components can often quite adequately be represented by a single fluid, with the various components contributing only via the phenomenological coefficients of the constitutive relations, for some processes it is essential to represent the components separately. Such processes are too rapid for the components to be in mutual equilibrium, the diffusion of one species through another, especially when chemical reactions between the species are possible, being an obvious example. To describe this situation the compound fluid must be treated as being a mixture of distinct fluids, homogeneous in each small fluid particle, having different fluid velocities, and, in some situations, involving very short time scales, even having different temperatures. For example, fully-ionized plasmas may sometimes be adequately described in terms of an ion fluid and an electron fluid, with distinct temperatures T_i and T_e that can be more than an order of magnitude different. In general, the different fluids can exchange mass, momentum, and energy, and phenomenological laws are required to describe these exchanges, which give rise to additional entropy production.

Even on the longest time-scales a compound fluid system may not be adequately represented as a single fluid, unless perhaps this fluid is given additional phenomenological properties having their physical origin in the compound nature of the system. A good example of this is the *one*-fluid magnetohydrodynamics of § 38.6, where electric currents, which have their origin in there really being *two* fluids (ions and electrons), are superimposed phenomenologically. Unless we wish to remain entirely at a long-time-scale phenomenological description, it is necessary to treat the various components as separate fluids and then to approach the long time scales asymptotically. This has the virtue both of exposing the limits of the one-fluid theory and of providing a many-fluid theory when a deeper description is unavoidable.

In Chapter 10 we shall apply mixture theory to two particularly interesting examples, namely to the fully-ionized plasma described above and to the two-fluid model of the 'superfluid' helium II.

51. The basic entropy inequality for a mixture

The inequality contained in § 37.1 will now be generalized to a mixture of several diffusing fluid components that are free to exchange mass, momentum, and energy.

51.1. Conservation of mass

Let $\mathscr{P}_i$ denote a fluid particle comprised entirely of the ith species, and suppose it has a density ρ_i and a velocity $\mathbf{v}_i$. The compound fluid particle $\mathscr{P}$ will then have a density ρ and a velocity $\mathbf{v}$ given by

$$\rho = \sum_i \rho_i, \qquad \rho\mathbf{v} = \sum_i \rho_i \mathbf{v}_i. \tag{51.1}$$

To describe the ith species it is convenient to introduce the concentration c_i, and the diffusion momentum $\mathbf{J}_i$ (or sometimes the diffusion velocity $\mathbf{w}_i$) defined by

$$c_i = \rho_i/\rho, \qquad \mathbf{J}_i = \rho_i(\mathbf{v}_i - \mathbf{v}) = \rho_i \mathbf{w}_i \qquad (\mathbf{w}_i = \mathbf{v}_i - \mathbf{v}), \tag{51.2}$$

which satisfy

$$\sum_i c_i = 1, \qquad \sum_i \mathbf{J}_i = 0. \tag{51.3}$$

The arguments given in § 38.1 that lead to

$$\mathrm{D}_i = \frac{\partial}{\partial t} + \mathbf{v}_i \,.\, \nabla = \mathrm{D} + \mathbf{w}_i \,.\, \nabla, \tag{51.4}$$

D being the material derivative for $\mathscr{P}$.

Let $\overset{+}{\rho}_i$ be the rate at which the ith species is being created per unit volume within $\mathscr{P}_i$, then on the assumption that the system $\mathscr{P}$ has constant mass we have

$$\sum_i \overset{+}{\rho}_i = 0. \tag{51.5}$$

The arguments given in § 38.1 that lead to

$$\begin{aligned} \mathrm{D}(\rho\, \mathrm{d}\tau) &= 0, \qquad & \rho \mathrm{D}(1/\rho) &= \nabla \,.\, \mathbf{v}, \\ \mathrm{D}\rho + \rho \nabla \,.\, \mathbf{v} &= 0, \qquad & \partial\rho/\partial t + \nabla \,.\, \rho\mathbf{v} &= 0, \end{aligned} \tag{51.6}$$

when applied to $\mathscr{P}_i$ will yield

$$\begin{aligned} \mathrm{D}_i(\rho_i\, \mathrm{d}\tau) &= \overset{+}{\rho}_i\, \mathrm{d}\tau, \qquad & \rho_i \mathrm{D}_i(1/\rho_i) &= \nabla \,.\, \mathbf{v}_i - \overset{+}{\rho}_i/\rho_i, \\ \mathrm{D}_i\rho_i + \rho_i \nabla \,.\, \mathbf{v}_i &= \overset{+}{\rho}_i, \qquad & \partial\rho_i/\partial t + \nabla \,.\, (\rho_i \mathbf{v}_i) &= \overset{+}{\rho}_i. \end{aligned} \tag{51.7}$$

By (51.4) and $(51.6)_2$ we can write $(51.7)_3$ in the form

$$(\mathrm{D} + \mathbf{w}_i \,.\, \nabla)\rho_i + \rho_i \rho \mathrm{D}(1/\rho) + \rho_i \nabla \,.\, (\mathbf{v}_i - \mathbf{v}) = \overset{+}{\rho}_i$$

or

$$\rho \mathrm{D} c_i + \nabla \,.\, \mathbf{J}_i = \overset{+}{\rho}_i, \tag{51.8}$$

and a comparison with (49.2) reveals that the external and internal contribution to $\rho \mathrm{D}c_i$ are

$$\rho \mathrm{D_e} c_i = -\nabla \cdot \mathbf{J}_i \tag{51.9}$$

and

$$\overset{+}{\rho}_i = \rho \sum_r \nu_{ir} \dot{\xi}_r. \tag{51.10}$$

Finally, let ζ_i denote some specific property of $\mathscr{P}_i$ with an average value ζ given by

$$\zeta = \sum_i c_i \zeta_i, \tag{51.11}$$

then by (51.4) and $(51.6)_3$

$$\begin{aligned} \rho \mathrm{D}\zeta &= \mathrm{D}(\rho\zeta) + \rho\zeta\, \nabla \cdot \mathbf{v} = \sum_i [\mathrm{D}(\rho_i \zeta_i) + \rho_i \zeta_i \nabla \cdot \mathbf{v}] \\ &= \sum_i [\mathrm{D}_i(\rho_i \zeta_i) - \mathbf{w}_i \cdot \nabla(\rho_i \zeta_i) + \rho_i \zeta_i \nabla \cdot \mathbf{v}]. \end{aligned}$$

Thus by $(51.7)_3$

$$\rho \mathrm{D}\zeta = \sum_i [\rho_i \mathrm{D}_i \zeta_i - \nabla \cdot (\zeta_i \mathbf{J}_i) + \overset{+}{\rho}_i \zeta_i], \tag{51.12}$$

which will be useful shortly.

51.2. Changes due to mass exchange between species

The creation or destruction of particles of the ith species will naturally affect the momentum and energy content of $\mathscr{P}_i$, as well as its mass. Eqn (51.10), which expresses the rate at which mass is created in $\mathscr{P}_i$ in terms of the independent reaction rates $\dot{\xi}_r$, does not indicate from which of the other fluid particle systems $\mathscr{P}_j$, $j \neq$ i, this mass has come or how much mass is lost from $\mathscr{P}_i$ to the other species. Such details are needed in order to write down momentum and energy balance equations.

Let $b_{ijr}\dot{\xi}_r$ ($\geqslant 0$) be the rate at which $\mathscr{P}_i$ *gains* mass from $\mathscr{P}_j$ in the rth reaction, and let

$$a_{ij} = \sum_r b_{ijr} \dot{\xi}_r \qquad (i, j = 1, 2, \ldots) \tag{51.13}$$

be the total rate at which $\mathscr{P}_i$ gains mass from $\mathscr{P}_j$, then (51.10) may be written

$$\overset{+}{\rho}_i = \rho \sum_r \nu_{ir} \dot{\xi}_r = \sum_{r,j} (b_{ijr} - b_{jir}) \dot{\xi}_r = \sum_j (a_{ij} - a_{ji}). \tag{51.14}$$

The number b_{ijr} can be obtained when the details of the reactions are known.

Now let $\overset{+}{X}_i$ be the rate at which some physical property of $\mathscr{P}_i$ per unit volume is created within $\mathscr{P}_i$ due solely to the appearance or disappearance of the ith species, then

$$\overset{+}{X}_i = \sum_j (a_{ij}X_j/\rho_j - a_{ji}X_i/\rho_i)$$

and

$$\overset{+}{X}_i - \overset{+}{\rho}_i X_i/\rho_i = \sum_j a_{ij}(X_j/\rho_j - X_i/\rho_i). \tag{51.15}$$

We note that

$$\sum_i \overset{+}{X}_i = 0. \tag{51.16}$$

Eqn (38.2), when applied to $\mathscr{P}_i$, does not allow for the effect just described and needs to be replaced by

$$\frac{\partial X_i}{\partial t} + \nabla\,.\,(X_i\mathbf{v}_i) = \rho_i \mathrm{D}_i\!\left(\frac{X_i}{\rho_i}\right) + \overset{+}{\rho}_i\frac{X_i}{\rho_i} = \mathscr{S}(X_i) + \overset{+}{X}_i, \tag{51.17}$$

where $\mathscr{S}(X_i)$ is the source term for the supply of X_i due to causes other than mass exchange.

51.3. The entropy balance equation

Take X_i in (51.17) to be the entropy density $\rho_i s_i$, then as $\mathscr{S}(X_i) = -\nabla\,.\,\mathbf{J}_{si} + \mathscr{S}_i$, where $\mathbf{J}_{si}$ is the entropy flux vector and $\mathscr{S}_i$ is the internal source strength due to dissipation,

$$\rho_i \mathrm{D}_i s_i + \nabla\,.\,\mathbf{J}_{si} = \mathscr{S}_i + \rho_i \overset{+}{s}_i - \overset{+}{\rho}_i s_i, \tag{51.18}$$

where for notational convenience we have written $\rho_i \overset{+}{s}_i$ for $(\rho_i s_i)^{+}$. By (51.12) with $\zeta_i = s_i$,

$$\sum_i \rho_i \mathrm{D}_i s_i = \rho \mathrm{D}s + \nabla\,.\sum_i s_i\mathbf{J}_i - \sum_i \overset{+}{\rho}_i s_i,$$

enabling the sum of (51.18) to be written

$$\rho \mathrm{D}s + \nabla\,.\sum_i (\mathbf{J}_{si} + s_i\mathbf{J}_i) = \sum_i \mathscr{S}_i, \tag{51.19}$$

where by (51.16)

$$\sum_i \rho_i \overset{+}{s}_i = 0. \tag{51.20}$$

Eqn (38.8), namely,

$$\rho \mathrm{D}s + \nabla\,.\left(\sum_i \mathbf{q}_i/T_i\right) = \sigma + \sum_i\left(\sum_j \varphi_{ij}/T_j + \psi_i/T_i\right),$$

is the special case of (51.19) for zero diffusion, that is, for $\mathbf{J}_i = 0$. Hence

$$\sum_i \mathscr{S}_i = \sigma + \sum_i \left(\sum_j \varphi_{ij}/T_j + \psi_i/T_i \right) \tag{51.21}$$

and

$$\mathbf{J}_{si} = \mathbf{q}_i/T_i.$$

Thus (51.19) yields the required generalization of (38.8),

$$\sigma = \rho \mathrm{D}s + \nabla \cdot \left[\sum_i (\mathbf{q}_i + T_i s_i \mathbf{J}_i)/T_i \right] - \sum_i \left(\sum_j \varphi_{ij}/T_j + \psi_i/T_i \right). \tag{51.22}$$

As $\sigma \geqslant 0$, the volume integral of (51.22) gives the inequality

$$\frac{\mathrm{d}S}{\mathrm{d}t} \geqslant - \int_\Sigma \sum_i \frac{\mathbf{n} \cdot (\mathbf{q}_i + T_i s_i \mathbf{J}_i)}{T_i} \, \mathrm{d}\Sigma + \int_V \sum_i \left(\sum_i \frac{\varphi_{ij}}{T_j} + \frac{\psi_i}{T_i} \right) \mathrm{d}\tau, \tag{51.23}$$

which replaces that given in § 37.1. The origin of the additional term in the surface integral is obvious enough.

In the special case of equal temperatures $T_i = T$, $i = 1, 2, \ldots$, we may use

$$T\mathrm{D}s = \mathrm{D}u + \mathbf{p} \cdot \mathrm{D}\mathbf{l} + \sum_i \frac{g_i}{\rho} \nabla \cdot \mathbf{J}_i + \sum_r A_r \dot{\xi}_r, \tag{51.24}$$

which follows from (38.9) and (51.9), to eliminate Ds from (51.22). By (18.4)

$$A_r = - \sum_i \nu_{ir} g_i,$$

hence

$$\sum_r A_r \dot{\xi}_r = - \sum_r \sum_i \nu_{ir} g_i \dot{\xi}_r = - \sum_i g_i \overset{+}{\rho}_i/\rho, \tag{51.25}$$

but it should be remembered that, whereas the reaction rates $\dot{\xi}_r$ are chosen to be independent, this is not true of the creation rates $\overset{+}{\rho}_i$.

An alternative expression for σ follows from (51.18), (51.20), and (51.21),

$$\sigma = \sum_i \left[\rho_i \mathrm{D}_i s_i + \nabla \cdot \frac{\mathbf{q}_i}{T_i} + \overset{+}{\rho}_i s_i - \sum_j \frac{\varphi_{ij}}{T_j} - \frac{\psi_i}{T_i} \right], \tag{51.26}$$

from which $\mathrm{D}_i s_i$ may be eliminated by (37.5),

$$T_i \mathrm{D}_i s_i = \mathrm{D}_i u_i + \mathbf{p}_i \cdot \mathrm{D}_i \mathbf{l}_i. \tag{51.27}$$

Let $\dot{\varphi}_i$ and $\dot{\omega}_i$ denote the rates at which unit mass of $\mathscr{P}_i$ receives heat and irreversible work, then writing (7.7) as a rate equation we obtain

$$T_i \mathrm{D}_i s_i = \dot{\varphi}_i + \dot{\omega}_i.$$

Now $-\nabla\,.\,\mathbf{q}_i\,d\tau$ is the rate at which $\mathscr{P}_i$ receives heat from contiguous phases, so by the definitions of φ_{ij} and ψ_i given in § 37.1, it follows that

$$\rho_i\dot{\varphi}_i = -\nabla\,.\,\mathbf{q}_i+\sum_j \varphi_{ij}+\psi_i.$$

These results plus (51.10) enable us to write (51.26) in the form

$$\sigma = \sum_i\left[\frac{\dot{\omega}_i}{T_i}+\mathbf{q}_i\,.\,\nabla\frac{1}{T_i}+\rho\sum_j s_i v_{ir}\dot{\xi}_r+\sum_j\varphi_{ij}\left(\frac{1}{T_i}-\frac{1}{T_j}\right)\right], \tag{51.28}$$

which exhibits the irreversible components of work transfer, heat conduction, mass transfer, and heat transfer between species.

52. The rate of entropy production in a mixture

52.1. Conservation of momentum

Before we are in a position to use (51.22) or (51.26) to derive bilinear forms for σ, we require expressions for the rates $\mathrm{D}u$ or $\mathrm{D}_i u_i$ appearing in (51.24) and (51.27). These we can deduce in the usual way from the conservation equations for momentum and energy.

The body force acting on unit mass of $\mathscr{P}_i$ will be written $(\mathbf{F}_i+\mathbf{R}_i)$, where $\mathbf{F}_i$ is due to electrical and gravitational fields, etc. and $\mathbf{R}_i$ is the friction force acting on $\mathscr{P}_i$ because of the relative motion of all the other species past $\mathscr{P}_i$. Hence when (51.15) and (51.17) are applied to the momentum density $X_i = \rho_i\mathbf{v}_i$ we find (cf. (38.4))

$$\frac{\partial}{\partial t}(\rho_i\mathbf{v}_i)+\nabla\,.\,(\rho_i\mathbf{v}_i\mathbf{v}_i+\mathbf{p}_i) = \rho_i(\mathbf{F}_i+\mathbf{R}_i)+\sum_j(a_{ij}\mathbf{v}_j-a_{ji}\mathbf{v}_i)$$

and

$$\rho_i\mathrm{D}_i\mathbf{v}_i+\nabla\,.\,\mathbf{p}_i = \rho_i(\mathbf{F}_i+\mathbf{R}_i)+\sum_j a_{ij}(\mathbf{v}_j-\mathbf{v}_i). \tag{52.1}$$

We shall assume in what follows that all the pressure tensors are symmetric.

Shortly we shall find it necessary to analyse the dissipative force density $\rho_i\mathbf{R}_i$ into components $\mathbf{R}_{ij}$, where $\mathbf{R}_{ij}$ is the force density acting on $\mathscr{P}_i$ due to $\mathscr{P}_j$, that is, the friction force between the ith and jth species. As

$$\mathbf{R}_{ij} = -\mathbf{R}_{ji}, \tag{52.2}$$

we have

$$\rho_i\mathbf{R}_i = \sum_j\mathbf{R}_{ij} = -\sum_j\mathbf{R}_{ji} \tag{52.3}$$

and

$$\sum_{i,j}\mathbf{R}_{ij} = \sum_i\rho_i\mathbf{R}_i = 0. \tag{52.4}$$

52.2. Conservation of energy

There are four sources for the accumulation of energy within $\mathscr{P}_i$ additional to those described in § 38.3, namely, (1) heat energies φ_{ij}, $j = 1, 2, \ldots$, due to a difference in the temperatures T_i and T_j, (2) dissipation energies ξ_{ij}, $j = 1, 2, \ldots$, generated by the friction forces $\mathbf{R}_{ij}$, (3) energy $\overset{+}{e}_i$ $(e_i \equiv \rho_i(u_i + \frac{1}{2}v_i^2))$ due to species creation, and (4) flow work $\overset{+}{p}_i$ done by the pressures to introduce the created species into their new systems. Hence on applying (51.15) and (51.17) to the energy density $\rho_i(u_i + \frac{1}{2}v_i^2)$ we obtain (cf. (38.5) and (38.6))

$$\frac{\partial}{\partial t}[\rho_i(u_i + \tfrac{1}{2}v_i^2)] + \nabla \cdot [\rho_i \mathbf{v}_i(u_i + \tfrac{1}{2}v_i^2) + \mathbf{p}_i \cdot \mathbf{v}_i + \mathbf{q}_i]$$
$$= \rho_i(\mathbf{F}_i + \mathbf{R}_i) \cdot \mathbf{v}_i + \psi_i + \sum_j [\varphi_{ij} + \xi_{ij} + a_{ij}(u_j + p_j/\rho_j + \tfrac{1}{2}v_j^2)$$
$$- a_{ji}(u_i + p_i/\rho_i + \tfrac{1}{2}v_i^2)] \tag{52.5}$$

and

$$\rho_i \mathrm{D}_i(u_i + \tfrac{1}{2}v_i^2) + \nabla \cdot (\mathbf{p}_i \cdot \mathbf{v}_i + \mathbf{q}_i)$$
$$= \rho_i(\mathbf{F}_i + \mathbf{R}_i) \cdot \mathbf{v}_i + \psi_i + \sum_j [\varphi_{ij} + \xi_{ij} + a_{ij}(u_j + p_j/\rho_j + \tfrac{1}{2}v_j^2 - u_i - \tfrac{1}{2}v_i^2) - a_{ji}p_i/\rho_i]. \tag{52.6}$$

We now introduce the vicous stress tensor $\boldsymbol{\pi}_i$ and the dissipation function Φ_i by

$$\mathbf{p}_i = p_i\mathbf{1} + \boldsymbol{\pi}_i \qquad (\boldsymbol{\pi}_i = \tilde{\boldsymbol{\pi}}_i) \tag{52.7}$$

and

$$\Phi_i \equiv -\boldsymbol{\pi}_i : \nabla \mathbf{v}_i, \tag{52.8}$$

then from $(51.7)_2$

$$\nabla \cdot (\mathbf{p}_i \cdot \mathbf{v}_i) - \mathbf{v}_i \cdot \nabla \cdot \mathbf{p}_i = p_i \rho_i \mathrm{D}_i(1/\rho_i) + \overset{+}{\rho}_i p_i/\rho_i - \Phi_i.$$

Consequently if the scalar product of $(52.1)_2$ with $\mathbf{v}_i$ is subtracted from (52.6), the result may be written

$$\rho_i\left[\mathrm{D}_i u_i + p_i \mathrm{D}_i\left(\frac{1}{\rho_i}\right)\right]$$
$$= \Phi_i + \psi_i - \nabla \cdot \mathbf{q}_i + \sum_j \left\{\varphi_{ij} + \xi_{ij} + a_{ij}\left[u_j + \frac{p_j}{\rho_j} - \left(u_i + \frac{p_i}{\rho_i}\right) + \tfrac{1}{2}(v_j - v_i)^2\right]\right\}. \tag{52.9}$$

The force density $\mathbf{R}_{ij}$ has been postulated to be dissipative. As $\mathscr{P}_j$ moves with a velocity $(\mathbf{v}_j - \mathbf{v}_i)$ relative to $\mathscr{P}_i$, it exerts a force $\mathbf{R}_{ij}$ on $\mathscr{P}_i$, and the positive work done, namely, $\mathbf{R}_{ij}\cdot(\mathbf{v}_j-\mathbf{v}_i)$, is initially dissipated directly into $\mathscr{P}_i$ and $\mathscr{P}_j$. Hence

$$\xi_{ij}+\xi_{ji} = \mathbf{R}_{ij}\cdot(\mathbf{v}_j-\mathbf{v}_i) = \mathbf{R}_{ij}\cdot\mathbf{v}_j+\mathbf{R}_{ji}\cdot\mathbf{v}_i = \mathbf{R}_{ji}\cdot(\mathbf{v}_i-\mathbf{v}_j), \tag{52.10}$$

and it remains to decide how this energy divides between $\mathscr{P}_i$ and $\mathscr{P}_j$. One reasonable hypothesis, based on the behaviour of isotropic elastic collisions, is that the energy is distributed inversely as the molecular masses, that is, $\xi_{ij}/\xi_{ji} = m_j/m_i$, whence (see § 71.4 for further justification)

$$\xi_{ij} = x_{ij}\mathbf{R}_{ij}\cdot(\mathbf{v}_j-\mathbf{v}_i) \qquad \left(x_{ij} \equiv \frac{m_j}{m_i+m_j}\right). \tag{52.11}$$

52.3. The bilinear form

With the assumption that the reversible work term in (51.27) is $p_i\mathrm{D}_i\rho_i^{-1}$, we can combine (51.20), (51.26), (51.27), and (52.9) to obtain

$$\sigma = \sum_i \left(\frac{\Phi_i}{T_i}-\frac{\mathbf{q}_i}{T_i^2}\cdot\nabla T_i+\sum_j\left\{\frac{\xi_{ij}}{T_i}+(\varphi_{ij}+a_{ij}T_js_j)\left(\frac{1}{T_i}-\frac{1}{T_j}\right)+\right.\right.$$
$$\left.\left.+\frac{a_{ij}}{T_i}[g_j-g_i+\tfrac{1}{2}(\mathbf{v}_j-\mathbf{v}_i)^2]\right\}\right), \tag{52.12}$$

where we have introduced the specific Gibbs function, $g_i = u_i+p_i/\rho_i-T_is_i$. Here the rates a_{ij} are not independent affinities (see remarks in § 44.2), being related by (51.14) to the reaction rates $\dot{\xi}_r$, $r = 1, 2, \dots$, which it will be assumed have been chosen so as to be independent. Our final expression may be written

$$\sigma = \sum_i\left\{\frac{\Phi_i}{T_i}-\frac{\mathbf{q}_i}{T_i^2}\cdot\nabla T_i+\sum_j\left[\frac{\xi_{ij}}{T_i}+\varphi_{ij}\left(\frac{1}{T_i}-\frac{1}{T_j}\right)\right]\right\}+\sum_r\mathscr{A}_r\dot{\xi}_r, \tag{52.13}$$

where

$$\mathscr{A}_r \equiv \sum_{i,j} b_{ijr}\left\{T_js_j\left(\frac{1}{T_i}-\frac{1}{T_j}\right)+\frac{1}{T_i}[g_j-g_i+\tfrac{1}{2}(\mathbf{v}_j-\mathbf{v}_i)^2]\right\}. \tag{52.14}$$

Comparing (52.13) with (38.13), we note that the entropy production rate in a mixture is not merely the sum of the production rates in each fluid taken separately.

By (52.11) the friction term in (52.13) may be expressed in the symmetrical form

$$\sum_{i,j}\xi_{ij}/T_i = \sum_{i,j}{}' \mathbf{R}_{ij}\cdot(\mathbf{v}_j-\mathbf{v}_i)/\overline{T}_{ij}, \tag{52.15}$$

where

$$\bar{T}_{ij} = (m_i + m_j)T_iT_j/(m_iT_i + m_jT_j)$$

and the dash on the summation sign is to restrict $i < j$ so that each type of term appears only once.

When the components all have the same temperature T, (51.14) and (51.25) enable us to reduce (52.14) to

$$T\mathscr{A}_r = \rho A_r + \sum_{i,j} \tfrac{1}{2} b_{ijr}(\mathbf{v}_j - \mathbf{v}_i)^2, \tag{52.16}$$

and further, by (52.2), (52.3), (52.4), and (52.10),

$$\sum_{i,j} \xi_{ij} = \sum_{i,j} \mathbf{R}_{ij} \,.\, (\mathbf{v}_i - \mathbf{v}_j) = -\sum_i \rho_i \mathbf{v}_i \,.\, \mathbf{R}_i = -\sum_i \rho_i(\mathbf{v}_i - \mathbf{v}) \,.\, \mathbf{R}_i \tag{52.17}$$

(which is independent of the hypothesis involved in (52.11)). Eqn (52.13) now takes the form

$$T\sigma = \sum_i \left[\Phi_i - \frac{\mathbf{q}_i}{T} \,.\, \nabla\mathbf{T} - \rho_i \mathbf{R}_i \,.\, (\mathbf{v}_i - \mathbf{v})\right] + \rho \sum_r \left[A_r + \tfrac{1}{2}\sum_{i,j} b_{ijr}(\mathbf{v}_j - \mathbf{v}_i)^2\right]\dot{\xi}_r. \tag{52.18}$$

53. One-fluid equations for a mixture

When complete details of the individual behaviour of the various fluid species $\mathscr{P}_i$, $i = 1, 2, \ldots$, in a mixture are not required, we may sum the component momentum equations to form a single momentum equation—and likewise a single energy equation—for the compound fluid $\mathscr{P}$. This approach is useful only when it may be assumed that all the components $\mathscr{P}_i$ are in mutual thermal equilibrium, that is, if $T_i = T$ for all i. This important reduction in the amount of detail required in the model means that we can use the Gibbs equation (51.24) to relate the thermodynamic properties of $\mathscr{P}$. Even with this restriction, difficulties arise involving the pressure tensor and the heat flux for the compound system that can be resolved only by adopting the 'weak-diffusion' approximation, which means the neglect of quadratic terms in the diffusion vectors $\mathbf{w}_i$ and $\mathbf{J}_i = \rho_i \mathbf{w}_i$. It will be instructive to defer making this second approximation until it appears essential to do so.

Following the usual practice, we shall cast the one-fluid equations obtained by summation over the components into the same form as the one-component fluid equations of § 38, although it is important to distinguish between 'one-fluid' and 'one-component' equations, save of course in the limit when *all* details of individual species behaviour are suppressed.

53.1. Fluid mass and momentum

By (51.5), $(51.7)_4$, and (51.10)

$$\frac{\partial \rho_i}{\partial t}+\nabla \cdot \rho_i \mathbf{v}_i = \overset{+}{\rho}_i = \rho \sum_r \nu_{ir}\dot{\xi}_r \qquad \left(\sum_i \overset{+}{\rho}_i = 0\right), \tag{53.1}$$

which sums into

$$\frac{\partial \rho}{\partial t}+\nabla \cdot \rho \mathbf{v} = 0. \tag{53.2}$$

However, we will not abandon the component details contained in (53.1), for later we shall need to calculate the diffusion momentum $\mathbf{J}_i$ and the concentration c_i which by (53.1) and (53.2) satisfy (51.8), namely,

$$\rho \mathrm{D} c_i+\nabla \cdot \mathbf{J}_i = \overset{+}{\rho}_i. \tag{53.3}$$

By (51.1) and (52.4) eqns (52.1) may be summed over the species to give the one-fluid momentum equations

$$\begin{aligned} \frac{\partial(\rho \mathbf{v})}{\partial t}+\nabla \cdot (\rho \mathbf{v}\mathbf{v}+\mathsf{p}) &= \rho \mathbf{F}, \\ \rho \mathrm{D}\mathbf{v}+\nabla \cdot \mathsf{p} &= \rho \mathbf{F}, \end{aligned} \tag{53.4}$$

provided we define the total pressure tensor p to be

$$\mathsf{p} \equiv \sum_i [\mathsf{p}_i+\rho_i(\mathbf{v}_i-\mathbf{v})(\mathbf{v}_i-\mathbf{v})] \tag{53.5}$$

and write

$$\rho \mathbf{F} = \sum_i \rho_i F_i.$$

In terms of the total and partial viscous stress tensors $\boldsymbol{\pi}$ and $\boldsymbol{\pi}_i$ eqn (53.5) may be written

$$\mathsf{p} = p\mathbf{1}+\boldsymbol{\pi} = \sum_i (p_i\mathbf{1}+\boldsymbol{\pi}_i+\rho_i \mathbf{w}_i \mathbf{w}_i), \tag{53.6}$$

where $\mathbf{w}_i = \mathbf{v}_i-\mathbf{v}$ is the diffusion velocity. Here p_i is defined to be the equilibrium or thermodynamic pressure in the component $\mathscr{P}_i$, while p is as yet undefined. For the compound system $\mathscr{P}$ equilibrium occurs when each component is in equilibrium, that is, when $\boldsymbol{\pi}_i = 0$ *and* when the diffusion velocities are zero. Then (53.6) yields Dalton's law for the total pressure

$$p \equiv \sum_i p_i, \tag{53.7}$$

which we take to be the definition of p. It now follows from (53.6) that

$$\boldsymbol{\pi} = \sum_i (\boldsymbol{\pi}_i+\rho_i \mathbf{w}_i \mathbf{w}_i). \tag{53.8}$$

The average value of the normal stresses in $\mathscr{P}_i$ taken over a vanishingly small sphere is $\frac{1}{3}\operatorname{tr}(\mathbf{p}_i) = \check{p}_i/3$ † The trace of $\mathbf{p}_i = p_i\mathbf{1} + \boldsymbol{\pi}_i$ yields $p_i = \frac{1}{3}\check{p}_i - -\frac{1}{3}\check{\pi}_i$, showing that only when $\check{\pi}_i$ happens to be zero, or in equilibrium when $\boldsymbol{\pi}_i = 0$, it is true that $p_i = \frac{1}{3}\check{p}_i$. Similarly from (53.6) and (53.8) the total pressure is related to $\check{p}$ by

$$p = \tfrac{1}{3}\check{p} - \tfrac{1}{3}\sum_i \check{\pi}_i - p' \qquad \left(p' \equiv \tfrac{1}{3}\sum_i \rho_i w_i^2\right), \tag{53.9}$$

where p' is the diffusion pressure.

On the basis of eqn (53.9) the diffusion pressure is often included as part of the total pressure, but to do this means a departure from the purely thermodynamic significance of pressure. The definition in (53.7) has the advantage that this pressure is the same as that occurring in Gibbs' equation (51.24). In any case, as we shall shortly see, the one-fluid model is practical only when the quadratic terms in (53.8) and (53.9) involving the diffusion velocity are negligible.

53.2. *Fluid energy*

The next task is to form a one-fluid energy equation by summing eqns (52.5) and (52.6) over i. From its definition φ_{ij} is zero when $T_i = T_j = T$, and the terms involving a_{ij} in (52.5) obviously cancel on summation over both i and j. Thus by (52.17) the right-hand side of (52.5) reduces to

$$\sum_i \rho_i \mathbf{F}_i \cdot \mathbf{v}_i + \psi \qquad \left(\psi = \sum \psi_i\right).$$

To deal with the left-hand side of (52.5) we need definitions for the internal energy density ρu and the heat flux vector $\mathbf{q}$ in the composite system. As we wish to present the summed equations in much the same form as the one-component energy equation given in (38.5), we write

$$\sum_i c_i(u_i + \tfrac{1}{2}v_i^2) = u + u' + \tfrac{1}{2}v^2$$

and

$$\sum_i [\mathbf{p}_i \cdot \mathbf{v}_i + \mathbf{q}_i + \rho_i \mathbf{w}_i(u_i + \tfrac{1}{2}v_i^2)] = \mathbf{p} \cdot \mathbf{v} + \mathbf{q},$$

† Let the sphere S have a radius vector $\mathbf{r}$, then the average value of the normal stress over S is

$$\frac{1}{4\pi r^2}\int_S \mathbf{n} \cdot \mathbf{p} \cdot \mathbf{n}\, \mathrm{d}S = \frac{1}{4\pi r^3}\int_S \mathbf{n} \cdot \mathbf{p} \cdot \mathbf{r}\, \mathrm{d}S = \frac{1}{4\pi r^3}\int_V \nabla \cdot (\mathbf{p} \cdot \mathbf{r})\, \mathrm{d}\tau \approx \tfrac{1}{3}(\nabla \cdot \mathbf{p}) \cdot \mathbf{r} + \tfrac{1}{3}\mathbf{p} : \nabla r.$$

Thus $\lim_{r\to 0}(1/4\pi r^2)\int_S \mathbf{n} \cdot \mathbf{p} \cdot \mathbf{n}\, \mathrm{d}S = \frac{1}{3}\mathbf{p} : \mathbf{1}$, as $\nabla \mathbf{r} = \mathbf{1}$. This limit is often wrongly taken to be the definition of the scalar pressure.

where

$$u \equiv \sum_i c_i u_i, \qquad u' \equiv \sum_i \tfrac{1}{2} c_i w_i^2 = \frac{3}{2}\frac{p'}{\rho}, \tag{53.10}$$

and

$$\mathbf{q} = \sum_i \left[\mathbf{q}_i + \rho_i \mathbf{w}_i(u_i + \tfrac{1}{2}w_i^2) + \mathbf{p}_i \cdot \mathbf{w}_i\right]. \tag{53.11}$$

Hence u' is the internal energy of diffusion and u is the thermodynamic internal energy of Gibb's equation. It is misleading to describe $(u+u')$ as being 'the internal energy', as the application of either heat or work affects u and u' quite differently. Eqn (53.11) reveals that the concepts of heat and work transfer are blurred in mixture theory, and that we must not take over the results of one-component theory without further consideration.

The sum of (52.5) is thus

$$\frac{\partial}{\partial t}[\rho(u+u'+\tfrac{1}{2}v^2)] + \nabla \cdot [\rho\mathbf{v}(u+u'+\tfrac{1}{2}v^2) + \mathbf{p}\cdot\mathbf{v} + \mathbf{q}] = \sum_i \rho_i \mathbf{F}_i \cdot \mathbf{v}_i + \psi. \tag{53.12}$$

Eqn (53.2) enables us to write (53.12) as

$$\rho \mathrm{D}(u+u'+\tfrac{1}{2}v^2) + \nabla \cdot (\mathbf{p}\cdot\mathbf{v}+\mathbf{q}) = \rho\mathbf{F}\cdot\mathbf{v} + \psi + \sum_i \mathbf{J}_i \cdot \mathbf{F}_i,$$

then from the definition of u'

$$\rho \mathrm{D}u' = \sum_i \mathbf{J}_i \cdot \mathrm{D}\mathbf{w}_i + \rho \sum_i \tfrac{1}{2} w_i^2 \mathrm{D}c_i, \tag{53.13}$$

hence

$$\rho \mathrm{D}(u+\tfrac{1}{2}v^2) + \nabla \cdot (\mathbf{p}\cdot\mathbf{v}+\mathbf{q}) = \rho\mathbf{F}\cdot\mathbf{v} + \psi + \sum_i [\mathbf{J}_i \cdot (\mathbf{F}_i - \mathrm{D}\mathbf{w}_i) - \tfrac{1}{2}\rho w_i^2 \mathrm{D}c_i]. \tag{53.14}$$

It is the last term that distinguishes the one-fluid energy equation from the one-component energy equation.

Finally reducing (53.14) to the intrinsic form with the help of (53.4) we obtain

$$\rho \mathrm{D}u + \rho p \mathrm{D}\rho^{-1} = -\boldsymbol{\pi} : \nabla\mathbf{v} + \sum_i [\mathbf{J}_i \cdot (\mathbf{F}_i - \mathrm{D}\mathbf{w}_i) - \tfrac{1}{2}\rho w_i^2 \mathrm{D}c_i] - \nabla \cdot \mathbf{q} + \psi, \tag{53.15}$$

which is the first law of thermodynamics for the composite system $\mathscr{P}$.

53.3. Entropy production rate

For the present application (51.22) may be written

$$\sigma = \rho \mathrm{D}s + \nabla \cdot \left[\sum_i (\mathbf{q}_i + T s_i \mathbf{J}_i)/T\right] - \psi/T. \tag{53.16}$$

On the other hand, the one-fluid heat flux vector, which we might have expected to appear in (53.16), is

$$\mathbf{q} = \sum_i [\mathbf{q}_i + \mathbf{J}_i(h_i + \tfrac{1}{2}w_i^2) + \boldsymbol{\pi}_i \cdot \mathbf{w}_i] \qquad (h_i = u_i + p_i/\rho_i),$$

so that

$$\sum_i (\mathbf{q}_i + Ts_i\mathbf{J}_i) = \mathbf{q} - \sum_i \tilde{g}_i\mathbf{J}_i - \boldsymbol{\varepsilon}, \tag{53.17}$$

where

$$\boldsymbol{\varepsilon} \equiv \sum_i \boldsymbol{\pi}_i \cdot \mathbf{w}_i, \qquad \tilde{g}_i = g_i + \tfrac{1}{2}w_i^2 \qquad (g_i = h_i - Ts_i).$$

We shall also need (51.24),

$$T\mathrm{D}s = \mathrm{D}u + \mathbf{p} \cdot \mathrm{D}\mathbf{l} + \sum_i \frac{g_i}{\rho}\nabla \cdot \mathbf{J}_i + \sum_r A_r\dot{\xi}_r, \tag{53.18}$$

in which $\mathbf{p} \cdot \mathrm{D}\mathbf{l}$ denotes the reversible work. As pointed out in §§ 38.4 and 38.6, it is necessary to consider the nature of the fluid system of interest to decide what work is actually reversible. It is reasonable to make the usual assumption that the work $p\mathrm{D}\rho^{-1}$ done by the normal pressures is reversible, and it remains to decide if any of the further work terms appearing on the right-hand side of (53.15) could be reversible.† The only possible candidates appear to be the components of $-\rho\mathrm{D}u'$, which by (51.9), (51.10), and (53.13) are

$$-\sum_i \mathbf{J}_i \cdot \mathrm{D}\mathbf{w}_i + \sum_i \tfrac{1}{2}w_i^2\nabla \cdot \mathbf{J}_i - \tfrac{1}{2}\rho\sum_i\sum_r w_i^2 v_{ir}\dot{\xi}_r.$$

Now the diffusion energy u' is different in character from the internal energy u; in particular it may be degraded irreversibly into u by internal friction between components. Work of certain kinds, for example, electrical work on a mixture of electrons and ions, may increase u', but eventually this energy flows into u and cannot be returned to u'. It is therefore reasonable to assume that $-\rho\mathrm{D}u'$ is not reversible.

On replacing $\mathbf{p} \cdot \mathrm{D}\mathbf{l}$ by $p\mathrm{D}\rho^{-1}$ and collecting (53.15)–(53.18) into an expression for σ, we arrive at

$$T\sigma = -\boldsymbol{\pi}:\nabla\mathbf{v} - \mathbf{q} \cdot \nabla T/T + \sum_i \mathbf{J}_i \cdot [\mathbf{F}_i - \mathrm{D}\mathbf{w}_i - T\nabla(\tilde{g}_i/T)] + \\ + \rho\sum_r \tilde{A}_r\dot{\xi}_r - T\nabla \cdot (\boldsymbol{\varepsilon}/T), \tag{53.19}$$

where (cf. $(51.25)_1$)

$$\tilde{A}_r \equiv -\sum_i v_{ir}\tilde{g}_i.$$

† In § 52.3 we chose this work to be $\sum_i p_i\mathrm{D}_i\rho_i^{-1}$, which is not quite the same as $p\mathrm{D}\rho^{-1}$, but the difference between the two expressions is easily traced to the additional term $\sum_i g_i \nabla \cdot \mathbf{J}_i/\rho$ in (53.18).

To this point the only property of the individual species that has been suppressed is their separate temperatures, so unless we intend to retain individual momentum equations, some further approximation is necessary. The obvious thing is to limit the theory to 'weak diffusion', which means that we neglect quadratic terms in the diffusion vectors $\mathbf{w}_i$ and $\mathbf{J}_i$. Further, as we may expect the viscosity stress tensor $\boldsymbol{\pi}_i$ to be small under normal flow conditions, the product $\boldsymbol{\pi}_i \cdot \mathbf{w}_i$ is also negligible. Such approximations reduce (53.19) to

$$T\sigma = -\boldsymbol{\pi}:\nabla\mathbf{v} - \mathbf{q}\cdot\nabla T/T + \sum_i \mathbf{J}_i\cdot[\mathbf{F}_i - T\nabla(g_i/T)] + \rho\sum_r A_r\dot{\xi}_r. \tag{53.20}$$

Dropping the component inertia $\mathrm{D}\mathbf{w}_i$ imposes a lower bound on the macroscopic time scale. To see this note from (53.19) that one of the constitutive equations for the first component will read

$$\mathbf{F}_1 - \mathrm{D}\mathbf{w}_1 - T\nabla(g_1/T) = \beta\mathbf{J}_1 + \text{various coupling terms},$$

where β is a positive phenomenological constant. Thus

$$\mathrm{D}\mathbf{w}_1 + \beta\rho\mathbf{w}_1 = \mathbf{F}_1 - T\nabla(g_1/T) + \dots,$$

and a comparison of the magnitudes of the left-hand terms show that $\mathrm{D}\mathbf{w}_1$ may be neglected for times $t \gg (\beta\rho)^{-1}$.

In the present approximation (53.5), (53.11), and (53.14) reduce to

$$\mathsf{p} = \sum_i \mathsf{p}_i, \qquad \mathbf{q} = \sum_i (\mathbf{q}_i + h_i\mathbf{J}_i) \tag{53.21}$$

and

$$\rho\mathrm{D}(u + \tfrac{1}{2}v^2) + \nabla\cdot(\mathsf{p}\cdot\mathbf{v} + \mathbf{q}) = \rho\mathbf{F}\cdot\mathbf{v} + \psi + \sum_i \mathbf{J}_i\cdot\mathbf{F}_i. \tag{53.22}$$

53.4. *Alternative forms for the entropy production rate*

The distinction made in (53.20) between dissipated work and dissipated heat is not unique, for with the definitions

$$\bar{\mathbf{q}} = \sum_i \mathbf{q}_i, \qquad \mathbf{J}_s = T^{-1}\sum_i (\mathbf{q}_i + Ts_i\mathbf{J}_i), \tag{53.23}$$

and the thermodynamic identities

$$T\,\mathrm{d}(g_i/T) = \mathrm{d}h_i - T\,\mathrm{d}s_i - (s_i\,\mathrm{d}T + g_i\,\mathrm{d}T/T) = (\mathrm{d}g_i)_T - h_i\,\mathrm{d}T/T,$$

the formula for σ may be written in the equivalent forms,

$$\begin{aligned} T\sigma &= -\boldsymbol{\pi}:\nabla\mathbf{v} - \mathbf{J}_s\cdot\nabla T + \sum_i \mathbf{J}_i\cdot(\mathbf{F}_i - \nabla g_i) + \rho\sum_r A_r\dot{\xi}_r, \\ T\sigma &= -\boldsymbol{\pi}:\nabla\mathbf{v} - \bar{\mathbf{q}}\cdot\nabla T/T + \sum_i \mathbf{J}_i\cdot(\mathbf{F}_i - (\nabla g_i)_T) + \rho\sum_r A_r\dot{\xi}_r. \end{aligned} \tag{53.24}$$

However, the non-uniqueness of the bilinear form for σ causes no difficulty, as the conditions of the theorem stated in § 44.1 have been met by the linear transformations applied to the fluxes in deriving (53.24).

We return now to the exact expression contained in (53.19). Written out in full it will be found to be a somewhat distorted formula that mixes up diffusion, heat transfer, and viscous effects in its various terms (cf. (53.5) and (53.11)). This is a consequence of our adopting the convention that the one-fluid equations should be cast into the same form as the one-component equations. When the weak diffusion approximation is valid this does not matter; otherwise we may disentangle the effects in (53.19) by rewriting it as

$$T\sigma = -\sum_i \boldsymbol{\pi}_i : \nabla \mathbf{v} - \mathbf{J}_s \cdot \nabla T + \sum_i \mathbf{J}_i \cdot (\mathbf{F}_i - \mathrm{D}_i \mathbf{v}_i - \nabla g_i) + \rho \sum_r \tilde{A}_r \dot{\xi}_r - \nabla \cdot \boldsymbol{\varepsilon},$$

which is still exact.

The approximations $\nabla \cdot \boldsymbol{\varepsilon} \approx 0$, $\tilde{A}_r \dot{\xi}_r \approx A_r \dot{\xi}_r$, which are rather less demanding than those used in deriving (53.20), now yield

$$T\sigma = -\bar{\boldsymbol{\pi}} : \nabla \mathbf{v} - \mathbf{J}_s \cdot \nabla T + \sum_i \mathbf{J}_i \cdot (\mathbf{F}_i - \mathrm{D}_i \mathbf{v}_i - \nabla g_i) + \rho \sum_r A_r \dot{\xi}_r, \qquad (53.25)$$

where

$$\bar{\boldsymbol{\pi}} \equiv \sum_i \boldsymbol{\pi}_i .$$

However, the retention of the term involving $\mathrm{D}_i \mathbf{v}_i$ does mean that separate species momentum equations are now necessary, so that (53.25) must be regarded as being an approximation to (52.18) rather than being a generalization of $(53.24)_1$.

54. Constitutive equations for mixtures

54.1. The one-fluid equations

We shall base the constitutive equations on $(53.24)_1$. With the assumption that the viscous stress tensor is symmetric we may write it in the form (cf. (38.17))

$$T\sigma = -\tfrac{1}{3}\check{\pi} \nabla \cdot \mathbf{v} + \rho \sum_r A_r \dot{\xi}_r + \sum_i \mathbf{J}_i \cdot (\mathbf{F}_i - \nabla g_i) - \mathbf{J}_s \cdot \nabla T - \mathring{\boldsymbol{\pi}} : \mathring{\nabla \mathbf{v}}. \qquad (54.1)$$

Suppose that the medium possesses lateral isotropy about unit vector $\mathbf{k}$ parallel to the $0z$-axis, then by a simple extension of the theory of § 50.1 we may infer from (54.1) that the constitutive relations for the scalar and second-order tensor processes have the appearance

$$\left.\begin{aligned}
\dot{\xi}_r &= \sum_s \gamma_{rs} A_s + \alpha_r \nabla \cdot \mathbf{v} + \beta_r (\mathbf{kk} - \tfrac{1}{3}\mathbf{1}) : \nabla \overset{\circ}{\mathbf{v}}, \\
\tfrac{1}{3}\overset{\times}{\pi} &= -\rho \sum_s \alpha_s A_s - \rho v' \nabla \cdot \mathbf{v} - \rho v'' (\mathbf{kk} - \tfrac{1}{3}\mathbf{1}) : \nabla \overset{\circ}{\mathbf{v}}, \\
\overset{\circ}{\boldsymbol{\pi}} &= -\rho \left(\sum_s \beta_s A_s + v'' \nabla \cdot \mathbf{v} \right) (\mathbf{kk} - \tfrac{1}{3}\mathbf{1}) - \rho \sum_{i=1}^{5} v_i \mathbf{N}_i : \nabla \overset{\circ}{\mathbf{v}},
\end{aligned}\right\} \quad (54.2)$$

where $\gamma_{rs}, \alpha_r, \beta_r, v', v'', v_1, v_2, v_3$ are *even* functions of $\mathbf{B}$; v_4, v_5 are *odd* functions of $\mathbf{B}$.

In writing down (54.2) we have incorporated the appropriate reciprocal relations (§ 45.2) except for

$$\gamma_{rs} = \gamma_{sr}. \quad (54.3)$$

We see from (53.1) that

$$\sum_{i,r} v_{ir} \dot{\xi}_r = 0,$$

consequently because of the independence of the affinities in $(54.2)_1$ the phenomenological coefficients must satisfy the constraints

$$\sum_{i,r} v_{ir} \gamma_{rs} = 0, \qquad \sum_{i,r} v_{ir} \alpha_r = 0, \qquad \sum_{i,r} v_{ir} \beta_r = 0. \quad (54.4)$$

For the vector processes we find

$$\left.\begin{aligned}
\mathbf{J}_i &= \sum_j \boldsymbol{\lambda}_{ij} \cdot (\mathbf{F}_j - \nabla g_j) + \boldsymbol{\delta}_i \cdot \nabla T, \\
\mathbf{J}_s &= -\sum_j \boldsymbol{\eta}_j \cdot (\mathbf{F}_j - \nabla g_j) - \boldsymbol{\kappa} \cdot \nabla T,
\end{aligned}\right\} \quad (54.5)$$

in which all the phenomenological tensors have the same structure as $\boldsymbol{\kappa}$, namely,

$$\boldsymbol{\kappa} = \kappa_{\parallel} \mathbf{kk} + \kappa_{\perp} (\mathbf{ii} + \mathbf{jj}) + \kappa_{\wedge} (\mathbf{ji} - \mathbf{ij}) \qquad \begin{pmatrix} \kappa_{\parallel}, \kappa_{\perp} & \textit{even} & \text{in } \mathbf{B} \\ \kappa_{\wedge} & \textit{odd} & \end{pmatrix}. \quad (54.6)$$

From (54.6) and similar expansions it follows that

$$\boldsymbol{\kappa} = \boldsymbol{\kappa}^*, \qquad \boldsymbol{\lambda}_{ij} = \boldsymbol{\lambda}^*_{ij}, \qquad \boldsymbol{\delta}_i = \boldsymbol{\delta}^*_i, \qquad \boldsymbol{\eta}_i = \boldsymbol{\eta}^*_i, \quad (54.7)$$

where by * we indicate the double operation of taking the transpose and of changing the sign of $\mathbf{B}$, that is,

$$\text{if} \quad \boldsymbol{\kappa} = \boldsymbol{\kappa}(\mathbf{B}), \qquad \boldsymbol{\kappa}^* = \tilde{\boldsymbol{\kappa}}(-\mathbf{B}). \quad (54.8)$$

Eqn $(54.7)_1$ is also a consequence of the reciprocal relation (45.5), which also yields

$$\boldsymbol{\lambda}_{ij} = \boldsymbol{\lambda}^*_{ji}, \qquad \boldsymbol{\delta}_i = \boldsymbol{\eta}^*_i,$$

hence by (54.7)

$$\boldsymbol{\lambda}_{ij} = \boldsymbol{\lambda}_{ji}, \qquad \boldsymbol{\delta}_i = \boldsymbol{\eta}_i. \tag{54.9}$$

From (54.1) and (54.2) we find that the entropy production rate due to the scalar and tensor processes is given by

$$T\rho^{-1}(\sigma_0+\sigma_2) = \sum_{r,s} A_r\gamma_{rs}A_s + \nu'(\nabla\cdot\mathbf{v})^2 + \sum_{i=1}^{5} \nu_i\overset{\circ}{\nabla\mathbf{v}}:\boldsymbol{N}_i:\overset{\circ}{\nabla\mathbf{v}} + 2\nabla\cdot\mathbf{v}\sum_r \alpha_r A_r +$$
$$+2\nu''\nabla\cdot\mathbf{v}(\mathbf{kk}-\tfrac{1}{3}\mathbf{1}):\overset{\circ}{\nabla\mathbf{v}} + 2\sum_r \beta_r A_r(\mathbf{kk}-\tfrac{1}{3}\mathbf{1}):\overset{\circ}{\nabla\mathbf{v}}. \tag{54.10}$$

By (46.4) this is a non-negative form which, by the independence of the affinities A_r, $\nabla\cdot\mathbf{v}$, and $\overset{\circ}{\nabla\mathbf{v}}$, imposes a variety of constraints on the coefficients. We shall not attempt a complete list. Some have been given in § 50.2; others are that the principal minors of $|\gamma_{rs}|$ must be positive, that $\gamma_{rr}\nu' \geqslant \alpha_r^2$, $r = 1, 2, \ldots$, and so on.

By (54.1) and (54.5) the vectorial entropy production rate σ_1, also non-negative, is given by

$$T\sigma_1 = \sum_{i,j}\mathbf{E}_i\cdot\boldsymbol{\lambda}_{ij}\cdot\mathbf{E}_j + \sum_i \mathbf{E}_i\cdot(\boldsymbol{\delta}_i+\tilde{\boldsymbol{\delta}}_i)\cdot\nabla T + \nabla T\cdot\boldsymbol{\kappa}\cdot\nabla T, \tag{54.11}$$

where

$$\mathbf{E}_i \equiv \mathbf{F}_i - \nabla g_i.$$

From (47.5) each tensor coefficient in (54.11) possesses an expansion like

$$\boldsymbol{\lambda} = \lambda_{\parallel}\mathbf{kk} + \lambda_{\perp}(\mathbf{1}-\mathbf{kk}) + \lambda_{\wedge}\mathbf{k}\times\mathbf{1}, \tag{54.12}$$

so that with the definitions

$$\mathbf{E}_{\parallel} \equiv \mathbf{kk}\cdot\mathbf{E}, \qquad \mathbf{E}_{\perp} \equiv \mathbf{E}-\mathbf{E}_{\parallel}, \tag{54.13}$$

we find

$$T\sigma_1 = \sum_{i,j}(\lambda_{ij\parallel}\mathbf{E}_{i\parallel}\cdot\mathbf{E}_{j\parallel} + \lambda_{ij\perp}\mathbf{E}_{i\perp}\cdot\mathbf{E}_{j\perp} + \lambda_{ij\wedge}\mathbf{k}\cdot\mathbf{E}_j\times\mathbf{E}_i) +$$
$$+2\sum_i(\delta_{i\parallel}\mathbf{E}_{i\parallel}\cdot\nabla_{\parallel}T + \delta_{i\perp}\mathbf{E}_{i\perp}\cdot\nabla_{\perp}T) +$$
$$+\kappa_{\parallel}(\nabla_{\parallel}T)^2 + \kappa_{\perp}(\nabla_{\perp}T)^2 \geqslant 0. \tag{54.14}$$

This form enables a large number of constraints on the scalar coefficients to be deduced, amongst which are

$$\lambda_{ii\parallel},\lambda_{ii\perp},\kappa_{\parallel},\kappa_{\perp}\geqslant 0,\qquad \lambda_{ii\parallel}\kappa_{\parallel}\geqslant\delta_{i\parallel}^{2},\qquad \lambda_{ii\perp}\kappa_{\perp}\geqslant\delta_{i\perp}^{2}. \tag{54.15}$$

54.2. The complete set of one-fluid equations

It may be helpful to the reader at this stage to see a complete list of the equations and constraints involved in the theory. These relations, which fall into four distinct groups, are set out in Table 8, and also listed are the basic variables. These have been placed in a one-to-one correspondence with the equations merely to make it immediately clear that the set is complete—their values are, of course, determined by all the equations. The body force **F** and radiation heat ψ are supposed given.

TABLE 8

Name	*Equations*	*Variables*
THE CONSERVATION EQUATIONS		
Mass	$\rho\mathrm{D}c_i+\nabla.\mathbf{J}_i=\rho\sum_i v_{ir}\dot{\xi}_r,\qquad \frac{\partial\rho}{\partial t}+\nabla.\rho\mathbf{v}=0$	$c_1,c_2,\ldots\rho$
Momentum	$\rho\mathrm{D}\mathbf{v}+\nabla.\mathbf{p}=\rho\mathbf{F}$	$\mathbf{v}$
Energy	$\rho\mathrm{D}u+\rho p\mathrm{D}\rho^{-1}=-\boldsymbol{\pi}:\nabla\mathbf{v}+\sum_i\mathbf{J}_i.\mathbf{F}_i-\nabla.\mathbf{q}+\psi$	u
Entropy	$T\mathrm{D}s=\mathrm{D}u+p\mathrm{D}\rho^{-1}+\sum_i g_i\nabla.\mathbf{J}_i/\rho+\sum_r A_r\dot{\xi}_r$	s
THE THERMODYNAMIC RELATIONS		
Gibbs	$T\mathrm{d}s=\mathrm{d}u+p\,\mathrm{d}\rho^{-1}-\sum_i g_i\,\mathrm{d}c_i+\sum_r A_r\,\mathrm{d}\xi_r$	s
State equations	$T=T(u,\rho,c_1,c_2,\ldots\xi_1,\xi_2,\ldots)$	T
	$p=p(u,\rho,c_1,c_2,\ldots\xi_1,\xi_2,\ldots)$	p
	$g_i=g_i(u,\rho,c_1,c_2,\ldots\xi_1,\xi_2,\ldots),\quad i=1,2,\ldots$	$g_1,g_2,\ldots$
	$A_r=A_r(u,\rho,c_1,c_2,\ldots\xi_1,\xi_2,\ldots),\quad r=1,2,\ldots$	$A_1,A_2,\ldots$
CONSTITUTIVE RELATIONS		
Kinetic	First of (54.2)	$\dot{\xi}_1,\dot{\xi}_2,\ldots$
Navier–Stokes	Second and third of (54.2); $\mathbf{p}=(p+\frac{1}{3}\overset{\times}{\pi})\mathbf{1}+\mathring{\boldsymbol{\pi}}$	$\mathbf{p}$
Fick	First of (54.5)	$\mathbf{J}_1,\mathbf{J}_2,\ldots$
Fourier	Second of (54.5); $\mathbf{q}=T\mathbf{J}_s+\sum_i g_i\mathbf{J}_i$	$\mathbf{q}$
CONSTRAINT RELATIONS		
(1) Thermodynamic inequalities, for example (54.15).		
(2) Material symmetries, for example (54.6).		
(3) Reciprocal relations, for example (54.3), the first of (54.7), and (54.9).		
(4) Mass conservation in reactions, for example (54.4).		

55. Minimum entropy production

Suppose a thermodynamic system extending over a volume V has time-independent boundary conditions applied to its surface. The total entropy production within the system is

$$P = \int_V \sigma \, \mathrm{d}\tau \tag{55.1}$$

per second and, under the rather restrictive circumstances to be described below, it is possible to show that

$$\frac{\partial P}{\partial t} \leqslant 0. \tag{55.2}$$

From this inequality we infer that the stationary state is characterized by a minimum of the entropy production, compatible with the imposed boundary conditions.

55.1. A basic inequality

As a first step towards our objective we shall prove the inequality

$$\mathrm{d}\left(\frac{1}{T}\right) \mathrm{d}(\rho u) - \sum_{i=1}^{l} \mathrm{d}\left(\frac{g_i}{T}\right) \mathrm{d}\rho_i \leqslant 0 \tag{55.3}$$

for a compound system of l components at a single temperature T. If eqns (17.10) and (15.6) are divided through by the constant total mass of the system, we have

$$\mathrm{d}\left(\frac{1}{T}\right) \mathrm{d}u + \mathrm{d}\left(\frac{p}{T}\right) \mathrm{d}\left(\frac{1}{\rho}\right) - \sum_i \mathrm{d}\left(\frac{g_i}{T}\right) \mathrm{d}\left(\frac{\rho_i}{\rho}\right) \leqslant 0 \tag{55.4}$$

and

$$s \, \mathrm{d}T - \rho \, \mathrm{d}p + \sum_i \frac{\rho_i}{\rho} \mathrm{d}g_i = 0, \tag{55.5}$$

as the volume V is the only deformation coordinate in the present application. Now (55.4) can be written

$$\mathrm{d}\left(\frac{1}{T}\right) \mathrm{d}(\rho u) - \sum_i \mathrm{d}\left(\frac{g_i}{T}\right) \mathrm{d}\rho_i - \left[u\mathrm{d}\left(\frac{1}{T}\right) + \frac{1}{\rho} \mathrm{d}\left(\frac{p}{T}\right) - \frac{1}{\rho} \sum_i \rho_i \, \mathrm{d}\left(\frac{g_i}{T}\right) \right] \mathrm{d}\rho \leqslant 0,$$

and on using $\sum_i \rho_i g_i = \rho g = \rho u + p - Ts$ and (55.5), the last term may be eliminated and hence (55.3) established.

The basic conservation laws for diffusing components given in § 54.2 may be written

$$\frac{\partial \rho_i}{\partial t} = -\nabla \, . \, \rho_i \mathbf{v}_i + \rho \sum_r \nu_{ir} \dot{\xi}_r, \tag{55.6}$$

$$\rho \frac{\partial \mathbf{v}}{\partial t} = -\rho \mathbf{v} \, . \, \nabla \mathbf{v} - \nabla \, . \, \mathbf{p} + \rho \mathbf{F}, \tag{55.7}$$

and

$$\frac{\partial(\rho u)}{\partial t} = -\nabla \cdot (\rho u \mathbf{v}) - \mathbf{p}:\nabla\mathbf{v} - \nabla \cdot \mathbf{q} + \sum_i \mathbf{J}_i \cdot \mathbf{F}_i. \tag{55.8}$$

Now in (55.3) we may replace the differentials by time derivatives and then, adding the negative quantity $-(\rho/T)(\partial\mathbf{v}/\partial t)^2$, arrive at

$$\frac{\partial}{\partial t}\left(\frac{1}{T}\right)\frac{\partial}{\partial t}(\rho u) - \frac{\rho}{T}\left(\frac{\partial \mathbf{v}}{\partial t}\right)^2 - \sum_i \frac{\partial}{\partial t}\left(\frac{g_i}{T}\right)\frac{\partial \rho_i}{\partial t} \leqslant 0. \tag{55.9}$$

It follows from eqns (55.6)–(55.9) that

$$\frac{\partial}{\partial t}\left(\frac{1}{T}\right)\left[-\nabla \cdot (\rho u \mathbf{v}) - \mathbf{p}:\nabla\mathbf{v} - \nabla \cdot \mathbf{q} + \sum_i \mathbf{J}_i \cdot \mathbf{F}_i\right] + \frac{1}{T}\frac{\partial \mathbf{v}}{\partial t} \cdot (\rho\mathbf{v} \cdot \nabla\mathbf{v} + \nabla \cdot \mathbf{p} - \rho\mathbf{F}) +$$
$$+ \sum_i \frac{\partial}{\partial t}\left(\frac{g_i}{T}\right)\left\{\nabla \cdot \rho_i \mathbf{v}_i - \rho \sum_i v_{ir}\dot{\xi}_r\right\} \leqslant 0,$$

which may be rearranged in the form

$$\nabla \cdot \left[-(\mathbf{q} + \rho u \mathbf{v})\frac{\partial}{\partial t}\left(\frac{1}{T}\right) + \frac{1}{T}\mathbf{p} \cdot \frac{\partial \mathbf{v}}{\partial t} + \sum_i \rho_i \mathbf{v}_i \frac{\partial}{\partial t}\left(\frac{g_i}{T}\right)\right] +$$
$$+ (\mathbf{q} + \rho u \mathbf{v} + \mathbf{p} \cdot \mathbf{v} + \rho\mathbf{v}\tfrac{1}{2}v^2) \cdot \frac{\partial}{\partial t}\left(\nabla \frac{1}{T}\right) - \sum_i \rho_i \mathbf{v}_i \cdot \frac{\partial}{\partial t}\left[\nabla\left(\frac{g_i}{T}\right) - \frac{\mathbf{F}_i}{T}\right] -$$
$$-(\mathbf{p} + \rho\mathbf{v}\mathbf{v}):\frac{\partial}{\partial t}\left(\nabla \frac{\mathbf{v}}{T}\right) + \rho\mathbf{v} \cdot \frac{\partial}{\partial t}\left[\nabla\left(\frac{v^2}{2T}\right)\right] -$$
$$-\rho\mathbf{F} \cdot \frac{\partial}{\partial t}\left(\frac{\mathbf{v}}{T}\right) - \sum_i \rho_i \mathbf{v}_i \cdot \frac{1}{T}\frac{\partial \mathbf{F}_i}{\partial t} + \sum_r \rho\dot{\xi}_r \frac{\partial}{\partial t}\left(\frac{\mathbf{A}_r}{T}\right) \leqslant 0. \tag{55.10}$$

Other arrangements are possible; this particular choice has the merit of producing the time derivatives of the thermodynamic processes appearing in (53.20).

55.2. *Theorems of Prigogine and Helmholtz*

If (55.10) is now integrated over the volume V the first term becomes a surface integral that vanishes owing to the assumed steady-state boundary conditions. The integral is non-positive and may be expressed

$$\int_V \sum_\alpha J'_\alpha \frac{\partial}{\partial t} X'_\alpha \, d\tau \leqslant 0, \tag{55.11}$$

where J'_α and X'_α denote the scalar components of the quantities appearing in (55.10); they may be thought of as being 'generalized' processes and affinities. When it is possible to extract the time derivative from the integral in (55.11) and write it in the form

$$\frac{\partial \varphi}{\partial t} \leqslant 0, \tag{55.12}$$

φ being an integral over V, then φ will decrease in time until the stationary state is achieved. However, in general this procedure is not possible, and we are confined to special cases.

First suppose that there is no mechanical convection in V, although diffusion of species is permitted, that is, that $\mathbf{v}$ is zero but $\mathbf{v}_i$ is not. For this case it follows from (55.10) that the integrand in (55.11) becomes

$$\mathbf{q}\cdot\frac{\partial}{\partial t}\nabla\left(\frac{1}{T}\right)-\sum_i \mathbf{J}_i\cdot\frac{\partial}{\partial t}\left[\nabla\left(\frac{g_i}{T}\right)-\frac{\mathbf{F}_i}{T}\right]+\sum_r \rho\dot{\xi}_r\frac{\partial}{\partial t}\left(\frac{A_r}{T}\right),$$

and comparison with (53.20) shows that the processes and affinities appearing here are those that occur in the bilinear form for the entropy production σ. Thus (55.11) now reads (in matrix form)

$$\int_V J\frac{\partial X}{\partial t} \leqslant 0, \tag{55.13}$$

where

$$\sigma = JX.$$

We now assume

(1) linear phenomenological laws, $J = LX$, and

(2) constant phenomenological coefficients in L.

All the forces have even parity, so in the absence of magnetic fields the reciprocal relations are $L = \tilde{L}$. Any variation $\delta\sigma$ in $\sigma = X\tilde{L}X$ may be written

$$\delta\sigma = \delta X\tilde{L}X + X\tilde{L}\,\delta X = 2XL\,\delta X = 2J\,\delta X,$$

and in particular

$$\frac{\partial \sigma}{\partial t} = 2J\frac{\partial X}{\partial t}. \tag{55.14}$$

Hence (55.13) may be written

$$\frac{\partial}{\partial t}\int_V \sigma\,\mathrm{d}\tau \leqslant 0,$$

so that in the stationary state the production of entropy is a minimum. This interesting, albeit somewhat restricted principle, is due to Prigogine (1945).

Secondly suppose there is just a single, incompressible, isothermal fluid, moving sufficiently slowly for the inertial terms to be dropped. For this case it follows from (55.10) that the integral in (55.11) reduces to I/T, where

$$I = -\mathbf{p}:\frac{\partial}{\partial t}\nabla\mathbf{v} = -(p\mathbf{1}+\boldsymbol{\pi}):\frac{\partial}{\partial t}\nabla\mathbf{v} = -\mathring{\boldsymbol{\pi}}:\frac{\partial}{\partial t}(\nabla\mathring{\mathbf{v}}),$$

as $\nabla\cdot\mathbf{v}$ is zero for an incompressible fluid. By $(49.8)_3$ and the assumption that ν is constant,

$$I = \frac{\partial}{\partial t}(\rho\nu\nabla\mathring{\mathbf{v}}:\nabla\mathring{\mathbf{v}}) = -\frac{1}{2}\frac{\partial}{\partial t}(\mathring{\boldsymbol{\pi}}:\nabla\mathring{\mathbf{v}}) = \frac{1}{2}\frac{\partial}{\partial t}\Phi, \tag{55.15}$$

and therefore (55.11) may be written

$$\frac{\partial}{\partial t}\int_V \Phi\,\mathrm{d}\tau \leqslant 0. \tag{55.16}$$

From this follows the classical Helmholtz theorem about the minimum of the dissipation function Φ occurring at the stationary state.

55.3. The local potential

In an attempt to overcome the very restrictive conditions necessary for the validity of the principle of minimum entropy production, Glansdorff and Prigogine (1971) have introduced the concept of the 'local potential' φ. First they eliminate $\dot{\xi}_r$, $\mathbf{q}$, $\mathbf{J}_i$, and $\mathbf{p}$ from (55.10) by adopting the usual (linear) phenomenological laws, a step that enables them to replace some of the time derivatives in (55.10) by time derivatives of quadratic terms (cf. (55.15)). Any remaining variables not in the operands of time derivatives are then replaced by their constant steady-state values, an approximation that allows the time derivative to be completely removed from the integrand in (55.11). Of course it remains to verify that the inequality is preserved by this modification, that is, that (55.11) can be replaced by (55.12) in which φ is a function of both the steady-state and time-dependent variables.

As a simple example, consider a problem in which only heat conduction is involved. By (55.10) and (55.11)

$$\int_V \mathbf{q}\cdot\frac{\partial}{\partial t}(\nabla T^{-1})\,\mathrm{d}\tau = \tfrac{1}{2}\int_V T^2\kappa(T)\frac{\partial}{\partial t}(\nabla T^{-1})^2\,\mathrm{d}\tau \leqslant 0, \tag{55.17}$$

where Fourier's law, $\mathbf{q} = -\kappa(T)\nabla T$, has been used. Let $T_0(\mathbf{r})$ be the time-independent solution of the heat-conduction equation

$$\nabla\cdot\kappa(T_0)\nabla T_0 = 0, \tag{55.18}$$

and replace $T^2\kappa(T)$ in (55.17) by $T_0^2\kappa(T_0)$. This yields the local potential

$$\varphi(T, T_0) = \tfrac{1}{2}\int_V T_0^2\kappa_0(\nabla T^{-1})^2\,\mathrm{d}\tau \qquad (\kappa_0 = \kappa(T_0)), \tag{55.19}$$

and it remains to verify that

$$\frac{\partial}{\partial t}\varphi(T, T_0) \leqslant 0. \tag{55.20}$$

Now

$$\delta\varphi \equiv \varphi(T, T_0) - \varphi(T_0, T_0) = \tfrac{1}{2}\int_V T_0^2\kappa_0[(\nabla T_0^{-1} + \nabla\theta)^2 - (\nabla T_0^{-1})^2]\,\mathrm{d}\tau,$$

where $\theta = T^{-1} - T_0^{-1}$. Expanding the integrand in a power series in $\nabla\theta$, and integrating by parts the linear term in $\nabla\theta$, which term then vanishes by (55.18), we arrive at

$$\delta\varphi = \tfrac{1}{2}\int_V T_0^2\kappa_0(\nabla\theta)^2\,\mathrm{d}\tau > 0, \tag{55.21}$$

for small deviations from the steady state. It follows that φ must decrease as it evolves to its steady value, that is, (55.20) is confirmed. For the steady state (55.21) yields the variational principle

$$\delta_T\varphi(T, T_0) = 0, \qquad T = T_0, \tag{55.22}$$

which can be made the basis of a method of computing $T_0(\mathbf{r})$, not unlike Galerkin's method (see Donnelly, Herman, and Prigogine 1966, p. 291). Indeed the local potential concept is more a computational device than a genuine evolutionary principle, and we shall not pursue it further.

10

MAGNETO-PLASMAS AND SUPERFLUIDS

MAGNETO-PLASMAS and superfluids have in common that they are both systems which can be represented in certain circumstances by two interpenetrating fluids, and so they provide excellent illustrations of the mixture theory developed in the previous chapter. Liquid helium II consists of a 'normal' fluid at temperature T ($<2{\cdot}17$ K) and a superfluid at $T = 0$ K. The electron and ion fluids comprising the magneto-plasma may also be at different temperatures. An important point of difference is that mass exchange is possible between the normal and superfluids, but of course impossible for the electric fluids. Process thermodynamics is particularly useful with systems like these, in which coupling between the fluids is important.

56. Fluid equations for a magneto-plasma

56.1. Validity of fluid equations

The general circumstances under which a system of atomic particles can be treated as a fluid continuum has been described in §§ 1.1 and 19.2. A fully ionized plasma has potentially two fluids, namely, an electron fluid and an ion fluid occupying the same space, so there will be at least three time scales and length scales to consider. The three collisional relaxation times of importance are those for establishing equilibrium velocity distributions (see final paragraph of § 25.6) in (1) the electron gas alone (τ_2), (2) the ion gas alone (τ_1), and (3) the electron–ion composite gas (τ_ε). Let the subscripts $_1$ and $_2$ denote ion and electron properties respectively, so that m_1 is the ion mass and m_2 is the electron mass, then when the temperatures T_1 and T_2 are of comparable magnitude and turbulence is absent it is approximately the case that

$$\tau_2 : \tau_1 : \tau_\varepsilon = 1 : (m_1/m_2)^{\frac{1}{2}} : (m_1/m_2), \quad \text{that is, } \tau_2 \ll \tau_1 \ll \tau_\varepsilon,$$

there being at least two orders of magnitude separating these times.

Let $\mathscr{T}$ denote the time scale of interest—the observer's time scale—then fluid descriptions are valid as indicated in Table 9.

Kinetic or 'non-equilibrium' temperatures have been defined in § 23; by 'ion particles' is implied that a fluid description is inadequate for the ions and that a phase-space description involving the velocity distribution function (§ 71.2) is unavoidable. The constraints on the observer's length scale L,

TABLE 9

$\tau_\varepsilon \leqslant \mathscr{T}$	$T_1 = T_2$	One-fluid model
$\tau_1 < \mathscr{T} < \tau_\varepsilon$	$T_1 \neq T_2$	Two-fluid model
$\tau_2 < \mathscr{T} < \tau_1$	T_1 kinetic	Electron fluid, ion particles
$\mathscr{T} < \tau_2$	T_1 and T_2 kinetic	Electron and ion particles

namely, that it should be much larger than the mean free path for the particle interactions under consideration, are usually less demanding than those just given for $\mathscr{T}$.

The introduction of a magnetic induction **B** into the plasma somewhat complicates the situation. Let e_1, e_2 denote the electric charges on the ions and electrons and suppose these particles move with velocities $\mathbf{V}_1, \mathbf{V}_2$ in the presence of an electric field **E** and an induction **B**, then their equations of motion in the absence of particle collisions are

$$m_i\dot{\mathbf{V}}_i = e_i(\mathbf{E}+\mathbf{V}_i \times \mathbf{B}) \qquad (i = 1, 2).$$

Thus motions parallel to **B** are unaffected by its presence, whereas those normal to **B**, with velocities $V_{\perp i}$, are circular with radii r_i and angular velocities ω_i given by

$$\omega_i = e_i B/m_i, \qquad r_i = V_{\perp i}/\omega_i \qquad (i = 1, 2). \tag{56.1}$$

Collisions will tend to inhibit these particle gyrations, and the dimensionless numbers $\omega_i\tau_i$, $i = 1, 2$, measure the extent to which the magnetic field affects the microscopic properties of the magneto-plasma. If $\omega_i\tau_i \ll 1$ the field has little effect as collisions dominate the particle motion; the fluid in question is then isotropic. On the other hand, if $\omega_i\tau_i \gg 1$ particle motion parallel to **B** is quite different from motion normal to **B**, a difference that renders the fluid anisotropic, although with transverse isotropy (§ 47) about unit vector **b** parallel to **B**.

By introducing the collisionless time scales ω_i^{-1} magnetic fields can in fact enhance the range of validity of the fluid descriptions. The field lines provide a means for groups of particles to interact collectively with each other without collisions, the Lorentz force $e_i\mathbf{V}_i \times \mathbf{B}$ playing an intermediatory role. Thus random collective motions of large groups of particles, that is, turbulent motions, interacting with **B**, can produce effective collisional time scales τ_i^* and length scales λ_i^* comparable with ω_2^{-1} and r_i and hence, in strong fields, somewhat smaller than the 'classical' values derived from kinetic theory (§ 76.4). These reductions in the scales will occur in directions normal to **B**, and by virtue of waves travelling along the field lines, may also occur to a lesser extent in directions parallel to **B**. Similarly groups of particles can interact via the electric field **E**, and likewise reduce the classical scales.

The great variety of instabilities known to occur in plasmas means that plasma turbulence is commonplace, and so the study of random collective motions and their influences on the time and length scales involved in the transport of macroscopic properties like heat, electric current, etc. has become a central problem in plasma physics. For our purposes it is sufficient to appreciate that turbulence somewhat enhances the utility of the fluid models of a magneto-plasma to be described below.

56.2. Entropy production in a magneto-plasma

As explained above, on a long enough time scale a one-fluid model of a magneto-plasma is adequate; we shall treat this case first. On the assumption that the ions are singly ionized and have the same number density as the electrons, we write $e_1 = e$, $e_2 = -e$, $n_1 = n$, $n_2 = n$.† The electric current density $\mathbf{j}$ and mean fluid velocity are

$$\mathbf{j} = en(\mathbf{v}_1 - \mathbf{v}_2), \qquad \mathbf{v} = (m_1\mathbf{v}_1 + m_2\mathbf{v}_2)/(m_1 + m_2),$$

$\mathbf{v}_1, \mathbf{v}_2$ being the fluid velocities of the components. By (11.8) the body forces acting on the component fluids are

$$\rho_i\mathbf{F}_i = e_i n(\mathbf{E} + \mathbf{v}_i \times \mathbf{B}) \qquad (\rho_i = nm_i). \tag{56.2}$$

The diffusion momenta are

$$\mathbf{J}_1 \equiv nm_1(\mathbf{v}_1 - \mathbf{v}) = \frac{m_1 m_2}{m_1 + m_2}\frac{\mathbf{j}}{e}, \qquad \mathbf{J}_2 \equiv nm_2(\mathbf{v}_2 - \mathbf{v}) = -\frac{m_1 m_2}{m_1 + m_2}\frac{\mathbf{j}}{e}.$$

It will be assumed that the ions and electrons comprise perfect gases, consequently the thermodynamic relations given in § 21.1 will apply to each component, with $\gamma = \frac{5}{3}$, its value for monatomic particles (§ 23.3). As $T_1 = T_2 = T$ the total pressure is $p = p_1 + p_2 = 2nkT$. Hence by (21.5)

$$T\nabla\left(\frac{g_i}{T}\right) = \frac{1}{m_i}\left(\frac{\nabla p}{2n} - \tfrac{5}{2}k\nabla T\right),$$

so that

$$\sum_{i=1}^{2} \mathbf{J}_i \cdot \left[\mathbf{F}_i - T\nabla\left(\frac{g_i}{T}\right)\right] = \sum_{i=1}^{2} \mathbf{J}_i \cdot \left[\frac{e_i}{m_i}(\mathbf{E} + \mathbf{v}_i \times \mathbf{B}) - \frac{1}{m_i}\left(\frac{\nabla p}{2n} - \tfrac{5}{2}k\nabla T\right)\right]$$
$$= \mathbf{j} \cdot \mathbf{E}' - \frac{5}{2}\frac{k}{e}\frac{m_1 - m_2}{m_1 + m_2}\mathbf{j} \cdot \nabla T,$$

where

$$\mathbf{E}' \equiv \mathbf{E} + \mathbf{v} \times \mathbf{B} + \frac{1}{en}\frac{m_1 - m_2}{m_1 + m_2}(\tfrac{1}{2}\nabla p - \mathbf{j} \times \mathbf{B}).$$

† It may be shown that the approximation $n_1 = n_2$ is satisfactory provided ω_2 and the macroscopic scales $\mathscr{T}$ and L satisfy

$$\omega_2 \ll \omega_{p2}, \qquad 1 \ll \mathscr{T}\omega_{p2}, \qquad \lambda_D \ll L, \tag{56.3}$$

where $\omega_{p2} \equiv (e^2 n/\varepsilon m_2)^{\frac{1}{2}}$ is the plasma frequency, and $\lambda_D \equiv (\varepsilon k T_2/e^2 n)^{\frac{1}{2}}$ is the Debye length.

As $m_1 \gg m_2$ an obvious simplification that we shall now adopt is to replace $(m_1 - m_2)/(m_1 + m_2)$ by unity.

There are no internal coordinates in the plasma, and viscosity has been fully discussed in § 50. Thus from $(46.4)_1$ and (53.20)

$$T\sigma_1 = -\mathbf{q}'.\nabla T/T + \mathbf{j}.\mathbf{E}' \geqslant 0, \tag{56.4}$$

where

$$\mathbf{q}' \equiv \mathbf{q} + \frac{5}{2}\frac{kT}{e}\mathbf{j}, \quad \mathbf{E}' \equiv \mathbf{E} + \mathbf{v}\times\mathbf{B} - \frac{1}{en}\mathbf{j}\times\mathbf{B} + \frac{1}{2en}\nabla p. \tag{56.5}$$

The difference between (56.4) and the result

$$T\sigma_1 = -\mathbf{q}.\nabla T/T + \mathbf{j}.(\mathbf{E} + \mathbf{v}\times\mathbf{B}) \tag{56.6}$$

contained in § 38.6 merits comment. The point is simply that in the earlier treatment the contribution to the entropy from $-\sum_i g_i\,\mathrm{d}c_i$ was ignored, which is equivalent to neglecting the gradient of (g_i/T) above. The approximation

$$\mathbf{E}' = \mathbf{E} + \mathbf{v}\times\mathbf{B} \tag{56.7}$$

is valid when both (see (57.31) below) the so-called 'Hall term' $\mathbf{j}\times\mathbf{B}/en = (\nabla\times\mathbf{B})\times\mathbf{B}/(en\mu)$ and $\nabla p/2en = \nabla(nkT)/en$ are small compared with $|\mathbf{v}\times\mathbf{B}| = v_\perp B$. It is easily verified that these constraints require a macroscopic length scale L such that

$$\delta_1 \mathscr{b}/v_\perp \ll L, \qquad r_1 v_{\theta 1}/v_\perp \ll L, \tag{56.8}$$

where

$$\delta_1 \equiv (m_1/e^2 n\mu)^{\frac{1}{2}}, \qquad v_{\theta 1} \equiv (kT/m_1)^{\frac{1}{2}}, \tag{56.9}$$

$$\mathscr{b} \equiv B/(\mu n m_1)^{\frac{1}{2}}, \tag{56.10}$$

and r_1 is the Larmor radius based on the ion thermal speed $v_{\theta 1}$, that is, $r_1 = v_{\theta 1}/\omega_1$. The length $\delta_1 = \mathscr{b}/\omega_1$ is termed the *ion inertial length*, and $\mathscr{b}$ is the *Alfvén speed*, the significance of which will be made clear later (§ 57.3).

56.3. *The one-fluid constitutive equations*

Let us choose ∇T and $\mathbf{E}'$ to be the independent affinities, then from (56.4) and the fact that the magneto-plasma will have lateral isotropy about unit vector $\mathbf{b}$ parallel to $\mathbf{B}$, we may infer constitutive equations of the form (cf. eqns (54.5)–(54.9))

$$\begin{aligned} \mathbf{j} &= \boldsymbol{\sigma}.\mathbf{E}' + \boldsymbol{\beta}.\nabla T, \\ \mathbf{q} + \frac{5kT}{2e}\mathbf{j} &= -T\boldsymbol{\beta}.\mathbf{E}' - T\boldsymbol{\alpha}.\nabla T, \end{aligned} \tag{56.11}$$

where

$$\boldsymbol{\sigma} = \boldsymbol{\sigma}^*, \qquad \mathbf{a} = \mathbf{a}^*, \qquad \boldsymbol{\beta} = \boldsymbol{\beta}^*, \tag{56.12}$$

and each phenomenological tensor has the same form as $\boldsymbol{\sigma}$, namely,

$$\boldsymbol{\sigma} = \sigma_{\|}\mathbf{bb} + \sigma_{\perp}(\mathbf{1} - \mathbf{bb}) + \sigma_{\wedge}\mathbf{b}\times\mathbf{1} \quad \left(\begin{matrix}\sigma_{\|}, \sigma_{\perp} & \textit{even} \\ \sigma_{\wedge} & \textit{odd}\end{matrix} \text{ in } \mathbf{B}\right). \tag{56.13}$$

Eqns (56.1.2) are consequences both of lateral isotropy—which leads to (56.13)—and of the reciprocal relations; the latter are also responsible for the appearance of $\boldsymbol{\beta}$ in each of eqns (56.11). The scalar conductivities $\sigma_{\|}$, $\sigma_{\perp}$, and $\sigma_{\wedge}$ depend on the local thermodynamic state and also on $B = |\mathbf{B}|$ via the numbers $\omega_1\tau_1$ and $\omega_2\tau_2$, and likewise for the other scalar coefficients in (56.11). The structure of the tensor in (56.13) can more readily be appreciated on choosing $\mathbf{b}$ parallel to $0z$ and setting the components out in a matrix array:

$$\boldsymbol{\sigma} = \begin{bmatrix} \sigma_{\perp} & \sigma_{\wedge} & 0 \\ -\sigma_{\wedge} & \sigma_{\perp} & 0 \\ 0 & 0 & \sigma_{\|} \end{bmatrix}.$$

By eqns $(56.11)_1$ and (56.13)

$$\mathbf{j} = \sigma_{\|}\mathbf{E}'_{\|} + \sigma_{\perp}\mathbf{E}'_{\perp} + \sigma_{\wedge}\mathbf{b}\times\mathbf{E}' + \beta_{\|}\nabla_{\|}T + \beta_{\perp}\nabla_{\perp}T + \beta_{\wedge}\mathbf{b}\times\nabla T, \tag{56.14}$$

and $(56.11)_2$ may be expanded similarly. The entropy production rate is given by (cf. (54.14))

$$T\sigma_1 = \sigma_{\|}(E'_{\|})^2 + \sigma_{\perp}(E'_{\perp})^2 + \alpha_{\|}(\nabla_{\|}T)^2 + \alpha_{\perp}(\nabla_{\perp}T)^2 + 2\beta_{\|}\mathbf{E}'_{\|}\cdot\nabla_{\|}T +$$
$$+ 2\beta_{\perp}\mathbf{E}'_{\perp}\cdot\nabla_{\perp}T \geqslant 0,$$

which requires that

$$\sigma_{\|}, \sigma_{\perp}, \alpha_{\|}, \alpha_{\perp} \geqslant 0, \qquad \sigma_{\|}\alpha_{\|} \geqslant \beta_{\|}^2, \qquad \sigma_{\perp}\alpha_{\perp} \geqslant \beta_{\perp}^2. \tag{56.15}$$

Notice that no constraints are imposed on the *odd* functions of $\mathbf{B}$, namely, $\sigma_{\wedge}, \alpha_{\wedge}, \beta_{\wedge}$.

As $\omega_1\tau_1, \omega_2\tau_2$ tend to zero, $\sigma_{\wedge} \to 0$ and $\sigma_{\perp} \to \sigma_{\|} = \sigma$, say, so that $\boldsymbol{\sigma} \to \sigma\mathbf{1}$; similar reductions occur for the other tensors, when (56.11) may be written in the isotropic form

$$\mathbf{j} = \sigma\mathbf{E}' + \beta\nabla T,$$

$$\mathbf{q} = -T\left(\beta + \frac{5k}{2e}\sigma\right)\mathbf{E}' - \kappa\nabla T,$$

where κ is a new coefficient termed the thermal conductivity.

56.4. The complete set of one-fluid magneto-plasma equations

For typical plasmas it is a good approximation to take the permittivity ε and the permeability μ to be constants. Also the electric charge density in Maxwell's equations (§ 11.1) may be set equal to zero when the constraints given in (56.3) are satisfied. These approximations lead to the equations set out under (1) in the following list (see §§ 38.6, 21.1, and 50.2).

(1) *The electromagnetic equations*

$$\nabla \times \mathbf{E} = -\frac{\partial \mathbf{B}}{\partial t}, \qquad \nabla . \mathbf{B} = 0,$$
$$\mu \mathbf{j} = \nabla \times \mathbf{B}, \qquad \nabla . \mathbf{j} = 0. \tag{56.16}$$

(2) *The conservation equations*

$$\begin{aligned} \mathrm{D}\rho + \rho\nabla . \mathbf{v} &= 0, \\ \rho\mathrm{D}\mathbf{v} + \nabla . \mathbf{p} &= \rho\mathbf{F} + \mathbf{j}\times\mathbf{B}, \\ \rho\mathrm{D}u + \rho p\mathrm{D}\rho^{-1} &= -\boldsymbol{\pi}:\nabla\mathbf{v} - \nabla . \mathbf{q} + \psi + \mathbf{j} . (\mathbf{E} + \mathbf{v}\times\mathbf{B}), \\ T\mathrm{D}s &= \mathrm{D}u + p\mathrm{D}\rho^{-1}.\dagger \end{aligned} \tag{56.17}$$

(3) *The thermodynamic equations*

$$p = 2nkT, \qquad u = \tfrac{3}{2}p/\rho = c_V T, \qquad \rho \approx m_1 n \qquad (c_V \approx 3k/m_1),$$
$$p = \text{constant} \times \rho^{\frac{5}{3}} \exp(s/c_V). \tag{56.18}$$

(4) *The constitutive equations*

$$\left.\begin{aligned} &\mathbf{j} = \boldsymbol{\sigma} . \mathbf{E}' + \boldsymbol{\beta} . \nabla T \qquad \left(\mathbf{E}' = \mathbf{E} + \mathbf{v}\times\mathbf{B} - \frac{1}{en}\mathbf{j}\times\mathbf{B} + \frac{1}{2en}\nabla p\right), \\ &\mathbf{q} = -T\left(\boldsymbol{\beta} + \frac{5k}{2e}\boldsymbol{\sigma}\right) . \mathbf{E}' - \boldsymbol{\kappa} . \nabla T, \\ &\mathbf{p} = p\mathbf{1} - \rho \sum_{i=1}^{5} \nu_i \boldsymbol{N}_i : \nabla\mathring{\mathbf{v}}, \end{aligned}\right\} \tag{56.19}$$

where $\boldsymbol{N}_i$ are defined in § 47.4. The bulk viscosities are zero in a plasma.

56.5. The two-fluid constitutive equations

On a shorter time scale $T_1 \neq T_2$ and a two-fluid model is essential. We shall not write down the complete list of the two-fluid equations for a plasma. They are similar to the one-fluid equations given above, except that $(56.17)_1$, $(56.17)_4$, (56.18), and $(56.19)_3$ are duplicated for the component fluids;

† The term $\sum_i g_i \mathrm{D}c_i = \sum_i g_i \nabla . \mathbf{J}_i/\rho$ in $T\mathrm{D}s$ vanishes as $\nabla . \mathbf{J}_i \propto \nabla . \mathbf{j} = 0$.

eqns $(56.17)_2$ and $(56.17)_3$ are replaced by (see (56.2), (52.1), and (52.9) and also the derivation from kinetic theory given in § 76.4)

$$\begin{aligned} \rho_i \mathrm{D}_i \mathbf{v}_i + \nabla \cdot \mathbf{p}_i &= \rho_i \mathbf{F}_i + e_i n(\mathbf{E} + \mathbf{v}_i \times \mathbf{B}) + \mathbf{R}_{ij}, \\ \rho_i \mathrm{D}_i u_i + \rho_i p_i \mathrm{D}_i \rho_i^{-1} &= -\boldsymbol{\pi}_i : \nabla \mathbf{v}_i - \nabla \cdot \mathbf{q}_i + \psi_i + \varphi_{ij} + \xi_{ij} \end{aligned} \qquad (i, j = 1, 2; i = j); \tag{56.20}$$

and eqns $(56.19)_1$ and $(56.19)_2$ need to be replaced by the constitutive relations (56.23) given below. Constitutive relations are also required for φ_{ij} and ξ_{ij}.

The vectorial terms in (52.13) yield

$$\sigma_1 = \mathbf{q}_1 \cdot \nabla T_1^{-1} + \mathbf{q}_2 \cdot \nabla T_2^{-2} + \mathscr{R} \cdot \mathbf{j} T_0^{-1}, \tag{56.21}$$

where by (52.15)

$$\mathscr{R} \equiv \frac{\mathbf{R}_{21}}{en}, \qquad T_0 = \frac{(m_1 + m_2) T_1 T_2}{m_1 T_1 + m_2 T_2} \sim T_2, \tag{56.22}$$

the approximate form for T_0 requiring $m_1 T_1 \gg m_2 T_2$. From (56.21) follows constitutive relations of the form

$$\begin{aligned} \mathbf{q}_1 &= \mathbf{L}_{11} \cdot \nabla T_1^{-1} + \mathbf{L}_{12} \cdot \nabla T_2^{-1} + \mathbf{L}_{13} \cdot \mathbf{j}, \\ \mathbf{q}_2 &= \mathbf{L}_{21} \cdot \nabla T_1^{-1} + \mathbf{L}_{22} \cdot \nabla T_2^{-1} + \mathbf{L}_{23} \cdot \mathbf{j}, \\ \mathscr{R} T_0^{-1} &= \mathbf{L}_{31} \cdot \nabla T_1^{-1} + \mathbf{L}_{32} \cdot \nabla T_2^{-1} + \mathbf{L}_{33} \cdot \mathbf{j}, \end{aligned}$$

where the phenomenological tensors satisfy the constraints (cf. (45.5) and (56.13))

$$\left.\begin{aligned} \mathbf{L}_{11} &= \mathbf{L}_{11}^* \\ \mathbf{L}_{22} &= \mathbf{L}_{22}^* \\ \mathbf{L}_{33} &= \mathbf{L}_{33}^* \end{aligned}\right\} \begin{array}{c} \text{lateral} \\ \text{isotropy} \\ \textit{and} \\ \text{reciprocity} \end{array}, \quad \left.\begin{aligned} \mathbf{L}_{12} &= \mathbf{L}_{12}^* \\ \mathbf{L}_{13} &= \mathbf{L}_{13}^* \\ \mathbf{L}_{23} &= \mathbf{L}_{23}^* \end{aligned}\right\} \begin{array}{c} \text{lateral} \\ \text{isotropy} \end{array}, \quad \left.\begin{aligned} \mathbf{L}_{12} &= \mathbf{L}_{21}^* \\ \mathbf{L}_{13} &= -\mathbf{L}_{31}^* \\ \mathbf{L}_{23} &= -\mathbf{L}_{32}^* \end{aligned}\right\} \text{reciprocity.}$$

The various signs in the last set arise because the temperature gradients have even parity and the electric current has odd parity.

The constitutive equations may be rewritten as

$$\left.\begin{aligned} \mathbf{q}_1 &= -\boldsymbol{\kappa}_1 \cdot \nabla T_1 - T_1 T_2^{-1} \boldsymbol{\varepsilon} \cdot \nabla T_2 - T_1 \boldsymbol{\gamma} \cdot \mathbf{j}, \\ \mathbf{q}_2 &= -T_2 T_1^{-1} \boldsymbol{\varepsilon} \cdot \nabla T_1 - \boldsymbol{\kappa}_2 \cdot \nabla T_2 - T_2 \boldsymbol{\delta} \cdot \mathbf{j}, \\ \mathscr{R} &= -T_0 T_1^{-1} \boldsymbol{\gamma} \cdot \nabla T_1 - T_0 T_2^{-1} \boldsymbol{\delta} \cdot \nabla T_2 + \boldsymbol{\eta} \cdot \mathbf{j}, \end{aligned}\right\} \tag{56.23}$$

where each of the six remaining tensor coefficients has the same structure as $\boldsymbol{\sigma}$ in (56.13). These (18 scalar) coefficients may be determined by kinetic theory.† An important simplification arises from the assumption that the ion distribution function in phase-space is determined only by ion–ion collisions; in this case $\boldsymbol{\varepsilon}$ and $\boldsymbol{\gamma}$ are zero. Using the approximation in $(56.22)_2$ and writing $\bar{\mathbf{q}} = \mathbf{q}_1 + \mathbf{q}_2$, we find

$$\bar{\mathbf{q}} = -\boldsymbol{\kappa}_1 \,.\, \nabla T_1 - \boldsymbol{\kappa}_2 \,.\, \nabla T_2 - T_2 \boldsymbol{\delta} \,.\, \mathbf{j}, \qquad \mathscr{R} = -\boldsymbol{\delta} \,.\, \nabla T_2 + \boldsymbol{\eta} \,.\, \mathbf{j}. \tag{56.24}$$

To illustrate the way in which the second law of thermodynamics in the form $\sigma \geqslant 0$ constrains the scalar coefficients, we shall consider the simple case of equal temperatures for which (56.21) and (56.24) reduce to

$$\begin{aligned} &\mathscr{R} = \boldsymbol{\eta} \,.\, \mathbf{j} - \boldsymbol{\delta} \,.\, \nabla T, \qquad \sigma_1 = \bar{\mathbf{q}} \,.\, \nabla T^{-1} + \mathscr{R} \,.\, \mathbf{j} T^{-1}, \\ &\bar{\mathbf{q}} = -T\boldsymbol{\delta} \,.\, \mathbf{j} - \boldsymbol{\kappa} \,.\, \nabla T \quad (\boldsymbol{\eta} = \eta^*, \boldsymbol{\kappa} = \boldsymbol{\kappa}^*, \boldsymbol{\delta} = \boldsymbol{\delta}^*) \end{aligned} \tag{56.25}$$

Thus, expanding each tensor according to (56.13) we obtain

$$T\sigma_1 = \eta_{\parallel} j_{\parallel}^2 + \eta_{\perp} j_{\perp}^2 + \frac{1}{T}\kappa_{\parallel}(\nabla_{\parallel} T)^2 + \frac{1}{T}\kappa_{\perp}(\nabla_{\perp} T)^2 + 2\delta_{\wedge} \mathbf{b} \,.\, \mathbf{j} \times \nabla T \geqslant 0,$$

which is satisfied for all possible temperature gradients and currents provided

$$\eta_{\parallel}, \eta_{\perp}, \kappa_{\parallel}, \kappa_{\perp} \geqslant 0, \qquad \eta_{\perp}\kappa_{\perp} \geqslant T\delta_{\wedge}^2.$$

In an isotropic plasma ($\omega_i \tau_i \ll 1$, $i = 1, 2$) (56.25) becomes

$$\begin{aligned} &\mathscr{R} = \eta \mathbf{j} - \delta \nabla T, \qquad T\sigma_1 = \eta j^2 + \kappa (\nabla T)^2 / T \geqslant 0, \\ &\bar{\mathbf{q}} = -T\delta \mathbf{j} - \kappa \nabla T, \qquad \eta, \kappa \geqslant 0. \end{aligned} \tag{56.26}$$

Kinetic theory (see Braginskii (1965) yields the approximate values

$$\delta = 0{\cdot}7k/e, \qquad \eta = m_2/(ne^2\tau_2), \qquad \kappa = 1{\cdot}6 p_2 \tau_2 k / m_2,$$

where τ_2 is the electron–ion collision time for momentum transfer and n is the electron number density.

Constitutive relations for ξ_{ij} and φ_{ij} may be deduced from (52.13) and (52.11). As $m_2 \ll m_1$ we find

$$\xi_{12} = 0, \qquad \xi_{21} = \mathbf{R}_{21} \,.\, (\mathbf{v}_1 - \mathbf{v}_2) = \mathscr{R} \,.\, \mathbf{j}, \qquad \varphi_{ij} = k_{ij}\,(T_i - T_j) \qquad (T_i > T_j) \tag{56.27}$$

where k_{ij} is a heat transfer coefficient.

56.6. The generalized Ohm's law

By (56.22), $en\mathscr{R}$ is the friction force acting on the electrons per unit volume and hence when $(56.24)_2$ is an adequate approximation to $\mathscr{R}$, it follows from

† Although this has not yet been achieved for turbulent plasmas.

$(52.1)_2$ and from (56.2) that the electron momentum equation reads

$$\rho_e \frac{\partial \mathbf{v}_e}{\partial t} + \rho_e \mathbf{v}_e \,.\, \nabla \mathbf{v}_e = -\nabla \,.\, \mathbf{p}_e - en(\mathbf{E} + \mathbf{v}_e \times \mathbf{B} - \boldsymbol{\eta} \,.\, \mathbf{j} + \boldsymbol{\delta} \,.\, \nabla T_e), \qquad (56.28)$$

where we have adopted the subscript $_e$ as being clearer than $_2$. For equal temperatures (56.28) may be rearranged to resemble the one-fluid Ohm's law $(56.19)_1$,

$$\boldsymbol{\eta} \,.\, \mathbf{j} = \mathbf{E} + \mathbf{v} \times \mathbf{B} - \frac{1}{en} \mathbf{j} \times \mathbf{B} + \frac{1}{2en} \nabla p + \frac{m_e}{e} \left(\frac{\partial \mathbf{v}_e}{\partial t} + \mathbf{v}_e \,.\, \nabla \mathbf{v}_e \right) + \boldsymbol{\delta} \,.\, \nabla T, \quad (56.29)$$

where we have neglected electron viscosity (momentum transfer is usually dominated by the heavy ions) and replaced the electron velocity $\mathbf{v}_e$ in one term by $\mathbf{v}_e \approx \mathbf{v} - \mathbf{j}/en$. Eqn (56.29) is often termed the 'generalized Ohm's law', and it differs from the one-fluid form of the law by the presence of a term accounting for the electron inertia.

The Hall term

$$\frac{\mathbf{j} \times \mathbf{B}}{en} = -\mathbf{b} \times \mathbf{j} \eta_{\parallel} \omega_e \tau_e ,$$

where ω_e is the electron cyclotron frequency and $\eta_{\parallel} = m_e/(ne^2\tau_e)$ is the 'parallel' resistivity, may be incorporated into the left-hand side of (56.29) provided $\boldsymbol{\eta}$ is replaced by

$$\boldsymbol{\eta}' = (\eta_{\parallel} - \eta_{\perp})\mathbf{bb} + \eta_{\perp}\mathbf{1} + (\eta_{\wedge} - \eta_{\parallel}\omega_e\tau_e)\mathbf{b} \times \mathbf{1}. \qquad (56.30)$$

However, while this simplifies the appearance of the generalized Ohm's law, the natural error of assuming $\eta_{\parallel}\omega_e\tau_e$ to vanish in the absence of dissipation needs to be resisted. It is clear from (56.30) that only the first two terms are dissipative, $\mathbf{b} \times \mathbf{1} \,.\, \mathbf{j}$ having negative parity under motion-reversal. More generally, if phenomenological coefficients $\alpha_{\parallel}, \alpha_{\perp}$ are associated with irreversible processes, then $\alpha_{\wedge}$ will be associated with a reversible process, and vice versa. Thus, for example, unlike $\eta_{\wedge}$, $\delta_{\wedge}$ is dissipative and can be expected to contribute to wave-damping.

57. Wave attenuation in a magneto-plasma

The simplest plasma problem of thermodynamic interest that we could consider is the way in which the various 'transport' coefficients appearing in the constitutive equations contribute to the attenuation of small-amplitude waves propagating through the anisotropic medium. This problem is readily solved in principle, being mathematically linear, but it must be admitted that the anisotropy does generate a lot of algebra. We shall set out the solution here mainly to illustrate a dissipative process of some importance in fluid dynamics; an account of viscous damping in an anisotropic plasma appears to be lacking in the literature.

57.1. Linearization of the thermodynamic relations

We shall adopt the one-fluid magneto-plasma equation, save that in place of (56.11) we shall use the equivalent formulae

$$\mathbf{E} = -\mathbf{v}\times\mathbf{B}+\frac{1}{en}\mathbf{j}\times\mathbf{B}-\frac{1}{2en}\nabla p+\boldsymbol{\eta}\,.\,\mathbf{j}+\boldsymbol{\gamma}\,.\,\nabla T, \tag{57.1}$$

$$\mathbf{q} = -\frac{5kT}{2e}\mathbf{j}+T\boldsymbol{\gamma}\,.\,\mathbf{j}-\boldsymbol{\kappa}\,.\,\nabla T, \tag{57.2}$$

based on $\mathbf{j}$ and ∇T as the independent affinities. The steady state is assumed to be completely uniform, with no gradients at all and zero velocity. Thus in the steady state (57.1) reduces to

$$\mathbf{E}+\mathbf{v}\times\mathbf{B} = 0. \tag{57.3}$$

In the perturbed state the variables $\mathbf{E}$, $\mathbf{j}$, p, etc. will be written

$$\mathbf{E}+\hat{\mathbf{E}}\,\mathrm{e}^{\mathrm{i}(\mathbf{k}.\mathbf{r}-\omega t)},\quad \hat{\mathbf{j}}\,\mathrm{e}^{\mathrm{i}(\mathbf{k}.\mathbf{r}-\omega t)},\quad p+\hat{p}\,\mathrm{e}^{\mathrm{i}(\mathbf{k}.\mathbf{r}-\omega t)},\quad \text{etc.,} \tag{57.4}$$

where the first term in each case refers to the uniform background state and the second to a small wave-like perturbation. Second-order terms like $\hat{\mathbf{v}}\times\hat{\mathbf{B}}$, $\hat{p}\hat{\rho}$, etc. will be ignored. The propagation vector $\mathbf{k}$ is not to be confused with unit vector along $0z$, which will now be written $\mathbf{b}$ to remind us that it is parallel to the magnetic field $\mathbf{B}$ for the uniform state.

By (57.4) we may adopt the transformation

$$\frac{\partial}{\partial t} \to -\mathrm{i}\omega, \qquad \nabla \to \mathrm{i}\mathbf{k},$$

so that Maxwell's equations become

$$\begin{aligned} \mu\hat{\mathbf{j}} &= \mathrm{i}\mathbf{k}\times\hat{\mathbf{B}}, \qquad \mathbf{k}\,.\,\hat{\mathbf{j}} = 0,\\ \omega\hat{\mathbf{B}} &= \mathbf{k}\times\hat{\mathbf{E}}, \qquad \mathbf{k}\,.\,\hat{\mathbf{B}} = 0. \end{aligned} \tag{57.5}$$

Eqn (57.2) linearizes to $\mathrm{i}\mathbf{k}\,.\,\hat{\mathbf{q}} = -T\mathbf{k}\,.\boldsymbol{\gamma}\,.\,\mathbf{k}\times\hat{\mathbf{B}}/\mu+\mathbf{k}\,.\,\boldsymbol{\kappa}\,.\,\mathbf{k}\hat{T}$, so that from (57.3) and the second pair of (56.17) we find

$$\omega\rho\hat{s} = \mathbf{k}\,.\,\hat{\mathbf{q}} = \mathrm{i}\mathbf{k}\,.\boldsymbol{\gamma}\times\mathbf{k}\,.\,\hat{\mathbf{B}}/\mu-\mathrm{i}\mathbf{k}\,.\,\boldsymbol{\kappa}\,.\,\mathbf{k}\hat{T}/T, \tag{57.6}$$

the omitted terms being of second order.

Eqn (57.6) reveals that in the absence of damping ($\boldsymbol{\gamma}, \boldsymbol{\kappa} = 0$) an entropy perturbation, or 'wave', will be convected with the fluid, unchanged in magnitude. We shall ignore this singular solution to the wave equation in the following, but it is helpful at times to remember that such a solution exists. In § 60.3, when considering waves in helium II, we shall find a similar entropy wave, enjoying the inept name of 'second sound', that does in fact propagate relative to the fluid.

In what follows it will be convenient to employ the variables

$$\hat{\mathbf{h}} \equiv \hat{\mathbf{B}}/\sqrt{(\mu\rho)}, \qquad \hat{\mathbf{J}} \equiv \mathrm{i}\mathbf{k}\times\hat{\mathbf{h}},$$

which satisfy (57.7)

$$\mathbf{k}\,.\,\hat{\mathbf{h}} = 0, \qquad \mathbf{k}\,.\,\hat{\mathbf{J}} = 0, \qquad k^2\hat{\mathbf{h}} = \mathrm{i}\mathbf{k}\times\hat{\mathbf{J}}.$$

The coordinate system is chosen so that $\mathbf{k}$ lies in the xz-plane (Fig. 15),

$$\mathbf{k} = \mathbf{b}\cos\theta + \mathbf{i}\sin\theta. \tag{57.8}$$

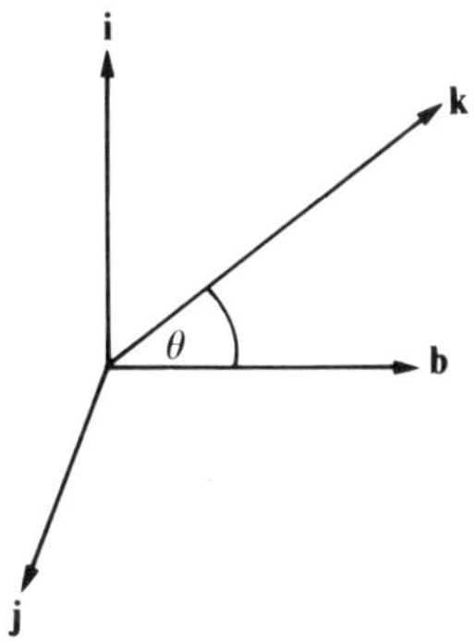

FIG. 15.

Now using an expansion for $\boldsymbol{\gamma}$ like that in (56.13) we find

$$\begin{aligned}\mathrm{i}\mathbf{k}\,.\boldsymbol{\gamma}\times\mathbf{k}\,.\,\hat{\mathbf{h}} &= \mathbf{k}\,.\,[(\gamma_{\parallel}-\gamma_{\perp})\mathbf{bb}+\gamma_{\perp}\mathbf{1}+\gamma_{\wedge}\mathbf{b}\times\mathbf{1}]\,.\,\hat{\mathbf{J}} \\ &= (\gamma_{\parallel}-\gamma_{\perp})k\cos\theta\hat{J}_{\parallel}+\mathrm{i}\gamma_{\wedge}k^2\hat{h}_{\parallel},\end{aligned} \tag{57.9}$$

where

$$\hat{J}_{\parallel} = \mathbf{b}\,.\,\hat{\mathbf{J}}, \qquad \hat{h}_{\parallel} = \mathbf{b}\,.\,\hat{\mathbf{h}}$$

are components along the magnetic field.

From (56.18) it follows that

$$\hat{p}/p = \hat{\rho}/\rho + \hat{T}/T, \qquad \hat{p} = a^2\hat{\rho} + \tfrac{2}{3}\rho T\hat{s}, \tag{57.10}$$

where

$$a \equiv (5p/3\rho)^{\frac{1}{2}}$$

is the speed of sound through an ideal gas. When (57.6), (57.9), and (57.10) are combined into a relation between $\hat{p}$ and $\hat{\rho}$, and quadratic terms involving the components of $\boldsymbol{\gamma}$ and $\boldsymbol{\kappa}$ are ignored, the result may be written

$$\hat{p} = a^2(1-\mathrm{i}c_\kappa)\hat{\rho} - \rho\ell c_\gamma \hat{\mathbf{J}}_{\|}/k + \mathrm{i}\rho\ell c'_\gamma \hat{h}_{\|}, \tag{57.11}$$

where

$$c_\kappa \equiv \frac{4T}{15p}\frac{k^2}{\omega}(\kappa_{\|}\cos^2\theta + \kappa_\perp \sin^2\theta),$$

$$c_\gamma \equiv \frac{2T}{3B}\frac{k^2}{\omega}(\gamma_\perp - \gamma_{\|})\cos\theta, \qquad c'_\gamma \equiv \frac{2T}{3B}\frac{k^2}{\omega}\gamma_\wedge,$$

and

$$\ell \equiv \frac{B}{\sqrt{(\mu\rho)}} \tag{57.12}$$

is a velocity whose physical significance will be identified shortly.

57.2. *Linearization of the equations for the velocity and the magnetic field*

Next consider the fluid equations; on linearizing and transforming the first pair of (56.17) we obtain

$$\omega\hat{\rho}/\rho - \mathbf{k}\,.\,\hat{\mathbf{v}} = 0 \tag{57.13}$$

and

$$\rho\omega\hat{\mathbf{v}} = \mathbf{k}\hat{p} + \mathrm{i}\hat{\mathbf{j}}\times\mathbf{B} - \mathrm{i}\rho k^2 \mathscr{L}\,.\,\hat{\mathbf{v}},$$

where from $(56.19)_3$ and results given in § 47.4 $\mathscr{L}$ is the second-order tensor

$$\begin{aligned}\mathscr{L} &\equiv k^{-2}\sum_i v_i \mathbf{k}\,.\,\boldsymbol{N}_i\,.\,\mathbf{k} \\ &= v_1(\mathbf{ii}+\mathbf{jj})\sin^2\theta + \tfrac{1}{3}v_2(2\mathbf{b}\cos\theta - \mathbf{i}\sin\theta)^2 + v_3[(\mathbf{i}\cos\theta + \mathbf{b}\sin\theta)^2 + \\ &\qquad + \mathbf{jj}\cos^2\theta] + v_4(\mathbf{ij}-\mathbf{ji})\sin^2\theta + v_5[(\mathbf{ij}-\mathbf{ji})\cos^2\theta + (\mathbf{bj}-\mathbf{jb})\sin\theta\cos\theta].\end{aligned} \tag{57.14}$$

By $(57.5)_1$, $(57.7)_1$, and (57.11) the momentum equation may be written

$$\omega\hat{\mathbf{v}} - \mathbf{k}a^2(1-\mathrm{i}c_\kappa)\hat{\rho}/\rho + \ell c_\gamma \hat{J}_{\|}\mathbf{k}/k - \mathrm{i}\ell c'_\gamma \hat{h}_{\|}\mathbf{k} + \ell k\cos\theta\;\hat{\mathbf{h}} - \mathbf{k}\ell\hat{\mathbf{h}}_{\|} + \mathrm{i}k^2\mathscr{L}\,.\,\hat{\mathbf{v}} = 0. \tag{57.15}$$

As will become apparent later, the most convenient scalar components into which to resolve the velocity perturbation are

$$\mathbf{k}\,.\,\hat{\mathbf{v}}, \qquad \hat{\zeta}_{\|} = \mathbf{b}\,.\,\hat{\boldsymbol{\zeta}} = \mathbf{b}\,.\,\mathrm{i}\mathbf{k}\times\hat{\mathbf{v}}, \qquad \hat{f}_{\|} = \mathbf{b}\,.\,\mathrm{i}\mathbf{k}\times\hat{\boldsymbol{\zeta}} = \mathbf{b}\,.\,\mathrm{i}\mathbf{k}\times(\mathrm{i}\mathbf{k}\times\hat{\mathbf{v}}), \tag{57.16}$$

and it is not difficult to show that

$$k^2 \sin^2\theta\,\hat{\mathbf{v}} = \sin^2\theta\,\mathbf{kk}\,.\,\hat{\mathbf{v}} + \mathrm{i}\mathbf{k}\times(\mathrm{i}\mathbf{k}\times\mathbf{b})\hat{f}_{\parallel}/k^2 + (\mathrm{i}\mathbf{k}\times\mathbf{b})\hat{\zeta}_{\parallel}. \tag{57.17}$$

We shall therefore obtain three scalar equations from (57.15) by multiplying it through by $\mathbf{k}\,.$, $\mathbf{b}\,.\,\mathrm{i}\mathbf{k}\times(\mathrm{i}\mathbf{k}\times$, and $\mathbf{b}\,.\,\mathrm{i}\mathbf{k}\times$ in turn. To deal with the awkward final term in (52.15) we replace $\hat{\mathbf{v}}$ by its value given in (57.17) and adopt the definitions

$$v_{11} = \frac{\mathbf{k}\,.\,\mathscr{L}\,.\,\mathbf{k}}{\omega}, \qquad v_{21} = \frac{-\mathbf{k}\,.\,\mathscr{L}\,.\,\mathrm{i}\mathbf{k}\times(\mathrm{i}\mathbf{k}\times\mathbf{b})}{\omega k\sin\theta},$$

$$v_{13} = \frac{\mathbf{k}\,.\,\mathscr{L}\,.\,\mathrm{i}\mathbf{k}\times\mathbf{b}}{\omega\sin^2\theta}, \qquad v_{21} = \frac{-\mathbf{b}\,.\,\mathrm{i}\mathbf{k}\times(\mathrm{i}\mathbf{k}\times\mathscr{L}\,.\,\mathbf{k})}{\omega k\sin\theta},$$

$$v_{22} = \frac{\mathbf{b}\,.\,\mathrm{i}\mathbf{k}\times[\mathrm{i}\mathbf{k}\times\mathscr{L}\,.\,\mathrm{i}\mathbf{k}\times(\mathrm{i}\mathbf{k}\times\mathbf{b})]}{\omega k^2\sin^2\theta}, \qquad v_{23} = \frac{-\mathbf{b}\,.\,\mathrm{i}\mathbf{k}\times(\mathrm{i}\mathbf{k}\times\mathscr{L}\,.\,\mathrm{i}\mathbf{k}\times\mathbf{b})}{\omega k\sin^2\theta},$$

$$v_{31} = \frac{\mathbf{b}\,.\,\mathrm{i}\mathbf{k}\times\mathscr{L}\,.\,\mathbf{k}}{\omega\sin\theta}, \qquad v_{32} = \frac{-\mathbf{b}\,.\,\mathrm{i}\mathbf{k}\times\mathscr{L}\,.\,\mathrm{i}\mathbf{k}\times(\mathrm{i}\mathbf{k}\times\mathbf{b})}{\omega k\sin^2\theta},$$

$$v_{33} = \frac{\mathbf{b}\,.\,\mathrm{i}\mathbf{k}\times\mathscr{L}\,.\,\mathrm{i}\mathbf{k}\times\mathbf{b}}{\omega\sin^2\theta}$$

From (57.8) and (57.14) we find the values

$$\left.\begin{aligned}
v_{11} &= \frac{k^2}{\omega}[v_1\sin^4\theta + \tfrac{1}{3}v_2(\cos 2\theta+\cos^2\theta)^2 + v_3\sin^2 2\theta], && \left(\frac{4k^2}{3\omega}v\right)\\
v_{12} = v_{21} &= \frac{k^2}{2\omega}\sin 2\theta[v_1\sin^2\theta - v_2(\cos 2\theta+\cos^2\theta) + \\
&\qquad + 2v_3\cos 2\theta], && (0)\\
v_{22} &= \frac{k^2}{\omega}[\tfrac{1}{4}(v_1+3v_2)\sin^2 2\theta + v_3\cos^2 2\theta], && \left(\frac{k^2}{\omega}v\right)\\
v_{23} = v_{32} &= -\mathrm{i}\frac{k^2}{\omega}\cos\theta(v_4\sin^2\theta + v_5\cos 2\theta), && (0)\\
v_{33} &= \frac{k^2}{\omega}(v_1\sin^2\theta + v_3\cos^2\theta), && \left(\frac{k^2}{\omega}v\right)\\
v_{31} = v_{13} &= -\mathrm{i}\frac{k^2}{\omega}\sin\theta(v_4\sin^2\theta + 2v_5\cos^2\theta), && (0)
\end{aligned}\right\} \tag{57.18}$$

where the bracketed expressions on the right-hand side apply in an isotropic medium, that is, when (see (50.11)) $v_1 = v_2 = v_3 = v$, $v_4 = v_5 = 0$. The three scalar equations derived from (57.15) appear in rows 2, 3, and 6 of the matrix equation (57.22) below.

On eliminating $\hat{\mathbf{E}}$ from the linearized form of (57.1) by $(57.5)_3$ we find

$$\omega\hat{B}-\mathbf{k}\,.\,\boldsymbol{\eta}\,.\,\hat{\mathbf{j}}+\mathbf{k}\,.\,\mathbf{B}\hat{\mathbf{v}}-\mathbf{Bk}\,.\,\hat{\mathbf{v}}-\frac{1}{en}\mathbf{k}\,.\,\mathbf{B}\hat{\mathbf{j}}-\mathrm{i}\,\frac{2T}{3\rho}\mathbf{k}\times\boldsymbol{\gamma}\,.\,\mathbf{k}\hat{\rho}=0,$$

the adiabatic approximation $\hat{T}=(2T/3\rho)\hat{\rho}$ being sufficiently accurate for the last term. In terms of the variables defined in (57.7) and the coefficients adopted in (57.11), this equation takes the form

$$\omega(1+\mathrm{i}c_\eta)\hat{\mathbf{h}}+\frac{\omega}{k^2}\left(\frac{\bar{c}_\eta-c_\eta}{\sin^2\theta}\right)\mathbf{b}\times\mathbf{k}\hat{J}_{\parallel}-k\ell\cos\theta\left(\delta_1-\frac{\eta_\wedge}{\mu\ell}\right)\hat{\mathbf{J}}+k\ell\cos\theta\hat{\mathbf{v}}-$$
$$-\frac{\ell}{k}\mathbf{k}\,.\,\hat{\mathbf{v}}[k\mathbf{b}(1+\mathrm{i}c'_\gamma)+\mathrm{i}c_\gamma\mathbf{b}\times\mathbf{k}-\mathrm{i}c'_\gamma\cos\theta\;\mathbf{k}]=0, \tag{57.19}$$

where

$$c_\eta\equiv\frac{k^2}{\omega\mu}\eta_\perp,\qquad \bar{c}_\eta\equiv\frac{k^2}{\omega\mu}(\eta_\perp\cos^2\theta+\eta_\parallel\sin^2\theta), \tag{57.20}$$

δ_1 is the length defined in (56.9), namely,

$$\delta_1=(m_1/e^2n\mu)^{\frac{1}{2}}, \tag{57.21}$$

and $\boldsymbol{\eta}$ has been expanded according to (56.13).

It only remains to form the scalar components of (57.19) by multiplying it through by $\mathbf{b}\,.\,\mathrm{i}\mathbf{k}\times(\mathrm{i}\mathbf{k}\times$ and $\mathbf{b}\,.\,\mathrm{i}\mathbf{k}\times$; the component in the $\mathbf{k}$-direction is identically zero. These equations appear in rows 4 and 5 of the matrix equation below, which sets out in a symmetric form eqn (57.13) and the five scalar equations derived from (57.15) and (57.19),

$$\begin{bmatrix} \tilde{u}(1+\mathrm{i}c_\kappa) & a & 0 & 0 & 0 & 0 \\ a & \tilde{u}(1+\mathrm{i}v_{11}) & \mathrm{i}\tilde{u}v_{12} & (1+\mathrm{i}c'_\gamma)\ell_x & c_\gamma\ell_x & \mathrm{i}\tilde{u}v_{13} \\ 0 & \mathrm{i}\tilde{u}v_{21} & \tilde{u}(1+\mathrm{i}v_{22}) & \ell_z & 0 & \mathrm{i}\tilde{u}v_{23} \\ 0 & (1+\mathrm{i}c'_\gamma)\ell_x & \ell_z & \tilde{u}(1+\mathrm{i}c_\eta) & X & 0 \\ 0 & c_\gamma\ell_x & 0 & X & \tilde{u}(1+\mathrm{i}\bar{c}_\eta) & \ell_z \\ 0 & \mathrm{i}\tilde{u}v_{31} & \mathrm{i}\tilde{u}v_{32} & 0 & \ell_z & \tilde{u}(1+\mathrm{i}v_{33}) \end{bmatrix}\begin{bmatrix} \hat{\psi} \\ -\mathbf{k}\,.\,\hat{\mathbf{v}}\sin\theta \\ \hat{f}_\parallel/k \\ \hat{h}_\parallel k \\ -\hat{J}_\parallel \\ -\hat{\zeta}_\parallel \end{bmatrix}=[0], \tag{57.22}$$

where for convenience the following notation is used

$$\tilde{u}=\omega/k,\qquad X=k\ell_z(\delta_1-\eta_\wedge/\mu\ell),\qquad \ell_x=\ell\sin\theta,\qquad \ell_z=\ell\cos\theta,$$

and (57.23)

$$\hat{\psi}=\frac{\hat{\rho}}{\rho}ka(1-\mathrm{i}c_\kappa)\sin\theta,$$

The merit of our particular choice of independent variables is now visible in the symmetry so produced in the square matrix operator.

57.3. *The low-frequency magneto-plasma waves*

It will be assumed that the frequency ω is a real number and that the amplitude k of the propagation vector is complex:

$$k = k_r(1+\mathrm{i}\alpha), \qquad \tilde{u} = u(1-\mathrm{i}\alpha), \qquad u \equiv \omega/k_r \qquad (\alpha \ll 1), \qquad (57.24)$$

where k_r is real, u is the phase-velocity, and the *attenuation* α (cf. (57.4)) is taken to be small compared with unity, without which constraint the wave would be extinguished in too few wavelengths ($\lambda = 2\pi/k_r$) to be of much interest.

The existence of non-zero solutions of (57.22) requires the determinant of the square matrix to vanish, that is by (57.24),

$$\begin{vmatrix} u(1+\mathrm{i}c_\kappa-\mathrm{i}\alpha) & a & 0 & 0 & 0 & 0 \\ a & u(1+\mathrm{i}v_{11}-\mathrm{i}\alpha) & \mathrm{i}uv_{12} & (1+\mathrm{i}c'_\gamma)\mathscr{b}_x & c_\gamma\mathscr{b}_x & \mathrm{i}uv_{13} \\ 0 & \mathrm{i}uv_{21} & u(1+\mathrm{i}v_{22}-\mathrm{i}\alpha) & \mathscr{b}_z & 0 & \mathrm{i}uv_{23} \\ 0 & (1+\mathrm{i}c'_\gamma)\mathscr{b}_x & \mathscr{b}_z & u(1+\mathrm{i}c_\eta-\mathrm{i}\alpha) & X & 0 \\ 0 & c_\gamma\mathscr{b}_x & 0 & X & u(1+\mathrm{i}\bar{c}_\eta-\mathrm{i}\alpha) & \mathscr{b}_z \\ 0 & \mathrm{i}uv_{31} & \mathrm{i}uv_{32} & 0 & \mathscr{b}_z & u(1+\mathrm{i}v_{33}-\mathrm{i}\alpha) \end{vmatrix} = 0, \qquad (57.25)$$

where in the definitions of c_κ, v_{ij}, c'_γ, etc. it is sufficiently accurate to replace k by k_r. Eqn (57.25) is a complex dispersion relation determining the possible values of u and α.

To appreciate its physical import we shall first consider the case when the wave is not attenuated, which requires all the imaginary numbers in (57.25) to vanish. Then u is the root of

Entropy wave

Acoustic wave

Magneto-acoustic wave

$$\begin{vmatrix} u & 0 & 0 & 0 & 0 & 0 & 0 \\ 0 & u & a & 0 & 0 & 0 & 0 \\ 0 & a & u & 0 & \mathscr{b}_x & c_\gamma\mathscr{b}_x & 0 \\ 0 & 0 & 0 & u & \mathscr{b}_z & 0 & 0 \\ 0 & 0 & \mathscr{b}_x & \mathscr{b}_z & u & X & 0 \\ 0 & 0 & c_\gamma\mathscr{b}_x & 0 & X & u & \mathscr{b}_z \\ 0 & 0 & 0 & 0 & 0 & \mathscr{b}_z & u \end{vmatrix} = 0, \qquad (57.26)$$

Alfvénic wave

where for completeness we have reintroduced the (stationary) entropy wave from eqn (57.6). The name of the waves associated with the various blocks of the array are shown. There are three types of propagating waves, namely, two magneto-acoustic waves, distinguished by the names 'slow' and 'fast' and the Alfvénic wave. It is interesting to note that in the limit as the magnetic field vanishes the electromagnetic variables $\hat{h}_{\|}$ and $\hat{J}_{\|}$ tend to zero, leaving just five waves, namely, the two propagating acoustic waves, $u = \pm a$, and three stationary waves, two of them associated with vorticity† and one with entropy.

Notice from (57.26) that the acoustic wave is not coupled with the remaining waves for propagation parallel to the field lines ($\mathscr{b}_x = 0$). Decoupling of the magneto-acoustic and Alfvénic branches requires both the Hall term X and the thermoelectric term $c_\gamma \mathscr{b}_x$ to vanish. The reason why the latter remains as a non-dissipative effect follows from the fact that the thermoelectric term $\boldsymbol{\gamma} . \nabla T$ in (57.1) has the opposite parity from the ohmic dissipation $\boldsymbol{\eta} . \mathbf{j}$ and hence is reversible.

When $c_\gamma \ll 1$, $\dot{\eta}_{\wedge}/\mu \mathscr{b} \ll \delta_1$, (57.26) gives

$$(u^2 - \mathscr{b}_z^2)[u^4 - (a^2 + \mathscr{b}^2)u^2 + a^2 \mathscr{b}_z^2] = u^2(u^2 - a^2)\mathscr{b}_z^2 k^2 \delta_1^2, \qquad (57.27)$$

showing that the pure Alfvénic wave

$$u = \pm \mathscr{b}_z \qquad (57.28)$$

and the pure magneto-acoustic waves

$$u = \pm c_{\mathrm{s}}, \qquad u = \pm c_{\mathrm{f}} \qquad (c_{\mathrm{s}} < c_{\mathrm{f}}), \qquad (57.29)$$

c_{s}^2, c_{f}^2 being the roots of

$$u^4 - (a^2 + \mathscr{b}^2)u^2 + a^2 \mathscr{b}_z^2 = (u^2 - c_{\mathrm{s}}^2)(u^2 - c_{\mathrm{f}}^2) = 0, \qquad (57.30)$$

can propagate only at wavelengths λ long enough to satisfy the decoupling condition

$$2\pi \cos\theta \, \delta_1/\lambda \ll 1. \qquad (57.31)$$

Alternatively we can use $\delta_1 = \mathscr{b}/\omega_1$ and $\omega = k_r u$ in the right-hand member of (57.27) to express this condition in the form

$$\cos\theta \, (\omega/\omega_1) \ll 1.$$

† This result is a form of Kelvin's theorem in classical fluid dynamics concerning the permanence of the vorticity of a given fluid particle.

Thus in the 'magnetohydrodynamic' limits,

$$2\pi\delta_1/\lambda \to 0, \qquad \omega/\omega_1 \to 0 \tag{57.32}$$

the roots of the dispersion relation yield the magnetohydrodynamic waves given in (57.28) and (57.29).

57.4. The attenuation coefficient

Now we return to eqn (57.25), the real part of which reduces to (57.27) and the imaginary part of which gives the value of the attenuation coefficient α. We ignore second-order terms in α, ν_{11}, c_κ, etc., and after some algebra find

$$\begin{aligned}2\alpha[(u^2-b_z^2)(2u^2-a^2-b^2)+(u^2-c_s^2)(u^2-c_f^2)-k^2\delta_1^2b_z^2(2u^2-a^2)]\\ =(u^2-b_z^2)[(c_\kappa+\nu_{11})(u^2-b_z^2)+(c_\eta+\nu_{22})(u^2-a^2)-(c_\kappa+\nu_{22}+2c'_\gamma)b_x^2+\\ +2\nu_{12}b_xb_z]+(\bar{c}_\eta+\nu_{33})(u^2-c_s^2)(u^2-c_f^2)-2k\delta_1b_z^2[ub_x\nu_{31}+\\ +u^{-1}b_z(u^2-a^2)\nu_{32}]-k^2\delta_1^2b_z^2[(c_\kappa+\nu_{11})u^2+(\nu_{22}+\nu_{33})(u^2-a^2)].\end{aligned} \tag{57.33}$$

In the magnetohydrodynamic limit $k\delta_1 = 0$, and for the fast magneto-acoustic wave (57.33) yields

$$\begin{aligned}\alpha = \frac{1}{2(2c_f^2-a^2-b^2)}[(c_\kappa+\nu_{11})(c_f^2-b_z^2)+(c_\eta+\nu_{22})(c_f^2-a^2)-\\ -(c_\kappa+\nu_{22}+2c'_\gamma)b_x^2+2\nu_{12}b_xb_z];\end{aligned} \tag{57.34}$$

a similar expression (replace c_f by c_s) holds for the slow wave. For the Alfvénic wave in the magnetohydrodynamic limit (57.33) gives

$$\alpha = \tfrac{1}{2}(\bar{c}_\eta+\nu_{33}) = \frac{1}{2}\frac{\omega}{b_z^2}\left[\left(\frac{\eta_\perp}{\mu}+\nu_3\right)\cos^2\theta+\left(\frac{\eta_\parallel}{\mu}+\nu_1\right)\sin^2\theta\right]. \tag{57.35}$$

Finally consider propagation parallel to **B** when $b_x = 0$ and $c_f = a$ for $a > b$, $c_f = b$ for $b > a$. In the first case (57.34) reduces to the classical result for the attenuation of sound waves,

$$\alpha = \tfrac{1}{2}(c_\kappa+\nu_{11}) = \frac{\omega}{a^2}\left(\frac{2}{3}\nu_2+\frac{2}{15}\frac{T}{p}\kappa_\parallel\right),$$

while in the second

$$\alpha = \tfrac{1}{2}(c_\eta+\nu_{22})+\frac{1}{2}\frac{\omega}{b^2}\left(\frac{\eta_\perp}{\mu}+\nu_3\right),$$

in agreement with the case $\theta = 0$ of (57.35).

58. Fluid equations for liquid helium II

58.1. The two-fluid model

Below a temperature of 2·17 K and over quite a wide range of pressures, helium 4 is a liquid with remarkable properties. Above 2·17 K the liquid is termed helium I and below 2·17 K, helium II. Helium I is entirely 'normal' in its behaviour, whereas helium II behaves as if it were a mixture of a normal fluid like helium I plus a strange superfluid to which is attributed the unusual properties of the fluid as a whole. It is not possible in a brief account of the subject to give adequate justification of the two-fluid model of helium II, which lacks the obvious physical reality of the two-fluid magneto-plasma model, although a few remarks may be helpful. But first we shall describe some of the liquid's unusual properties.

First there is the thermomechanical or 'fountain' effect, in which a very small temperature difference is able to produce substantial pressure gradients. Thus, for example, radiant heat impinging on the lower part of an open tube packed with a fine powder and immersed in a bath of helium II will result in a strong ejection of fluid from the end of the tube clear of the bath. This is due to the flow of the superfluid from the bath to the heated region of the tube, the normal fluid being impeded by the powder. The propagation of temperature (or entropy) waves, a phenomenon described by the curious name 'second sound', is an interesting consequence of the thermomechanical effect and will be discussed in § 60. Other phenomena are associated with the absence of any viscosity in the superfluid, which allows this component of helium II to flow quite frictionless through capillaries, past obstacles, and even up the sides and over the edge of a test tube should there be any difference in level between the helium II inside the tube and that outside it.

The two-fluid model postulates that helium II *behaves* as if it were a mixture of a normal fluid of density ρ_n and a superfluid of density ρ_s, such that the total density is $\rho = \rho_s + \rho_n$. The ratio ρ_n/ρ is a function of temperature and may be deduced directly by measuring the period of oscillation P of a stack of closely spaced discs immersed in helium II and rigidly fastened on a torsion fibre. Only the normal fluid between the discs takes part in their oscillatory motion, consequently P is a function of ρ_n. The experimental data closely fit the empirical relation

$$c_n \equiv \rho_n/\rho = \begin{cases} (T/T_\lambda)^{5\cdot 6} & (T < T_\lambda) \\ 1 & (T > T_\lambda) \end{cases} \qquad (T_\lambda \approx 2\cdot 17\ \text{K}).$$

The usual physical interpretation of ρ_n is that it is the density associated with the mean drift momentum of collective motions of the helium atoms known as 'excitations'. There are believed to be two types of excitations comprising the normal fluid of helium II, namely, 'phonons' (§ 34.1) and 'rotons', and the statistical average of their velocities of propagation is taken to be the normal fluid velocity $\mathbf{v}_n$. The total fluid density ρ and mass flux $\rho\mathbf{v}$ have their normal physical meanings, so that the superfluid properties ρ_s and $\mathbf{v}_s$ may be obtained via the definitions $\rho_s = \rho - \rho_n$, $\rho_s\mathbf{v}_s = \rho\mathbf{v} - \rho_n\mathbf{v}_n$. Another theory associates the superfluid with helium atoms in the Bose–Einstein condensed state (§ 33.6); while this theory is attractive in offering a direct physical interpretation of ρ_s and $\mathbf{v}_s$, some of its predictions are at variance with experiment. It appears that *all* the helium atoms are involved in the excitations, making it impossible to assign them individually to one fluid or the other.

For details of the various kinetic theories of helium II, which aim *inter alia* at underpinning the two-fluid model, we must refer the reader to the literature cited in the text by J. Wilks (1967). For present purposes, namely, to use helium II as an interesting example in process thermodynamics, it is sufficient to accept the model as a phenomenological construction that is very convincingly supported by a variety of experimental evidence. The following treatment differs in several respects from the standard account although the final results are much the same. The differences are discussed in § 59.5.

58.2. Superfluid thermodynamics

To the superfluid component of helium II will now be attributed two basic thermodynamic properties, namely, (1) that it possesses no entropy and (2) that under steady conditions it is in material equilibrium with the normal component.† The first condition requires the superfluid to be energetically at absolute zero, that is,

$$T_s = 0, \qquad s_s = 0, \qquad \rho_s = \rho_n s_n, \tag{58.1}$$

s being the specific entropy of the composite system helium II.

For the normal fluid, at temperature T, the specific Gibbs function satisfies

$$g_n = u_n + p_n\rho_n^{-1} - Ts_n, \qquad dg_n = \rho_n^{-1}\,dp_n - s_n\,dT, \tag{58.2}$$

the second being the Gibbs–Duhem relation (§ 15.1), which follows from $T\,ds_n = du_n + p_n\,d\rho_n^{-1}$ and the differential of eqn $(58.2)_1$. Without commitment on the question of superfluid internal energy, corresponding to $(58.2)_2$ we may write

$$dg_s = \rho_s^{-1}\,dp_s, \tag{58.3}$$

which proves basic to the theory.

† Here 'material' equilibrium is to be understood not in terms of atoms of helium II, but as a thermodynamic equilibrium between the density of excitations ρ_n and the complementary density $(\rho - \rho_n)$.

For the composite helium II there are the thermodynamic relations (see § 14)

$$\left.\begin{aligned} T\,\mathrm{d}s &= \mathrm{d}u+p\,\mathrm{d}\rho^{-1}-(g_{\mathrm{n}}-g_{\mathrm{s}})\,\mathrm{d}c_{\mathrm{n}}, \\ Ts &= u+p/\rho-g, \qquad g = c_{\mathrm{n}}g_{\mathrm{n}}+c_{\mathrm{s}}g_{\mathrm{s}}, \\ \mathrm{d}g &= \rho^{-1}\,\mathrm{d}p-s\,\mathrm{d}T+(g_{\mathrm{n}}-g_{\mathrm{s}})\,\mathrm{d}c_{\mathrm{n}}, \end{aligned}\right\} \tag{58.4}$$

where

$$c_{\mathrm{n}} = \rho_{\mathrm{n}}/\rho, \qquad c_{\mathrm{s}} = \rho_{\mathrm{s}}/\rho, \qquad c_{\mathrm{n}}+c_{\mathrm{s}} = 1,$$

and

$$p = p_{\mathrm{n}}+p_{\mathrm{s}}.$$

The second thermodynamic property mentioned above means that there exists a 'density equilibrium' between the fluids, and that near this equilibrium, at constant total density ρ, 'exchanges' $\mathrm{d}\rho_{\mathrm{s}} = -\mathrm{d}\rho_{\mathrm{n}}$ will occur to restore the balance. Suppose this density exchange process is associated with a finite relaxation time τ_{ns}, then on a time scale much greater than τ_{ns} the two fluids are always in density equilibrium, which by the theory of § 17.3 and $(58.4)_1$ requires that

$$g_{\mathrm{s}} = g_{\mathrm{n}}. \tag{58.5}$$

Later we shall consider non-equilibrium processes on a time scale comparable with τ_{ns}, for which (58.5) is only an approximation. It follows from $(58.4)_4$, (58.5), and

$$g_{\mathrm{s}}-g = c_{\mathrm{n}}(g_{\mathrm{s}}-g_{\mathrm{n}}), \tag{58.6}$$

that

$$\rho_{\mathrm{s}}^{-1}\,\mathrm{d}p_{\mathrm{s}} = \mathrm{d}g_{\mathrm{s}} = \mathrm{d}g = \rho^{-1}\,\mathrm{d}p-s\,\mathrm{d}T, \tag{58.7}$$

which relation leads to the fountain effect as will next be shown.

58.3. The equations of fluid motion

By an application of eqn $(51.7)_4$ and (51.14) to the super and normal fluids we find

$$\frac{\partial\rho_{\mathrm{s}}}{\partial t}+\nabla\cdot\rho_{\mathrm{s}}\mathbf{v}_{\mathrm{s}} = a_{\mathrm{sn}}-a_{\mathrm{ns}}, \qquad \frac{\partial\rho_{\mathrm{n}}}{\partial t}+\nabla\cdot\rho_{\mathrm{n}}\mathbf{v}_{\mathrm{n}} = a_{\mathrm{ns}}-a_{\mathrm{sn}}, \tag{58.8}$$

where a_{sn} is the rate at which the superfluid gains density from the normal fluid and conversely for a_{ns}. Likewise $(52.2)_2$ yields

$$\left.\begin{aligned} \rho_{\mathrm{s}}\mathrm{D}_{\mathrm{s}}\mathbf{v}_{\mathrm{s}}+\nabla p_{\mathrm{s}} &= \rho_{\mathrm{s}}\mathbf{F}_{\mathrm{s}}+a_{\mathrm{sn}}(\mathbf{v}_{\mathrm{n}}-\mathbf{v}_{\mathrm{s}}), \\ \rho_{\mathrm{n}}\mathrm{D}_{\mathrm{n}}\mathbf{v}_{\mathrm{n}}+\nabla\cdot\mathbf{p}_{\mathrm{n}} &= \rho_{\mathrm{n}}\mathbf{F}_{\mathrm{n}}+a_{\mathrm{ns}}(\mathbf{v}_{\mathrm{s}}-\mathbf{v}_{\mathrm{n}}), \end{aligned}\right\} \tag{58.9}$$

where we have anticipated the facts to be justified in § 59.1 that the superfluid possesses no viscosity and that the friction force between the components is zero.

As

$$\nabla p_s = \rho_s \nabla g_s$$

and

$$\nabla \cdot \mathbf{p}_n = \nabla(p - p_s) + \nabla \cdot \boldsymbol{\pi}_n = \nabla p - \rho_s \nabla g_s + \nabla(\tfrac{1}{3}\check{\pi}_n) + \nabla \cdot \mathring{\boldsymbol{\pi}}_n, \tag{58.10}$$

(58.9) may be written

$$\begin{aligned} \rho_s \mathrm{D}_s \mathbf{v}_s &= -\rho_s \nabla g_s + a_{sn}(\mathbf{v}_n - \mathbf{v}_s), \\ \rho_s \mathrm{D}_n \mathbf{v}_n &= -\nabla p + \rho_s \nabla g_s - \nabla(\tfrac{1}{3}\check{\pi}_n) - \nabla \cdot \mathring{\boldsymbol{\pi}}_n + a_{ns}(\mathbf{v}_s - \mathbf{v}_n), \end{aligned} \tag{58.11}$$

on assuming no body forces are involved. Constitutive equations are required for g_s, $\check{\pi}_n$, $\mathring{\boldsymbol{\pi}}_n$, a_{sn}, and a_{ns}.

It is at times convenient to replace one of (58.11) by the total momentum equation

$$\mathrm{D}\mathbf{v} = \frac{\partial \mathbf{v}}{\partial t} + \mathbf{v} \cdot \nabla \mathbf{v} = -\frac{1}{\rho} \nabla \cdot \mathbf{p} \qquad (\rho \mathbf{v} = \rho_s \mathbf{v}_s + \rho_n \mathbf{v}_n), \tag{58.12}$$

when by (53.5) and the definition

$$\mathbf{w} \equiv \mathbf{v}_n - \mathbf{v}_s, \tag{58.13}$$

the total pressure tensor is

$$\mathbf{p} = \mathbf{p}_n + p_s \mathbf{1} + \frac{\rho_s \rho_n}{\rho} \mathbf{w}\mathbf{w} = (p + \tfrac{1}{3}\check{\pi}_n)\mathbf{1} + \mathring{\boldsymbol{\pi}}_n + \frac{\rho_s \rho_n}{\rho} \mathbf{w}\mathbf{w}. \tag{58.14}$$

As will be shown shortly, the drag force on the superfluid due to density transfer, namely, $a_{sn}(\mathbf{v}_n - \mathbf{v}_s)$, is usually quite small when eqn $(58.11)_1$ gives

$$\frac{\partial \mathbf{v}_s}{\partial t} + \mathbf{v}_s \cdot \nabla \mathbf{v}_s = -\nabla g_s, \tag{58.15}$$

and further if the superfluid is in density equilibrium with the normal fluid we may adopt (58.7) and obtain the ideal flow equation

$$\frac{\partial \mathbf{v}_s}{\partial t} + \mathbf{v}_s \cdot \nabla \mathbf{v}_s = -\frac{1}{\rho} \nabla p + s \nabla T. \tag{58.16}$$

Similarly, when dissipation in the normal fluid can be neglected, (58.7) and $(58.11)_2$ give

$$\frac{\partial \mathbf{v}_n}{\partial t} + \mathbf{v}_n \cdot \nabla \mathbf{v}_n = -\frac{1}{\rho} \nabla p - \frac{\rho_s s}{\rho_n} \nabla T. \tag{58.17}$$

Finally, we note that when the superfluid is at rest (58.16) reduces to London's famous relation

$$\nabla p = \rho s \nabla T, \tag{58.18}$$

which provides an immediate explanation of the fountain effect.

58.4. Superfluid vorticity

Eqn (58.15) may be written

$$\frac{\partial \mathbf{v}_s}{\partial t} + \boldsymbol{\zeta}_s \times \mathbf{v}_s + \nabla(g_s + \tfrac{1}{2}v_s^2) = 0, \tag{58.19}$$

where $\boldsymbol{\zeta}_s = \nabla \times \mathbf{v}_s$ is the vorticity vector. Taking the curl of this equation we find

$$\frac{\partial \boldsymbol{\zeta}_s}{\partial t} + \mathbf{v}_s \cdot \nabla \boldsymbol{\zeta}_s = \boldsymbol{\zeta}_s \cdot \nabla \mathbf{v}_s - \boldsymbol{\zeta}_s \nabla \cdot \mathbf{v}_s,$$

an equation that is known to possess a unique solution with specified initial conditions. From this fact it follows that if a fluid particle of superfluid has zero vorticity at some instant, it always has zero vorticity thereafter. Consequently if at $t = 0$ the flow is spacewise uniform, thereafter

$$\nabla \times \mathbf{v}_s = 0 \tag{58.20}$$

throughout the domain of flow. Of course this no longer holds when density transfer applies a drag on the superfluid.

It has been often suggested that superfluidity is to be viewed as being a *consequence* of zero vorticity, however, as remarked by Lin (1963), irrotationality should not be given a fundamental status in the theory. That it has been is no doubt because of the close analogy between (58.20) and the Meissner effect in superconductivity. When (58.20) does apply it is no more than a deduction from the basic fluid equations with prescribed initial conditions.

59. Dissipative processes in helium II

59.1. The rate of entropy production

Our only basic assumption about the superfluid will be that it is always at absolute zero temperature, even when subject to dissipative processes like fluid drag. Thus any 'incipient' dissipation in the superfluid must be removed via a mass transfer from it into the normal component.

To obtain an expression for the entropy production rate we shall adopt the general formula for mixtures given in (52.12), in which the subscript i

is s for the superfluid and n for the normal fluid. Initially we shall take the temperature T_s and entropy s_s to be non-zero and subsequently take the limits $T_s \to 0$, $s_s \to 0$.† Thus

$$\sigma = T_n^{-1}[-\boldsymbol{\pi}_n:\nabla\mathbf{v}_n - T_n^{-1}\mathbf{q}_n\cdot\nabla T_n + \xi_{ns} + \varphi_{ns} + a_{ns}T_s s_s - \varphi_{sn} - a_{sn}T_n s_n + \\ + a_{ns}(g_s - g_n + \tfrac{1}{2}w^2)] + T_s^{-1}[-\boldsymbol{\pi}_s:\nabla\mathbf{v}_s - T_s^{-1}\mathbf{q}_s\cdot\nabla T_s + \xi_{sn} + \varphi_{sn} + \\ + a_{sn}T_n s_n - \varphi_{ns} - a_{ns}T_s s_s + a_{sn}(g_n - g_s + \tfrac{1}{2}w^2)]. \quad (59.1)$$

In the limit $T_s \to 0$, σ will become infinite unless the second term in (59.1) vanishes identically, that is, vanishes for all possible dissipative processes. We shall impose this physically reasonable constraint, which is equivalent to excluding dissipative processes from the superfluid component. As the processes due to viscosity $\boldsymbol{\pi}_s$, heat conduction $\mathbf{q}_s$, energy transfer from the normal fluid to the superfluid φ_{sn}, and mutual friction ξ_{sn} are independent, in the limit we must have

$$\boldsymbol{\pi}_s = 0, \qquad \mathbf{q}_s = 0, \qquad \varphi_{sn} = 0, \qquad \xi_{sn} = 0,$$

leaving (59.2)

$$\varphi_{ns} = a_{sn}(g_n + Ts_n - g_s + \tfrac{1}{2}w^2),$$

where $T_n = T$ may now be interpreted as being the temperature of the composite helium II.

On the tentative assumption that it is possible for mutual friction to occur between the components even when the resulting dissipation occurs *only* in the normal component, we have from (52.10)

$$\xi_{ns} = \mathbf{R}_{sn}\cdot(\mathbf{v}_n - \mathbf{v}_s),$$

$\mathbf{R}_{sn}$ being the force density acting on the superfluid due to the relative motion of the normal fluid. Thus (59.1) reduces to

$$T\sigma = -\boldsymbol{\pi}_n:\nabla\mathbf{v}_n - T^{-1}\mathbf{q}_n\cdot\nabla T + \mathbf{R}_{sn}\cdot(\mathbf{v}_n - \mathbf{v}_s) + (a_{sn} - a_{ns})(g_n - g_s) + \\ + \tfrac{1}{2}(a_{sn} + a_{ns})w^2, \quad (59.3)$$

on using $(59.2)_5$ to eliminate φ_{ns}.

At first sight eqn $(59.2)_5$ appears to require $a_{sn} = 0$ for φ_{ns} should be zero since heat cannot flow from the colder superfluid into the normal fluid. However, there is another interpretation of this equation namely, that when a mass $a_{sn}\,dt$ is transferred into the superfluid, the normal fluid must simultaneously retain the incremental energy density $a_{sn}\,dt(h_n - g_s + \tfrac{1}{2}w^2)$ associated with this mass to prevent the superfluid being heated. In this view

† We expect $s_s = 0$ to follow from $T_s = 0$ by the third law of thermodynamics (§ 25.4).

$\varphi_{ns}\,dt$ is heat deposited in the normal fluid *as if* it came from the superfluid, and as a consequence of interactions with the superfluid.

There remains the question of the mutual friction term in (59.3), based on the hypothesis that the two fluids can interstream with a non-zero friction $\mathbf{R}_{sn}$ that produces dissipation in only one of them. This implies an improbable exchange process between fluid elements in which momentum is transferred but energy is not. Hence we shall adopt

$$\mathbf{R}_{sn} = 0, \tag{59.4}$$

which is confirmed by the experimental evidence. It is important to note that in this discussion we have tacitly assumed that the work $\mathbf{R}_{sn} \cdot (\mathbf{v}_n - \mathbf{v}_s)$ is entirely dissipated in the usual way. The possibility that it could contribute irreversibly to the energy in *non-dissipative* superfluid eddies and vortices will be taken up in § 61.2.

Now $(58.8)_2$ can be written

$$a_{ns} - a_{sn} = c_n\left(\frac{\partial\rho}{\partial t} + \mathbf{v}_n \cdot \nabla\rho + \rho\nabla \cdot \mathbf{v}_n\right) + \rho D_n c_n,$$

so that by conservation of mass for helium II, that is,

$$\frac{\partial\rho}{\partial t} + \nabla \cdot \rho\mathbf{v} = 0, \tag{59.5}$$

we find

$$a_{ns} - a_{sn} = -c_n\nabla \cdot \rho(\mathbf{v} - \mathbf{v}_n) + \rho D_n c_n. \tag{59.6}$$

By (59.3) the second right-hand term in (59.6) contributes to $T\sigma$ an amount

$$T\sigma' = \rho(g_s - g_n)D_n c_n,$$

where by $(58.4)_1$

$$TD_n s = D_n u + pD_n\rho^{-1} + (g_s - g_n)D_n c_n.$$

Thus c_n is an internal variable with affinity $(g_s - g_n)$ and whose relaxation towards equilibrium can result in bulk viscosity. With approximations similar to those adopted in § 49.5, we obtain

$$\left.\begin{aligned}
T\sigma' &= (g_s - g_n)\varepsilon D_n\rho = \varepsilon(g_n - g_s)[\rho\nabla \cdot \mathbf{v}_n + \nabla \cdot \rho(\mathbf{v} - \mathbf{v}_n)] \\
&\text{and} \\
a_{sn} - a_{ns} &= \varepsilon\rho\nabla \cdot \mathbf{v}_n + (c_n + \varepsilon)\nabla \cdot \rho(\mathbf{v} - \mathbf{v}_n), \\
&\text{where} \\
\varepsilon &\equiv (\partial c_n/\partial\rho)_{s, g_s = g_n}.
\end{aligned}\right\} \tag{59.7}$$

Thus when (59.6) is substituted into (59.3), by (59.4) the result may be written

$$T\sigma = [-\tfrac{1}{3}\overset{\times}{\pi}_n + \rho\varepsilon(g_n - g_s)]\nabla \cdot \mathbf{v}_n + (c_n + \varepsilon)(g_n - g_s)\nabla \cdot \rho(\mathbf{v} - \mathbf{v}_n) + \\ + \tfrac{1}{2}(a_{sn} + a_{ns})w^2 - T^{-1}\mathbf{q}_n \cdot \nabla T - \mathring{\boldsymbol{\pi}}_n : \mathring{\nabla \mathbf{v}}_n, \tag{59.8}$$

on setting $\boldsymbol{\pi}^a = 0$, that is, ignoring the possibility of rotational viscosity (§ 49.2).

It is a simple matter to generalize (59.3) to the case of several normal fluids present. We find

$$\sigma = \sigma_n + \sum_i T_i^{-1}\{a_{si}[g_i - g_s + \tfrac{1}{2}(\mathbf{v}_i - \mathbf{v}_s)^2] + a_{is}[g_s - g_i + \tfrac{1}{2}(\mathbf{v}_s - \mathbf{v}_i)^2]\},$$

the summation being over the normal fluids; σ_n is an expression like (52.12) and involves the normal fluids only.

59.2. *Constitutive equations*

For the present we shall ignore the possibility of rotational viscosity and of anisotropy, so that by the theory of § 48.3 we may split (59.8) into three components,

$$\left.\begin{aligned} T\sigma_0 &= [-\tfrac{1}{3}\overset{\times}{\pi} + \rho\varepsilon(g_n - g_s)]\nabla \cdot \mathbf{v}_n + (c_n + \varepsilon)(g_n - g_s)\nabla \cdot \rho(\mathbf{v} - \mathbf{v}_n) + \tfrac{1}{2}(a_{sn} + a_{ns})w^2 \\ &\geqslant 0, \\ T\sigma_1 &= -T^{-1}\mathbf{q}_n \cdot \nabla T \geqslant 0, \\ T\sigma_2 &= -\mathring{\boldsymbol{\pi}}_n : \mathring{\nabla \mathbf{v}}_n \geqslant 0. \end{aligned}\right\} \tag{59.9}$$

From these forms we deduce the linear constitutive relations

$$\left.\begin{aligned} -\tfrac{1}{3}\overset{\times}{\pi}_n &= \rho_n \nu_2 \nabla \cdot \mathbf{v}_n + c_n \nu_1 \nabla \cdot \rho(\mathbf{v} - \mathbf{v}_n) + \rho_n \alpha w^2, \\ g_n - g_s &= \nu_4 \nabla \cdot \mathbf{v}_n + \rho^{-1}\nu_3 \nabla \cdot \rho(\mathbf{v} - \mathbf{v}_n) + \beta w^2, \\ \tfrac{1}{2}(a_{sn} + a_{ns}) &= -\rho_n \alpha' \nabla \cdot \mathbf{v}_n - c_n \beta' \nabla \cdot \rho(\mathbf{v} - \mathbf{v}_n) + \rho_s \rho_n \gamma w^2, \end{aligned}\right\} \tag{59.10}$$

$$\mathbf{q}_n = -\kappa \nabla T, \tag{59.11}$$

$$\mathring{\boldsymbol{\pi}}_n = -2\rho_n \nu \mathring{\nabla \mathbf{v}}_n \tag{59.12}$$

and the reciprocal relations

$$c_n \nu_1 = (c_n + \varepsilon)\nu_4 - \varepsilon \nu_3, \qquad c_n \alpha' = c_n \alpha + \varepsilon \beta, \qquad c_n \beta' = (c_n + \varepsilon)\beta. \tag{59.13}$$

The bulk viscosity ν_2 is due to a relaxation process entirely within the normal fluid (§ 49.5), specifically resulting from the finite time required for the normal fluid excitations to achieve mutual equilibrium. As $(59.10)_2$ is the generalization of (58.5) to rapid processes in which density equilibrium between the two fluids is not quite achieved, we see that the bulk viscosity ν_3 is due to density relaxation between the two components. The 'cross' viscosities ν_1 and ν_4 then result from cross-coupling between the two relaxation processes. The inequalities in (59.9) impose the constraints

$$\nu \geqslant 0, \qquad (c_n\nu_2+\varepsilon\nu_4)\nu_3 \geqslant (c_n+\varepsilon)\nu_4^2, \qquad \kappa \geqslant 0, \qquad \gamma \geqslant 0. \quad (59.14)$$

By (59.7) and $(59.10)_3$

$$\begin{matrix} a_{ns} \\ a_{sn} \end{matrix} = -\rho(c_n\alpha' \pm \tfrac{1}{2}\varepsilon)\nabla \cdot \mathbf{v}_n - [c_n\beta' \pm \tfrac{1}{2}(c_n+\varepsilon)]\nabla \cdot \rho(\mathbf{v}-\mathbf{v}_n) + \rho_s\rho_n\gamma w^2, \qquad (59.15)$$

while from (58.6), $(58.4)_4$, and $(59.10)_2$,

$$\nabla g_s = \nabla g + \nabla[c_n(g_s-g_n)] = \frac{1}{\rho}\nabla p - s\nabla T + c_n\nabla(g_s-g_n),$$

that is,

$$\nabla g_s = \frac{1}{\rho}\nabla p - s\nabla T - c_n\nabla[\nu_4\nabla \cdot \mathbf{v}_n + \rho^{-1}\nu_3\nabla \cdot \rho(\mathbf{v}-\mathbf{v}_n) + \beta w^2]. \qquad (59.16)$$

59.3. The momentum balance equations

Eqns $(58.11)_1$, $(59.15)_2$, and (59.16) yield

$$\begin{aligned} \frac{\partial \mathbf{v}_s}{\partial t} + \mathbf{v}_s \cdot \nabla \mathbf{v}_s = {} & -\frac{1}{\rho}\nabla p + s\nabla T + c_n\nabla[\nu_4\nabla \cdot \mathbf{v}_n + \rho^{-1}\nu_3\nabla \cdot \rho(\mathbf{v}-\mathbf{v}_n) + \beta w^2] + \\ & + \frac{\mathbf{w}}{\rho_s}\{\rho_n\rho_s\gamma w^2 + [\tfrac{1}{2}(c_n+\varepsilon) - c_n\beta']\nabla \cdot \rho(\mathbf{v}-\mathbf{v}_n) + \\ & + \rho(\tfrac{1}{2}\varepsilon - c_n\alpha')\nabla \cdot \mathbf{v}_n\} \end{aligned} \qquad (59.17)$$

for the superfluid acceleration. In writing down the corresponding equation for the normal fluid we shall introduce a slight approximation, namely, to regard the coefficients of $\nabla \cdot \mathbf{v}_n$, $\nabla \cdot \rho(\mathbf{v}-\mathbf{v}_n)$, and w^2 in (59.10) as being constants. Then $(58.11)_2$, $(59.10)_1$, (59.12), and $(59.15)_1$ may be combined to give

$$\begin{aligned} \frac{\partial \mathbf{v}_n}{\partial t} + \mathbf{v}_n \cdot \nabla \mathbf{v}_n = {} & -\frac{1}{\rho}\nabla p - \frac{\rho_s s}{\rho_n}\nabla T + \nabla[(\tfrac{1}{3}\nu + \nu_2 - c_s\nu_4)\nabla \cdot \mathbf{v}_n + \\ & + \rho^{-1}(\nu_1 - c_s\nu_3)\nabla \cdot \rho(\mathbf{v}-\mathbf{v}_n)] + (\alpha - c_s\beta)\nabla w^2 + \nu\nabla^2\mathbf{v}_n + \\ & + \frac{\mathbf{w}}{\rho_n}\{-\rho_n\rho_s\gamma w^2 + [\tfrac{1}{2}(c_n+\varepsilon) + c_n\beta']\nabla \cdot \rho(\mathbf{v}-\mathbf{v}_n) + \\ & + \rho(\tfrac{1}{2}\varepsilon + c_n\alpha')\nabla \cdot \mathbf{v}_n\}. \end{aligned} \qquad (59.18)$$

Similarly, (58.12) yields

$$\frac{\partial \mathbf{v}}{\partial t}+\mathbf{v}\,.\,\nabla\mathbf{v} = -\frac{1}{\rho}\nabla p+c_{\mathrm{n}}\nabla[(\tfrac{1}{3}v+v_2)\nabla\,.\,\mathbf{v}_{\mathrm{n}}+\rho^{-1}v_1\nabla\,.\,\rho(\mathbf{v}-\mathbf{v}_{\mathrm{n}})+\alpha w^2]+$$
$$+c_{\mathrm{n}}v\nabla^2\mathbf{v}_{\mathrm{n}}-\nabla\,.\left(\frac{\rho_{\mathrm{s}}\rho_{\mathrm{n}}}{\rho}\mathbf{w}\mathbf{w}\right). \tag{59.19}$$

The terms involving w in these equations are usually much smaller than the others and may be omitted with little error. The resulting equations then agree with those usually presented in texts on superfluids, with the exception that the standard treatment neglects the derivative ε, so that the first of (59.13) reads $v_1 = v_4$. The notation generally adopted is $\eta = \rho_{\mathrm{n}}v$, $\zeta_1 = c_{\mathrm{n}}v_1$, $\zeta_4 = \rho_{\mathrm{n}}v_4$, $\zeta_2 = \rho_{\mathrm{n}}v_2$, $\zeta_3 = v_3/\rho$, and $\mathbf{j} = \rho\mathbf{v}$.

59.4. The normal fluid energy equation

We now apply eqn (52.9) to the normal fluid component and use (59.2) and (59.3) to obtain the energy balance equation

$$\rho_{\mathrm{n}}\mathrm{D}_{\mathrm{n}}u_{\mathrm{n}}+\rho_{\mathrm{n}}p_{\mathrm{n}}\mathrm{D}_{\mathrm{n}}\rho_{\mathrm{n}}^{-1} = -\boldsymbol{\pi}_{\mathrm{n}}:\nabla\mathbf{v}_{\mathrm{n}}-\nabla\,.\,\mathbf{q}_{\mathrm{n}}+\psi_{\mathrm{n}}+a_{\mathrm{sn}}(h_{\mathrm{n}}-g_{\mathrm{s}}+\tfrac{1}{2}w^2)+$$
$$+a_{\mathrm{ns}}(g_{\mathrm{s}}-h_{\mathrm{n}}+\tfrac{1}{2}w^2),$$

where ψ_{n} is the heat density received per second from external radiative sources. By the Gibbs relation this is equivalent to

$$T\rho_{\mathrm{n}}\mathrm{D}_{\mathrm{n}}s_{\mathrm{n}} = -\boldsymbol{\pi}_{\mathrm{n}}:\nabla\mathbf{v}_{\mathrm{n}}-\nabla\,.\,\mathbf{q}_{\mathrm{n}}+\psi_{\mathrm{n}}+a_{\mathrm{sn}}(g_{\mathrm{n}}-g_{\mathrm{s}}+\tfrac{1}{2}w^2)+$$
$$+a_{\mathrm{ns}}(g_{\mathrm{s}}-g_{\mathrm{n}}+\tfrac{1}{2}w^2)+(a_{\mathrm{sn}}-a_{\mathrm{ns}})Ts_{\mathrm{n}}.$$

By $(58.8)_2$

$$\rho_{\mathrm{n}}\mathrm{D}_{\mathrm{n}}s_{\mathrm{n}}+(a_{\mathrm{ns}}-a_{\mathrm{sn}})s_{\mathrm{n}} = \frac{\partial}{\partial t}(\rho_{\mathrm{n}}s_{\mathrm{n}})+\nabla\,.\,(\rho_{\mathrm{n}}s_{\mathrm{n}}\mathbf{v}_{\mathrm{n}}),$$

so that with (59.3) defining σ and $\rho_{\mathrm{n}}s_{\mathrm{n}} = \rho s$ we arrive at

$$\frac{\partial}{\partial t}(\rho s)+\nabla\,.\left[\frac{\rho sT\mathbf{v}_{\mathrm{n}}+\mathbf{q}_{\mathrm{n}}}{T}\right]-\frac{\psi_{\mathrm{n}}}{T} = \sigma. \tag{59.20}$$

The quantity $\mathbf{Q} \equiv \rho sT\mathbf{v}_{\mathrm{n}}$ is known as the 'heat current' and, under certain flow conditions, like $\mathbf{q}_{\mathrm{n}}$, it is proportional to ∇T. Consider the steady flow of a normal fluid through a narrow capillary tube of radius a. From the well-known Poisseuille formula (for example, see Batchelor (1967, p. 180)) for the volume flux of such a flow,

$$\mathbf{Q} = \rho sT\frac{\pi a^2}{8\rho v}(-\nabla p) = -(\rho s)^2T\frac{\pi a^2}{8\rho v}\nabla T, \tag{59.21}$$

by the London relation (58.18). However, for sufficiently large values of the relative velocity $\mathbf{w}$ it is an experimental observation that (59.21) is replaced by $|\mathbf{Q}| \propto |\nabla T|^{\frac{1}{3}}$. Such a result can be deduced from (59.17) and (59.18), which reduce to

$$0 = -\frac{1}{\rho}\nabla p + s\nabla T + \rho_n \gamma w^2 \mathbf{w}, \qquad 0 = -\frac{1}{\rho}\nabla p - \frac{\rho_s s}{\rho_n}\nabla T - \rho_s \gamma w^2 \mathbf{w},$$

when viscosity and axial velocity gradients are negligible. With $\mathbf{v} = 0$, $\mathbf{w} = \rho \mathbf{v}_n/\rho_s$, and so subtracting these equations we find

$$|\mathbf{Q}| = \rho_s s T \left| \frac{s}{\rho_n \gamma} \nabla T \right|^{\frac{1}{3}}. \tag{59.22}$$

It is usual to explain this result phenomenologically as a consequence of fluid turbulence, consisting of an irregular mass of tangled vortex lines and which occurs when the relative velocity $\mathbf{w}$ exceeds a critical value. On our model it is a consequence of the assumed irreversibility of the density exchange between the fluid components, but of course such an exchange could be enhanced by turbulence. That some form of turbulence is involved in many of the experiments seems clear from the existence of a critical velocity at which the heat-current law changes suddenly from (59.21) to (59.22).

59.5. Khalatnikov's theory

The standard treatment of dissipative effects in the two-fluid model is due in substance to Khalatnikov (1965), and it differs in several respects from that given above. The first point to note is that Khalatnikov incorporates the diffusion energy $\frac{1}{2}\rho_n w^2$ into the thermodynamics of helium II by taking

$$e \equiv \rho u + \tfrac{1}{2}\rho_n w^2$$

to be the 'internal' energy per unit volume. This is of course the total energy in a frame moving with the superfluid. Eqn $(58.4)_1$ may be written

$$\mathrm{d}(\rho u) = T\,\mathrm{d}(\rho s) + g\,\mathrm{d}\rho + \rho(g_n - g_s)\,\mathrm{d}c_n$$

and adding

$$\mathrm{d}(\tfrac{1}{2}\rho_n w^2) = \mathbf{w}\,.\,\mathrm{d}(\rho_n \mathbf{w}) - \tfrac{1}{2}c_n w^2\,\mathrm{d}\rho - \tfrac{1}{2}\rho w^2\,\mathrm{d}c_n$$

to it, we obtain

$$\mathrm{d}e = T\,\mathrm{d}\mathscr{s} + \mu\,\mathrm{d}\rho + \mathbf{w}\,.\,\mathrm{d}\mathbf{j}_0 + \rho(g_n - g_s - \tfrac{1}{2}w^2)\,\mathrm{d}c_n, \tag{59.23}$$

where

$$\mathscr{s} \equiv \rho s, \qquad \mu \equiv g - \tfrac{1}{2}c_n w^2, \qquad \mathbf{j}_0 \equiv \rho_n \mathbf{w}.$$

From $(58.4)_2$

$$p = -\rho u + T\mathfrak{s} + \rho g$$

or

$$p = -e + T\mathfrak{s} + \mu\rho + \mathbf{w} \cdot \mathbf{j}_0. \tag{59.24}$$

In fact Khalatnikov's thermodynamics does not include c_n among the state coordinates, so in place of (59.23) he has

$$\mathrm{d}e = T\,\mathrm{d}\mathfrak{s} + \mu\,\mathrm{d}\rho + \mathbf{w} \cdot \mathrm{d}\mathbf{j}_0, \tag{59.25}$$

that is, it is implicit in this theory that

$$\textit{either} \quad \mathrm{d}c_n = 0 \quad (c_n \text{ frozen}) \quad \textit{or} \quad g_s = g_n - \tfrac{1}{2}w^2 = \mu \quad (c_n \text{ relaxed}). \tag{59.26}$$

Addition of the differential of (59.24) to (59.25) gives $\mathrm{d}\mu = \rho^{-1}\,\mathrm{d}p - s\,\mathrm{d}T - \frac{1}{2}c_n\,\mathrm{d}w^2$, consequently by (58.15) and $(59.26)_2$ we arrive at the result (originally due to Landau)

$$\frac{\partial \mathbf{v}_s}{\partial t} + \mathbf{v}_s \cdot \nabla \mathbf{v}_s = -\frac{1}{\rho}\nabla p + s\nabla T + \tfrac{1}{2}c_n \nabla w^2. \tag{59.27}$$

On the other hand, with dissipation included, Khalatnikov obtains the special case

$$\alpha' = \gamma = \varepsilon = 0, \qquad \beta' = \beta = \tfrac{1}{2} \tag{59.28}$$

of (59.17), which in fact requires one to adopt the *first* of (59.26), albeit in the form $\varepsilon = 0$.

Another point of difference is that the standard theory omits the effect of density transfer on each of the momentum equations, an omission that is really possible only if both a_{sn} and a_{ns} are zero.† From (59.15) we see that (59.28) makes a_{sn} zero but not a_{ns}. The particular value $\beta = \frac{1}{2}$ arises from the inclusion of the diffusion energy in the Gibbs equation, a choice that gives this energy a reversible character, indistinguishable from the internal energy u, and hence in contrast to the result given in (59.3), it cannot contribute to the entropy production rate (see also the remarks in § 53.3). Eqn (59.22), which is qualitatively confirmed by experiment, is some support for our view that the diffusion energy should not be part of the reversible thermodynamics.

† F. London (1954) includes the effect of an increase in the normal fluid density, but does not develop the consequences of this. He argues that density transfers cannot affect the superfluid velocity, which by the first of (58.9) requires $a_{sn} = 0$ and so leads to (59.27).

60. Waves in liquid helium II

60.1. Linearization of the thermodynamic relations

We start with the Gibbs relation $(58.4)_1$ written as

$$du = p\rho^{-2}\,d\rho + T\,ds + (g_n - g_s)\,dc_n, \tag{60.1}$$

and consider small perturbations $(\hat{\rho}, \hat{s}, \hat{c}_n)$ from a time-independent equilibrium state (ρ, s, c_n). The fluid velocities will also be subject to small perturbations $\hat{\mathbf{v}}_n$, $\hat{\mathbf{v}}$ about their zero equilibrium values. As products of these perturbations will be neglected, we can omit the final term in (60.1), it being clear from the discussion following eqn (59.6)—and also from (59.10)—that this term would contribute quadratically in the perturbations. The linearized forms of the state equations based on (60.1) may be written

$$\hat{p} = a_1^2\hat{\rho} + \rho^2 b\hat{s}, \qquad \hat{T} = b\hat{\rho} + c\hat{s}, \tag{60.2}$$

where (see § 10)

$$a_1^2 = \left(\frac{\partial p}{\partial \rho}\right)_s = \frac{c_p}{\rho\kappa_T c_V}, \qquad b = \left(\frac{\partial T}{\partial \rho}\right)_s = \rho^{-2}\left(\frac{\partial p}{\partial s}\right)_\rho = \frac{T\beta}{\rho^2\kappa_T c_V},$$
$$c = \left(\frac{\partial T}{\partial s}\right)_\rho = \frac{T}{c_V}. \tag{60.3}$$

Later we shall require

$$\frac{\rho^2 b^2}{c a_1^2} = \frac{T\beta^2}{\rho\kappa_T c_p} = \frac{c_p - c_V}{c_p} \equiv \chi, \tag{60.4}$$

and the definition

$$a_2^2 \equiv \frac{\rho_s s^2}{\rho_n}\left(\frac{\partial T}{\partial s}\right)_\rho = \frac{\rho_s s^2 T}{\rho_n c_V}. \tag{60.5}$$

The reader will recognize a_1 as being the normal sound speed, and shortly it will be shown that a_2 is the speed at which temperature waves propagate—the so-called 'second sound' of helium II.

60.2. The dispersion relation

We shall take the perturbations $\hat{\rho}$, $\hat{\mathbf{v}}$, etc. to be the amplitudes of simple harmonic waves, so that the density, velocity, etc. have the form

$$\rho + \hat{\rho}\,e^{i(\mathbf{k}.\mathbf{r}-\omega t)}, \qquad \hat{\mathbf{v}}\,e^{i(\mathbf{k}.\mathbf{r}-\omega t)}$$

whence

$$\frac{\partial}{\partial t} \to -i\omega, \qquad \nabla \to i\mathbf{k}.$$

Then the linearized forms of eqns (59.5), (59.18), (59.19), and (59.20) are

$$\left.\begin{aligned} \omega\hat{\rho} &= \rho\mathbf{k}\,.\,\hat{\mathbf{v}}, \\ (\omega+\mathrm{i}k^2\nu)\hat{\mathbf{v}}_\mathrm{n} &= \mathbf{k}\left[\frac{1}{\rho}\hat{p}+\frac{\rho_\mathrm{s}s}{\rho_\mathrm{n}}\hat{T}-\mathrm{i}(\tfrac{1}{3}\nu+\nu_2-\nu_1+c_\mathrm{s}\nu_3-c_\mathrm{s}\nu_4)\mathbf{k}\,.\,\hat{\mathbf{v}}_\mathrm{n}-\right. \\ &\qquad\left.-\mathrm{i}(\nu_1-c_\mathrm{s}\nu)\mathbf{k}\,.\,\hat{\mathbf{v}}\right], \\ \omega\hat{\mathbf{v}} &= \mathbf{k}[\hat{p}/\rho-\mathrm{i}c_\mathrm{n}(\tfrac{1}{3}\nu+\nu_2-\nu_1)\mathbf{k}\,.\,\hat{\mathbf{v}}_\mathrm{n}-\mathrm{i}c_\mathrm{n}\nu_1\mathbf{k}\,.\,\hat{\mathbf{v}}]-\mathrm{i}c_\mathrm{n}\nu k^2\hat{\mathbf{v}}_\mathrm{n}, \\ 0 &= \omega(\hat{\rho}s+\rho\hat{s})-\rho s\mathbf{k}\,.\,\hat{\mathbf{v}}_\mathrm{n}+\mathrm{i}k^2\kappa\hat{T}/T, \end{aligned}\right\} \tag{60.6}$$

in the last of which we have used (59.11) to eliminate the heat flux vector.

The difference of $(60.6)_2$ and $(60.6)_3$ gives

$$[\omega+\mathrm{i}k^2c_\mathrm{s}(\tfrac{4}{3}\nu+\nu_2+\nu_3-\nu_1-\nu_4)]\mathbf{k}\,.\,\hat{\mathbf{v}}_\mathrm{n}-[\omega+\mathrm{i}k^2c_\mathrm{s}(\nu_3-\nu_1)]\mathbf{k}\,.\,\hat{\mathbf{v}} = k^2\frac{\rho_\mathrm{s}s}{\rho_\mathrm{n}}\hat{T}, \tag{60.7}$$

and using (60.2), $(60.6)_1$ and $(60.6)_4$ to eliminate $\hat{p}$, $\hat{T}$, $\mathbf{k}\,.\,\hat{\mathbf{v}}$, and $\mathbf{k}\,.\,\hat{\mathbf{v}}_\mathrm{n}$ from $(60.6)_3$ and (60.7) and ignoring quadratic terms in the dissipations we find

$$[\tilde{u}^2-a_1^2+\mathrm{i}\omega c_\mathrm{n}(\tfrac{4}{3}\nu+\nu_2)]\frac{\hat{\rho}}{\rho}-[\rho bs-\mathrm{i}\omega c_\mathrm{n}(\tfrac{4}{3}\nu+\nu_2-\nu_1)]\frac{\hat{s}}{s} = 0,$$

$$-\left[\frac{\rho_\mathrm{s}}{\rho_\mathrm{n}}\rho sb-\mathrm{i}\omega c_\mathrm{s}\left(\tfrac{4}{3}\nu+\nu_2-\nu_4+\frac{\kappa b}{c_\mathrm{s}sT}\right)\right]\frac{\hat{\rho}}{\rho}+$$
$$+\left[\tilde{u}^2-a_2^2+\mathrm{i}\omega c_\mathrm{s}\left(\tfrac{4}{3}\nu+\nu_2+\nu_3-\nu_1-\nu_4+\frac{\kappa c}{T\rho_\mathrm{s}}\right)\right]\frac{\hat{s}}{s} = 0,$$

where $\tilde{u} = \omega/k$. Hence for non-zero $\hat{\rho}$, $\hat{s}$ we require $\tilde{u} = u(1-\mathrm{i}\alpha)$ to satisfy the dispersion relation

$$\begin{aligned} &[u^2(1-2\mathrm{i}\alpha)-a_1^2(1-2\mathrm{i}\alpha_1)][u^2(1-2\mathrm{i}\alpha)-a_2^2(1-2\mathrm{i}\alpha_2)]- \\ &\qquad -\chi a_1^2a_2^2+2\mathrm{i}\chi^{\frac{1}{2}}a_1^2a_2^2\alpha_3+\mathrm{i}\omega\rho\kappa b^2/T = 0, \end{aligned} \tag{60.8}$$

where u is the (real) phase velocity, α is the attenuation (see (57.24)),

$$\left.\begin{aligned} &\alpha_1 \equiv \frac{\omega c_\mathrm{n}}{2a_1^2}(\tfrac{4}{3}\nu+\nu_2), \qquad \alpha_2 \equiv \frac{\omega c_\mathrm{s}}{2a_2^2}\left(\tfrac{4}{3}\nu+\nu_2+\nu_3-\nu_1-\nu_4+\frac{\kappa c}{T\rho_\mathrm{s}}\right), \\ &\alpha_3 \equiv \frac{\omega}{2a_1a_2}(c_\mathrm{n}c_\mathrm{s})^{\frac{1}{2}}(\tfrac{8}{3}\nu+2\nu_2-\nu_1-\nu_4), \end{aligned}\right. \tag{60.9}$$

and we have used (60.4) and (60.5).

60.3. *First and second sound*

As c_p is very close to c_V for helium II, we may neglect χ and b in (60.8) to obtain the solutions

$$u = a_1, \qquad \alpha = \alpha_1; \qquad u = a_2, \qquad \alpha = \alpha_2. \tag{60.10}$$

The first of these is ordinary or 'first' sound and the second is the entropy wave known as 'second' sound. In the limit as ρ_s tends to zero (60.8) and (60.9) give

$$u = a_1, \qquad \alpha = \frac{\omega}{2a_1^2}\left(\tfrac{4}{3}\nu + \nu_2 + \frac{\rho\kappa b^2}{a_1^2 T}\right),$$

(cf. a similar result given in § 57.4), while the entropy wave becomes stationary.

The experimental evidence suggests that the attenuation is dominated by the normal fluid bulk viscosity ν_2 for ordinary sound, and by the thermal conductivity κ for second sound.

61. Rotating superfluids

61.1. *Introduction*

The early two-fluid theory of liquid helium implied that in a bucket of helium II rotating with angular velocity Ω only the normal fluid would partake in the rotation, so that the surface parabola would be

$$z = c_n \Omega^2 r^2 / 2g.$$

At $T = 1{\cdot}5$ K, $c_n \sim 0{\cdot}1$, so that the meniscus should attain only 10 per cent of its 'normal' ($T > 2{\cdot}17$ K) height. When this reduction was not observed, it was realized that rotating superfluids presented a special difficulty to the two-fluid theory. Experiments involving rotating discs lead to the same general outcome, namely, as far as the rotation is concerned the superfluidity seems to disappear. On the other hand, both the fountain phenomenon and the velocity of second sound are found to be unchanged in the rotating system, showing that the distinction between the normal and superfluid components still holds (for details see Andronikashvili and Mamaladze (1966).

Hall and Vinen (1956) made the particularly important observation that the attenuation of second sound suffered an additional anisotropic increment that is proportional to the angular velocity Ω. For propagation vectors normal to the axis of rotation α_2 is increased by as much as 30 per cent in some cases, whereas for axial propagation the increase in α_2 is quite small. Similar results have been obtained for heat currents.

The generally accepted explanation of these experimental results, and of many others of a similar nature, is that the superfluid vorticity ζ_s is zero everywhere (see § 58.4), save at a finite number of line vortices, which behave like bound vortices in aerodynamics, that is move relative to the superfluid. These vortices are supposed to be quantized in strength, having a circulation $\kappa = nh/m$, where n is an integer, h is Planck's constant, and m is the mass of a helium atom. The ground state $n = 1$ is shown to be thermodynamically the most probable in the sense of minimizing the energy for given angular momentum. For solid-body rotation

$$\Omega = \tfrac{1}{2}(\text{average vorticity}) = \tfrac{1}{2}\kappa n_{\mathrm{L}} = \tfrac{1}{2}(h/m)nn_{\mathrm{L}} \qquad (n = 1, 2, \dots), \quad (61.1)$$

where n_{L} is the number of vortex lines per unit area. At $n = 1$ this gives $n_{\mathrm{L}} \approx 2{\cdot}2 \times 10^3$ lines/cm^{-2} for $\Omega = 1$ rad/s^{-1}, which is large enough for one to adopt the averaged value

$$\bar{\zeta}_{\mathrm{s}} = \kappa n_{\mathrm{L}} = hnn_{\mathrm{L}}/m \quad (61.2)$$

for the superfluid vorticity.

This vortex line model, due originally to Feynman, raises as many questions as it resolves, for example, questions about the vortex creation, stability, and spatial distribution. It also seems to have been forgotten that vortex curvature induces in the vortex a large velocity along the *binormal*, this being responsible for the progress of vortex rings. Another difficulty concerns the kinetic energy of a line vortex, which for unit length is taken in the model to be

$$\varepsilon = \int_{a_0}^{b} \tfrac{1}{2}\rho_{\mathrm{s}} v_{\mathrm{s}}^2 2\pi r\,\mathrm{d}r = \int_{a_0}^{b} \tfrac{1}{2}\rho_{\mathrm{s}}\left(\frac{\kappa}{2\pi r}\right)^2 2\pi r\,\mathrm{d}r = \frac{\rho_{\mathrm{s}}\kappa^2}{4\pi}\ln\left(\frac{b}{a_0}\right), \quad (61.3)$$

where a_0 is the vortex core radius and b is an arbitrary outer radius less than the inter-vortex spacing $n_{\mathrm{L}}^{-\frac{1}{2}}$; the total energy per unit volume is then assumed to be $n_{\mathrm{L}}\varepsilon$. The arbitrariness introduced into the model via the 'cut-offs' $r = a_0$, $r = b$ is unimportant compared with the neglect of the *mutual* energy of the vortices, which from the standard expression for the kinetic energy of an array of vortex lines (Batchelor 1967) may be shown to be proportional to n_{L}^2 and for $\Omega = 1$ rad/s^{-1} at least two orders of magnitude greater than $n_{\mathrm{L}}\varepsilon$.

Notwithstanding these difficulties, the vortex-line model does provide an immediate physical explanation, both the anisotropy of rotating helium II and of the additional mutual friction between the normal and super components, the latter being ascribed to the plausible hypothesis that the normal fluid excitations are able to interact with a vortex structure, especially in the core neighbourhood. However, as Lin (1963) has pointed out, any model possessing a granular structure oriented preferentially along the axis of rotation would serve to reconcile the macroscopic fluid theory with the experimental results.

61.2. The mutual friction force due to superfluid vorticity

For a reason given in § 59.1, namely, to avoid dissipation within the superfluid, the mutual friction forces $\mathbf{R}_{sn}$ and $\mathbf{R}_{ns}$ were omitted from the right-hand sides of eqns (58.9). The neglected possibility of a friction force generating non-dissipative small-scale vortex motions in the superfluid must now be considered.

It appears that superfluid vorticity, whether distributed in line vortices or in any other *granular array*, is a form of excitation able to interact with the normal fluid excitations, consequently when there is substantial vorticity in the superfluid a friction force is possible. In this case $\mathbf{R}_{sn} \cdot (\mathbf{v}_n - \mathbf{v}_s)$ would be partly dissipated in the normal fluid, with the remainder contributing to the energy of the superfluid vorticity. Owing to the absence of superfluid viscosity, this latter energy would remain undissipated in the macroscopic fluid motions, but the large number of possible vortex modes would mean that the energy transfer is essentially irreversible.

In rotating helium II density exchange will certainly create superfluid vorticity, for if we add the friction force $\mathbf{R}_{sn}$ to eqn $(58.11)_1$ and take its curl we obtain

$$\frac{\partial \boldsymbol{\zeta}_s}{\partial t} + \nabla \times (\boldsymbol{\zeta}_s \times \mathbf{v}_s) = \frac{a_{sn}}{\rho_s}(\boldsymbol{\zeta}_n - \boldsymbol{\zeta}_s) + \nabla \times \left(\frac{\mathbf{R}_{sn}}{\rho_s}\right), \tag{61.4}$$

the right-hand side of which acts as a source term for $\boldsymbol{\zeta}_s$. When there is no rotation both $\boldsymbol{\zeta}_s$ and $\boldsymbol{\zeta}_n$ (the normal fluid vorticity) will remain negligible and $\mathbf{R}_{sn}$ will be zero as assumed earlier, but in rotating helium II $\boldsymbol{\zeta}_n$ is generated at the walls and diffused throughout the fluid in the usual way, when it provides a source for $\boldsymbol{\zeta}_s$ according to (61.4). The friction force $\mathbf{R}_{sn}$ then follows and accelerates the creation of superfluid vorticity. Of course, if we take the view that superfluid vorticity is entirely a quantum phenomenon, a mechanical explanation of its origin, such as that just offered, is unnecessary.

The entropy production rate will now contain a term $x\mathbf{R}_{sn} \cdot (\mathbf{v}_n - \mathbf{v}_s)$ (cf. (59.3)), where x is a number less than unity, whose precise value is not important. And if thermo-diffusive coupling between $(\mathbf{v}_n - \mathbf{v}_s)$ and ∇T is neglected, we can conclude in the usual way that

$$\mathbf{R}_{sn} = a\mathbf{D} \cdot (\mathbf{v}_n - \mathbf{v}_s), \tag{61.5}$$

where a is a scalar to be specified shortly and $\mathbf{D}$ is a phenomenological tensor possessing lateral isotropy about unit vector $\boldsymbol{\varpi}$ parallel to the axis of rotation, that is, having the form (see § 47.2)

$$\mathbf{D} = D_{\parallel}\boldsymbol{\varpi}\boldsymbol{\varpi} + D_{\perp}(\mathbf{1} - \boldsymbol{\varpi}\boldsymbol{\varpi}) + D_{\wedge}\boldsymbol{\varpi} \times \mathbf{1} \qquad (D_{\parallel}, D_{\perp} \geqslant 0). \tag{61.6}$$

The anisotropy assumed here is a consequence of the axial orientation of the vorticity generated at the walls of the rotating bucket.

In the light of the dependence of $\mathbf{R}_{sn}$ on the existence of superfluid vorticity, it is tempting to assume that the scalar a depends on $|\zeta_s|$. Indeed the experimental evidence that the incremental attenuation of second sound is proportional to the angular velocity Ω of the bucket suggests a linear dependence of a on the magnitude of ζ_s, which in the steady state has the average value 2Ω. However, such an assumption would be at variance with the principle of frame-indifference discussed in § 40.4: if ζ_s is the vorticity measured in the laboratory frame, in a frame rotating with the bucket the vorticity is $(\zeta_s - 2\Omega\bar{\varpi})$, which vanishes in a uniform state, and we cannot eliminate or reduce dissipation merely by choice of reference frame.

An escape from this difficulty lies in the hypothesis that the vorticity is concentrated in a large but finite number of distinct regions, say N per unit volume, and that N is proportional to Ω. To this we add the plausible assumption that the friction force acting on the normal fluid excitations depends linearly on the number density N of 'scattering centres', which is, of course, frame-indifferent. Thus we may set $a = \rho_s\rho_n\Omega/\rho$ in (61.5), that is,

$$\mathbf{R}_{sn} = \frac{\rho_s\rho_n}{\rho}\Omega\mathbf{D}\,.\,(\mathbf{v}_n - \mathbf{v}_s) = \rho_n\Omega\mathbf{D}\,.\,(\mathbf{v}_n - \mathbf{v}), \tag{61.7}$$

where Ω is to be interpreted as the positive angular velocity of the helium II *relative* to a frame in which N is found to be zero.

Notice from (61.1) that the line-vortex model is a simple example having the granular structure required by our theory; other distributions would also serve, but before speculating on this, we would need to know just how the vorticity enters the superfluid, an aspect on which very little has been written.

61.3. The attenuation of small-amplitude waves

First we shall note the modifications required in the basic equation of § 59 to allow for the effects of rotation. The anisotropy ought to extend to all the transport coefficients, for example, (59.11) should be replaced by

$$\mathbf{q}_n = -\boldsymbol{\kappa}\,.\,\nabla T, \tag{61.8}$$

where $\boldsymbol{\kappa}$ has a structure similar to $\mathbf{D}$. The theory given in § 50.1 shows how eqns $(59.10)_1$ and (59.12) should be modified, but no doubt the changes required in the viscous terms will be negligible in practice, and we shall ignore them. Another approximation we have made is to omit the thermo-diffusive coupling between ∇T and $(\mathbf{v}_n - \mathbf{v}_s)$, but as the omitted terms are reversible they will not contribute to wave-damping. The friction term defined in (61.7) will add $\mathbf{R}_{sn}$ and $-\mathbf{R}_{sn}/\rho_n$ to the right-hand sides of (59.17) and (59.18), and finally the accelerations on the left-hand sides of these equations should be increased by $2\Omega\bar{\varpi}\times\mathbf{v}_n + \Omega^2\bar{\varpi}\times(\bar{\varpi}\times\mathbf{r})$ and $2\Omega\bar{\varpi}\times\mathbf{v} + \Omega^2\bar{\varpi}\times(\bar{\varpi}\times\mathbf{r})$

respectively. These additional accelerations due to the rotation contribute effects of order $(\omega/\Omega)^2$ to the dispersion relation and can be ignored.

The changes to the theory of § 60.2 caused by adopting (61.8) and of adding $\Omega\mathbf{D}\,.\,(\mathbf{v}-\mathbf{v}_n)$ to the right-hand side of (59.18) are easily calculated. Referring to eqn (60.8), we find that α_1 is unchanged, α_2 and α_3 are replaced by

$$\alpha_2 = \frac{\omega c_s}{2a_2^2}\left(\tfrac{4}{3}\nu+\nu_2+\nu_3-\nu_1-\nu_4+\frac{\bar{\kappa}c}{T\rho_s}\right)+\frac{\Omega}{2\omega}\bar{D}, \tag{61.9}$$

$$\alpha_3 = \frac{\omega}{2a_1a_2}(c_nc_s)^{\frac{1}{2}}\left(\tfrac{8}{3}\nu+2\nu_2-\nu_1-\nu_4+\frac{\rho^2}{\rho_n}\frac{\Omega}{\omega^2}b\bar{D}\right),$$

and κ replaced by $\bar{\kappa}$, where

$$\bar{\kappa} \equiv \kappa_{\parallel}\cos^2\theta+\kappa_{\perp}\sin^2\theta, \qquad \bar{D} \equiv D_{\parallel}\cos^2\theta+D_{\perp}\sin^2\theta, \tag{61.10}$$

θ being the angle between $\mathbf{k}$ and the axis $\bar{\boldsymbol{\omega}}$. The incremental damping of second sound is then $\Omega\bar{D}/2\omega$, and the experimental evidence shows that $D_{\perp} \gg D_{\parallel}$, that is, this increment is a maximum for waves propagating normal to the axis of rotation.

61.4. Theories based on the vortex-line model

The model given above ignores the vortex energy density εn_L, which is given some prominence in the usual treatments of rotating helium II. There are two approaches, both based on the vortex-line model, but employing quite different concepts and leading to apparently similar results.

The first that we shall outline (due to Khalatnikov (1965)) is entirely phenomenological in character, being based on the simple idea of adding a term $\varepsilon\,\mathrm{d}n_L = \lambda\,\mathrm{d}\zeta_s$ $(\lambda = \varepsilon/\kappa)$ to the right-hand side of (59.25) to account for changes in the vortex energy density. The move amounts to adopting a frame-dependent internal coordinate to complete the specification of the thermodynamic state of the rotating superfluid. The dissipation resulting from this step is stated by Khalatnikov to amount to

$$T\sigma' = -[\mathbf{f}+\boldsymbol{\zeta}_s\times(\mathbf{v}_n-\mathbf{v}_s)]\,.\,[\rho(\mathbf{v}-\mathbf{v}_n)+\nabla\times(\lambda\hat{\boldsymbol{\zeta}}_s)], \tag{61.11}$$

where $\hat{\boldsymbol{\zeta}}_s$ is unit vector parallel to $\boldsymbol{\zeta}_s$ and $\mathbf{f} \equiv \mathbf{R}_{sn}/\rho_s$. From this he infers the constitutive equation

$$\mathbf{f}+\boldsymbol{\zeta}_s\times(\mathbf{v}_n-\mathbf{v}_s) = -\tfrac{1}{2}\zeta_s\frac{\rho_n}{2\rho_s\rho}\bar{\mathbf{D}}\,.\,[\rho(\mathbf{v}-\mathbf{v}_n)+\nabla\times(\lambda\hat{\boldsymbol{\zeta}}_s)],$$

or, as

$$\rho(\mathbf{v}-\mathbf{v}_n) = \rho_s(\mathbf{v}_s-\mathbf{v}_n),$$

$$\mathbf{R}_{sn} = (\tfrac{1}{2}\zeta_s)\frac{\rho_n}{\rho}\bar{\mathbf{D}}\,.\,[\rho_s(\mathbf{v}_n-\mathbf{v}_s)-\nabla\times(\lambda\hat{\boldsymbol{\zeta}}_s)]+\rho_s(\mathbf{v}_n-\mathbf{v}_s)\times\boldsymbol{\zeta}_s. \tag{61.12}$$

The phenomenological tensor $\overline{\mathbf{D}}$ is assumed to have lateral isotropy about $\hat{\zeta}_s$; in Khalatnikov's notation

$$\overline{\mathbf{D}} = \frac{2\rho\rho_s}{\rho_n}[-\alpha\hat{\zeta}_s\times\mathbf{1}-\beta\hat{\zeta}_s\times(\hat{\zeta}_s\times\mathbf{1})+\gamma\hat{\zeta}_s\hat{\zeta}_s]$$

$$= \frac{2\rho\rho_s}{\rho_n}[(\gamma-\beta)\hat{\zeta}_s\hat{\zeta}_s-\beta\mathbf{1}-\alpha\hat{\zeta}_s\times\mathbf{1}] \qquad (\beta,\gamma \geqslant 0). \tag{61.13}$$

Let $\mathbf{v}_{s\infty}$ denote the uniform superfluid velocity a great distance from a collection of vortex lines that move with a velocity $\mathbf{v}_L$, then the Magnus or 'lift' force acting per unit length of line is $\rho_s(\mathbf{v}_{s\infty}-\mathbf{v}_L)\times\boldsymbol{\kappa}$. Hence the force on the superfluid per unit mass is

$$\mathbf{F}_s = (\mathbf{v}_L-\mathbf{v}_{s\infty})\times\boldsymbol{\zeta}_s \qquad (\boldsymbol{\zeta}_s = n_L\boldsymbol{\kappa}), \tag{61.14}$$

which suggests an *approximate* interpretation for the final term in (61.12), namely, that it is the reversible force on the superfluid by vortex lines moving with the normal fluid velocity.

There are several difficulties in the theory: first there is the tacit assumption that the irreversible force $[\mathbf{f}+\boldsymbol{\zeta}_s\times(\mathbf{v}_n-\mathbf{v}_s)]$ depends on $\boldsymbol{\zeta}_s$—eqn (61.11) does not imply this at all. Secondly, if the dependence on $\boldsymbol{\zeta}_s$ is accepted, the result then violates frame-indifference, a problem that could be resolved only by providing kinetic details that 'explain' the dependence on ζ_s via some frame-indifferent scalar (cf. discussion in § 61.2). Finally there is the perplexing question of how we can possibly divide a velocity field into two parts such that the kinetic energy of one part, neighbouring the vortices, is reckoned as being internal thermodynamic energy, while that of the other part is taken to be ordinary fluid kinetic energy.

Hall and Vinen (1956) also use the vortex-line model, making rather more assumptions about the kinetics of the interactions between the lines and the two fluids. With each line is associated a tension

$$\int_0^\infty (p_\infty-p)2\pi r\,\mathrm{d}r \simeq \int_{a_0}^b \tfrac{1}{2}\rho_s v_s^2 2\pi r\,\mathrm{d}r = \varepsilon, \tag{61.15}$$

p_∞ being the pressure at $r=\infty$, and the force acting on unit volume due to this tension is

$$\mathbf{R}_T = -n_L\varepsilon\,\hat{\zeta}_s\times(\nabla\times\hat{\zeta}_s)+\hat{\zeta}_s\hat{\zeta}_s\cdot\nabla(n_L\varepsilon)-\nabla(n_L\varepsilon), \tag{61.16}$$

of which the first term is due to the straightening force in each line ($|\hat{\zeta}_s\times(\nabla\times\hat{\zeta}_s)|$ is the line curvature), the second to the axial changes in tension, and the third is a result of the pressure induced by the vortices in the region between them. This additional pressure arises from the incremental internal energy density $n_L\varepsilon$ and $g_s = u_s+p_s/\rho_s = (p_s+n_L\varepsilon)/\rho_s$. Writing $n_L\varepsilon = \lambda\zeta_s$, we find that (61.16) may be expressed

$$R_{\mathrm{T}} = -\hat{\zeta}_{\mathrm{s}} \times [\nabla \times (\lambda \zeta_{\mathrm{s}})], \tag{61.17}$$

but as Hall and Vinen have neglected the effects contained in the last two terms of (61.16), they find

$$R_{\mathrm{T}} = -\lambda \zeta_{\mathrm{s}} \times (\nabla \times \hat{\zeta}). \tag{61.18}$$

Other forces assumed to be acting on the vortex lines are the Magnus force (61.14)—with change of sign—and a drag force due to the relative motion of the normal fluid, say

$$\mathbf{F}_{\mathrm{Ln}} = \mathbf{D}' \cdot (\mathbf{v}_{\mathrm{L}} - \mathbf{v}_{\mathrm{n}}), \tag{61.19}$$

where $\mathbf{D}'$ has lateral isotropy about $\hat{\zeta}_{\mathrm{s}}$. Hall and Vinen have developed a kinetic theory from which estimates of $D'_{\perp}$ and $D'_{\wedge}$ may be made; $D'_{\parallel}$ they assumed to be zero. On the assumption that the three forces acting on the vortex lines balance, they obtain

$$(\mathbf{v}_{\mathrm{L}} - \mathbf{v}_{\mathrm{s}} - \lambda \nabla \times \hat{\zeta}_{\mathrm{s}}) \times \zeta_{\mathrm{s}} + \mathbf{D}' \cdot (\mathbf{v}_{\mathrm{n}} - \mathbf{v}_{\mathrm{L}}) = 0, \tag{61.20}$$

from which a value for $\mathbf{v}_{\mathrm{L}}$ can be obtained. Finally this value of $\mathbf{v}_{\mathrm{L}}$ is substituted into (61.19) to yield the required drag force $\mathbf{R}_{\mathrm{sn}} = \mathbf{F}_{\mathrm{Ln}}$ acting on the superfluid in terms of $(\mathbf{v}_{\mathrm{n}} - \mathbf{v}_{\mathrm{s}})$. The final result given by Hall and Vinen may be expressed

$$\mathbf{R}_{\mathrm{sn}} = (\tfrac{1}{2}\zeta_{\mathrm{s}}) \frac{\rho_{\mathrm{n}}}{\rho} \mathbf{D} \cdot [\rho_{\mathrm{s}}(\mathbf{v}_{\mathrm{n}} - \mathbf{v}_{\mathrm{s}}) - \lambda \nabla \times \hat{\zeta}_{\mathrm{s}}], \tag{61.21}$$

which should be compared with (61.7) and (61.12). These three formulae differ only in the term involving λ, and even if such a term could be given a convincing treatment, it is probably negligible anyway.

This quite ingenious theory is not without defects. First there are difficulties in the model itself, already discussed, secondly, there is the problem of frame-dependence, which could no doubt be overcome by reinterpreting ζ_{s}. Thirdly, we could ask if the Magnus force used in (61.20) makes sense in view of the fact that $(\mathbf{v}_{\mathrm{L}} - \mathbf{v}_{\mathrm{s}})$ is a rapidly changing function of position in the vortex array. The velocity $(\mathbf{v}_{\mathrm{L}} - \mathbf{v}_{\mathrm{s}\infty})$ that should be adopted in lifting vortex-line theory is not part of the physical model as it is deployed in the rotating bucket experiment. Finally, the approximations involved in (61.15) are as difficult to justify as those adopted by Khalatnikov in dividing the velocity field into two in his vortex-energy picture.

These criticisms apart, the vortex-line model has been employed to good effect by many besides the authors whose important contributions we have outlined here. It has, at the least, provided a clear physical picture in a field where the difficulties of physical description are notorious.

11

INTERNAL PROCESSES

UNTIL now our concern has been with 'external' processes, by which we mean phenomena like heat flux, electric currents, and fluid stress that are directly influenced by boundary conditions on the system. While internal processes may depend indirectly on boundary conditions, they are driven by affinities or 'forces' internally developed, and their importance hinges on the time scales on which such forces can relax. We have already encountered one example of an internal process, namely, the generation of bulk viscosity by thermal affinities in § 49 and by density transfers between the normal fluids and superfluids of helium II in § 59. Chemical reactions between fluid components and the relaxation of velocity distribution functions in the kinetic theory of gases are two more important examples of internal processes.

In this chapter we shall give an account in general terms of the theory of internal processes, and in Chapter 13 we shall deal with applications of this theory to gas kinetics. Below it will be shown that the internal affinities A and their processes $\dot{\xi}$ can interact with the external affinities X and their processes J to produce the phenomenon of time-integrated responses as represented in the phenomenological relation

$$J(t) = \int_{-\infty}^{t} \psi(t-t')X(t')\,\mathrm{d}t' \qquad (\eta\psi = \psi^*\eta).$$

The response function $\psi(t)$ now replaces the (instantaneous) phenomenological coefficient L and satisfies the same reciprocal relation. Likewise the 'static' state equations of equilibrium thermodynamics, which include the progress variables ξ as independent coordinates, can be converted to time-dependent 'dynamic' forms by eliminating ξ. The propagation of sound through a 'relaxing' gas illustrates some of the consequences of time delays in the various thermodynamic relations.

When, as a consequence of random fluctuations, it is necessary to treat ξ as being a stochastic variable, it is found that the spectral density of these fluctuations is related to the response function ψ giving $\xi(t)$ as a functional of the external affinity $X(t')$. This relation is the famous 'dissipation–fluctuation' theorem, and it offers a method of determining ψ via measurements of correlations of the internal coordinates under stationary conditions. This theorem will be the subject of the next chapter.

62. Relaxation phenomena

62.1. 'Static' constitutive relations

Consider a thermodynamic (fluid-particle) system, whose state is defined by 'external' coordinates ζ_j, $j = 1, 2, \ldots$, and internal coordinates ξ_r, $r = 1, 2, \ldots R$. Let $x(\zeta_1, \zeta_2, \ldots, \xi_1, \xi_2, \ldots)$ denote the thermodynamic potential, then in matrix notation the fundamental equation of thermodynamics may be written

$$\mathrm{d}x = Z\,\mathrm{d}\zeta - A\,\mathrm{d}\xi,$$

with (62.1)

$$Z = Z(\zeta, \xi), \qquad A = A(\zeta, \xi)$$

representing the state equations. Here A is the affinity vector; a typical choice for x is the specific energy u, for which

$$Z = \{T, -p, g_1, g_2, \ldots\}, \qquad \zeta = \{s, \rho^{-1}, c_1, c_2, \ldots\}.$$

Of course in one sense the concentrations $c_1, c_2, \ldots$ behave more like internal than external coordinates, but in (62.1) it is only the changes in $c_1, c_2, \ldots$ due to *external* causes (cf. (37.4)) that are included in $Z\,\mathrm{d}\zeta$—other changes are represented in $A\,\mathrm{d}\xi$. The entropy production, as given for example by (52.13), may be written

$$\sigma = JX + \frac{\rho}{T}\dot{\xi}A \geqslant 0, \tag{62.2}$$

where JX is the contribution from 'external' processes like viscous dissipation and heat conduction, and $\rho\dot{\xi}A/T$ is the contribution from internal relaxations.

Let (ζ^0, ξ^0) denote a reference state; choose ξ^0 to be ξ_{eq}, the value of ξ when the internal processes are in equilibrium; then as $A(\zeta^0, \xi_{\text{eq}}) = 0$, we can determine $\xi_{\text{eq}} = \xi_{\text{eq}}(\zeta^0)$. For small departures $(\Delta\zeta, \Delta\xi)$ from the reference state it follows from (62.1) that

$$\Delta Z = a\,\Delta\zeta + b\,\Delta\xi, \qquad A = -\tilde{b}\,\Delta\zeta - g\,\Delta\xi, \tag{62.3}$$

where

$$a = \frac{\partial Z}{\partial \zeta} = \frac{\partial^2 x}{\partial\zeta\,\partial\zeta}, \qquad b = \frac{\partial Z}{\partial \xi} = \frac{\partial^2 x}{\partial\xi\,\partial\zeta} = -\frac{\partial \tilde{A}}{\partial \zeta}, \qquad g = -\frac{\partial A}{\partial \xi} = \frac{\partial^2 x}{\partial\xi\,\partial\xi}.$$

The symmetric, positive definite character of the matrix g was established in § 18.3; notice that a is also symmetric.

From the entropy production (62.2) we may infer linear constitutive relations of the form

$$\begin{bmatrix} J \\ f\dot{\xi} \end{bmatrix} = \begin{bmatrix} L & M \\ fN & fP \end{bmatrix} \begin{bmatrix} X \\ A \end{bmatrix} \qquad \left(f \equiv \frac{\rho}{T}\right),$$

or

$$J = LX + MA, \qquad \dot{\xi} = NX + PA. \tag{62.4}$$

The phenomenological matrices introduced here satisfy the reciprocal relations (see (44.1))

$$\begin{bmatrix} \eta_e & \cdot \\ \cdot & \eta_i \end{bmatrix}\begin{bmatrix} L & M \\ fN & fP \end{bmatrix} = \begin{bmatrix} L & M \\ fN & fP \end{bmatrix}^*\begin{bmatrix} \eta_e & \cdot \\ \cdot & \eta_i \end{bmatrix}$$
$$= \begin{bmatrix} L^* & fN^* \\ M^* & fP^* \end{bmatrix}\begin{bmatrix} \eta_e & \cdot \\ \cdot & \eta_i \end{bmatrix},$$

or

$$\eta_e L = L^*\eta_e, \qquad \eta_e M = \frac{\rho}{T}N^*\eta_i, \qquad \eta_i P = P^*\eta_i, \tag{62.5}$$

where the asterisk denotes the transposed matrix with the sign of **B** changed, and η_i, η_e are the diagonal parity matrices for the internal (A) and the external (X) affinities respectively, that is, by (42.3) written in matrix form

$$A(\tau) = \eta_i A(-\tau), \qquad X(\tau) = \eta_e X(-\tau), \tag{62.6}$$

where τ is an infinitesimal microscopic time interval. On assuming that each external coordinate ζ_j has positive parity† and noting that, as $\dot{\xi}_r A_r$ has negative parity, the internal coordinate ξ_r must have the same parity as A_r, we find on applying $(62.6)_1$ to (62.3) that

$$\begin{gathered} a(\mathbf{B}) = a(-\mathbf{B}) = \tilde{a}(-\mathbf{B}), \qquad b(\mathbf{B}) = b(-\mathbf{B})\eta_i, \\ \eta_i g(\mathbf{B}) = g(-\mathbf{B})\eta_i = \tilde{g}(-\mathbf{B})\eta_i. \end{gathered} \tag{62.7}$$

62.2. 'Dynamic' constitutive relations

From $(62.3)_2$, $(62.4)_2$, and the assumption that the equilibrium state (ζ^0, ξ_{eq}) changes on a much longer time scale than $(\Delta\zeta, \Delta\xi)$ we have

$$(\mathrm{D} + Pg)\Delta\xi = NX - P\tilde{b}\,\Delta\zeta$$

or

$$\Delta\xi = \mathscr{L}(NX - P\tilde{b}\,\Delta\xi), \tag{62.8}$$

where $\mathrm{D} = U\,\mathrm{d}/\mathrm{d}t$, $\mathscr{L} = (\mathrm{D} + Pg)^{-1}$, and U denotes unit matrix.

Hence, on eliminating $\Delta\xi$ from $(62.3)_1$ and $(62.4)_1$ we may write formally

$$\Delta Z = a\,\Delta\zeta - b\mathscr{L}(P\tilde{b}\,\Delta\zeta) + b\mathscr{L}(NX), \tag{62.9}$$

$$J = LX - Mg\mathscr{L}(NX) - M\tilde{b}\,\Delta\zeta + Mg\mathscr{L}(P\tilde{b}\,\Delta\zeta). \tag{62.10}$$

† This is almost always so; for an exceptional case see the final term of eqn (59.25).

Now the matrices L, M, N, P, a, b, g are functions of the thermodynamic state (ζ, ξ), and as this state changes with time these matrices also depend on t. However, for small departures from the quasi-static reference state $(\zeta^0, \xi_{\text{eq}})$, the changes in the coefficient matrices may be neglected and we shall take them to be independent of time. Hence the solution of (62.8) is

$$\Delta\xi = e^{-Pgt}(\Delta\xi)_0 + \int_0^t e^{-(t-t')Pg}[NX(t') - P\tilde{b}\,\Delta\zeta(t')]\,dt', \tag{62.11}$$

$(\Delta\xi)_0$ being the value of $\Delta\xi$ at $t = 0$. Alternatively, on the assumption that $\Delta\xi$ was zero in the distant past, we may write

$$\Delta\xi = \xi - \xi_{\text{eq}} = \int_{-\infty}^t e^{-(t-t')Pg}[NX(t') - P\tilde{b}\,\Delta\zeta(t')]\,dt'. \tag{62.12}$$

Eqn (62.9) may be regarded as being a 'dynamic' equation of state replacing the static version in eqn $(62.3)_1$. The internal coordinates have been replaced by an integral over time: a differential form can be obtained by applying the operator $(b\mathscr{L})^{-1}$ to (62.9). Likewise, (62.10) can be regarded as being a 'dynamic' constitutive relation in which the *history* of the affinity plays a role through the integral operator $\mathscr{L}$ in determining the process J. We shall return to this point in § 63.

Suppose that the conditions are such that the external coordinates are held constant, that is, that $\Delta\zeta = 0$, then from $(62.3)_2$ and (62.4)

$$J = LX - Mg\,\Delta\xi, \qquad \dot{\xi} = NX - Pg\,\Delta\xi \qquad (A = -g\,\Delta\xi), \tag{62.13}$$

and $\Delta\xi$ may be regarded as a new affinity replacing A. However, the new phenomenological coefficients Mg, Pg do not satisfy the reciprocal relations (62.5). To overcome this we make the transformation

$$\xi' = \beta\xi, \qquad A' = \tilde{\beta}^{-1}A, \tag{62.14}$$

which preserve the bilinear form in (62.2) for σ, and from the theory of § 44 also leave the reciprocal relations (62.5) intact provided the transformation matrix β satisfies

$$\eta_i\beta(\mathbf{B}) = \beta(-\mathbf{B})\eta_i. \tag{62.15}$$

Eqns (62.13) now appear as

$$J = LX - M'g'\,\Delta\xi', \qquad \dot{\xi}' = N'X - P'g'\,\Delta\xi', \qquad (A' = -g'\,\Delta\xi'), \tag{62.16}$$

where

$$M' = M\tilde{\beta}, \qquad N' = \beta N, \qquad P' = \beta P\tilde{\beta}, \qquad g' = \tilde{\beta}^{-1}g\beta^{-1}. \tag{62.17}$$

Now as g is a positive definite matrix, we are able to choose a real matrix β such that the congruent matrix g' is the unit matrix. With this choice, the coefficients $M'g'$, $P'g'$ in (62.16) reduce to M', P', which can be readily verified to satisfy the same reciprocal relations as M and P, namely $(62.5)_2$ and $(62.5)_3$.

62.3. Relaxation times

For the special case $X = 0$ (62.2) and (62.14) give $A'\dot{\xi}' \geqslant 0$, or by (62.16) with $g' = U$, $\Delta\xi' P' \Delta\xi' \geqslant 0$. When this is added to its transpose, we obtain

$$\Delta\xi'(P' + \tilde{P}')\Delta\xi' \geqslant 0, \tag{62.18}$$

which holds for all real R-dimensional vectors $\Delta\xi'$. The (complex) eigenvectors x of P' satisfy $P'x = \nu x$, where ν is one of the eigenvalues $\nu_1, \nu_2, \ldots \nu_R$. Hence $x^*P'x = \nu x^* x$, where x^* is the complex conjugate of x. Now add the transpose of the complex conjugate of this last equation to itself to find

$$x^*(P' + \tilde{P}')x = (\nu + \nu^*)x^* x,$$

as P' is real. Set $x = \eta + i\chi$, where η, χ are real R-dimensional vectors, then

$$\eta(P' + \tilde{P}')\eta + \chi(P' + \tilde{P}')\chi = (\nu + \nu^*)(\eta^2 + \chi^2).$$

By (62.18) the left-hand side is non-negative, and the conclusion follows that the real part of ν is also non-negative. As

$$P' = P'g' = \beta P g \beta^{-1},$$

that is, P' is derived from Pg by a *collineatory* transformation, the eigenvalues $\nu_1, \nu_2, \ldots \nu_R$ of P' are also the eigenvalues of Pg. We have therefore proved that the matrix Pg in (62.11) and (62.12) has eigenvalues whose real parts are non-negative, which means that $\Delta\xi$ remains bounded as $t \to \infty$.

This last conclusion can be seen more clearly on applying a further transformation

$$\xi'' = \alpha\xi' = \alpha\beta\xi, \qquad A'' = \tilde{\alpha}^{-1}A' = (\widetilde{\alpha\beta})^{-1}A,$$

where α satisfies a relation similar to that given in (62.15) for β. Then, as

$$P'' = \alpha P' \alpha^{-1} = \alpha\beta P g(\alpha\beta)^{-1}, \tag{62.19}$$

P'', P', and Pg possess the same eigenvalues, and provided these are distinct, the collineatory transformation (62.19) can be chosen to render P'' diagonal with elements $\nu_1, \nu_2, \ldots \nu_R$. With this (62.11) transforms into (recall that at present X and $\Delta\zeta$ are zero).

$$\Delta\xi''_r = (\Delta\xi'')_{r0} \exp(-\nu_r t) \qquad (r = 1, 2, \ldots R), \tag{62.20}$$

so that the relaxation times (see (19.2)) τ_r, $r = 1, 2, \ldots R$, are identified as the reciprocals of the real parts of the frequencies ν_r.

While it follows from $(62.5)_3$ and (62.15) that $P' = \beta P \tilde{\beta}$ satisfies $\eta_i P' = (P')^* \eta_i$, the diagonal matrix P'' satisfies a like relation only when $\alpha^{-1} = \tilde{\alpha}$, that is, when the transformation from P' to P'' is both collineatory and congruent. Such a constraint on α is possible only if P' is a symmetric matrix, that is, if the internal coordinates all have the same parity ($\eta_i = U$) and the magnetic field is negligible. In this special and important case the eigenvalues are real and P'' has positive diagonal elements $1/\tau_1$, $1/\tau_2$, etc.

The conclusion reached above about the nature of the eigenvalues of Pg, the replacement of g by unit matrix, and the reduction of P to diagonal form are independent of the values of X and $\Delta\zeta$—at least for a linear theory—and so may be applied to the general theory contained in eqns (62.6)–(62.10).

Finally, on the question of how many internal coordinates ξ_r, $r = 1, 2, \ldots R$, are useful in the theory, that is, are worth retaining in (62.2), it is worth observing that if dt denotes the macroscopic time scale of interest, then only those processes whose relaxation times are *about the same order* as dt are important. Label the times so that $\tau_1 \geqslant \tau_2 \geqslant \ldots \geqslant \tau_R$, then the situation may be summarized as follows (cf. discussion in § 3.2),

$$\begin{array}{ccccc} \tau_m & \gg & \mathrm{d}t & \gg & \tau_n \\ A'_m, \Delta\xi'_m \neq 0, \Delta\dot{\xi}'_m \simeq 0 & & & & A'_n, \Delta\xi'_n \approx 0, \Delta\dot{\xi}'_n \neq 0 \\ \text{frozen equilibrium for } \xi'_r, r \leqslant m & & & & \text{relaxed equilibrium for } \xi'_r, r \geqslant n \\ \text{(`persisting' memory)} & & & & \text{(`fading' memory)} \end{array} \qquad (62.21)$$

and the product $\dot{\xi}'_r A'_r$ is negligible for both types of equilibrium.

63. Dynamic state and constitutive equations

63.1. The admittance matrix

We start with the dynamic equation of state in the form given in (62.9),

$$\Delta Z = a\,\Delta\zeta - b\mathscr{L}(P\tilde{b}\,\Delta\zeta) + b\mathscr{L}(NX) \qquad (\mathscr{L} \equiv (\mathrm{D}+Pg)^{-1}), \qquad (63.1)$$

where

$$\eta_i P = P^*\eta_i. \qquad (63.2)$$

As shown in § 62.2, we may replace g by the unit matrix without affecting the reciprocal relation (63.2), but if we also wish to write P as a diagonal matrix, (63.2) no longer applies, save if it reduces to $P = \tilde{P}$, when P takes the simple form

$$P = \left[\begin{array}{ccc|c} \tau_1^{-1} & \cdot & \cdot & \cdot \\ \cdot & \tau_2^{-1} & & \cdot \\ \hline \cdot & \cdot & \cdot & \tau_R^{-1} \end{array}\right] \qquad (\tau_r > 0). \qquad (63.3)$$

To simplify the discussion, we shall first suppose that the last term in (13.1) is zero. This will be so if in physical space all the quantities represented by X are *vectors* (for example, temperature and concentration gradients) and those represented by ξ are *scalars* (for example, isotropic relaxation phenomena), for then, by the theory of § 46, the coupling matrices M and N in (62.4) must vanish under central symmetry. If this does not apply, setting the term in question equal to zero restricts us to the case of vanishing external affinities.† The general case will be considered shortly.

Corresponding to the two limits contained in (62.21), eqn (63.1) yields

$$\begin{array}{cc} \mathrm{d}t\text{ 'small' (D large)} & \mathrm{d}t\text{ 'large' (D small)} \\ \Delta Z = a\,\Delta\zeta\,, & \Delta Z = (a-bg^{-1}\tilde{b})\,\Delta\zeta\,, \end{array} \tag{63.4}$$

where by (62.3)

$$a = \left(\frac{\partial^2 x}{\partial\zeta\,\partial\zeta}\right)_{\Delta\xi=0}, \qquad a-bg^{-1}\tilde{b} = \left(\frac{\partial^2 x}{\partial\zeta\,\partial\zeta}\right)_{A=0}.$$

$$\text{(frozen)} \qquad\qquad\qquad \text{(relaxed)}$$

Suppose that the time dependence of ΔZ and $\Delta\zeta$ is harmonic, that is, proportional to $\exp(-\mathrm{i}\omega t)$, then on replacing D by $-\mathrm{i}\omega U$ we find that (63.1) reads

$$\Delta Z = \hat{\psi}(\omega, \mathbf{B})\,\Delta\zeta, \tag{63.5}$$

where

$$\hat{\psi}(\omega, \mathbf{B}) \equiv a - b(g - \mathrm{i}\omega P^{-1})^{-1}\tilde{b} \tag{63.6}$$

is termed an 'admittance' matrix. Other such admittances will be introduced shortly, distinguished by various subscripts. Notice that (63.4) contains the limiting values $\hat{\psi}(\infty, \mathbf{B})$ and $\hat{\psi}(0, \mathbf{B})$. From eqns $(62.5)_3$, (62.47), and $\eta_{\mathrm{i}}^2 = U$ it is easily shown that $\hat{\psi}$ satisfies the reciprocal relation

$$\hat{\psi}(\omega, \mathbf{B}) = \tilde{\hat{\psi}}(\omega, -\mathbf{B}) \qquad (\hat{\psi} = \hat{\psi}^*). \tag{63.7}$$

When the internal coordinates have even parity we may set $g = U$ and assume P to have the form given in (63.3). In this case the components of the admittance are

$$\hat{\psi}_{ij} = a_{ij} - \sum_{k=1}^{R} \frac{b_{ik}b_{jk}}{1-\mathrm{i}\omega\tau_k}.$$

At low frequencies, by expanding $(U - \mathrm{i}\omega g^{-1}P^{-1})^{-1}$ in powers of ω, we may express $\hat{\psi}$ thus

$$\hat{\psi}(\omega) = \hat{\psi}(0) - \mathrm{i}\omega H,$$

† The theory of bulk viscosity presented in § 49.5 is an important example of coupling between external irreversible phenomena and internal relaxation processes.

where

$$\hat{\psi}(0) = a - bg^{-1}\tilde{b}, \qquad H = bg^{-1}P^{-1}g^{-1}\tilde{b}, \qquad (H = H^*).$$

In this approximation (63.5) may be written

$$\mathscr{I} \equiv \Delta Z - \Delta Z_{\text{eq}} = -\mathrm{i}\omega H\,\Delta\zeta = H\,\Delta\dot{\zeta} \qquad (\Delta Z_{\text{eq}} \equiv \hat{\psi}(0)\,\Delta\zeta),$$

where the subscript eq refers to *relaxed* equilibrium, that is, $\mathrm{d}t$ is 'large'. This relation shows that $\mathscr{I}$ are processes dependent on the affinities $\Delta\dot{\zeta}$; the internal processes are submerged into the phenomenological matrix H and need not be made explicit.

63.2. *The equations of state*

When the affinities $\Delta\zeta$ have a general time dependence, (63.1) and (62.12) yield the dynamic state equation

$$\Delta Z(t) = \int_{-\infty}^{t} [a\,\delta(t-t') + \varphi_{22}(t-t')]\,\Delta\zeta(t')\,\mathrm{d}t', \tag{63.8}$$

where $\delta(t)$ is the delta function and

$$\varphi_{22}(t) \equiv -b\,\mathrm{e}^{-Pgt}P\tilde{b} \qquad (t \geqslant 0), \tag{63.9}$$

the components of which, for obvious reasons, are called *after-effect* functions. On integrating (63.8) by parts we find the alternative form

$$\mathscr{I}(t) \equiv \Delta Z(t) - \Delta Z_{\text{eq}}(t) = \int_{-\infty}^{t} \psi_{22}(t-t')\,\Delta\dot{\zeta}(t')\,\mathrm{d}t', \tag{63.10}$$

where

$$\psi_{22}(\tau) \equiv bg^{-1}\,\mathrm{e}^{-\tau gP}\tilde{b} \qquad (\tau \geqslant 0).$$

Functions like ψ_{22} that relate a process to the history of an affinity are called *response* functions; they are generalizations of the phenomenological matrices of earlier chapters.

Now

$$\mathrm{e}^{-\tau Pg}P = P - \tau PgP + \frac{1}{2!}\tau^2 PgPgP - \frac{1}{3!}\tau^3 PgPgPgP + \ldots$$

has the transpose $\tilde{P}\exp(-\tau\tilde{g}\tilde{P})$, so that on changing the sign of $\mathbf{B}$, taking the transpose of (63.9), and making use of $(62.5)_3$ and (62.7) we find

$$\varphi^* = -\tilde{b}^*P^*\,\mathrm{e}^{-\tau g^*P^*}b^* = -\tilde{b}^*\eta_i\eta_i P^*\,\mathrm{e}^{-\tau g^*P^*}\eta_i\eta_i b^* = -b\,\mathrm{e}^{-\tau Pg}P\tilde{b} = \varphi,$$

that is, $\varphi(\tau, \mathbf{B}) = \tilde{\varphi}(\tau, -\mathbf{B})$. Similarly (cf. (63.7))

$$\psi_{22}(\tau, \mathbf{B}) = \tilde{\psi}_{22}(\tau, -\mathbf{B}).$$

If the term containing X in (63.1) is retained, (63.10) is replaced by

$$\mathscr{J}(t) = \int_{-\infty}^{t} [\psi_{21}(t-t')X(t') + \psi_{22}(t-t')\,\Delta\dot{\zeta}(t')]\,\mathrm{d}t',$$

where

$$\psi_{21}(\tau) \equiv b\,\mathrm{e}^{-\tau Pg}N \qquad (\tau \geqslant 0).$$

The reciprocal relation satisfied by ψ_{21} will be given shortly.

63.3. The constitutive relation

Turning now to the dynamic constitutive relation (62.10), we may treat it similarly to the dynamic state equation, and we shall just outline the results.

For harmonic time dependence it becomes

$$J = \hat{\psi}_{11}(\omega)X - h(\omega)\,\Delta\zeta,$$

where

$$\hat{\psi}_{11}(\omega) \equiv L - M(P - \mathrm{i}\omega g^{-1})^{-1}N, \qquad h(\omega) \equiv M[U - (U - \mathrm{i}\omega P^{-1}g^{-1}]\tilde{b} \tag{63.11}$$

and $\hat{\psi}_{11}$ satisfies the same reciprocal relations as L, namely, $\eta_{\mathrm{e}}\hat{\psi}_{11} = \hat{\psi}_{11}^{*}\eta_{\mathrm{e}}$. With internal coordinates of even parity it may be cast into a form with components

$$(\hat{\psi}_{11})_{ij} = L_{ij} - \sum_{k=1}^{R} \frac{M_{ik}N_{kj}}{(\nu_k - \mathrm{i}\omega)}.$$

The integral form of the constitutive relation is (after an integration by parts of the integral containing $\Delta\zeta$)

$$J(t) = \int_{-\infty}^{t} [\psi_{11}(t-t')X(t') + \psi_{12}(t-t')\,\Delta\dot{\zeta}(t')]\,dt'$$

where

$$\psi_{11}(\tau) \equiv L\,\delta(\tau) - Mg\,\mathrm{e}^{-\tau Pg}N,\ \psi_{12}(\tau) \equiv -M\,\mathrm{e}^{-\tau gP}\tilde{b} \qquad (\tau \geqslant 0), \tag{63.12}$$

which may be shown to satisfy the reciprocal relations

$$\eta_{\mathrm{e}}\psi_{11}(\tau, \mathbf{B}) = \tilde{\psi}_{11}(\tau, -\mathbf{B})\eta_{\mathrm{e}}, \qquad \eta_{\mathrm{e}}\psi_{12}(\tau, \mathbf{B}) = \frac{\rho}{T}\tilde{\psi}_{21}(\iota, -\mathbf{B}).$$

63.4. Combined state and constitutive equations

Collecting the results given above into one generalized 'constitutive' relation, we have

$$\begin{bmatrix} J(t) \\ \mathscr{J}(t) \end{bmatrix} = \int_{-\infty}^{\infty} \begin{bmatrix} \psi_{11}(\tau) & \psi_{12}(\tau) \\ \psi_{21}(\tau) & \psi_{22}(\tau) \end{bmatrix} \begin{bmatrix} X(t-\tau) \\ \Delta\dot{\zeta}(t-\tau) \end{bmatrix} \mathrm{d}\tau, \tag{63.13}$$

where

$$\begin{bmatrix} \eta_{\mathrm{e}} & \cdot \\ \cdot & U \end{bmatrix} \begin{bmatrix} \psi_{11} & \psi_{12} \\ f\psi_{21} & f\psi_{22} \end{bmatrix} = \begin{bmatrix} \psi_{11}^{*} & f\psi_{21}^{*} \\ \psi_{12}^{*} & f\psi_{22}^{*} \end{bmatrix} \begin{bmatrix} \eta_{\mathrm{e}} & \cdot \\ \cdot & U \end{bmatrix} \qquad (f \equiv \rho/T), \tag{63.14}$$

and we have defined the response functions $\psi_{ij}(\tau)$, $i, j = 1, 2$, to be zero for $\tau < 0$, that is,

$$\psi_{ij}(t-t') = 0 \quad \text{for } t < t'. \tag{63.15}$$

Eqn (63.15) is an expression of the *principal of causality*, namely, that the response (process) may not preceed its cause (affinity); its consequences will be considered further below. Comparing (63.14) with the corresponding 'static' constraint (62.5), we notice the disappearance of the internal parities in the conversion to dynamic constitutive relations.

Eqn (63.13) shows that on short enough time scales the constitutive relations necessarily involve the material time derivatives of the external *independent* thermodynamic variables—typically $\dot{s}$, $\dot{\rho}$, $\dot{\rho}_i$, $i = 1, 2, \ldots$ (see first paragraph of § 62.1)—and it is therefore necessary to prescribe these quantities. In some physical circumstances we may be able to choose as independent variables thermodynamic properties that remain constant, in which case the external processes are related to the external affinities by

$$J(t) = \int_{-\infty}^{\infty} \psi(t-\tau)X(\tau)\,\mathrm{d}\tau = \int_{-\infty}^{\infty} \psi(\tau)X(t-\tau)\,\mathrm{d}\tau, \tag{63.16}$$

where $\psi = \psi_{11}$.

While the various phenomenological matrix functions $\psi_{ij}(\tau)$ in (63.13) may be interpreted in terms of internal coordinates and the associated spectrum of relaxation times, we may be content to accept them as functions to be supplied directly from experiment or from the statistical mechanics of the fluid involved. On the latter view (63.13) and (63.16) are generalized constitutive relations in which the history of the process is accounted for through the response functions $\psi_{ij}(t)$, which must at all times satisfy constraints similar to those given in § 40 for the matrix coefficient L. In particular, in an isotropic system material invariance will eliminate coupling between processes of different tensorial order. It also follows that if internal relaxation processes in an isotropic system are to influence external processes, they must possess the same tensorial order as the processes in question.

64. Linear response theory

64.1. *The generalized admittance*

In § 63 we obtained several constitutive relations for a process vector $J(t)$ of the general form

$$J(t) = \int_{-\infty}^{\infty} \psi(t-\tau)X(\tau)\,\mathrm{d}\tau, \tag{64.1}$$

in which the linear response matrix $\psi(t)$ satisfies

$$\eta\psi = \psi^*\eta, \qquad \psi(t) = 0 \quad \text{for } t < 0, \tag{64.2}$$

η being the (diagonal) parity matrix of the affinity vector $X(\tau)$. The general theory of such constitutive relations is best developed with the aid of Fourier transforms.

Let $w = \omega + \mathrm{i}v$ be a complex variable with ω and v its real and imaginary parts, and let $\psi(t)$ be a response function such that

$$|\psi(t)| \leqslant \begin{cases} A \exp(v_- t) & \text{as } t \to \infty \\ A \exp(v_+ t) & \text{as } t \to -\infty \end{cases} \qquad (v_- < v_+), \tag{64.3}$$

where A is a positive real constant. Then the generalized admittance is the Fourier transform $\hat{\psi}(w)$, namely,†

$$\hat{\psi}(w) \equiv \int_{-\infty}^{\infty} \mathrm{e}^{\mathrm{i}wt}\psi(t)\,\mathrm{d}t, \tag{64.4}$$

which is an analytic function of w (that is, does not become infinite and is differentiable) in the strip $-\infty < \omega < \infty$, $v_- < v < v_+$. Causality as expressed in eqn $(64.2)_2$ enables us to replace the lower limit in (64.4) by zero, and it also permits us to choose v_+ to be as large as we please. Thus if $\psi(t)$ is bounded as $t \to \infty$, $v_- = 0$, so that

$$\hat{\psi}(w) = \int_0^{\infty} \mathrm{e}^{\mathrm{i}wt}\psi(t)\,\mathrm{d}t \tag{64.5}$$

is analytic in the open upper half plane $0 < v < \infty$.

For the dissipative systems with which we are concerned the response will decay exponentially to zero with time, which by (64.1) means that the distant past contributes negligibly to the present value of a process. In this case $v_- < 0$ and $\hat{\psi}(w)$ is certainly analytic in the *whole* of the upper half plane. A purely oscillatory response function like $\psi(t) = \exp(-\mathrm{i}\omega_0 t)$, with

$$\hat{\psi}(w) = \int_0^{\infty} \exp[\mathrm{i}(w-\omega_0)t]\,\mathrm{d}t = 2\pi\,\delta(w-\omega_0), \tag{64.6}$$

where $\delta(w)$ is the delta function (see the inversion integral (64.11)), is associated with a reversible process but may be included in the general theory of dissipative systems as a limiting case,

$$\hat{\psi}(w) = \lim_{\varepsilon\to 0}\int_0^{\infty} \exp[\mathrm{i}(w-\omega_0+\mathrm{i}\varepsilon)t]\,\mathrm{d}t \qquad (\varepsilon > 0).$$

† Strictly, an element $\hat{\psi}_{ij}$ is an 'admittance' only when J_i is *irreversibly* related to X_j, that is (see (64.14) below), $\eta_i\eta_j = \pm 1$ when $\hat{\psi}_{ij}(\mathbf{B}) = \pm\hat{\psi}_{ji}(-\mathbf{B})$ (cf. (42.16)); otherwise $\hat{\psi}_{ij}$ is a 'susceptibility', but within our generalized treatment such precise terminology would be pedantic.

Suppose that $\psi(t)$ can be written as the sum of an instantaneous response $L\,\delta(t)$ and an after-effect function $\varphi(t)$,

$$\psi(t) = L\,\delta(t)+\varphi(t), \qquad \hat{\psi}(w) = L+\hat{\varphi}(w), \tag{64.7}$$

then as $w \to \infty$ in the upper half plane it follows from (64.5) written for φ that $\hat{\varphi}(\infty) = 0$, leaving

$$\hat{\psi}(\infty) = L.$$

When φ is negligible, that is, when the system has no 'memory' for the affinities under consideration (64.1) and $(64.7)_2$ reduce to

$$J = LX \qquad (\eta L = L^*\eta),$$

and we recover the constitutive relation of earlier chapters. It should be noted that the division in (64.7) depends on the observer's time scale $\mathscr{T}$ and that a small enough $\mathscr{T}$ will eliminate any instantaneous response. As it is convenient in the following theory to have $\hat{\psi}(\infty) = 0$, we shall assume $\mathscr{T}$ small enough to render L negligible; alternatively we could suppose that this theory applies to the function φ rather than to ψ.

The matrix $\hat{\psi}(0)$ also has a special significance in the theory, for it follows from (64.1) and (64.4) that constant affinities applied over a very long period will yield the steady-state process

$$J_0 = \hat{\psi}(0)X_1. \tag{64.8}$$

As a typical example take the response function in eqn $(63.12)_1$, namely,

$$\psi(t) = L\,\delta(t) - \begin{cases} Mg\,\mathrm{e}^{-tPg}N & (t > 0) \\ 0 & (t < 0). \end{cases}$$

Its transform is (cf. $(63.11)_1$)

$$\hat{\psi}(w) = L - M(P-\mathrm{i}wg^{-1})^{-1}N,$$

whence

$$\hat{\psi}(\infty) = L, \qquad \hat{\psi}(0) = L - MP^{-1}N. \tag{64.9}$$

64.2. The inversion formula

Cauchy's integral formula applies to a function $f(w)$ continuous within and on a contour C and analytic within C, and relates the value $f(w)$ at a point w within C to the values $f(u)$ on C by (Fig. 16)

$$f(w) = \frac{1}{2\pi\mathrm{i}}\int_C \frac{f(u)\,\mathrm{d}u}{u-w}.$$

We shall choose $f(w) = \hat{\psi}(w)$ and take C to be the semi-circular contour comprised of the straight line $v = v'$, $-R < \omega < R$ and the circle $(v^2+\omega^2)^{\frac{1}{2}} = R$,

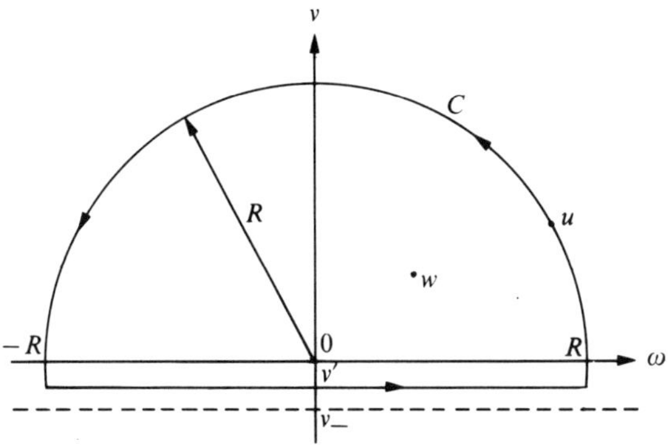

FIG. 16. The contour C.

for, provided $v' > v_-$, $\hat{\psi}(w)$ is analytic within C. As $\hat{\psi}(u)$ is assumed to tend to zero as $u \to \infty$, the contribution to the integral from the circular path will vanish in the limit $R = \infty$, which leaves

$$\hat{\psi}(w) = \frac{1}{2\pi \mathrm{i}} \int_{-\infty + \mathrm{i}v'}^{\infty + \mathrm{i}v'} \frac{\hat{\psi}(u)\,\mathrm{d}u}{u - w} \qquad (v' > v_-). \tag{64.10}$$

The value of this integral is unaltered by changes in v' so long as $\operatorname{Im} w > v' > v_-$, and as we are dealing with dissipative systems, $v_- < 0$ so we may choose $v' \leqslant 0$.

Since $\operatorname{Im} w > v' = \operatorname{Im} u$, we may write (64.10) as

$$\begin{aligned}\hat{\psi}(w) &= \frac{1}{2\pi} \int_{-\infty + \mathrm{i}v'}^{\infty + \mathrm{i}v'} \left(\int_0^\infty \mathrm{e}^{\mathrm{i}(w-u)t}\,\mathrm{d}t \right) \hat{\psi}(u)\,\mathrm{d}u \\ &= \int_0^\infty \mathrm{e}^{\mathrm{i}wt} \left(\frac{1}{2\pi} \int_{-\infty + \mathrm{i}v'}^{\infty + \mathrm{i}v'} \mathrm{e}^{-\mathrm{i}ut} \hat{\psi}(u)\,\mathrm{d}u \right) \mathrm{d}t,\end{aligned}$$

and by comparison with (64.5) conclude that

$$\psi(t) = \frac{1}{2\pi} \int_{-\infty + \mathrm{i}v'}^{\infty + \mathrm{i}v'} \mathrm{e}^{-\mathrm{i}wt} \hat{\psi}(w)\,\mathrm{d}w \qquad (v' > v_-), \tag{64.11}$$

provided, of course that it is true that $\hat{\psi}(w)$ *uniquely* determines the value of $\psi(t)$. A discussion of this question would be out of place here; in fact for given $\hat{\psi}(w)$, $\psi(t)$ is uniquely determined, save for a pathological additive function that differs from zero only on a point set of measure zero and which may be completely ignored in physical applications (see, for example, Pol and Bremner (1950)). We shall take uniqueness for granted in what follows. Eqn (64.11) is the required Fourier inversion formula. Notice that the choice $v' = 0$ and $\hat{\psi}(w) = 2\pi\,\delta(w - \omega_0)$ yields $\psi(t) = \exp(-\mathrm{i}\omega_0 t)$, in accordance with (64.6).

We shall now use (64.11) to prove the convolution theorem,

$$\int_{-\infty}^{\infty} \psi(t-\tau)X(\tau)\,\mathrm{d}\tau = \int_{-\infty}^{\infty}\left(\frac{1}{2\pi}\int_{-\infty+\mathrm{i}v'}^{\infty+\mathrm{i}v'} \mathrm{e}^{-\mathrm{i}w(t-\tau)}\hat{\psi}(w)\,\mathrm{d}w\right)X(\tau)\,\mathrm{d}\tau$$

$$= \frac{1}{2\pi}\int_{-\infty+\mathrm{i}v'}^{\infty+\mathrm{i}v'} \mathrm{e}^{-\mathrm{i}wt}\hat{\psi}(w)\left(\int_{-\infty}^{\infty} \mathrm{e}^{\mathrm{i}w\tau}X(\tau)\,\mathrm{d}\tau\right)\mathrm{d}w$$

$$= \frac{1}{2\pi}\int_{-\infty+\mathrm{i}v'}^{\infty+\mathrm{i}v'} \mathrm{e}^{-\mathrm{i}wt}\hat{\psi}(w)\hat{X}(w)\,\mathrm{d}w,$$

which may be inverted to read

$$\int_{0}^{\infty} \mathrm{e}^{\mathrm{i}wt}\left(\int_{-\infty}^{\infty} \psi(t-\tau)X(\tau)\,\mathrm{d}\tau\right)\mathrm{d}t = \hat{\psi}(w)\hat{X}(w). \tag{64.12}$$

Applying this theorem to (64.1), we obtain the transformed constitutive equation

$$\hat{J}(w) = \hat{\psi}(w)\hat{X}(w), \tag{64.13}$$

where by the transform of (64.2) the admittance satisfies the reciprocal relation

$$\eta\hat{\psi} = \hat{\psi}^*\eta. \tag{64.14}$$

64.3. *The Kramers–Krönig relations*

Returning to (64.10) let us put $v' = 0$ and deduce the value of

$$\hat{\psi}(w) = \frac{1}{2\pi\mathrm{i}}\int_{-\infty}^{\infty} \frac{\hat{\psi}(u)\,\mathrm{d}u}{u-w} \tag{64.15}$$

in the limit as w tends to a point ω on the real axis. Since

$$\frac{1}{u-w} = \frac{1}{(u-\omega)-\mathrm{i}v} = \frac{(u-\omega)}{(u-\omega)^2+v^2} + \mathrm{i}\pi\,\delta(u-\omega, v),$$

where

$$\delta(u-\omega, v) \equiv \frac{1}{\pi}\frac{v}{(u-\omega)^2+v^2}, \qquad \int_{-\infty}^{\infty} \delta(u-\omega, v)\,\mathrm{d}\omega = 1,$$

we see that as $v \to +0$, $\delta(u-\omega, v)$ becomes sharply peaked about $u = \omega$ and in the limit becomes the delta function $\delta(u-\omega)$. The limit

$$\lim_{v\to 0}\frac{1}{2\pi\mathrm{i}}\int_{-\infty}^{\infty} \hat{\psi}(u)\frac{u-\omega}{(u-\omega)^2+v^2}\,\mathrm{d}u = \frac{1}{2\pi\mathrm{i}}\int_{-\infty}^{\infty} \frac{\hat{\psi}(u)\,\mathrm{d}u}{(u-\omega)}$$

is termed the 'Cauchy principal value' of the improper integral on the right, and can be shown to exist (see, for example, Woods (1971)) provided $\hat{\psi}(u)$ satisfies the Hölder condition at $u = \omega$, namely,

$$|\hat{\psi}(u)-\hat{\psi}(\omega)| \leqslant A|u-\omega|^{\mu} \qquad (A, \mu \text{ real}, 0 < A < \infty, 0 < \mu).$$

In what follows all improper integrals shall be understood to be Cauchy principal values.

Thus in the limit $v = +0$, we may write formally

$$\frac{1}{u-w} \to \frac{1}{u-\omega} + \mathrm{i}\pi\,\delta(u-\omega). \tag{64.16}$$

Taking this limit in (64.15) we obtain

$$\hat{\psi}(\omega) = \tfrac{1}{2}\hat{\psi}(\omega) + \frac{1}{2\pi\mathrm{i}}\int_{-\infty}^{\infty} \frac{\hat{\psi}(u)\,\mathrm{d}u}{u-\omega},$$

that is,

$$\hat{\psi}(\omega) = \frac{1}{\pi\mathrm{i}}\int_{-\infty}^{\infty} \frac{\hat{\psi}(u)\,\mathrm{d}u}{u-\omega}. \tag{64.17}$$

Let $\hat{\psi}'$, $\hat{\psi}''$ denote the real and imaginary parts of the admittance, then the real and imaginary parts of (64.17) are

$$\hat{\psi}'(\omega) = \frac{1}{\pi}\int_{-\infty}^{\infty} \frac{\hat{\psi}''(u)}{u-\omega}\,\mathrm{d}u, \qquad \hat{\psi}''(\omega) = -\frac{1}{\pi}\int_{-\infty}^{\infty} \frac{\hat{\psi}'(u)}{u-\omega}\,\mathrm{d}u, \tag{64.18}$$

which are the Kramers–Krönig relations for $\psi(t)$.

From

$$\hat{\psi}(\omega) + \mathrm{i}\hat{\psi}''(\omega) = \int_{0}^{\infty} (\cos\omega t + \mathrm{i}\sin\omega t)\psi(t)\,\mathrm{d}t$$

and the reality of the matrix $\psi(t)$ it follows that

$$\hat{\psi}'(\omega) = \hat{\psi}'(-\omega), \qquad \hat{\psi}''(\omega) = -\hat{\psi}''(-\omega), \tag{64.19}$$

so that (64.18) may be written as

$$\hat{\psi}'(\omega) = \frac{2}{\pi}\int_{0}^{\infty} \frac{u\hat{\psi}''(u)}{u^2-\omega^2}\,\mathrm{d}u, \qquad \hat{\psi}''(\omega) = -\frac{2\omega}{\pi}\int_{0}^{\infty} \frac{\hat{\psi}'(u)\,\mathrm{d}u}{u^2-\omega^2}. \tag{64.20}$$

These relations enable us to find the real part of the admittance given the imaginary part as a function of frequency, and vice versa.

If $\hat{\psi}(\infty)$ is not zero, it is necessary to replace $\hat{\psi}(\omega)$ in (64.18) and (64.20) by $\hat{\varphi}(\omega) = \hat{\psi}(\omega) - \hat{\psi}(\infty)$.

64.4. *The relaxation function*

Apart from the steady-state affinity used in obtaining the process J_0 in (64.8), there are two other special affinities of importance in the general theory. First there is harmonic time dependence, that is,

$$X(t) = X_0 \, e^{-i\omega t},$$

when (64.21)

$$J(t) = \int_{-\infty}^{t} \psi(t-\tau) X_0 \, e^{-i\omega\tau} \, d\tau = \hat{\psi}(\omega) X_0 \, e^{-i\omega t} = \hat{\psi}(\omega) X,$$

(cf. (63.5)). Secondly there is the 'switched-off' affinity, that is, one that is held constant in $-\infty < t < 0$ and then suddenly removed at $t = 0$. For this affinity

$$J(t) = \int_{-\infty}^{tV(-t)} \psi(t-\tau) X_1 \, d\tau \qquad \left(V(t) = \begin{matrix} 1, & t > 0 \\ 0, & t < 0 \end{matrix} \right)$$

that is,

$$J(t) = \Phi(t) X_1 \qquad \left(\Phi(t) \equiv \int_t^{\infty} \psi(\tau) \, d\tau \right), \tag{64.22}$$

where $\Phi(t)$ is known as the 'relaxation' function. For $t < 0$, $\Phi(t)$ has the constant value $\Phi(0)$.

From the definition of $\Phi(t)$ it follows that $\psi(t) = -\Phi'(t)$, whence on integrating by parts,

$$\hat{\psi}(w) = \Phi(0) + iw \int_0^{\infty} e^{iwt} \Phi(t) \, dt. \tag{64.23}$$

Hence

$$\Phi(t) = \Phi(0) = \hat{\psi}(0) \qquad (t \leqslant 0), \tag{64.24}$$

which is also a consequence of causality and the result contained in (64.8). Let $\Phi_0(t) = \Phi(t)$ for $t \geqslant 0$ and $\Phi_0(t) = 0$ for $t < 0$, then (64.23) shows that $\hat{\Phi}_0(w) = [\hat{\psi}(w) - \hat{\psi}(0)]/iw$, which may be inverted by (64.11) with v' zero to give

$$\Phi(t) = \frac{1}{2\pi i} \int_{-\infty}^{\infty} e^{-i\omega t} [\hat{\psi}(\omega) - \hat{\psi}(0)] \frac{d\omega}{\omega} \qquad (t \geqslant 0)$$

or

$$\Phi(t) = \tfrac{1}{2} \hat{\psi}(0) + \frac{1}{2\pi i} \int_{-\infty}^{\infty} e^{-i\omega t} \hat{\psi}(\omega) \frac{d\omega}{\omega} \qquad (\text{all } t), \tag{64.25}$$

as

$$i \int_{-\infty}^{\infty} e^{-i\omega t} \frac{d\omega}{\omega} = \int_{-\infty}^{\infty} \sin \omega t \frac{d\omega}{\omega} = \pi,$$

and (see (64.26) below) (64.25) reduces correctly to (64.24) for negative t.

To find a convenient expression for $\hat{\psi}(0)$ we consider the function

$$f(w) = e^{\pm iwt}\hat{\psi}(w) \qquad \begin{matrix}(+ \text{ if } t > 0)\\ (- \text{ if } t < 0)\end{matrix}$$

which is analytic in the whole of the upper half plane. By the argument applied to $f(w)$ that lead to (64.17) for $\hat{\psi}$ we find (interchanging u and ω)

$$e^{\pm iut}\hat{\psi}(u) = \frac{1}{\pi i}\int_{-\infty}^{\infty} e^{\pm i\omega t}\hat{\psi}(\omega)\frac{d\omega}{\omega - u},$$

whence

$$\hat{\psi}(0) = \frac{1}{\pi i}\int_{-\infty}^{\infty} e^{\pm i\omega t}\hat{\psi}(\omega)\frac{d\omega}{\omega} \qquad \begin{matrix}(+ \text{ if } t > 0)\\ (- \text{ if } t < 0).\end{matrix} \tag{64.26}$$

Then (64.25) and (64.26) give

$$\Phi(t) = \begin{cases} \hat{\psi}(0) & (t \leqslant 0) \\ \dfrac{1}{\pi i}\displaystyle\int_{-\infty}^{\infty} \cos \omega t\, \hat{\psi}(\omega)\frac{d\omega}{\omega} & (t \geqslant 0). \end{cases} \tag{64.27}$$

Eqns (64.21) and (64.23) suggest experimental methods of finding the response function $\psi(t)$. Measurements of the process amplitude over a spectrum of frequencies will yield the admittance $\hat{\psi}(\omega)$ directly, and hence $\psi(t)$ by inversion, or alternatively the process relaxation following switch-off will lead to $\hat{\psi}(\omega)$—and hence $\psi(t)$—via $\Phi(t)$ and eqn (64.25).

64.5. Relation between dissipation and admittance

Consider the contribution to σ due only to the external processes J in (62.2),

$$\begin{aligned}\sigma_e(t) = JX &= \int_{-\infty}^{\infty} X(t)\psi(t-\tau)X(\tau)\, d\tau \\ &= X(t)\frac{1}{2\pi}\int_{-\infty}^{\infty} e^{-i\omega t}\hat{\psi}(\omega)\hat{X}(\omega)\, d\omega.\end{aligned}$$

We shall need (64.19) and the similar result for $\hat{X}(\omega)$,

$$\hat{\psi}(\omega) = \hat{\psi}_c(-\omega), \qquad \hat{X}(\omega) = \hat{X}_c(-\omega), \tag{64.28}$$

where the subscript c denotes complex conjugates, that is, $\hat{\psi}_c = \hat{\psi}' - i\hat{\psi}''$, etc. Thus

$$\begin{aligned}\sigma_e(t) &= \frac{1}{2\pi}\int_{-\infty}^{\infty} e^{i\omega t}X(t)\hat{\psi}_c(\omega)\hat{X}_c(\omega)\, d\omega \\ &= \frac{1}{4\pi^2}\int_{-\infty}^{\infty}\int_{-\infty}^{\infty} e^{i(\omega-\omega')t}\hat{X}(\omega')\hat{\psi}_c(\omega)\hat{X}_c(\omega)\, d\omega\, d\omega'.\end{aligned}$$

Now suppose that the affinities act for a finite time, then the total dissipation $\mathscr{D}$ over that time is given by

$$\mathscr{D} \equiv \int_{-\infty}^{\infty} \sigma(t)\,\mathrm{d}t = \frac{1}{2\pi}\int_{-\infty}^{\infty} \hat{X}(\omega)\hat{\psi}_{\mathrm{c}}(\omega)\hat{X}_{\mathrm{c}}(\omega)\,\mathrm{d}\omega, \tag{64.29}$$

in the evaluation of which we have used (64.6). It is easily verified from (64.28) that the integral in (64.29) is real—as of course it should be—and may be expressed as

$$\mathscr{D} = \frac{1}{4\pi}\int_{-\infty}^{\infty} \hat{X}_{\mathrm{c}}(\omega)[\hat{\psi}(\omega) + \tilde{\hat{\psi}}_{\mathrm{c}}(\omega)]\hat{X}(\omega)\,\mathrm{d}\omega$$

or (64.30)

$$\mathscr{D} = \frac{1}{4\pi}\int_{-\infty}^{\infty} \hat{X}_{\mathrm{c}}(\omega)[\hat{\psi}(\omega, \mathbf{B}) + \eta\hat{\psi}_{\mathrm{c}}(\omega, -\mathbf{B})\eta]\hat{X}(\omega)\,\mathrm{d}\omega$$

by (64.14). When there is no magnetic field and the affinities all have the same parity, it follows from the second of (64.30) that the dissipation depends on the real part of the admittance.

65. Propagation of sound waves through a relaxing gas

65.1. Basic equations

To illustrate the theory given above we shall consider the simplest example of importance, namely, a small amplitude disturbance propagating through a gas possessing one scalar relaxation mechanism, such as energy distribution between molecular modes (§ 49.4), characterized by a progress variable ξ. We shall take the internal energy u to be the potential x in (62.1), then the static equations (62.3) are

$$\begin{bmatrix} -\Delta p \\ \Delta T \\ -A \end{bmatrix} = \begin{bmatrix} a_1 & a_2 & b_1 \\ a_2 & a_3 & b_2 \\ b_1 & b_2 & g \end{bmatrix} \begin{bmatrix} \Delta v \\ \Delta s \\ \Delta \xi \end{bmatrix} \qquad (a_1, a_3, g > 0), \tag{65.1}$$

where $v = 1/\rho$ is the specific volume and the constraints on the constants a_1, a_3, g are a consequence of thermodynamic stability (§ 17.4).

The entropy production rate is (see § 49.1)

$$T\sigma = (-\tfrac{1}{3}\overset{\times}{\pi}\nabla\,.\,\mathbf{v} + \rho\dot{\xi}A) - T^{-1}\mathbf{q}\,.\,\nabla T - \mathring{\boldsymbol{\pi}}:\nabla\mathring{\mathbf{v}} \geqslant 0,$$

and hence in an isotropic fluid the constitutive relations take the form (§ 48.3)

$$-\tfrac{1}{3}\overset{\times}{\pi} = \rho v'\nabla\,.\,\mathbf{v} - \rho\gamma A, \qquad \dot{\xi} = \gamma\nabla\,.\,\mathbf{v} + \beta A \qquad (v', \beta > 0), \tag{65.2}$$

$$\mathbf{q} = -\kappa\nabla T, \qquad \mathring{\boldsymbol{\pi}} = -2\rho v\nabla\mathring{\mathbf{v}} \qquad (\kappa, v > 0). \tag{65.3}$$

It follows from $(65.1)_3$ and $(65.2)_2$ that, provided $\Delta\xi = 0$ at $t = -\infty$,

$$\Delta\xi = \int_{-\infty}^{t} \exp[-(t-t')/\tau_1]\{\gamma\nabla \cdot \mathbf{v}(t') - \beta[b_1\,\Delta v(t') + b_2\,\Delta s(t')]\}\,\mathrm{d}t', \tag{65.4}$$

where

$$\tau_1 \equiv \frac{1}{\beta} = -\frac{1}{\beta}\left(\frac{\partial\xi}{\partial A}\right)_{v,s} \qquad (\tau_1 > 0) \tag{65.5}$$

is the relaxation time. It is convenient to integrate the terms containing $\Delta v(t')$ and $\Delta s(t')$ by parts; doing this and using the result to eliminate $\Delta\xi$ from (65.1) we arrive at the dynamic state equations

$$\begin{aligned}\Delta p &= \Delta p_{\mathrm{eq}} - \int_{-\infty}^{t} \exp[-(t-t')/\tau_1]\{g^{-1}b_1[b_1\,\Delta\dot{v}(t') + b_2\,\Delta\dot{s}(t')] + \\ &\qquad\qquad + \gamma b_1\nabla \cdot \mathbf{v}(t')\}\,\mathrm{d}t' \\ \Delta T &= \Delta T_{\mathrm{eq}} + \int_{-\infty}^{t} \exp[-(t-t')/\tau_1]\{g^{-1}b_2[b_1\,\Delta\dot{v}(t') + b_2\,\Delta\dot{s}(t')] + \\ &\qquad\qquad + \gamma b_2\nabla \cdot \mathbf{v}(t')\}\,\mathrm{d}t',\end{aligned} \tag{65.6}$$

where

$$\begin{aligned}\Delta p_{\mathrm{eq}} &= -(a_1 - g^{-1}b_1^2)\,\Delta v - (a_2 - g^{-1}b_1b_2)\,\Delta s \qquad (a_1 - g^{-1}b_1^2 > 0), \\ \Delta T_{\mathrm{eq}} &= (a_2 - g^{-1}b_1b_2)\,\Delta v + (a_3 - g^{-1}b_2^2)\,\Delta s \qquad (a_3 - g^{-1}b_2^2 > 0)\end{aligned} \tag{65.7}$$

are the 'relaxed' values, appropriate to $A = 0$. The stability inequalities here could, of course, have been inferred directly from (65.1).

From (65.1) and (65.7) we find that the non-negative derivatives

$$\left(\frac{\partial p}{\partial \rho}\right)_{s,\xi} \equiv a_\infty^2, \qquad \left(\frac{\partial p}{\partial \rho}\right)_{s,A} \equiv a_0^2$$

are related by (65.8)

$$a_0^2 = a_\infty^2 - b_1^2/(\rho^2 g).$$

It is convenient to rewrite (65.7) in the form (cf. (60.2))

$$\begin{aligned}\Delta p_{\mathrm{eq}} &= a_0^2\,\Delta\rho + \rho^2 b_0\,\Delta s \\ \Delta T_{\mathrm{eq}} &= b_0\,\Delta\rho + c_0\,\Delta s\end{aligned} \qquad (c_0 > 0) \tag{65.9}$$

Eqns $(65.2)_2$, $(65.1)_3$, and (65.4) yield the dynamic constitutive relation

$$-\tfrac{1}{3}\breve{\pi} = \rho v'\nabla \cdot \mathbf{v} + \int_{-\infty}^{t} \exp[-(t-t')/\tau_1]\rho\gamma[\gamma g\nabla \cdot \mathbf{v}(t') + b_1\,\Delta\dot{v}(t') + b_2\,\Delta\dot{s}]\,\mathrm{d}t'. \tag{65.10}$$

The rates $\Delta\dot{v}$, $\Delta\dot{s}$ may be calculated from the mass and entropy balances

$$\rho\,\Delta\dot{v} = -\rho^{-1}\,\Delta\dot{\rho} = \nabla\,.\,\mathbf{v}, \qquad \rho\,\Delta\dot{s} = \nabla\,.\,(T^{-1}\kappa\nabla T) \simeq T^{-1}\kappa\nabla^2 T, \quad (65.11)$$

the second of which is a consequence of $(65.3)_1$ and (38.8), which for the present application reduces to

$$\rho\,\Delta\dot{s}+\nabla\,.\,(T^{-1}\mathbf{q}) = \sigma \simeq 0,$$

σ being quadratic in perturbation amplitudes.

The perturbation in the pressure tensor

$$\Delta\mathbf{p} = (\Delta p+\tfrac{1}{3}\overset{\times}{\pi})\mathbf{1} + \overset{\circ}{\boldsymbol{\pi}}$$

is given by $(65.3)_2$, $(65.6)_1$, (65.10), and (65.11),

$$\Delta\mathbf{p} = (\Delta p_{\mathrm{eq}}-\rho v'\nabla\,.\,\mathbf{v})\mathbf{1} - 2\rho v\overset{\circ}{\nabla\mathbf{v}} - \mathbf{1}\int_{-\infty}^{t}\exp\left[-(t-t')/\tau_1\right]\times$$
$$\times\,\rho[\varepsilon\nabla\,.\,\mathbf{v}(t')-\eta\kappa T^{-1}\nabla^2 T(t')]\,\mathrm{d}t', \quad (65.12)$$

where

$$\varepsilon \equiv \frac{b_1^2}{\rho^2 g}+\frac{2\gamma b_1}{\rho}+\gamma^2 g \quad \text{and} \quad \eta \equiv -\frac{b_2}{\rho}\left(\frac{b_1}{\rho g}+\gamma\right). \quad (65.13)$$

To complete the set of equations we require the perturbation form of momentum balance, namely,

$$\rho\frac{\partial\mathbf{v}}{\partial t} = -\nabla\,.\,(\Delta\mathbf{p}). \quad (65.14)$$

65.2. The dispersion relation

Write $\tilde{\rho}$, $\tilde{\mathbf{v}}$, etc. for the amplitudes of simple harmonic waves,

$$\Delta\rho = \tilde{\rho}\,\mathrm{e}^{\mathrm{i}(\mathbf{k}.\mathbf{r}-\omega t)}, \qquad \mathbf{v} = \tilde{\mathbf{v}}\,\mathrm{e}^{\mathrm{i}(\mathbf{k}.\mathbf{r}-\omega t)},$$

then the transformation $\Delta\rho \to \tilde{\rho}$, $\mathbf{v} \to \tilde{\mathbf{v}}$, ...

$$\frac{\partial}{\partial t} \to -\mathrm{i}\omega, \qquad \int_{-\infty}^{t}\exp\left[-(t-t')/\tau_1\right]\ldots\,\mathrm{d}t \to \frac{\tau_1}{1-\mathrm{i}\omega\tau_1}, \qquad \nabla \to \mathrm{i}\mathbf{k},$$

applied to (65.12) and (65.14) gives

$$\omega\rho\tilde{\mathbf{v}} = \mathbf{k}\,.\,\tilde{\mathbf{p}} = k\left[a_0^2\tilde{\rho}+\rho^2 b_0\tilde{s}-\mathrm{i}\rho\mathbf{k}\,.\,\tilde{\mathbf{v}}\left(v'+\tfrac{1}{3}v+\frac{\tau_1\varepsilon}{1-\mathrm{i}\omega\tau_1}\right)-\right.$$
$$\left.-\frac{\tau_1\eta\rho\kappa k^2}{(1-\mathrm{i}\omega\tau_1)T}\tilde{T}\right]-\mathrm{i}\rho v k^2\tilde{\mathbf{v}}. \quad (65.15)$$

Likewise $(65.6)_2$ and (65.11) become

$$\tilde{T} = b_0\tilde{\rho}+c_0\tilde{s}-\frac{\mathrm{i}\tau_1}{1-\mathrm{i}\omega\tau_1}\left(\rho\eta\mathbf{k}\,.\,\tilde{\mathbf{v}}-\frac{\mathrm{i}b_2^2\kappa k^2}{g\rho T}\tilde{T}\right)$$

and (65.16)

$$\omega\tilde{\rho} = \rho\mathbf{k}\,.\,\tilde{\mathbf{v}}, \qquad \omega\tilde{s} = -\frac{\mathrm{i}\kappa k^2}{\rho T}\tilde{T}.$$

The dispersion relation now follows in the usual way (cf. § 57.3) on eliminating $\mathbf{k}\,.\,\tilde{\mathbf{v}}$, $\tilde{\rho}$, $\tilde{s}$, and $\tilde{T}$ from (65.15) and (65.16). The conditions for a non-zero solution in the perturbations is

$$\left(\frac{\omega}{k}\right)^2-a^2+\mathrm{i}\omega\bar{\nu}+$$
$$+\mathrm{i}\kappa\frac{\rho}{T}\frac{k^2}{\omega}\left(b_0-\frac{\mathrm{i}\omega\tau_1\eta}{1-\mathrm{i}\omega\tau_1}\right)^2\Big/\left[1+\mathrm{i}\frac{\kappa}{\rho T}\frac{k^2}{\omega}\left(c_0-\frac{\mathrm{i}\omega\tau_1 b_2^2g^{-1}}{1-\mathrm{i}\omega\tau_1}\right)\right] = 0, \tag{65.17}$$

where

$$a^2 \equiv a_0^2+\frac{(\omega\tau_1)^2}{1+(\omega\tau_1)^2}\varepsilon \quad\text{and}\quad \bar{\nu} \equiv \nu'+\tfrac{4}{3}\nu+\frac{\tau_1\varepsilon}{1+(\omega\tau_1)^2}.$$

This is the required dispersion relation.

Let ω be real and k complex, then

$$\frac{\omega}{k} = v_p(1-\mathrm{i}\alpha) \qquad \left(\alpha = \frac{\mathrm{Im}\,k}{\mathrm{Re}\,k}\right), \tag{65.18}$$

where v_p is the phase velocity and α is the attenuation factor, which shall be assumed to be small compared with unity.

First consider adiabatic conditions, $\kappa = 0$. If it is also valid to ignore the coupling term γ in (65.2), it follows from $(65.8)_2$ and (65.13) that $\varepsilon = a_\infty^2-a_0^2$, whence (65.17) and (65.18) yield

$$v_\mathrm{p}^2 = a_0^2+\frac{(\omega\tau_1)^2}{1+(\omega\tau_1)^2}(a_\infty^2-a_0^2), \qquad \alpha = \frac{\omega}{2v_\mathrm{p}^2}\left[\nu'+\tfrac{4}{3}\nu+\frac{\tau_1(a_\infty^2-a_0^2)}{1+(\omega\tau_1)^2}\right]. \tag{65.19}$$

Notice that α is a maximum when $\omega \simeq \tau_1^{-1}$, and that v_p increases steadily from a_0 to a_∞ with increasing frequency. When $\omega\tau_1 \ll 1$ it follows from (65.13) and (65.15) that the relaxation process is contributing a bulk viscosity of $\nu'' = \tau_1 b_1^2/(\rho^2 g)$, a result in accord with the theory of § 49.5.

Thermal conductivity introduces the possibility of an additional wave, as may be seen from the form of the term containing κ in (65.17). This wave would have a small velocity proportional to $\kappa^{\frac{1}{2}}/a$, were it to propagate, but it appears that the wave is evanescent unless

$$2\rho\eta(b_0+\eta) > a^2 b_2^2/g.$$

Both

$$b_0 = \left(\frac{\partial T}{\partial \rho}\right)_{s,A}, \qquad \eta = \left(\frac{\partial T}{\partial \rho}\right)_{s,\xi} - \left(\frac{\partial T}{\partial \rho}\right)_{s,A} \qquad (\gamma = 0) \tag{65.20}$$

are positive, and so it may be possible to satisfy the inequality in some cases; indeed the second sound of helium II could be interpreted in this fashion.

Neglecting terms of order κ^2 we find from (65.17) that thermal conductivity increases the attenuation factor by an amount (cf. result given in § 60.3)

$$\alpha_\kappa = \frac{\omega}{2v_p^2}\frac{\kappa\rho}{Tv_p^2}\left[b_0 + \frac{\omega\tau_1(\omega\tau_1-1)}{1+(\omega\tau_1)^2}\eta\right]\left[b_0 + \frac{\omega\tau_1(\omega\tau_1+1)}{1+(\omega\tau_1)^2}\eta\right], \tag{65.21}$$

which has the limiting values

$$\lim_{\omega\tau_1\to 0}\alpha_\kappa = \frac{\omega}{2v_p^2}\frac{\kappa\rho}{Tv_p^2}\left(\frac{\partial T}{\partial \rho}\right)_{s,A}^2,$$

$$\lim_{\omega\tau_1\to \infty}\alpha_\kappa = \frac{\omega}{2v_p^2}\frac{\kappa\rho}{Tv_p^2}\left(\frac{\partial T}{\partial \rho}\right)_{s,\xi}^2,$$

as might have been anticipated.

65.3. General theory using response functions

The single relaxation time introduced in (65.5) corresponds to the single internal variable ξ. Complicated internal processes will require several internal variables to be adequately represented, but rather than attempting to identify these separately for specific mechanisms we may adopt a 'black-box' approach and write

$$\begin{aligned}\Delta p &= \int_{-\infty}^{t}[\psi_p(t-t')\,\Delta\rho(t')+\varphi_p(t-t')\,\Delta s(t')]\,\mathrm{d}t',\\ \Delta T &= \int_{-\infty}^{t}[\psi_T(t-t')\,\Delta\rho(t')+\varphi_T(t-t')\,\Delta s(t')]\,\mathrm{d}t',\end{aligned} \tag{65.22}$$

in place of (65.6). Any dependence on $\nabla\,.\,\mathbf{v}(t')$ may be incorporated in the terms containing $\Delta\rho(t')$. The four phenomenological response functions introduced here may be supposed known from measurements of pressure and temperature made under a variety of changing conditions (see remarks at end of § 64.4), or perhaps from measurements of the velocity and attenuation of the waves under present consideration (see eqns (65.26) below).

We now introduce the four-fold Fourier transforms

$$\Delta p(\mathbf{r}, t) = \left(\frac{1}{2\pi}\right)^4 \int_{-\infty}^{\infty} \int_{-\infty}^{\infty} e^{i(\mathbf{k}.\mathbf{r}-\omega t)} \hat{p}(k, \omega)\, d\mathbf{k}\, d\omega,$$
$$\hat{p}(\mathbf{k}, \omega) = \int_{-\infty}^{\infty} \int_{-\infty}^{\infty} e^{-i(\mathbf{k}.\mathbf{r}-\omega t)}\, \Delta p(\mathbf{r}, t)\, d\mathbf{r}\, dt, \tag{65.23}$$

in which ω is a real scalar and $\mathbf{k}$ is a complex vector. Under this transformation (65.22) and $(65.11)_2$ become

$$\hat{p} = \hat{\psi}_p \hat{\rho} + \hat{\varphi}_p \hat{s}, \qquad \hat{T} = \hat{\psi}_T \hat{\rho} + \hat{\varphi}_T \hat{s}, \qquad \omega \hat{s} = -i \frac{\kappa k^2}{\rho T} \hat{T},$$

whence

$$\hat{p} = \hat{\psi} \hat{\rho},$$

where

$$\hat{\psi} \equiv \hat{\psi}_p - i\kappa \frac{\rho k^2}{T\omega} \frac{\hat{\psi}_T^2}{1 + i(\kappa k^2/\rho T\omega)\hat{\varphi}_T}, \tag{65.24}$$

and we have used

$$\hat{\varphi}_p = \rho^2 \hat{\psi}_T,$$

which follows from (65.22) and the Maxwell relation $(\partial p/\partial s)_\rho = \rho^2 (\partial T/\partial \rho)_s$.

We shall ignore any coupling between $\mathring{\pi}$ and the histories $\Delta\rho(t')$, $\Delta s(t')$, so that the transform of the momentum equation will be

$$\omega \rho \hat{\mathbf{v}} = \mathbf{k}\,.\,\mathbf{p} = \mathbf{k}[\hat{p} - i(\nu' + \tfrac{1}{3}\nu)\rho \mathbf{k}\,.\,\hat{\mathbf{v}}] - i\rho\nu k^2 \hat{\mathbf{v}},$$

whence from mass conservation

$$\omega^2 \hat{\rho} = \omega \rho \mathbf{k}\,.\,\hat{\mathbf{v}} = k^2[\hat{\psi} - i\omega(\nu' + \tfrac{4}{3}\nu)]\hat{\rho}.$$

The dispersion relation is

$$(\omega/k)^2 - \hat{\psi}(\mathbf{k}, \omega) + i\omega(\nu' + \tfrac{4}{3}\nu) = 0, \tag{65.25}$$

and comparing this and (65.24) with (65.17) we can identify the special values of $\hat{\psi}_p$, $\hat{\psi}_T$, and $\hat{\varphi}_T$ appropriate to the theory of the previous article.

Eqns (65.18) and (65.25) yield

$$v_p^2 = \hat{\psi}'(\mathbf{k}, \omega), \qquad \alpha = \frac{1}{2v_p^2}[\omega(\nu' + \tfrac{4}{3}\nu) - \hat{\psi}''(\mathbf{k}, \omega)], \tag{65.26}$$

where $\hat{\psi}'$, $\hat{\psi}''$ are the real and imaginary parts of $\hat{\psi}$. In an isotropic system $\mathbf{k}$ can occur only as $\mathbf{k}\,.\,\mathbf{k} = k^2$ in (65.26) as v_p and α must be independent of the direction of propagation. Finally, note that if the velocity $[\hat{\psi}'(\mathbf{k}, \omega)]^{\frac{1}{2}}$ of small amplitude waves is obtained experimentally, the damping function $\hat{\psi}''(\mathbf{k}, \omega)$ can be calculated directly by the Kramers–Krönig relation $(64.19)_2$.

66. Processes in an 'internal' space

66.1. Functional constitutive relations

Let the internal coordinates ξ be continuous functions of some internal variable y as described in § 21.3, then the Gibbs relation will have the form given in (21.16). In matrix notation we can write it as (cf. (62.1))

$$dx = Z\,d\zeta - \int A\,d\xi\,dy, \tag{66.1}$$

where the functions $\xi(y)$ are subject to the constraints

$$\int \xi(y)\,dy = 1. \tag{66.2}$$

The state equations now express Z and $A(y)$ as functions of ζ and as functionals of ξ. Thus for small departures from a reference state $(\zeta^0, \xi_{\text{eq}})$ we have, corresponding to (62.3)

$$\begin{aligned} \Delta Z &= a\Delta\zeta + \int b(y)\,\Delta\xi(y)\,dy, \\ A(y) &= -\tilde{b}(y)\,\Delta\zeta - \int g(y, y')\,\Delta\xi(y')\,dy'. \end{aligned} \tag{66.3}$$

As eqns (66.1) and $(66.3)_2$ are the limit forms of

$$dx = Z\,d\zeta - \sum_k A_k\,d\xi_k, \qquad A_k = -\tilde{b}_k\,\Delta\zeta - \sum_{k'} g_{kk'}\,\Delta\xi_{k'},$$

where

$$g_{kk'} = -\frac{\partial A_k}{\partial \xi_{k'}} = \frac{\partial^2 x}{\partial \xi_{k'}\,\partial \xi_k},$$

it is clear that $g_{kk'} = \tilde{g}_{k'k}$, which in $(66.3)_2$ takes the form

$$g(y, y') = \tilde{g}(y', y). \tag{66.4}$$

The entropy production is given by

$$\sigma = JX + \frac{\rho}{T}\int A(y)\dot{\xi}(y)\,dy \geqslant 0, \tag{66.5}$$

so that the linear constitutive relations have the appearance (cf. (62.4))

$$J = LX + \int M(y)A(y)\,\mathrm{d}y$$

and (66.6)

$$\dot{\xi}(y) = N(y)X + \int P(y, y')A(y')\,\mathrm{d}y',$$

where $\dot{\xi}(y)$ is the rate of change of $\xi(y)$ in a frame moving with the fluid. By first considering the discrete forms of (66.6) in which the integrals are replaced by sums, we infer that the phenomenological matrices appearing in (66.6) satisfy (cf. (62.5))

$$\eta_e L = L^*\eta_e, \qquad \eta_e M(y) = \frac{\rho}{T}N^*(y)\eta_i, \qquad \eta_i P(y, y') = P^*(y, y')\eta_i$$

and that further (66.7)

$$P(y, y') = P^*(y', y).$$

From $(66.3)_2$ and $(66.6)_2$

$$\dot{\xi}(y) = N(y)X + F(y)\,\Delta\zeta - \int H(y, y')\,\Delta\xi(y')\,\mathrm{d}y', \tag{66.8}$$

where

$$F(y) = -\int P(y, y')\tilde{b}(y')\,\mathrm{d}y', \qquad H(y, y') = \int P(y, y'')g(y'', y')\,\mathrm{d}y''.$$

The solution of the integral equation (66.8) may be used to eliminate $\Delta\xi(y)$ from (66.3), and the general theory—which we will not pursue further—proceeds along the same lines as that given in § 62 and § 63.

From § 21.3 we see that the components of ξ are scalars—density ratios per unit y—and so if the components of X are either vectors or traceless tensors in physical space, it follows from the theory given in § 46 that (66.5) may be replaced by

$$\sigma = \sigma_{\mathrm{ex}} + \sigma_{\mathrm{in}}, \qquad \sigma_{\mathrm{ex}} = JX \geqslant 0; \qquad \sigma_{\mathrm{in}} = \frac{\rho}{T}\int A(y)\dot{\xi}(y)\,\mathrm{d}y \geqslant 0. \tag{66.9}$$

The coupling matrices M and N now vanish, leaving

$$J = LX$$

(66.10)

$$\dot{\xi}(y) = F(y)\,\Delta\zeta - \int H(y, y')\,\Delta\xi(y')\,\mathrm{d}y'.$$

Thus there is no *phenomenological* relation between J and ξ, but as we have seen in § 23 this does not exclude the possibility of a quite different type of relationship holding between external fluxes and internal coordinates.

66.2. The Fokker–Planck equation

In many cases the integral equation $(66.10)_2$ may be simplified. If both $g(y, y')$ and $P(y, y')$ are sharply peaked about $y = y'$, it reduces to

$$\dot{\xi} = -P\tilde{b}\,\Delta\zeta - Pg\,\Delta\xi = PA, \tag{66.11}$$

where P and g are the functions $P(y, y)$, $g(y, y)$. Hence (cf. (62.12))

$$\Delta\xi = -\int_{-\infty}^{t} \mathrm{e}^{-(t-t')Pg}(P\tilde{b}\,\Delta\zeta(t') + \dot{\xi}_{\mathrm{eq}}(t'))\,\mathrm{d}t'. \tag{66.12}$$

Notice that we could have inferred a constitutive relation like (66.11) simply by supposing the *integrand* in $(66.9)_3$ to be positive for all values of y.

$$A(y)\dot{\xi}(y) \geqslant 0. \tag{66.13}$$

This means that $\dot{\xi}$ is determined *locally* in y-space.

Now define a new variable φ by

$$\dot{\xi} = -\frac{\partial\varphi}{\partial y}, \tag{66.14}$$

so that up to an arbitrary additive constant, each component of φ may be interpreted as being a flux in the space of its internal variable, then (66.13) may be expressed

$$-\frac{\partial}{\partial y}(A\varphi) + \frac{\partial A}{\partial y}\varphi \geqslant 0. \tag{66.15}$$

The time derivative of (66.2) gives

$$0 = \int \dot{\xi}(y)\,\mathrm{d}y = -\int \mathrm{d}\varphi(y),$$

so that we may choose the arbitrary constants in φ to render it zero at the limits of the y-domain. Hence on integration (66.15) becomes

$$\int \frac{\partial A}{\partial y}\varphi\,\mathrm{d}y \geqslant 0. \tag{66.16}$$

If φ is also determined locally in y-space, (66.16) may be reduced to $(\partial A/\partial y)\varphi \geqslant 0$, and the constitutive relation is simply $\varphi = D\partial A/\partial y$, where D

is a phenomenological matrix. Then by (66.14)

$$\dot{\xi} = -\frac{\partial}{\partial y}\left(D\frac{\partial A}{\partial y}\right). \tag{66.17}$$

For an ideal system (21.18), namely,

$$A_i(y) = -\frac{kT}{m}\ln\frac{\xi_i(y)}{\xi_i^{\text{eq}}(y)},$$

applies, when the components of (66.17) are

$$\dot{\xi}_i = \frac{kT}{m}\sum_i \frac{\partial}{\partial y} D\left(\frac{1}{\xi_i}\frac{\partial \xi_i}{\partial y} - \frac{1}{\xi_i^{\text{eq}}}\frac{\partial \xi_i^{\text{eq}}}{\partial y}\right). \tag{66.18}$$

This type of constitutive equation, known as a Fokker–Planck equation, is important in kinetic theory.

12

THE FLUCTUATION-DISSIPATION THEOREM

THE genuinely original and inherent concepts of process thermodynamics are contained in just two theorems, namely, the reciprocal theorem of Thomson, Onsager, and Casimir (described in Chapter 7) and the fluctuation–dissipation theorem to be explained in this chapter. When as a consequence of random fluctuations in a system it is necessary to treat the internal coordinates ξ as being stochastic variables, it is found that the spectral density of these fluctuations is related to the response function $\psi(t)$ that gives the mean value $\bar{\xi}(t)$ as a functional of the affinity $\mathscr{A}(t)$. This relation is one of several forms that the fluctuation–dissipation theorem can take, and it offers a method of determining ψ via measurements of correlations of the internal coordinates under stationary conditions.

The theorem started its career with Einstein's work (1905) on Brownian motion and his famous relation between diffusivity and mobility is a special case of it. Another special case was obtained by Nyquist (1928) in his study of Johnson noise in resistors, and somewhat later the theorem was given a generalized form in the work of Callen and his colleagues (Callen and Welton 1951; Callen and Greene 1951, 1952; Callen, Barasch, and Jackson 1952), whose contributions owe much to Onsager's hypothesis (§ 41.2). In fact one version of the theorem, in its general form, is implicit in Onsager's work on the reciprocal relations. Among other contributors, the work of R. Kubo (Kubo 1957; Kubo and Yokota 1957), especially his treatment of adiabatic systems using Hamiltonians, is noteworthy.

67. Fluctuating internal variables

67.1. Constitutive relations

In the previous chapter the theory of internal processes was developed on the supposition that the internal coordinate vector ξ has small but determinate departures from its equilibrium value ξ_{eq}. By (62.2), (62.4), and $(62.3)_2$, namely,

$$\sigma = JX + \frac{\rho}{T}\dot{\xi}A \geqslant 0, \tag{67.1}$$

$$J = LX + MA, \qquad \dot{\xi} = NX + PA, \tag{67.2}$$

and

$$A = -\tilde{b}\,\Delta\zeta - g\,\Delta\xi, \tag{67.3}$$

it follows that these irreversible departures are due to non-zero values for X and $\Delta\zeta$, both of which are generated by causes external to the fluid-particle system (cf. (65.6), (65.10), and (65.11), which show that Δp, ΔT, $\frac{1}{3}\mathring{\pi}$ depend on volume changes and temperature gradients). If both X and $\Delta\zeta$ are zero, $\dot{\xi} = -Pg\,\Delta\xi$ and any non-zero $\Delta\xi$ would decay rapidly to zero and on a deterministic theory remain zero.

However, we shall now admit the possibility that, even with $(X, \Delta\zeta)$ zero, non-zero values of $\Delta\xi$ occur due to small spontaneous internal changes nearer the microscopic level of description than the macroscopic (for example, as in Brownian motion). On the usual ergodic hypothesis (§ 26.2) the interaction may be represented by an ensemble of systems over which at any fixed instant of time $\Delta\xi$ is a random variable. It is convenient to choose origins to make ξ_{eq} zero, then $\xi(t)$ is a stochastic process with a mean value $\bar{\xi}(t)$ that depends on $(X, \Delta\zeta)$, vanishing when it is zero.

The entropy production rate for a particular system is given by (67.1), which may be cast into the form

$$\sigma = JX + \frac{\rho}{T}\dot{\bar{\xi}}\bar{A} + \dot{\jmath}, \tag{67.4}$$

where

$$\bar{A} = -\tilde{b}\,\Delta\zeta - g\bar{\xi} \tag{67.5}$$

and

$$\jmath \equiv -\frac{\rho}{2T}(\xi - \bar{\xi})(\xi - \bar{\xi}). \tag{67.6}$$

The entropy fluctuations represented by $\jmath$ are now assumed to be rapid, small changes, essentially reversible in character and not governed by macroscopic phenomenological equations like (67.2). This means that $\dot{\jmath}$ will be both positive and negative, so that for the ensemble average we may write

$$\dot{\bar{\jmath}} \approx 0. \tag{67.7}$$

If $(X, \Delta\zeta)$ has been constant for a long period, the fluctuations will have the equilibrium distribution described in § 43.1, in which case, as noted after (43.5), the approximation (67.7) will hold exactly. In fact adapting the result established in § 28.3 for the variance of thermodynamically conjugate variables to the present problem, we find for equilibrium or 'stationary' fluctuations that

$$\int_V \overline{(\xi - \bar{\xi})(\xi - \bar{\xi})}\rho\,\mathrm{d}\tau = kTg^{-1}, \tag{67.8}$$

the integral being over the volume of the whole system. Hence $\dot{\bar{\jmath}}$ is zero in this case.

If $(X, \Delta\zeta)$ changes, then it will do so on a macroscopic time scale $\mathscr{T}$, and the fluctuations will occur about a mean $\bar{\xi}$ that changes similarly, and provided the time scale τ_F for the fluctuations is much smaller than $\mathscr{T}$, it is reasonable to suppose that they remain normally distributed about $\bar{\xi}$, with a variance whose rate of change is negligible.† Then the ensemble average of (67.4) may be written

$$\bar{\sigma} = JX + \frac{\rho}{T}\dot{\bar{\xi}}\bar{A} \geqslant 0. \tag{67.9}$$

It follows that the constitutive equations in (67.2) must be replaced by

$$J = LX + M\bar{A}, \qquad \dot{\bar{\xi}} = NX + P\bar{A}. \tag{67.10}$$

67.2. The rate of entropy production

It is convenient to represent the external thermodynamic 'force' $(X, \Delta\zeta)$ by the affinity

$$\mathscr{A} \equiv P^{-1}NX - \tilde{b}\,\Delta\zeta, \tag{67.11}$$

which is related to $\bar{\xi}$ by an expression of the form (cf. (62.12))

$$\bar{\xi}(t) = \int_{-\infty}^{t} \psi(t-t').\mathscr{A}(t')\,\mathrm{d}t', \tag{67.12}$$

where the response function satisfies the reciprocal relation

$$\eta_{\mathrm{i}}\psi = \psi^{*}\eta_{\mathrm{i}}. \tag{67.13}$$

By (67.5) and $(67.10)_2$

$$\dot{\bar{\xi}} = P(\mathscr{A} - g\bar{\xi}), \tag{67.14}$$

whence

$$\psi(t) = \mathrm{e}^{-tPg}P, \tag{67.15}$$

but we shall regard this form as simply a particular case of the general response function appearing in (67.12).

Under steady-state conditions (67.12) and (67.14) yield (see (64.8))

$$\bar{\xi} = \bar{\xi}_0 = \hat{\psi}(0).\mathscr{A}_1 = g^{-1}\mathscr{A}_1 \qquad (\dot{\bar{\xi}} = 0),$$

so

$$\hat{\psi}(0) = g^{-1} = \tilde{\hat{\psi}}(0), \tag{67.16}$$

† It is of interest to remark that, if the time scale for the very slow reversible volume changes mentioned in § 3.3 is τ, then

$$\underset{\text{reversible}}{\tau_F} \ll \underset{\text{irreversible}}{\mathscr{T}} \ll \underset{\text{reversible}}{\tau},$$

showing the close connection between time scales and the concept of irreversibility (see also the discussion in § 20).

as g is a symmetric matrix. This result also follows immediately from the transform of (67.15), but the method just adopted has the virtue of requiring no assumption about the functional form of $\psi(t)$.

Now use $(67.10)_2$ to eliminate $\bar{A}$ from $(67.10)_1$ and (67.9),

$$J = J_0 + MP^{-1}\dot{\bar{\xi}} \qquad (J_0 = (L - MP^{-1}N)X)$$

and (67.17)

$$\bar{\sigma} = \sigma_0 + \sigma_\xi \qquad (\sigma_0 = J_0 X),$$

with

$$\sigma_\xi \equiv \frac{\rho}{T}\dot{\bar{\xi}}\left(\frac{T}{\rho}\tilde{P}^{-1}\tilde{M}X - \tilde{b}\,\Delta\zeta - g\bar{\xi}\right).$$

This separates $\bar{\sigma}$ into a steady state component σ_0 and a transient component σ_ξ. Provided

$$P^{-1}(\mathbf{B})N(\mathbf{B}) = \eta_i P^{-1}(-\mathbf{B})N(-\mathbf{B})\eta_e \tag{67.18}$$

then by $(62.5)_3$

$$\frac{T}{\rho}\tilde{P}^{-1}(\mathbf{B})\tilde{M}(\mathbf{B}) = \eta_i P^{-1}(-\mathbf{B})N(-\mathbf{B})\eta_e = P^{-1}(\mathbf{B})N(\mathbf{B}),$$

so that

$$\sigma_\xi = \frac{\rho}{T}\dot{\bar{\xi}}(\mathscr{A} - g\bar{\xi}). \tag{67.19}$$

Note from (67.14) that σ_ξ is quadratically small near steady-state conditions.

By integrating (67.12) by parts and using the convolution theorem (64.12) we find

$$\dot{\bar{\xi}}(t) = \int_{-\infty}^{t} \psi(t-t')\dot{\mathscr{A}}(t')\,\mathrm{d}t' = \frac{1}{2\pi \mathrm{i}}\int_{-\infty}^{\infty} \omega\,\mathrm{e}^{-\mathrm{i}\omega t}\hat{\psi}(\omega)\hat{\mathscr{A}}(\omega)\,\mathrm{d}\omega,$$

and following an analysis like that given in § 64.5 we arrive at

$$\mathscr{D}_\xi = \frac{\mathrm{i}\rho}{2\pi T}\int_{-\infty}^{\infty} \omega[\hat{\mathscr{A}}(\omega)\hat{\psi}_c(\omega)\hat{\mathscr{A}}_c(\omega) - \hat{\mathscr{A}}(\omega)\tilde{\hat{\psi}}(\omega)g\hat{\psi}_c(\omega)\hat{\mathscr{A}}_c(\omega)]\,\mathrm{d}\omega$$

for the total transient dissipation per unit volume

$$\mathscr{D}_\xi \equiv \int_{-\infty}^{\infty} \sigma_\xi(t)\,\mathrm{d}t \qquad \geqslant 0.$$

As

$$\hat{\psi}(\omega) = \hat{\psi}_c(-\omega), \qquad \hat{\mathscr{A}}(\omega) = \hat{\mathscr{A}}_c(-\omega), \tag{67.20}$$

$g = \tilde{g}$, and the transform of a scalar leaves it unchanged, $\hat{\mathscr{A}}(\omega)\hat{\psi}(\omega)g\hat{\psi}_c(\omega)\hat{\mathscr{A}}_c(\omega)$ is an even function of ω, so that the formula for $\mathscr{D}_\xi$ reduces to

$$\mathscr{D}_\xi = \frac{\mathrm{i}\rho}{2\pi T}\int_{-\infty}^{\infty} \omega\hat{\mathscr{A}}(\omega)\hat{\psi}_c(\omega)\hat{\mathscr{A}}_c(\omega)\,\mathrm{d}\omega,$$

or

$$\mathscr{D}_\xi = \frac{\mathrm{i}\rho}{4\pi T}\int_{-\infty}^{\infty} \omega\hat{\mathscr{A}}(\omega)[\hat{\psi}_c(\omega)-\tilde{\hat{\psi}}(\omega)]\hat{\mathscr{A}}_c(\omega)\,\mathrm{d}\omega \geqslant 0. \tag{67.21}$$

The form of the integrand here differs from that appearing in eqn $(64.30)_1$ because there (for $\eta = U$ and $\mathbf{B} = 0$) $\hat{\psi}$ is an admittance, whereas here it is really a susceptibility (see footnote on p. 271).

67.3. The spectral density of a stochastic process

To this stage we have merely replaced ξ and A in the earlier theory by their ensemble averages $\bar{\xi}$ and $\bar{A}$. To understand how spontaneous internal fluctuations contribute to the transient dissipation, we shall require the concept of the spectral density of a stochastic process. Here we shall be concerned with the *stationary* stochastic process $\xi(t)$ that results when the driving affinity $\mathscr{A}$ has been zero for a long time; in this case (67.14) yields $\bar{\xi} = 0$.

Let $f_m(\xi_1, \xi_2, \ldots \xi_m;\ t_1, t_2, \ldots t_m)\,\mathrm{d}\xi_1\,\mathrm{d}\xi_2 \ldots \mathrm{d}\xi_m$, $m = 1, 2, \ldots$, be the joint probability of finding ξ in the ranges $(\xi_1, \xi_1+\mathrm{d}\xi_1)$, $(\xi_2, \xi_2+\mathrm{d}\xi_2)$, ... $(\xi_m, \xi_m+ +\mathrm{d}\xi_m)$ at times $t_1, t_2, \ldots t_m$ respectively, then if for any time interval τ and all $m \leqslant k$,

$$f_m(\xi_1, \xi_2, \ldots \xi_m; t_1, t_2, \ldots t_m) = f_m(\xi_1, \xi_2, \ldots \xi_m; t_1+\tau, t_2+\tau, \ldots t_m+\tau),$$

the stochastic process $\xi(t)$ is said to be 'stationary to order k'.

The correlation matrix $\varphi(t_1, t_2)$ is defined by†

$$\varphi(t_1, t_2) \equiv \langle \xi(t_1)\xi(t_2)\rangle = \iint \xi_1\xi_2 f_2(\xi_1, \xi_2; t_1, t_2)\,\mathrm{d}\xi_1\,\mathrm{d}\xi_2,$$

the integration being over the complete range. If the process is stationary to second order, then $\varphi(t_1, t_2)$ depends only on the difference (t_2-t_1), and we may write

$$\varphi(\tau) = \langle \xi(t)\xi(t+\tau)\rangle = \langle \xi(t-\tau)\xi(t)\rangle = \tilde{\varphi}(-\tau). \tag{67.22}$$

Let η_α, η_β be the parities of the coordinates ξ_α, ξ_β, then (see § 43.2) $\varphi_{\alpha\beta}(\tau; \mathbf{B}) = \eta_\alpha\eta_\beta\varphi_{\beta\alpha}(\tau; -\mathbf{B})$, that is, in terms of the diagonal parity matrix η_i for the internal coordinates,

$$\eta_\mathrm{i}\varphi = \varphi^*\eta_\mathrm{i}. \tag{67.23}$$

† It is convenient here to adopt the conventional notation $\langle\ \ \rangle$ for ensemble averages.

The Fourier transform $\hat{\varphi}(\omega)$ of the correlation function is called the *spectral density* $S(\omega)$,

$$S(\omega) \equiv \hat{\varphi}(\omega) = \int_{-\infty}^{\infty} e^{i\omega\tau}\varphi(\tau)\,d\tau. \tag{67.24}$$

It follows from (67.22), (67.23), and (67.24) that

$$S(\omega) = S_c(-\omega) = \tilde{S}_c(\omega); \qquad \eta_i S = S^*\eta_i, \tag{67.25}$$

where c denotes the complex conjugate and * the operations of changing the sign of **B** and transposing.

Provided the ensemble may be assumed to be ergodic (§ 26.2), the functions φ and S may be estimated from a single history. We approximate $\xi(t)$ by the cut-off process $\xi(t;t_0)$, which equals $\xi(t)$ for $|t| < t_0$ and is zero for $t_0 < |t|$. Then

$$\varphi(\tau) = \lim_{t_0\to\infty} \frac{1}{2t_0}\int_{-t_0}^{t_0} \xi(t;t_0)\xi(t+\tau;t_0)\,dt$$

and

$$S(\omega) = \lim_{t_0\to\infty} \frac{1}{2t_0}\left|\int_{-t_0}^{t_0} e^{i\omega t}\xi(t;t_0)\,dt\right|^2 = \lim_{t_0\to\infty} \frac{1}{2t_0}\hat{\xi}(\omega;t_0)\hat{\xi}_c(\omega;t_0),$$

the second of which follows from the Fourier transform of the first and the substitution $t+\tau = t'$.

A relationship exists between the value of the correlation at $\tau = 0$ and that of the admittance at $\omega = 0$ that will be required shortly. This relation is

$$\int_V \varphi(0)\rho\,d\tau = kTg^{-1} = kT\hat{\psi}(0), \tag{67.26}$$

which follows from (67.8) in which $\bar{\xi}$ is zero, and (67.16).

67.4. The correlation function and conditional expectations

The correlation function $\varphi(\tau)$ may be written in the form

$$\varphi(\tau) = \int \xi' E\{\xi, \tau|\xi'\} f_1(\xi')\,d\xi', \tag{67.27}$$

where $f_1(\xi')$ is the stationary probability density, and $E\{\xi, \tau|\xi'\}$ is the mean or expectation value of ξ at time t given that its value was precisely ξ' at time $t-\tau$. To establish (67.27) we need the general notion of conditional expectations.

Let ξ have the precise value ξ' at a time t, then owing to the fluctuations in the system, at a later time $(t+\tau)$ it will be a random variable with an expectation value $E\{\xi(t+\tau)|\xi'(t)\}$. In the absence of a driving 'force' $\mathscr{A}$ this conditional expectation will be stationary, and can be adequately denoted by $E\{\xi, \tau|\xi'\}$. Let $f_1(\xi), f_2(\xi', \xi; \tau)$ be the corresponding stationary probability densities, then

$$E\{\xi, \tau|\xi'\} = \int \xi[f_2(\xi', \xi; \tau)/f_1(\xi')]\,\mathrm{d}\xi,$$

$[f_2(\xi', \xi; \tau)/f_1(\xi')]\,\mathrm{d}\xi$ being the probability that at a time τ after a measurement that gave ξ', we find ξ in $(\xi, \xi+\mathrm{d}\xi)$. Thus the correlation function

$$\varphi(\tau) = \iint \xi'\xi f_2(\xi', \xi; \tau)\,\mathrm{d}\xi'\,\mathrm{d}\xi$$

may be written in the form given in (67.27).

Mathematical note. Let the random variables x and z be jointly normal with zero means, then they are independent if they are uncorrelated, that is, if $E\{xz\} = 0$, in which case $E\{z|x\} = E\{z\}$. Now suppose x and y are jointly normal with zero means then, defining a constant a by $a \equiv E\{xy\}/E\{x^2\}$, we have $E\{(y-ax)x\} = 0$, whence $E\{y-ax|x\} = E\{y-ax\} = E\{y\}-aE\{x\} = 0$. As $E\{ax|x\} = ax$, it follows that

$$E\{y|x\} = ax \qquad (a = E\{xy\}/E\{x^2\}). \tag{67.28}$$

Now the best mean square linear estimate of y in terms of x is obtained by minimizing $E\{(y-ax\}^2) = E\{y^2\}-2aE\{xy\}+a^2E\{x^2\}$ with respect to a, which yields the value of a given in (67.28). It can also be shown (see, for example, Papoulis (1965)) that, whether x and y are jointly normal or not, $E\{y|x\}$ is the best non-linear mean square estimate of y in terms of x. The important conclusion is that for normal distributions (67.28) provides the best mean square estimate (linear or non-linear) of y in terms of x.

Applying (67.28) to the stationary expectation introduced above we have

$$E\{\xi, \tau|\xi'\} = l(\tau)\xi' \qquad (\tau > 0), \tag{67.29}$$

where $l(\tau)$ is a matrix coefficient whose value may be found by substituting (67.29) into (67.27),

$$\varphi(t) = \int \xi' l(t)\xi' f_1(\xi')\,\mathrm{d}\xi' = \varphi(0)\tilde{l}(t).$$

By (67.26) the volume integral of this relation is

$$\int_V \varphi(t)\rho\,\mathrm{d}\tau = kT\hat{\psi}(0)\tilde{l}(t) \qquad (t > 0). \tag{67.30}$$

68. The fluctuation-dissipation theorem

68.1. Spontaneous and driven fluctuations

In the last two sections our concern was with spontaneous fluctuations in the absence of a driving force $\mathscr{A}$. This meant that $\bar{\xi}$ was zero and that there was no dissipation in the system. Our present task is to escape from these constraints and to obtain results for dissipative processes; the method we shall adopt is similar to that used by Onsager in his derivation of his reciprocal relations (§ 43). Indeed we can already see a connection between our present study and that of § 43 through the variable $\overline{\mathbf{X}}'(t+\tau)$ appearing in (43.6), which is just the conditional expectation $E\{\mathbf{X}(t+\tau)|\mathbf{X}'(t)\}$.

Consider the situation that arises when $\mathscr{A}$ is given by

$$\mathscr{A} = \begin{cases} \mathscr{A}_1 & (-\infty < t < 0) \\ 0 & (0 \leqslant t < \infty), \end{cases}$$

where $\mathscr{A}_1$ is a constant. This is the 'switched-off' affinity described in § 64.4. For $t < 0$, $\xi(t)$ will be a stationary stochastic process, possessing a mean value

$$\bar{\xi}_0 = \hat{\psi}(0)\mathscr{A}_1, \tag{68.1}$$

as shown in § 67.2. Let $f_1^0(\xi)$ be the corresponding distribution density. At $t = 0$ the affinity is suddenly removed, after which $\xi(t)$ will be stationary no longer, and its mean value $\bar{\xi}(t)$ will regress steadily towards the equilibrium value $\bar{\xi} = 0$ appropriate to zero $\mathscr{A}$.

Suppose that at the instant $t = 0$, that is, the moment at which $\mathscr{A}_1$ is removed, ξ has the value ξ', then as the state of the system is determined entirely by $(\Delta\zeta, \xi') = (0, \xi')$, irrespective of whether $\mathscr{A}_1$ was present prior to $t = 0$ or not, the decay towards $\bar{\xi} = \xi_{\text{eq}} = 0$ will be just as if the state $(0, \xi')$ was achieved spontaneously in a fluctuation. Therefore the mean $\bar{\xi}(t)$ can be calculated by averaging the conditional expectation $E\{\xi, t|\xi'\}$ over the initial ($t = 0$) distribution for ξ',

$$\bar{\xi}(t) = \int f_1^0(\xi')E\{\xi, t|\xi'\}\, \mathrm{d}\xi' \qquad (t > 0).$$

Hence by (67.29) and (68.1) $\bar{\xi}(t) = l(t)\bar{\xi}_0 = l(t)\hat{\psi}(0)\mathscr{A}_1$.

It follows from the theory of § 64.4 that $\bar{\xi}(t)$ is also given by

$$\bar{\xi}(t) = \int_{-\infty}^{0} \psi(t-t')\,\mathrm{d}t'\mathscr{A}_1 = \Phi(t)\mathscr{A}_1,$$

where $\Phi(t)$ is the relaxation function for $\psi(t)$, that is, $\dot{\Phi}(t) = -\psi(t)$. Thus

$$\Phi(t) = l(t)\hat{\psi}(0) = l(t)\Phi(0),$$

and therefore (67.30) may be written

$$\int_V \tilde{\varphi}(t)\rho \, d\tau = kT\Phi(t),$$

or $$(t > 0) \qquad (68.2)$$

$$\int_V \dot{\tilde{\varphi}}(t)\rho \, d\tau = -kT\psi(t).$$

This connection between the spontaneous fluctuations represented by $\varphi(t)$ and the driven fluctuations, whose average regression is described by $\Phi(t)$, is one form of the *fluctuation–dissipation* theorem.

By stationarity $\dot{\varphi}(t-t') = \langle\xi(t')\dot{\xi}(t)\rangle = -\langle\dot{\xi}(t')\xi(t)\rangle$, so that by $(68.2)_2$ we can write (67.12) as

$$\bar{\xi}(t) = \beta\int_{-\infty}^{t}\int_V \rho\langle\xi(t)\dot{\xi}(t')\rangle\mathscr{A}(t')\, d\tau\, dt' \qquad (\beta = 1/kT), \qquad (68.3)$$

and for its rate of change

$$\dot{\bar{\xi}}(t) = \beta\int_{-\infty}^{t}\int_V \rho\langle\dot{\xi}(t)\dot{\xi}(t')\rangle\mathscr{A}(t')\, d\tau\, dt' + \beta\int_V \rho\langle\xi(t)\dot{\xi}(t)\rangle\mathscr{A}(t)\, d\tau. \qquad (68.4)$$

Another form of the theorem involves the spectral density (67.24),

$$S(\omega) = \int_0^{\infty} e^{i\omega t}\varphi(t)\, dt + \int_{-\infty}^{0} e^{i\omega t}\varphi(t)\, dt = \int_0^{\infty}[e^{i\omega t}\varphi(t) + e^{-i\omega t}\tilde{\varphi}(t)]\, dt,$$

the second form of which is a consequence of (67.22). Hence by $(68.2)_1$, (64.23), (64.28), and (67.16)

$$\int_V S(\omega)\rho\, d\tau = \frac{kT}{i\omega}\{\hat{\tilde{\psi}}(\omega) + \hat{\psi}(-\omega)\} = \frac{kT}{i\omega}\{\hat{\tilde{\psi}}(\omega) - \hat{\psi}_c(\omega)\}$$

$$= \frac{2kT}{\omega}\{[\hat{\psi}''(\omega)]^s + i[\hat{\psi}'(\omega)]^a\}. \qquad (68.5)$$

Thus knowledge of the admittance enables us to find the spectral density and thence the correlation matrix.

The total entropy dissipated per unit volume is given by (67.21), which may now be written

$$\mathscr{D}_\xi = \frac{\rho}{4\pi kT^2}\int_V \rho\int_{-\infty}^{\infty}\omega^2\hat{\mathscr{A}}(\omega)S(\omega)\hat{\mathscr{A}}_c(\omega)\, d\omega\, d\tau \geqslant 0, \qquad (68.6)$$

and the total energy dissipated is $T\mathscr{D}_\xi$. Oscillatory driving affinities are of some importance,

$$\mathscr{A}(t) = \tfrac{1}{2}\mathscr{A}_0 \exp(-i\omega_0 t) + \tfrac{1}{2}\mathscr{A}_{0c}\exp(i\omega_0 t),$$

$$\hat{\mathscr{A}}(\omega) = \pi\mathscr{A}_0\,\delta(\omega-\omega_0) + \pi\mathscr{A}_{0c}\,\delta(\omega+\omega_0),$$

but from (68.6) this choice gives $T\mathscr{D}_\xi = \infty$. We shall find instead the finite energy dissipated during one period $2\pi/\omega_0$. Returning to eqn (67.19) and transforming only one of the affinities occurring in the bilinear form for $\mathscr{D}_\xi$, we find after some algebra

$$T\mathscr{D}_\xi(\omega_0) = \frac{\pi\rho\omega_0}{2kT}\mathscr{A}_0 \int_V S(\omega_0)\rho \, d\tau \, \mathscr{A}_{0c} \tag{68.7}$$

for the energy dissipated per cycle. This result suggests an experimental method of determining the spectral density matrix—we measure the energy dissipated in the system for affinities of given amplitude and frequency ω_0, then vary ω_0 to cover the whole spectrum, or at least the range for which $T\mathscr{D}_\xi(\omega_0)$ is significant.

68.2. Linear regression

To illustrate the various functions arising in the theory, we shall take the linear regression law

$$\dot{\bar{\xi}} + Pg\bar{\xi} = 0, \tag{68.8}$$

which follows from (67.14) at $\mathscr{A} = 0$. For this case

$$\bar{\xi}(\tau) = e^{-\tau Pg}\bar{\xi}(0), \quad \text{or} \quad E\{\xi, \tau|\xi'\} = e^{-\tau Pg}\xi' \qquad (\tau \geqslant 0), \tag{68.9}$$

so by (67.22), (67.27), and the symmetry of $\varphi(0)$,

$$\varphi(t) = \varphi(0)\begin{cases} e^{-|t|Pg} & (t \leqslant 0) \\ e^{-|t|g\tilde{P}} & (t \geqslant 0). \end{cases} \tag{68.10}$$

The spectral density is

$$S(\omega) = \varphi(0)\int_0^\infty [e^{(i\omega U - g\tilde{P})t} + e^{-(i\omega U + Pg)t}] \, dt,$$

that is, (68.11)

$$S(\omega) = \varphi(0)[(Pg + i\omega U)^{-1} + (g\tilde{P} - i\omega U)^{-1}],$$

and from $(68.4)_1$ and (68.10) the relaxation function is

$$\Phi(t) = e^{-tPg}g^{-1} \qquad (t > 0).$$

The admittance now follows from (64.23),

$$\hat{\psi}(\omega) = g^{-1} + i\omega(Pg - i\omega U)^{-1}g^{-1} = (Pg - i\omega U)^{-1}P, \tag{68.12}$$

and finally, in agreement with (67.15), we arrive at the response function

$$\psi(t) = e^{-tPg}P. \tag{68.13}$$

In the simplest case of a single internal degree of freedom Pg is a scalar, when (67.26), (68.10), and (68.11) give

$$\varphi(t) = \frac{kT}{Mg}\mathrm{e}^{-Pg|t|}, \qquad S(\omega) = \frac{kT}{M}\frac{2P}{(Pg)^2+\omega^2}, \tag{68.14}$$

where M is the total mass of the system and spatial uniformity has been assumed.

68.3. Spectral density of thermodynamic noise

Eqn $(67.2)_2$, which may be written $\dot{\xi}+Pg\xi = P\mathscr{A}$, has been replaced by the average regression law given in (67.14). Let us introduce a stochastic force $F(t)$ with zero mean and defined by

$$\dot{\xi}(t)+Pg\xi(t) = P\mathscr{A}(t)+F(t)$$

or (68.15)

$$\dot{\eta}(t)+Pg\eta(t) = F(t) \qquad (\eta \equiv \xi-\bar{\xi}),$$

and ask the question: What can be said about the statistics of $F(t)$?

To answer this we need the relationship between the spectral density $S_\eta(\omega)$ of the 'output' η and the spectral density $S_F(\omega)$ of the 'input' F, namely (see *Mathematical note* on the next page),

$$S_\eta(\omega) = H_{\mathrm{c}}(\omega)S_F(\omega)\tilde{H}(\omega), \tag{68.16}$$

where $H(\omega)$ is the Fourier transform of the response to a delta function (impulse) input ($H(-\mathrm{i}\omega)$ is called the transfer function of the system). From $(68.15)_2$ and (68.13)

$$\eta(t) = \int_{-\infty}^{t} \exp[-Pg(t-t')]F(t')\,\mathrm{d}t' = \int_{-\infty}^{\infty} \psi(t-t')P^{-1}F(t')\,\mathrm{d}t',$$

and setting $F(t') = \delta(t')$ we obtain the impulse response $\psi(t)P^{-1}$, whence $H(\omega) = \hat{\psi}(\omega)P^{-1}$. Thus (68.16) may be inverted to give

$$S_F = P\hat{\psi}_{\mathrm{c}}^{-1}S_\eta\tilde{\hat{\psi}}^{-1}\tilde{P}. \tag{68.17}$$

Eqn (68.5) and the assumption of spatial uniformity give

$$S_\eta = \frac{\mathrm{i}kT}{\omega M}(\hat{\psi}_{\mathrm{c}}-\tilde{\hat{\psi}}),$$

so (68.12) and (68.17) lead to the spectral density S_F and (by inversion) the correlation function φ_F for the stochastic process $F(t)$. We obtain

$$\varphi_F(t) = S_F\,\delta(t), \qquad S_F = \frac{kT}{M}(P+\tilde{P}). \tag{68.18}$$

Such a process, in which S_F is independent of ω and φ_F is a delta function, is called 'white noise'.

Mathematical note. If

$$y(t_2) = \int_{-\infty}^{t_2} h(\tau)x(t_2-\tau)\,\mathrm{d}\tau \equiv h(t_2)^* x(t_2),$$

where the definition of the convolution operator * is provided by the integral, then

$$\langle x(t_1)y(t_2)\rangle = \int_{-\infty}^{\infty} \langle x(t_1)x(t_2-\tau)\rangle \tilde{h}(\tau)\,\mathrm{d}\tau,$$

that is,

$$\varphi_{xy}(t_1, t_2) = \varphi_{xx}(t_1, t_2)^*\tilde{h}(t_2).$$

Similarly

$$\varphi_{yy}(t_1, t_2) = h(t_1)^*\varphi_{xx}(t_1, t_2) \qquad (68.19)$$

so

$$\varphi_{yy}(t_1, t_2) = h(t_1)^*\varphi_{xx}(t_1, t_2)^*\tilde{h}(t_2).$$

When the process is stationary, the Fourier transforms of the first and second of (68.19) are

$$S_{xy}(\omega) = S_{xx}(\omega)\tilde{H}(\omega), \qquad S_{yy}(\omega) = H_c(\omega)S_{xy}(\omega),$$

where $H(\omega)$ is the transform of $h(t)$ and $H_c(\omega)$ is its complex conjugate. Hence

$$S_{yy}(\omega) = H_c(\omega)S_{xx}(\omega)\tilde{H}(\omega). \qquad (68.20)$$

In the case of a single internal variable, P and g are scalars, when (68.15), (68.18), and (68.14) may be written

$$m\dot{\eta}(t) + \beta\eta(t) = mF(t) \qquad (m \equiv Mg, \beta \equiv Pgm), \qquad (68.21)$$

$$\varphi_F(t) = S_F\,\delta(t), \qquad S_F = \frac{2kT\beta}{m^2}$$

and (68.22)

$$\varphi_\eta(t) = \frac{kT}{m}\mathrm{e}^{-\beta|t|/m}, \qquad S_\eta = \frac{2kT\beta}{\beta^2 + m^2\omega^2}.$$

If $\eta(t)$ is assumed to be a normal process, as its variance is $\varphi_\eta(0) = kT/m$, its one-dimensional probability density is (see (29.1))

$$f_1(\eta) = \left(\frac{m}{2\pi kT}\right)^{\frac{1}{2}} \exp(-m\eta^2/2kT). \qquad (68.23)$$

Notice that (68.21) is in the form of an equation of motion, with η, m, and β to be interpreted as a velocity, a mass, and a friction coefficient respectively.

By (67.6) and (67.7) the reversible fluctuations in internal energy occur at a rate

$$T\int_V \dot{\jmath}\,\mathrm{d}\tau = -\frac{\mathrm{d}}{\mathrm{d}t}(\tfrac{1}{2}m\eta^2)$$

and have zero ensemble average.

69. Special cases of the theorem

69.1. Brownian motion

Direct evidence of the existence of molecular motion is provided by the ceaseless small movements of very small particles suspended in a liquid, a phenomenon known as Brownian motion. If the thermodynamic system is taken to be a small element of fluid containing at least one Brownian particle $\mathscr{P}$ of mass m, then we may adopt the velocity $\mathbf{v}(t)$ of $\mathscr{P}$ relative to the fluid as an internal variable that satisfies an equation of motion

$$m\dot{\mathbf{v}}(t)+\beta\mathbf{v}(t) = m\mathbf{F}(t), \tag{69.1}$$

where β is due to viscous drag and $\mathbf{F}(t)$ is random acceleration caused by molecular impacts on $\mathscr{P}$. We shall assume $\mathbf{F}(t)$ to be thermodynamic noise, normally distributed and with zero mean.

The internal energy contains a term $\frac{1}{2}mv^2$,

$$\mathrm{d}U = T\,\mathrm{d}S+\ldots+\mathrm{d}(\tfrac{1}{2}mv^2), \quad \text{or} \quad T\,\mathrm{d}S = \mathrm{d}U+\ldots-m\mathbf{v}\,.\,\mathrm{d}\mathbf{v},$$

whence $\mathbf{v}$, qua internal coordinate, has the affinity $-m\mathbf{v}$, that is, g in the relation $A = -g\xi$ is m. The ensemble average of (69.1) is the constitutive relation $\dot{\bar{\mathbf{v}}} = P(-m\bar{\mathbf{v}}) = -\beta\bar{\mathbf{v}}/m$. It is clear that we can take over to each scalar component of $\mathbf{v}(t)$ the results given for $\eta(t)$ at the end of § 68.3.

It is of interest to investigate the statistics of the position $\mathbf{r}(t)$ of a Brownian particle, each component of which by (69.1) satisfies an equation like

$$m\ddot{x}(t)+\beta\dot{x}(t) = m\left[\int_0^t F(t)\,\mathrm{d}t+v_0(t)\right], \tag{69.2}$$

where we have assumed the particle to have a velocity component v_0 along $0x$ at $\mathbf{r} = 0, t = 0$. We shall be interested in $x(t)$ only after long times $t \gg m/\beta$, when both the initial velocity v_0 and the acceleration $m\ddot{x}$ in (69.2) may be neglected. Thus

$$\beta x(t) = m\int_0^t F(t')\,\mathrm{d}t' = m\int_{-\infty}^t V(t')F(t')\,\mathrm{d}t', \tag{69.3}$$

where $V(t)$ is the unit function. Clearly $x(t)$ has zero mean. We may find its variance from $(68.19)_3$, $(68.22)_1$, and (69.3), which give

$$\varphi_x(t_1, t_2) = \int_{-\infty}^{t_1} \left[\int_{-\infty}^{t_2} h(\alpha)\varphi_F(t_1-\alpha, t_2-\beta)h(\beta)\,\mathrm{d}\beta \right] \mathrm{d}\alpha,$$

$$\varphi_F(t_1, t_2) = S_F\,\delta(t_2-t_1), \qquad h(t) = \frac{m}{\beta}V(t).$$

Thus

$$\varphi_x(t_1, t_2) = \frac{m^2}{\beta^2} S_F \begin{cases} t_2 & (t_1 > t_2 > 0) \\ t_1 & (t_2 > t_1 > 0), \end{cases}$$

from which it follows that the variance $\langle x^2 \rangle = \varphi_x(t, t)$ is given by

$$\langle x^2 \rangle = 2Dt \qquad \left(D = \frac{m^2}{2\beta^2} S_F = \frac{kT}{\beta} \right). \tag{69.4}$$

This result for the 'diffusivity' D is due to Einstein; it is often expressed as

$$D = kT \times \text{mobility},$$

the 'mobility' being the average steady-state drift velocity per unit force, which by (69.1) equals β^{-1}.

If $x(t)$ is assumed to be a normal process, the probability density is

$$f_1(x, t) = (4\pi Dt)^{-\frac{1}{2}} \exp(-x^2/4Dt),$$

with similar expressions for $f_1(y, t)$ and $f_1(z, t)$. Thus N Brownian particles initially ($t = 0$) at the origin ($\mathbf{r} = 0$) result in a number density

$$n(\mathbf{r}, t) = N(4\pi Dt)^{-\frac{3}{2}} \exp(-r^2/4Dt), \tag{69.5}$$

from which formula it follows that $n(\mathbf{r}, t)$ satisfies the diffusion equation

$$D\nabla^2 n = \partial n/\partial t. \tag{69.6}$$

Conservation of particle numbers requires the particle current density $\mathbf{J}_n$ to satisfy

$$\partial n/\partial t + \nabla \cdot \mathbf{J}_n = 0,$$

so (69.7)

$$\mathbf{J}_n = -D\nabla n,$$

which is known as *Fick's law* of diffusion.

69.2. The Nyquist theorem

Next we consider a situation rather similar to Brownian motion, namely, the random motion of electrons in a resistor across which the average voltage

$\overline{V}$ is zero. The electrons produce fluctuations in V, known as 'Johnson noise', and if these come from a frequency range Δf Hz, the Nyquist theorem states that the variance $\overline{V^2}$ in the voltage is given by

$$\varphi_V(0) \equiv \overline{V^2} = 4kTR\,\Delta f \quad (f \text{ positive only}). \tag{69.8}$$

An equivalent statement of the theorem is that the spectral density S_V of this voltage is given by

$$S_V = 2kTR, \tag{69.9}$$

the connection between (69.8) and (69.9) following from

$$\varphi_V(t) = \frac{1}{2\pi}\int_{-\infty}^{\infty} \mathrm{e}^{-\mathrm{i}\omega t} S_V(\omega)\,\mathrm{d}\omega$$

with $t = 0$ and $\omega = 2\pi f$.

Nyquist's method of establishing this result involved counting the standing electromagnetic waves on an open-circuited, loss-free transmission line and associating an energy kT with each of these degrees of freedom ($\frac{1}{2}kT$ for each of the potential and kinetic energies). We shall give a different derivation based on the equation of motion for the electrons, namely,

$$m\dot{w}+\beta w = e\,\partial V/\partial x \qquad (\beta = e^2 n\eta = m/\tau_c), \tag{69.10}$$

where m is the electron mass, w is the average or drift velocity of the electrons in the direction $0x$ along the resistor, and η is the resistivity. This expression follows from eqns $(56.27)_2$ and (56.28) with some obvious simplifications. The average w is given by

$$w = \frac{1}{N}\sum_{i=1}^{N} w_i \qquad (N = ALn),$$

where N is the total number of electrons in the resistor of length L and cross-sectional area A, and w_i is the velocity of a particular electron. We now assume spatial uniformity and write

$$m\dot{w}_i+\beta w_i = mF_i \qquad \left(F_i \equiv \frac{eN}{Lm}V_i\right), \tag{69.11}$$

where V_i is the voltage due to the ith electron. This equation describes the 'Brownian motion' of a single electron.

An application of eqn $(68.22)_2$ gives $S_{F_i} = 2kT\beta/m^2$, and we require the spectral density of the voltage due to all of the electrons,

$$V = \frac{Lm}{eN}\sum_{i=1}^{N} F_i.$$

The electrons are assumed to move independently so that the cross-correlations $\varphi_{F_iF_j}$, $i \neq j$, are zero, and

$$S_V = \left(\frac{Lm}{eN}\right)^2 \sum_{i=1}^{N} S_{F_i} = \left(\frac{Lm}{eN}\right)^2 N \frac{2kT\beta}{m^2} = 2kT\frac{L\eta}{A}$$

and as $L\eta/A$ is the resistance R, eqn (69.9) is established.

Finally we remark that the electron diffusivity D and the conductivity σ are connected by the Einstein relation

$$\sigma = e^2nD/kT, \tag{69.12}$$

which follows from (69.4) and (69.10).

70. The fluctuation-dissipation theorem for adiabatic systems

70.1. Mechanical disturbances

Unlike thermal disturbances of a system, which are essentially statistical in nature, mechanical disturbances, being adiabatic, can be explicitly defined in terms of an interaction Hamiltonian, say $H_{\text{ext}} = -AF(t)$, where $F(t)$ is the force acting on the system and $A = A(\mathbf{p}, \mathbf{q})$ is the dynamic quantity thermodynamically conjugate to F. The total Hamiltonian Γ-space (§ 30) is expressed thus,

$$H(\mathbf{p}, \mathbf{q}, t) = H_0(\mathbf{p}, \mathbf{q}) - A(\mathbf{p}, \mathbf{q})F(t), \tag{70.1}$$

H_0 being the equilibrium Hamiltonian. It will be assumed that the perturbing force $F(t)$ is small enough to satisfy $H_{\text{ext}} \ll H_0$.

By (30.1) the total rate of change of a function $f(\mathbf{p}, \mathbf{q}, t)$ is

$$\frac{\mathrm{d}f}{\mathrm{d}t} = \frac{\partial f}{\partial t} + \dot{\mathbf{p}}\cdot\frac{\partial f}{\partial \mathbf{p}} + \dot{\mathbf{q}}\cdot\frac{\partial f}{\partial \mathbf{q}} = \frac{\partial f}{\partial t} + \frac{\partial H}{\partial \mathbf{p}}\cdot\frac{\partial f}{\partial \mathbf{q}} - \frac{\partial H}{\partial \mathbf{q}}\cdot\frac{\partial f}{\partial \mathbf{p}},$$

or

$$\frac{\mathrm{d}f}{\mathrm{d}t} = \frac{\partial f}{\mathrm{d}t} + \mathscr{H}f, \tag{70.2}$$

where $\mathscr{H}$ is the Hamiltonian operator defined via the Poisson bracket

$$\mathscr{H}f \equiv [f, H] = \frac{\partial f}{\partial \mathbf{q}}\cdot\frac{\partial H}{\partial \mathbf{p}} - \frac{\partial f}{\partial \mathbf{p}}\cdot\frac{\partial H}{\partial \mathbf{q}}.$$

Let $\xi(\mathbf{p}, \mathbf{q}, t)\,\mathrm{d}\Gamma$ be the probability that, in the ensemble of systems distributed over Γ-space, a particular system lies in $\mathrm{d}\Gamma$, then by (70.2) and (30.5)

$$\frac{\mathrm{d}\xi}{\mathrm{d}t} = \frac{\mathrm{d}\xi}{\partial t} + \mathscr{H}\xi = 0. \tag{70.3}$$

In the equilibrium state $\xi = \xi_0(\mathbf{p}, \mathbf{q})$, $H = H_0(\mathbf{p}, \mathbf{q})$, for which (70.3) gives

$$\mathscr{H}_0 \xi_0 = [\xi_0, H_0] = 0. \tag{70.4}$$

Dynamic variables like $A(\mathbf{p}, \mathbf{q})$ change along phase point trajectories according to

$$\frac{\mathrm{d}A}{\mathrm{d}t} = \mathscr{H} A = [A, H],$$

which has the formal solution

$$A(t) = \exp(t\mathscr{H})A(0), \tag{70.5}$$

where by $A(t)$ is meant $A(p_t, q_t)$ with (p_t, q_t) a point moving through phase-space, following the dynamical motion.

Now suppose that in the disturbed state

$$\xi(\mathbf{p}, \mathbf{q}, t) = \xi_0(\mathbf{p}, \mathbf{q}) + \Delta\xi(\mathbf{p}, \mathbf{q}, t),$$

where $\Delta\xi$ is small compared with ξ_0, then by (70.1), (70.3), and (70.4) and neglect of a second-order term,

$$\frac{\partial \Delta\xi}{\partial t} + \mathscr{H}_0 \, \Delta\xi = [\xi_0, A]F \qquad (\mathscr{H}_0 \cdot\cdot = [\cdot\cdot, H_0]).$$

We shall assume the system to be in equilibrium at $t = -\infty$, when the force $F(t)$ was switched on, then (cf. (62.12))

$$\Delta\xi(t) = \int_{-\infty}^{t} \exp[(t'-t)\mathscr{H}_0][\xi_0, A]F(t')\,\mathrm{d}t'. \tag{70.6}$$

70.2. The response function

The change $\Delta\langle B(t)\rangle$ in the ensemble average of a dynamic variable $B(\mathbf{p}, \mathbf{q})$ may be calculated from

$$\Delta\langle B(t)\rangle = \int \Delta\xi(t) B(\mathbf{p}, \mathbf{q})\,\mathrm{d}\Gamma = \int_{-\infty}^{t} \psi_{BA}(t-t')F(t')\,\mathrm{d}t', \tag{70.7}$$

where by (70.6) the response function is

$$\psi_{BA}(t-t') = \int \exp[(t'-t)\mathscr{H}_0][\xi_0, A]B\,\mathrm{d}\Gamma. \tag{70.8}$$

As ψ_{BA} appears in a product with the small force F, it is sufficient to calculate its value to first order, which permits us to use

$$A(t) = \exp(t\mathscr{H}_0)A(0) \qquad B(t') = \exp(t'\mathscr{H}_0)B(0),$$

in (70.8). Also, provided dynamic variables A, B, and C vanish for infinite

p and **q**, we may integrate by parts and show that

$$\int [A, B]C \, \mathrm{d}\Gamma = \int [C, A]B \, \mathrm{d}\Gamma = \int [B, C]A \, \mathrm{d}\Gamma.$$

Hence by (70.4)

$$\psi_{BA}(t-t') = \int [\xi_0, \exp(t'\mathscr{H}_0)A] \exp(t\mathscr{H}_0)B \, \mathrm{d}\Gamma = \int [A(t'), B(t)]\xi_0 \, \mathrm{d}\Gamma,$$

that is,

$$\psi_{BA}(t-t') = -\langle [B(t), A(t')] \rangle. \tag{70.9}$$

Another form for the response function follows on noting that with $\xi_0 \propto \exp(-\beta H_0)/Z$ (§ 30.3),

$$[\xi_0, A(t')] = -\beta[H_0, A(t')]\xi_0 = \beta \dot{A}(t')\xi_0,$$

whence

$$\psi_{BA}(t-t') = \beta \int B(t)\dot{A}(t')\xi_0 \, \mathrm{d}\Gamma,$$

that is,

$$\psi_{BA}(t-t') = \beta \langle B(t)\dot{A}(t') \rangle. \tag{70.10}$$

Defining the correlation

$$\varphi_{AB}(t) \equiv \langle A(0)B(t) \rangle$$

and employing stationarity to show that (§ 43.2) $\dot{\varphi}_{AB}(t) = -\dot{\varphi}_{BA}(-t)$, we can rewrite (70.10) as

$$\dot{\varphi}_{AB}(t) = -kT\psi_{AB}(t). \tag{70.11}$$

As here φ refers to the whole system, and not unit mass as in earlier sections, this form of the fluctuation–dissipation theorem agrees with that given in eqn $(68.2)_2$. It is thus a neat proof of the theorem for the special case of adiabatic systems.

Finally we note by (70.10) that the time derivative of (70.7) yields (cf. (68.4))

$$\Delta\langle \dot{B}(t) \rangle = \beta \int_{-\infty}^{t} \langle \dot{B}(t)\dot{A}(t') \rangle F(t') \, \mathrm{d}t', \tag{70.12}$$

provided $\dot{A}(t)$ and $B(t)$ are uncorrelated (otherwise an additional term $\beta\langle B(t)\dot{A}(t)\rangle F(t)$ is required).

For thermal disturbances Kubo adopted Onsager's theory for the regression of fluctuations (§ 43) and arrived at various formulae equivalent to

(68.3). A special case is

$$L_{\alpha\beta} = k^{-1}\langle x_\beta(t)\dot{x}_\alpha(t+\mathrm{d}t)\rangle,$$

for the coefficient $L_{\alpha\beta}$ appearing in (43.6), a result that is readily derived by multiplying (43.6) by $x'_\beta(t)$, averaging over all initial states, and then taking the macroscopic limit $\tau \to \mathrm{d}t \sim 0$.

70.3. Electrical conductivity

The simplest example of an adiabatic disturbance is provided by an electric field $\mathbf{E}(t)$ which generates a current $\mathbf{J}(t)$. The perturbation energy is $H_{\mathrm{ext}} = -\sum_i e_i\mathbf{r}_i \cdot \mathbf{E}(t)$, where e_i is the charge on the ith particle of the system and $\mathbf{r}_i$ is its position vector. The dynamic variable A and its time derivative are the vectors

$$\mathbf{A} = \sum_i e_i\mathbf{r}_i, \qquad \dot{\mathbf{A}} = \mathbf{J},$$

so taking B to be a similar variable, we obtain from (70.12)

$$\langle \mathbf{J}(t)\rangle = \beta \int_{-\infty}^{t} \langle \mathbf{J}(t)\mathbf{J}(t')\rangle \cdot \mathbf{E}(t')\,\mathrm{d}t'. \tag{70.13}$$

Suppose the field is periodic, then $\mathbf{E}(t) = \mathbf{E}_0 \exp(-\mathrm{i}\omega t)$, and it follows from (70.13) that the conductivity tensor for such a field is

$$\boldsymbol{\sigma}(\omega) = \beta \int_0^\infty \langle \mathbf{J}(0)\mathbf{J}(t)\rangle\, \mathrm{e}^{\mathrm{i}\omega t}\,\mathrm{d}t,$$

provided the correlations are over unit volume of the system. For steady-state conditions

$$\boldsymbol{\sigma} = \beta \int_0^\infty \langle \mathbf{J}(0)\mathbf{J}(t)\rangle\,\mathrm{d}t. \tag{70.14}$$

If the carriers are all electrons, then as the motions of the ith and jth electrons are uncorrelated,

$$\langle \mathbf{J}(0)\mathbf{J}(t)\rangle = \left\langle \sum_{i,j} e_i e_j \dot{\mathbf{r}}_i(0)\dot{\mathbf{r}}_j(t)\right\rangle = \left\langle \sum_i e_i^2 \dot{\mathbf{r}}_i(0)\dot{\mathbf{r}}_j(t)\right\rangle = ne^2\langle \mathbf{v}(0)\mathbf{v}(t)\rangle,$$

where $\mathbf{v}$ is the electron drift velocity. Hence (70.14) becomes

$$\boldsymbol{\sigma} = \frac{ne^2}{kT}\int_0^\infty \langle \mathbf{v}(0)\mathbf{v}(t)\rangle\,\mathrm{d}t.$$

A generalization of the Einstein relation (69.12) can be obtained by noting that

$$\begin{aligned}\int_0^\infty \langle \mathbf{v}(0)\mathbf{v}(t)\rangle \, \mathrm{d}t &= \int_0^\infty \langle \mathbf{v}(t')\mathbf{v}(t''-t')\rangle \, \mathrm{d}t'' \\ &= \lim_{t\to\infty} \frac{1}{2t}\int_0^t\int_0^t \langle \mathbf{v}(t')\mathbf{v}(t'')\rangle \, \mathrm{d}t' \, \mathrm{d}t'' \\ &= \lim_{t\to\infty} \frac{1}{2t}\langle(\mathbf{x}(t)-\mathbf{x}(0))(\mathbf{x}(t)-\mathbf{x}(0))\rangle = \mathsf{D},\end{aligned}$$

where D is the diffusion tensor (cf. (69.4). Thus

$$\boldsymbol{\sigma} = \frac{ne^2}{kT}\mathsf{D}.$$

13

KINETIC PROCESSES

THIS final chapter deals with processes that are the time derivatives of distribution functions—'kinetic processes' as we have called them. Such processes can be very rapid, with the distribution function always quite close to its relaxed equilibrium form. The constitutive relations, usually termed 'kinetic equations', specify the irreversible evolution of one-particle distributions in phase-space, and from their solutions we can evaluate fluid properties such as thermal conductivity, viscosity, etc. that on a more elementary theory are simply assigned as phenomenological constants. To obtain the kinetic equations it is necessary to take the details of particle interactions into account; we shall confine attention only to the simplest case in which the particles are represented by featureless spherical force fields. It is possible, but not usual, to derive the kinetic equations from an expression for the rate of entropy production, and in other ways, too, kinetic processes have much in common with the fluid processes of earlier chapters.

The most famous of all kinetic equations was established over a 100 years ago by L. Boltzmann, and extended 70 years later by N. N. Bogoluibov by a method that has the virtue of making clear the nature of the approximations involved. We show that Boltzmann's equation also follows from a constitutive relation that can be deduced from the rate of entropy production in a two-particle description of the system. The existence of several distinct time scales of quite different magnitudes proves to be the key (as indeed it is in all branches of thermodynamics) by providing a small dimensionless parameter to enable asymptotic expansions to be made.

71. Conservation laws

71.1. Liouville's theorem in μ-space

The starting point for our study of kinetic processes will be a conservation law for the velocity distribution function $f_i(\mathbf{r}, \mathbf{w}, t)$ defined in § 21.5, and to find this we shall consider the fate of a cloud of particles moving in 'μ-space', that is, the six-dimensional phase-space that is the union of physical and velocity-space. The first step is to obtain a form of Liouville's theorem (§ 30.2) for this space.

Let $\mathbf{F}$ denote the *macroscopic* force per unit mass acting on the particles, so that the short-range forces occurring between particles during collisions are not included in $\mathbf{F}$. Likewise, by $\dot{\mathbf{w}}$ is meant the acceleration of a bunch of particles due to **F** alone, that is, $\dot{\mathbf{w}} = \mathbf{F}$. Now μ-space has a position vector $(\mathbf{r}, \mathbf{w})$, a 'velocity' vector $(\dot{\mathbf{r}}, \dot{\mathbf{w}}) = (\mathbf{w}, \mathbf{F})$, and a gradient operator $(\partial/\partial\mathbf{r}, \partial/\partial\mathbf{w})$. Thus the time rate of change of some function $\varphi(\mathbf{r}, \mathbf{w}, t)$ in a frame following the motion of a bunch of particles through μ-space may be written

$$\mathrm{D}\varphi \equiv \frac{\partial\varphi}{\partial t}+(\mathbf{w}, \mathbf{F}) \,.\left(\frac{\partial}{\partial\mathbf{r}}, \frac{\partial}{\partial\mathbf{w}}\right)\varphi = \frac{\partial\varphi}{\partial t}+\mathbf{w} \,.\frac{\partial\varphi}{\partial\mathbf{v}}+\mathbf{F} \,.\frac{\partial\varphi}{\partial\mathbf{w}}. \tag{71.1}$$

We shall show that, under certain restrictions, a volume element $\mathrm{d}\mu \equiv \mathrm{d}\mathbf{r}\,\mathrm{d}\mathbf{w}$ of μ-space is convected incompressibly, that is, that

$$\mathrm{D}(\mathrm{d}\mu) = 0, \tag{71.2}$$

which is a special case of the similar result given in § 30.2. Now adapting the result $D(\mathrm{d}\mathbf{r}) = \nabla \,.\, \mathbf{v}\,\mathrm{d}\mathbf{r}$ (given in § 38 for physical space) to μ-space, we have

$$\frac{1}{\mathrm{d}\mu}\mathrm{D}(\mathrm{d}\mu) = \left(\frac{\partial}{\partial\mathbf{r}}, \frac{\partial}{\partial\mathbf{w}}\right) .\, (\mathbf{w}, \mathbf{F}) = 0+\frac{\partial}{\partial\mathbf{w}} \,.\, \mathbf{F},$$

as the independence of **w** and **r** in μ-space means that $(\partial/\partial\mathbf{r}) \,.\, \mathbf{w} = 0$. Now we restrict **F** to be either independent of **w** or to have the Lorentz force dependence $\mathbf{w} \times \mathbf{B}$ (see (11.8)). In either case $(\partial/\partial\mathbf{w}) \,.\, \mathbf{F}$ is zero and (71.2) is established.

71.2. The transport equation for $f_i(\mathbf{r}, \mathbf{w}, t)$

First consider just one component, with velocity distribution function $f(\mathbf{r}, \mathbf{w}, t)$. The total number of particles in the volume element $\mathrm{d}\mu$ is $f\,\mathrm{d}\mu$, and the macroscopic rate of change of this number as $\mathrm{d}\mu$ is convected with the particles through μ-space, namely, $\mathrm{D}(f\,\mathrm{d}\mu) = \mathrm{d}\,\mu\mathrm{D}f$, must equal the rate of change, say $(\partial f/\partial t)_{\mathrm{c}}$, caused by the scattering of the particles in and out of $\mathrm{d}\mu$ by short-range microscopic forces, that is, collision forces. This balance is the content of the transport equation for f,

$$\mathrm{D}f = \frac{\partial f}{\partial t}+\mathbf{w} \,.\frac{\partial f}{\partial\mathbf{r}}+\mathbf{F} \,.\frac{\partial f}{\partial\mathbf{w}} = \left(\frac{\partial f}{\partial t}\right)_{\mathrm{c}}, \tag{71.3}$$

the essential feature of which is the division of the causes of change in f into two distinct types, namely, macroscopic (left-hand side) and microscopic (right-hand side) causes.

It should be remarked that, while the division is easily effected for neutral gases, in which the collision forces are very short-range, if not impulsive, it is rather difficult to achieve in a plasma, where the Coulomb forces of particle interaction have considerable range and no sharp distinction between 'macroscopic' and 'microscopic' properties is possible (Wu 1966, Chapter 5). The separation of the rates of change of f into $\mathrm{D}f$ and $(\partial f/\partial t)_{\mathrm{c}}$ may also be viewed as being that between reversible ($\mathrm{D}f$) and irreversible ($(\partial f/\partial t)_{\mathrm{c}}$) rates of change, for as remarked in § 40.2, particle collisions are irreversible on a macroscopic time scale. Our problem is to find a constitutive relation for $(\partial f/\partial t)_c$.

By applying (71.3) to the ith component of a compound system we find

$$\mathrm{D}_i f_i = \frac{\partial f_i}{\partial t} + \mathbf{w}_i \cdot \frac{\partial f_i}{\partial \mathbf{r}} + \mathbf{F}_i \cdot \frac{\partial f_i}{\partial \mathbf{w}_i} = \left(\frac{\partial f_i}{\partial t}\right)_c = \sum_j C_{ij} \qquad (i, j = 1, 2, ...), \qquad (71.4)$$

in which C_{ij} is the contribution to $(\partial f_i/\partial t)_c$ due to collisions between particles of the ith and jth components. We shall refer to the right-hand side of (71.4) as being the 'collision term', and the various constitutive relations that have been developed for this term will be described later. Meanwhile there are conservation laws for mass, momentum, and energy that must hold for all forms of the collision term.

71.3. The transport equation for a property $\varphi_i(\mathbf{r}, \mathbf{w}, t)$

First we recall the definition given in § 22 for the average value $\bar{\varphi}_i(r, t)$ of a function $\bar{\varphi}_i(\mathbf{r}, \mathbf{w}_i, t)$ associated with the ith component of the system,

$$n_i \bar{\varphi}_i \equiv \int \varphi_i f_i \, \mathrm{d}\mathbf{w}_i, \qquad (71.5)$$

where

$$n_i = \int f_i \, \mathrm{d}\mathbf{w}_i \qquad (71.6)$$

is the number density. In particular we have the average velocity $\mathbf{v}_i = \bar{\mathbf{w}}_i$.

Omitting the subscript i *pro tempore*, we may write for each component

$$\int \varphi \mathbf{F} \cdot \frac{\partial f}{\partial \mathbf{w}} \, \mathrm{d}\mathbf{w} = \int \frac{\partial}{\partial \mathbf{w}} \cdot (\mathbf{F}\varphi f) \, \mathrm{d}\mathbf{w} - \int f \frac{\partial}{\partial \mathbf{w}} \cdot (\mathbf{F}\varphi) \, \mathrm{d}\mathbf{w}.$$

And by the divergence theorem, the first right-hand integral can be transformed into a surface integral over the infinite sphere in velocity-space that vanishes provided that as $|\mathbf{w}| \to \infty$, $|\varphi \mathbf{F} f| \to 0$ rapidly enough. Assuming this to be so, and using the result $(\partial/\partial \mathbf{w}) \cdot \mathbf{F} = 0$ given in § 71.1, we have

$$\int \varphi \mathbf{F} \cdot \frac{\partial f}{\partial \mathbf{w}} \, \mathrm{d}\mathbf{w} = -n \overline{\mathbf{F} \cdot \frac{\partial \varphi}{\partial \mathbf{w}}}.$$

Further

$$\int \varphi \mathbf{w} \cdot \frac{\partial f}{\partial \mathbf{r}} \, \mathrm{d}\mathbf{w} = \frac{\partial}{\partial \mathbf{r}} \cdot (n \overline{\mathbf{w}\varphi}) = \nabla \cdot (n \mathbf{v} \bar{\varphi}) + \nabla \cdot (n \overline{\boldsymbol{c}\varphi}),$$

on replacing $\mathbf{w}$ by $\mathbf{v} + \boldsymbol{c}$ and writing $\partial/\partial \mathbf{r}$ as ∇. Hence if (71.3) is multiplied by φ

and then integrated over velocity space, the outcome is

$$\frac{\partial}{\partial t}(n\bar{\varphi})+\nabla\,.\,(n\mathbf{v}\bar{\varphi}+n\overline{\boldsymbol{c}\varphi})-n\mathbf{F}\,.\,\overline{\frac{\partial\varphi}{\partial\mathbf{w}}} = \int \varphi\left(\frac{\partial f}{\partial t}\right)_{\mathrm{c}}\mathrm{d}\mathbf{w}. \tag{71.7}$$

Thus, when $(\partial f/\partial t)_{\mathrm{c}}$ is known, (71.7) enables the transport of the property φ through the fluid to be determined.

For the ith component (71.4) and (71.7) give

$$\frac{\partial}{\partial t}(n_i\bar{\varphi}_i)+\nabla\,.\,(n_i\mathbf{v}_i\bar{\varphi}_i+n_i\overline{\boldsymbol{c}_i\varphi_i})-n_i\mathbf{F}_i\,.\,\overline{\frac{\partial\varphi_i}{\partial\mathbf{w}_i}} = \sum_j \int \varphi_i C_{ij}\,\mathrm{d}\mathbf{w}, \tag{71.8}$$

where a typical right-hand-side term represents the change in the property φ_i owing to collisions between particles of the ith and jth components. It is of particular importance to know under what circumstances the right-hand side of (71.8) vanishes.

71.4. Conservation of mass, momentum, and energy

Individual particle mass is unchanged at each collision, consequently with the choice $\varphi_i = m_i$ the collision terms in (71.8) separately vanish, that is,

$$\int C_{ij}\,\mathrm{d}\mathbf{w}_i = 0. \tag{71.9}$$

On the other hand, momentum and energy are transferred between the interacting particles, although for the smooth, structureless particles of our kinetic model, collisions between the like particles of a given component conserve the momentum and energy of that component in any reference frame, that is,

$$\int m_i\mathbf{w}_i C_{ii}\,\mathrm{d}\mathbf{w}_i = 0, \qquad \int m_i\boldsymbol{c}_i C_{ii}\,\mathrm{d}\boldsymbol{c}_i = 0, \tag{71.10}$$

$$\int \tfrac{1}{2}m_i w_i^2 C_{ii}\,\mathrm{d}\mathbf{w}_i = 0, \qquad \int \tfrac{1}{2}m_i \boldsymbol{c}_i^2 C_{ii}\,\mathrm{d}\boldsymbol{c}_i = 0, \tag{71.11}$$

where $\boldsymbol{c}_i \equiv \mathbf{w}_i - \mathbf{v}_i$.

To deal with collisions between unlike particles it is convenient to define quantities $\mathbf{R}_{ij}$, φ_{ij}, ξ_{ij} (see § 52) by

$$\mathbf{R}_{ij} \equiv \int m_i\mathbf{w}_i C_{ij}\,\mathrm{d}\mathbf{w}_i = \int m_i\boldsymbol{c}_i C_{ij}\,\mathrm{d}\mathbf{w}_i, \tag{71.12}$$

$$\varphi_{ij} \equiv \int \tfrac{1}{2}m_i(\mathbf{w}_i-\mathbf{V}_{ij})^2 C_{ij}\,\mathrm{d}\mathbf{w}_i \qquad (\mathbf{V}_{ij} \equiv (m_i\mathbf{v}_i+m_j\mathbf{v}_j)/(m_i+m_j)), \tag{71.13}$$

$$\xi_{ij} = \int m_i\mathbf{w}_i\,.\,(\mathbf{V}_{ij}-\mathbf{v}_i)C_{ij}\,\mathrm{d}\mathbf{w}_i = \frac{m_j}{m_i+m_j}\mathbf{R}_{ij}\,.\,(\mathbf{v}_j-\mathbf{v}_i), \tag{71.14}$$

where $\mathbf{V}_{ij}$ is the mean centre of mass motion of the two components in question. As $\mathbf{V}_{ij}$ is unaffected by elastic collisions, we see that φ_{ij} is the energy *transferred* by collisions between $\mathscr{P}_j$ (the jth component fluid-particle system) and $\mathscr{P}_i$, whereas ξ_{ij} represents energy dissipated in $\mathscr{P}_i$ as a result of the friction force $\mathbf{R}_{ij}$ between $\mathscr{P}_i$ and $\mathscr{P}_j$. The total force density on $\mathscr{P}_i$ is

$$\rho_i \mathbf{R}_i = \sum_j \mathbf{R}_{ij} = \sum_j \int m_i \mathbf{w}_i C_{ij}\, d\mathbf{w}_i. \tag{71.15}$$

By (71.9) and the definitions (71.12)–(71.13) we find that

$$\mathbf{R}_{ij}\cdot \mathbf{v}_i + \varphi_{ij} + \xi_{ij} = \int \tfrac{1}{2} m_i w_i^2 C_{ij}\, d\mathbf{w}_i. \tag{71.16}$$

As $\mathbf{R}_{ij} = -\mathbf{R}_{ji}$,

$$\int m_i \mathbf{w}_i C_{ij}\, d\mathbf{w}_i + \int m_j \mathbf{w}_j C_{ji}\, d\mathbf{w}_j = 0, \tag{71.17}$$

and further, conservation of the total momentum and energy of all components during collisional transfers requires that C_{ij} satisfy the constraints

$$\sum_{i,j} \mathbf{R}_{ij} = \sum_{i,j} \int m_i \mathbf{w}_i C_{ij}\, d\mathbf{w}_i = \sum_{i,j} \int m_i \boldsymbol{c}_i C_{ij}\, d\mathbf{w}_i = 0$$

and (71.18)

$$\sum_{i,j} (\mathbf{R}_{ij}\cdot \mathbf{v}_i + \varphi_{ij} + \xi_{ij}) = \sum_{i,j} \int \tfrac{1}{2} m_i w_i^2 C_{ij}\, d\mathbf{w}_i = \sum_{i,j} \int \tfrac{1}{2} m_i \boldsymbol{c}_i^2 C_{ij}\, d\mathbf{w}_i = 0.$$

Now take $\varphi_i = m_i$ in (71.8), use (71.9) and the fact that $\bar{\boldsymbol{c}}_i = 0$, to obtain the equation of continuity for $\mathscr{P}_i$,

$$\frac{\partial \rho_i}{\partial t} + \nabla \cdot (\rho_i \mathbf{v}_i) = 0. \tag{71.19}$$

This differs from eqn $(51.7)_4$ as no chemical or physical transformations of $\mathscr{P}_i$ into another species is possible with our simple kinetic model.

To derive the momentum equation for $\mathscr{P}_i$ we let $\varphi_i = m_i \mathbf{w}_i$ in (71.8). For this case we need

$$n_i \bar{\varphi}_i = \rho_i \mathbf{v}_i, \qquad n_i \overline{\mathbf{F}_i \cdot \frac{\partial \varphi_i}{\partial \mathbf{w}_i}} = \rho_i \mathbf{F}_i \cdot \mathbf{1} = \rho_i \mathbf{F}_i,$$

and

$$n_i \overline{\boldsymbol{c}_i \varphi_i} = \rho_i \overline{\boldsymbol{c}_i(\mathbf{v}_i + \boldsymbol{c}_i)} = \rho_i \overline{\boldsymbol{c}_i \boldsymbol{c}_i} = \mathbf{p}_i$$

by the definition for the pressure tensor given in $(23.2)_2$. Then by (71.8) and

(71.15) we arrive at (cf. (52.1))

$$\frac{\partial}{\partial t}(\rho_i \mathbf{v}_i)+\nabla\,.\,(\rho_i \mathbf{v}_i \mathbf{v}_i+\mathbf{p}_i) = \rho_i(\mathbf{F}_i+\mathbf{R}_i). \tag{71.20}$$

Finally the choice $\varphi_i = \frac{1}{2}m_i \mathbf{w}_i\,.\,\mathbf{w}_i = \frac{1}{2}m_i(v_i^2+2\mathbf{v}_i\,.\,\boldsymbol{c}_i+c_i^2)$ will lead to the energy equation for $\mathscr{P}_i$. Then by (22.5)

$$n_i\bar{\varphi}_i = \tfrac{1}{2}\rho_i v_i^2+\tfrac{1}{2}\rho_i\overline{c_i^2} = \tfrac{1}{2}\rho_i v_i^2+\rho_i u_i$$

and

$$n_i\overline{\boldsymbol{c}_i\varphi_i} = 0+\rho_i\overline{\boldsymbol{c}_i\boldsymbol{c}_i}\,.\,\mathbf{v}_i+\tfrac{1}{2}\rho_i\overline{\boldsymbol{c}_i c_i^2} = \mathbf{p}_i\,.\,\mathbf{v}_i+\mathbf{q}_i,$$

where

$$\mathbf{q}_i \equiv \tfrac{1}{2}\rho_i\overline{\boldsymbol{c}_i c_i^2} \tag{71.21}$$

defines the heat flux vector for our kinetic model. We also need

$$n_i\overline{\mathbf{F}_i\,.\,\frac{\partial\varphi_i}{\partial\mathbf{w}_i}} = \rho_i\overline{\mathbf{F}_i\,.\,\mathbf{1}\,.\,\mathbf{w}_i} = \rho_i\mathbf{F}_i\,.\,\mathbf{v}_i.$$

Using these results plus (71.15) and (71.16) in (71.8) we arrive at (cf. (52.5))

$$\frac{\partial}{\partial t}[\rho_i(u_i+\tfrac{1}{2}v_i^2)]+\nabla\,.\,[\rho_i\mathbf{v}_i(u_i+\tfrac{1}{2}v_i^2)+\mathbf{p}_i\,.\,\mathbf{v}_i+\mathbf{q}_i]$$

$$= \rho_i(\mathbf{F}_i+\mathbf{R}_i)\,.\,\mathbf{v}_i+\sum_j(\varphi_{ij}+\xi_{ij})+\psi_i, \tag{71.22}$$

where we have added a further heat source in ψ_i, the supply due to radiation.

Once f_i and C_{ij} have been calculated—this is the principal task of kinetic theory—expressions for $\mathbf{R}_{ij}$, φ_{ij}, $\mathbf{p}_i$, and $\mathbf{q}_i$ can be written down directly from the definitions given above. Thus kinetic theory enables us to replace the whole set of constitutive relations of fluid mechanics by a single scalar constitutive relation for the distribution function f, a gain to be set against the rather limited nature of the model. However, when this model is extended to allow for some molecular structure, the kinetic model successfully reproduces a wide range of fluid behaviour.

72. The *H*-theorem for kinetic processes

72.1. Kinetic entropy

Quoting results given in § 25.1, we have

$$\rho s = \sum_i \rho_i s_i = -k\sum_i\int f_i(\ln f_i-1)\,\mathrm{d}\mathbf{w}_i$$

or (72.1)

$$\rho s = \rho s^{\mathrm{eq}}-k\sum_i\int f_i\ln(f_i/f_i^{\mathrm{eq}})\,\mathrm{d}\mathbf{w}_i,$$

where ρs^{eq} is the 'equilibrium' entropy density, which evidently satisfies

$$\rho s^{\mathrm{eq}} = -k\sum_i \int f_i(\ln f_i^{\mathrm{eq}} - 1)\,\mathrm{d}\mathbf{w}_i = -k\sum_i \int f_i^{\mathrm{eq}}(\ln f_i^{\mathrm{eq}} - 1)\,\mathrm{d}\mathbf{w}_i.$$

No committment is yet made as to the precise value of f_i^{eq}, which as we indicated in § 24, depends on the variables appearing in the equilibrium (that is, $f_i = f_i^{\mathrm{eq}}$) form of the Gibbs equation. For mixtures of perfect gases it must satisfy the constraints (§§ 23.5 and 24)

$$\int f_i\,\mathrm{d}\mathbf{w}_i = \int f_i^{\mathrm{eq}}\,\mathrm{d}\mathbf{w}_i = n, \qquad \int m_i\mathbf{w}_i f_i\,\mathrm{d}\mathbf{w}_i = \int m_i\mathbf{w}_i f_i^{\mathrm{eq}}\,\mathrm{d}w_i = \rho_i\mathbf{v}_i, \tag{72.2}$$

$$\text{plus}\begin{cases}\text{either} & \displaystyle\int \tfrac{1}{2}m_i c_i^2 f_i\,\mathrm{d}\mathbf{w}_i = \int \tfrac{1}{2}m_i c_i^2 f_i^{\mathrm{eq}}\,\mathrm{d}\mathbf{w}_i = \rho_i u_i & \begin{pmatrix}\text{component}\\ \text{equilibrium}\end{pmatrix}\\ \text{or} & \displaystyle\sum_i\int \tfrac{1}{2}m_i \mathscr{C}_i^2 f_i\,\mathrm{d}\mathbf{w}_i = \sum_i\int \tfrac{1}{2}m_i \mathscr{C}_i^2 f_i^{\mathrm{eq}}\,\mathrm{d}\mathbf{w}_i = \rho u_{\mathrm{m}} & \begin{pmatrix}\text{mixture}\\ \text{equilibrium}\end{pmatrix}\end{cases} \tag{72.3}$$

The first choice leads to $f_i^{\mathrm{eq}} = f_{i\mathrm{c}}^{\mathrm{eq}}$, where

$$f_{i\mathrm{c}}^{\mathrm{eq}} = \exp\left[m_i(g_i - \tfrac{1}{2}c_i^2)/kT_i\right] = n_i\left(\frac{m_i}{2\pi kT_i}\right)^{\frac{3}{2}} \exp(-\tfrac{1}{2}m_i c_i^2/kT_i), \tag{72.4}$$

and the second to $f_i^{\mathrm{eq}} = f_{im}^{\mathrm{eq}}$, which is a Maxwellian with T_{m} and $\mathscr{C}_i$ in place of T_i and c_i in (72.4).

For the entropy production rate the substitutions $\xi_i = f_i/n$, $y_i = \mathbf{w}_i$ into (66.5) plus (22.1) give

$$\sigma = JX - k\sum_i \int \dot{f}_i \ln(f_i/f_i^{\mathrm{eq}})\,\mathrm{d}\mathbf{w}_i \geqslant 0 \tag{72.5}$$

in the integral of which n is held constant in accordance with eqn $(72.2)_1$ (see observation at the end of § 21.4). The distinction in (72.5) between the 'external' processes J and the 'internal' processes $\dot{f}_i$ depends on the choice made for the equilibrium distribution, as we shall now show.

The time derivative of the entropy density may be written

$$\begin{aligned}\frac{\partial(\rho s)}{\partial t} &= -k\sum_i \int \frac{\partial f_i}{\partial t}\ln f_i\,\mathrm{d}\mathbf{w}_i\\ &= -k\sum_{i,j}\int C_{ij}\ln f_i\,\mathrm{d}\mathbf{w}_i + k\sum_i\int\left(\mathbf{w}_i\cdot\frac{\partial f_i}{\partial \mathbf{r}} + \mathbf{F}_i\cdot\frac{\partial f_i}{\partial \mathbf{w}_i}\right)\ln f_i\,\mathrm{d}\mathbf{w}_i,\end{aligned}$$

by (71.4). As $(\partial/\partial\mathbf{w}_i)\cdot\mathbf{F}_i = 0$ (§ 71.1), then

$$\int \mathbf{F}_i\cdot\frac{\partial f_i}{\partial \mathbf{w}_i}\ln f_i\,\mathrm{d}\mathbf{w}_i = \int\frac{\partial}{\partial \mathbf{w}_i}\cdot[\mathbf{F}_i f_i(\ln f_i - 1)]\,\mathrm{d}\mathbf{w}_i = 0,$$

for f_i tends rapidly to zero at the limits of integration. Also with the definitions

$$\mathbf{J}_{si} \equiv -k\int(\mathbf{w}_i-\mathbf{v}_i)f_i\,(\ln f_i-1)\,d\mathbf{w}_i, \tag{72.6}$$

$$\mathbf{J}_i \equiv \rho_i(\mathbf{v}_i-\mathbf{v}) \qquad \left(\rho\mathbf{v} = \sum_i \rho_i\mathbf{v}_i\right), \tag{72.7}$$

we have

$$k\sum_i\int \mathbf{w}_i\,.\,\frac{\partial f_i}{\partial \mathbf{r}}\ln f_i\,d\mathbf{w}_i = k\sum_i\int(\mathbf{w}_i-\mathbf{v}_i)\,.\,\frac{\partial}{\partial \mathbf{r}}[f_i\;(\ln f_i-1)]\,d\mathbf{w}_i$$

$$= -\frac{\partial}{\partial \mathbf{r}}\,.\,\sum_i(\mathbf{J}_{si}+s_i\mathbf{J}_i)-\frac{\partial}{\partial \mathbf{r}}\,.\,(\rho s\mathbf{v}).$$

Consequently the expression for $\partial(\rho s)/\partial t$ may be written

$$\frac{\partial(\rho s)}{\partial t}+\nabla\,.\,(\rho s\mathbf{v})+\nabla\,.\,\sum_i(\mathbf{J}_{si}+s_i\mathbf{J}_i) = -k\sum_{i,j}\int C_{ij}\ln f_i\,d\mathbf{w}_i, \tag{72.8}$$

or (cf. (51.22) with $T_i = T$ and $\psi_i = 0$),

$$\rho Ds+\nabla\,.\,\sum_i(\mathbf{J}_{si}+s_i\mathbf{J}_i) = \sigma,$$

where

$$\sigma = -k\sum_{i,j}\int C_{ij}\ln f_i\,d\mathbf{w}_i \tag{72.9}$$

is the entropy source strength.

By (71.9), $(71.18)_2$, and (72.4)

$$kT\sum_{i,j}\int C_{ij}\ln f_{ic}^{eq}\,d\mathbf{w}_i = \sum_i \rho_i\mathbf{v}_i\,.\,\mathbf{R}_i = \sum_i\rho_i(\mathbf{v}_i-\mathbf{v})\,.\,\mathbf{R}_i;$$

similarly

$$kT\sum_{i,j}\int C_{ij}\ln f_{im}^{eq}\,d\mathbf{w}_i = 0.$$

Hence

$$\begin{aligned}\sigma &= \frac{1}{T}\sum_i\rho_i(\mathbf{v}_i-\mathbf{v})\,.\,\mathbf{R}_i-k\sum_{i,j}\int C_{ij}\ln(f_i/f_{ic}^{eq})\,d\mathbf{w}_i,\\ \sigma &= -k\sum_{i,j}\int C_{ij}\ln(f_i/f_{im}^{eq})\,d\mathbf{w}_i,\end{aligned} \tag{72.10}$$

or by (71.4)

$$\begin{aligned}\sigma &= \frac{1}{T}\sum_i\rho_i(\mathbf{v}_i-\mathbf{v})\,.\,\mathbf{R}_i-k\sum_i\int D_if_i\ln(f_i/f_{ic}^{eq})\,d\mathbf{w}_i\\ &= -k\sum_i\int D_if_i\ln(f_i/f_{im}^{eq})\,d\mathbf{w}_i.\end{aligned} \tag{72.11}$$

Comparing (72.11) with (72.5), in which $\dot{f}_i = \mathrm{D}_i f_i$, we see that the choice $f_i^{\mathrm{eq}} = f_{i\mathrm{m}}^{\mathrm{eq}}$ in (72.5) eliminates the processes termed 'external' by turning them into 'internal' processes.

72.2. Constraints on the collision term

If it is accepted that the essential nature of the distinction between the left and right-hand sides of the transport equation,

$$\mathrm{D}_i f_i = \frac{\partial f_i}{\partial t} + \mathbf{w}_i \cdot \frac{\partial f_i}{\partial \mathbf{r}} + \mathbf{F}_i \cdot \frac{\partial f_i}{\partial \mathbf{w}_i} = \sum_j C_{ij}, \tag{72.12}$$

is that described in § 71.2, namely, that the left-hand side is reversible while the right-hand side is not, then it follows that, as $\mathrm{D}_i f_i$ has negative parity under motion reversal, $\sum_j C_{ij}$ must have positive parity. Eqn (72.10) then shows that σ has the positive parity required of the entropy production rate (§ 40.2); when using (72.11), which appears to give σ the wrong parity, we should interpret $\mathrm{D}_i f_i$ being an irreversible time derivative.

The other constraint on $\sum_j C_{ij}$ is that it must render σ non-negative,

$$\sigma = -k \sum_{i,j} \int C_{ij} \ln f_i \, \mathrm{d}\mathbf{w}_i \geqslant 0. \tag{72.13}$$

This follows from (72.5) and is, of course, a direct consequence of the fact that thermodynamic systems evolve towards equilibrium states (if this were not so, we would scarcely be in a position to discuss the question!).

The constraint in (72.13) is usually expressed in terms of Boltzmann's H-function,

$$H \equiv \sum_i \iint f_i \ln f_i \, \mathrm{d}\mathbf{w}_i \, \mathrm{d}\mathbf{r} = -S/k + N, \tag{72.14}$$

where N is the total number of particles within the volume of integration. If (72.8) is integrated over a closed volume, it reduces to

$$\frac{\mathrm{d}S}{\partial t} + \int_\Sigma \sum_i (\mathbf{n} \cdot \mathbf{J}_{\mathrm{s}i}) \, \mathrm{d}\Sigma = \int \sigma \, \mathrm{d}\mathbf{r} \geqslant 0,$$

as $\mathbf{n} \cdot \mathbf{v}_i = 0$ on the surface Σ. Further if Σ is 'smooth', that is, the reflection of the incident particles is specular, it is easy to show from the symmetry of the reflection that $\mathbf{n} \cdot \boldsymbol{c}_i \ln f_i$, and hence $\mathbf{n} \cdot \mathbf{J}_{\mathrm{s}i}$, is zero on Σ. The conclusion may be expressed

$$\frac{\mathrm{d}H}{\mathrm{d}t} = \sum_{i,j} \iint C_{ij} \ln f_i \, \mathrm{d}\mathbf{w}_i \, \mathrm{d}\mathbf{r} \leqslant 0; \tag{72.15}$$

when specified collision terms C_{ij} are found to satisfy this inequality they are said to 'satisfy an H-theorem'.

Boltzmann's famous integro-differential equation for f_i (see § 75.1) has a collision integral C_{ij} based on the assumption that $(\partial f_i/\partial t)_c$ results from binary collisions between particles uncorrelated in velocity-space. Thus the probability that a molecule of $\mathscr{P}_i$ collides with one of $\mathscr{P}_j$ is supposed proportional to the product $f_i f_j$ of the one-particle distribution functions, and the neglect of possible correlations between the particles is termed the 'hypothesis of molecular chaos'. Soon after Boltzmann published the H-theorem for his collision term (1872), Loschmidt objected that it introduces irreversibility into the kinetic model, the individual particles of which certainly obey the reversible equations of dynamics.

The resolution of this paradox, presented in general terms in § 20, is that irreversible processes are merely *statistically* much more likely than reversible ones; thus (72.15) holds only in a statistical sense. When equilibrium is attained, the theorem requires H to have a constant minimum value for an isolated system, whereas small variations in f_i due to thermodynamic noise (§68.3) will cause H to fluctuate above this value. However, on a large enough time scale such fluctuations will be unimportant. The situation can be understood on appreciating that Boltzmann's collision integral is itself statistical, depending as it does on the hypothesis of moleclar chaos. For a system in a state of chaos, the binary collision represented in the collision term does cause H to decrease, but on a dynamical view—as distinct from a statistical one—the collision creates a correlation between the velocities of the two particles involved (for they are now certainly receding from each other), thereby destroying the initial state of molecular chaos. Hence on the dynamical theory H no longer necessarily decreases in the next instant, and a gas near the equilibrium state may experience an increase in its H-function For a gas very far from the equilibrium state, the statistical view predicts an overwhelmingly large probability that H will decrease.

72.3. *Alternative formulation of the H-theorem*

Consider a system comprised of a single component of N particles occupying a volume V. The kinetic entropy S_1 of this system is given by the volume integral of (72.1), which in terms of the one-particle distribution function $F_1(\mathbf{x}_1, t)$ defined in § 25.2 may be written

$$S_1(t) = -kN \int F_1(\mathbf{x}_1, t)[\ln F_1(\mathbf{x}_1, t) - 1]\, d\mathbf{x}_1 + \text{constant}, \qquad (72.16)$$

where $\mathbf{x}_1 = (\mathbf{p}_1, \mathbf{q}_1)$ is a position vector in six-dimensional phase-space and $d\mathbf{x}_1 = d\mathbf{p}_1\, d\mathbf{q}_1$. As $F_1\, d\mathbf{x}_1$ is the probability that particle 1 lies in $d\mathbf{x}_1$,

$$\int F_1\, d\mathbf{x}_1 = 1. \qquad (72.17)$$

We shall need the convective (in μ-space) time derivative $\mathscr{D}_1$, defined by

$$\mathscr{D}_1\varphi = \frac{\partial\varphi}{\partial t}+\dot{\mathbf{p}}_1\cdot\frac{\partial\varphi}{\partial\mathbf{p}_1}+\dot{\mathbf{q}}_1\cdot\frac{\partial\varphi}{\partial\mathbf{q}_1} = \frac{\partial\varphi}{\partial t}+[\varphi, H_1]$$

or

$$\mathscr{D}_1\varphi = \frac{\partial\varphi}{\partial t}+\mathscr{H}_1\varphi = \frac{\partial}{\partial t}[\exp(t\mathscr{H}_1)\varphi], \tag{72.18}$$

in which H_1 is the single-particle Hamiltonian and $\mathscr{H}_1$ is the corresponding operator (cf. (70.2)). The Liouville theorem (71.2) now reads

$$\mathscr{D}_1(\mathrm{d}\mathbf{x}_1) = 0. \tag{72.19}$$

If the system is isolated, so that neither mass nor energy cross the boundary surface Σ of the system, then by (72.16) and (72.19) entropy is created within the volume at the rate

$$\frac{\mathrm{d}S_1}{\mathrm{d}t} = -kN\int\mathscr{D}_1[F_1(\ln F_1-1)\,\mathrm{d}\mathbf{x}_1],$$

that is,

$$\frac{\mathrm{d}S_1}{\mathrm{d}t} = -kN\int\mathscr{D}_1F_1\ln F_1\,\mathrm{d}\mathbf{x}_1 \geqslant 0, \tag{72.20}$$

the constraint $\geqslant 0$ being a consequence of the law of entropy increase for an isolated system (§ 13.3). The H-theorem now follows on replacing $\mathscr{D}_1F_1$ by the collision term $(\partial F_1/\partial t)_\mathrm{c}$ and using (72.14). By an argument similar to that leading from (72.9) to $(72.11)_1$ we may rewrite (72.20) as

$$\frac{\mathrm{d}S_1}{\mathrm{d}t} = -kN\int\mathscr{D}_1F_1\ln\left(\frac{F_1}{F_1^{\mathrm{eq}}}\right)\mathrm{d}\mathbf{x}_1, \tag{72.21}$$

where F_1^{eq} is the Maxwellian distribution function.

72.4. Generalized H-theorem

To this stage we have been concerned with the kinetic or 'one-particle' entropy S_1. Now consider the r-particle entropy S_r, introduced in § 25.6, namely,

$$S_r = -\frac{kN}{r}\int F_r(\ln F_r-1)\,\mathrm{d}\Gamma_r+\text{constant}, \tag{72.22}$$

where

$$S_r = S_r(t), \qquad F_r = F_r(\mathbf{x}_1, \mathbf{x}_2 \ldots \mathbf{x}_r, t), \qquad \mathrm{d}\Gamma_r = \mathrm{d}\mathbf{x}_1 \ldots \mathrm{d}\mathbf{x}_r,$$

and

$$\int F_r\,\mathrm{d}\Gamma_r = 1.$$

Denote by H_r the Hamiltonian for a system of r particles, then

$$H_r = \sum_{i=1}^{r}\left[\frac{p_i^2}{2m}+\Phi(\mathbf{q}_i)\right]+\sum_{1\leqslant i<j\leqslant r}\varphi_{ij} \qquad (\varphi_{ij}=\varphi(|\mathbf{q}_i-\mathbf{q}_j|)), \tag{72.23}$$

$\Phi(\mathbf{q}_i)$ being the external field on particle i and φ_{ij} the intermolecular potential. Let $\mathscr{D}_r$ denote the convective time derivative in Γ_r-space, then

$$\mathscr{D}_r\varphi = \left[\frac{\partial}{\partial t}+\sum_{i=1}^{r}\left(\dot{\mathbf{q}}_i\cdot\frac{\partial}{\partial \mathbf{q}_i}+\dot{\mathbf{p}}_i\cdot\frac{\partial}{\partial \mathbf{p}_i}\right)\right]\varphi$$
$$= \left[\frac{\partial}{\partial t}+\sum_{i=1}^{r}\left(\frac{\partial H_r}{\partial \mathbf{p}_i}\cdot\frac{\partial}{\partial \mathbf{q}_i}-\frac{\partial H_r}{\partial \mathbf{q}_i}\cdot\frac{\partial}{\partial \mathbf{p}_i}\right)\right]\varphi,$$

that is,

$$\mathscr{D}_r\varphi = \left(\frac{\partial}{\partial t}+\mathscr{H}_r\right)\varphi,$$

where

$$\mathscr{H}_r \equiv \sum_{i=1}^{r}\left(\frac{\mathbf{p}_i}{m}\cdot\frac{\partial}{\partial \mathbf{q}_i}+\mathbf{K}_i\cdot\frac{\partial}{\partial \mathbf{p}_i}\right)-\sum_{1\leqslant i<j\leqslant r}\Theta_{ij},$$

$$K_i \equiv -\frac{\partial \Phi}{\partial \mathbf{q}_i},$$

and

$$\Theta_{ij} \equiv \frac{\partial \varphi_{ij}}{\partial \mathbf{q}_i}\cdot\frac{\partial}{\partial \mathbf{p}_i}+\frac{\partial \varphi_{ij}}{\partial \mathbf{q}_j}\cdot\frac{\partial}{\partial \mathbf{p}_j}. \tag{72.24}$$

Here $\mathscr{H}_r$ is the r-particle Hamiltonian operator, $\mathbf{K}_i$ is the external force on particle i, and Θ_{ij} is the molecular interaction operator.

As (cf. proof of (71.2))

$$\mathscr{D}_r(\mathrm{d}\Gamma_r) = \mathrm{d}\Gamma_r\sum_{i=1}^{r}\left(\frac{\partial}{\partial \mathbf{q}_i},\frac{\partial}{\partial \mathbf{p}_i}\right)\cdot(\mathbf{p}_i,\mathbf{K}_i)$$

and $\mathbf{p}_i$, $\mathbf{K}_i$ are independent of $\mathbf{q}_i$ and $\mathbf{p}_i$ respectively, it follows that

$$\mathscr{D}_r(\mathrm{d}\Gamma_r) = 0. \tag{72.25}$$

It remains only to differentiate (72.22) to obtain

$$\frac{\mathrm{d}S_r}{\mathrm{d}t} = -\frac{kN}{r}\int \mathscr{D}_rF_r \ln F_r\,\mathrm{d}\Gamma_r \geqslant 0, \tag{72.26}$$

for an isolated system; here the derivative $\mathscr{D}_rF_r$ must be interpreted as having even parity.

Since Liouville's theorem (§ 30.2) may be written

$$\mathscr{D}_N F_N = 0, \tag{72.27}$$

the limiting case $r = N$ of (72.26) requires the equality, that is,

$$\frac{\mathrm{d}S_N}{\mathrm{d}t} = -k \int \mathscr{D}_N F_N \ln F_N \,\mathrm{d}\Gamma_N = 0. \tag{72.28}$$

Comparing (72.26) and (72.28), we observe that anything less than the complete knowledge of a system embodied in F_N yields irreversibility.

There is another form of (72.26) that will be useful later. We define a 'frozen equilibrium' r-particle distribution F_r^0 by requiring it to satisfy

$$\mathscr{D}_r F_r^0 = 0, \tag{72.29}$$

which ensures that the corresponding entropy is stationary. Now (72.26) may be written

$$\frac{\mathrm{d}S_r}{\mathrm{d}t} = -\frac{kN}{r} \int \mathscr{D}_r F_r \ln (F_r/F_r^0) \,\mathrm{d}\Gamma_r - \frac{kN}{r} \int \mathscr{D}_r F_r \ln F_r^0 \,\mathrm{d}\Gamma_r \geqslant 0.$$

When F_r is close to F_r^0 in value, the first right-hand member is evidently quadratically small whereas the second is not, and the latter therefore dominates the expression. Both signs are available to the second term, and in view of the constraint $\geqslant 0$, it must be identically zero. Hence

$$\frac{\mathrm{d}S_r}{\mathrm{d}t} = -\frac{kN}{r} \int \mathscr{D}_r F_r \ln (F_r/F_r^0) \,\mathrm{d}\Gamma_r \geqslant 0; \tag{72.30}$$

the case $r = 1$ of this formula should not be confused with (72.21), as F_1^0 is not necessarily the same as F_1^{eq}.

73. Simple kinetic constitutive relations

73.1. The BGK relaxation model

In the process thermodynamics of earlier chapters the method of devising constitutive relations was to construct a bilinear form $\sigma = JX$ for the rate of irreversible entropy production, and from $\sigma \geqslant 0$ to infer a linear approximation $J = LX$ to the required relation. And as mentioned occasionally (§§ 38.4, 38.6, 53.3) it is necessary to exercise some physical intuition as to which are the reversible processes, so that these can then be excluded from contributing to σ. As described in § 71.2, in kinetic theory this separation of processes into reversible and irreversible parts is made at the outset in writing down the form of the basic transport equation, and it is made possible by the assumed existence of disparate time scales on which changes can occur (see footnote, p. 290).

On the assumption that the entropy is sufficiently specified by the one-particle entropy S_1 (see remarks in § 25.6 and § 74.3), we may follow the prescription just described and infer from (72.10) a linear constitutive equation for $\mathrm{D}_i f_i$ of the general form

$$\mathrm{D}_i f_i = \frac{\partial f_i}{\partial t} + \mathbf{w}_i \cdot \frac{\partial F}{\partial \mathbf{r}} + \mathbf{F}_i \cdot \frac{\partial f_i}{\partial \mathbf{w}_i}$$

$$= \sum_j \iint L_{ij}(\mathbf{r}_i, \mathbf{w}_i; \mathbf{r}_j, \mathbf{w}_j) \ln \frac{f_j(\mathbf{r}_j, \mathbf{w}_j, t)}{f_j^{\mathrm{eq}}(\mathbf{r}_j, \mathbf{w}_j)} \,\mathrm{d}\mathbf{w}_j \,\mathrm{d}\mathbf{r}_j, \qquad (73.1)$$

where L_{ij} is a phenomenological function. Such a closed relation for one-particle distribution functions is called a *kinetic equation*.

Now let us suppose that only those particles moving through phase-space with about the same velocity and about the same position can interact, that is, that $(\partial f_i/\partial t)_{\mathrm{c}}$ is determined *locally* in μ-space. Then L_{ij} may be written

$$L_{ij}(\mathbf{r}_i, \mathbf{w}_i; \mathbf{r}_j, \mathbf{w}_j) = \delta(\mathbf{r}_i - \mathbf{r}_j)\delta(\mathbf{w}_i - \mathbf{w}_j)\nu_{ij}(\mathbf{w}_i) f_j^{\mathrm{eq}}(\mathbf{r}_j, \mathbf{w}_j),$$

where $\nu_{ij}(\mathbf{w}_i)$ has the dimensions of a frequency. With the further assumption that the deviations from equilibrium are small, that is, $(f_i - f_i^{\mathrm{eq}})/f_i^{\mathrm{eq}} \ll 1$, eqn (73.1) reduces to

$$\mathrm{D}_i f_i = \sum_j C_{ij}, \qquad C_{ij} \equiv \nu_{ij}(f_j^{\mathrm{eq}} - f_j), \qquad (73.2)$$

which is the BGK relaxation model (Bhathager, Gross, and Krook 1954). It is of course necessary to ensure that f_j and ν_{ij} possess sufficient degrees of freedom to satisfy the various conservation laws imposed on C_{ij} as given in § 71.4.

For a single species (73.2) and $(72.10)_1$ yield

$$\frac{\partial f}{\partial t} + \mathbf{w} \cdot \frac{\partial f}{\partial \mathbf{r}} + \mathbf{F} \cdot \frac{\partial f}{\partial \mathbf{w}} = \nu(f^{\mathrm{eq}} - f) \qquad (73.3)$$

and

$$\sigma = k \int \nu(f - f^{\mathrm{eq}}) \ln (f/f^{\mathrm{eq}}) \,\mathrm{d}\mathbf{w}.$$

As $\ln x$ always has the same sign as $(x-1)$, we see that this model satisfies an H-theorem (that is, $\sigma \geqslant 0$) provided $\nu \geqslant 0$.

73.2. The Fokker–Planck model

For simplicity consider a one-component system and define $\mathbf{J}$ by

$$\mathrm{D}f = \left(\frac{\partial f}{\partial t}\right)_{\mathrm{c}} \equiv -\frac{\partial}{\partial \mathbf{w}} \cdot \mathbf{J}. \qquad (73.4)$$

Then, on supposing $\ln(f/f^{\text{eq}})$ to vanish at large $|\mathbf{w}|$, we have

$$\sigma = k\int \frac{\partial}{\partial \mathbf{w}}\,.\,\mathbf{J}\ln\left(\frac{f}{f^{\text{eq}}}\right)\mathrm{d}\mathbf{w} = -k\int \mathbf{J}\,.\,\frac{\partial}{\partial \mathbf{w}}\left[\ln\left(\frac{f}{f^{\text{eq}}}\right)\right]\mathrm{d}\mathbf{w} \geqslant 0.$$

Now take $\mathbf{J}f^{-\frac{1}{2}}$ to be the process and $f^{\frac{1}{2}}(\partial/\partial\mathbf{w})[\ln(f/f^{\text{eq}})]$ to be the affinity, and make the assumption that $\mathbf{J}f^{-\frac{1}{2}}$ is determined locally in μ-space; this leads to a linear constitutive relation of the appearance

$$\mathbf{J} = -\tfrac{1}{2}f\mathsf{B}\,.\,\frac{\partial}{\partial \mathbf{w}}[\ln(f/f^{\text{eq}})], \tag{73.5}$$

where in general the phenomenological tensor B depends on $(\mathbf{r}, \mathbf{w}, t)$ and is a functional of f. We rewrite (73.5) in the form

$$\mathbf{J} = -\frac{1}{2}\frac{\partial}{\partial \mathbf{w}}\,.\,(\mathsf{B}f) + \mathbf{A}f, \tag{73.6}$$

where

$$2A \equiv \frac{\partial}{\partial \mathbf{w}}\,.\,\tilde{\mathsf{B}} + \mathsf{B}\,.\,\frac{\partial(\ln f^{\text{eq}})}{\partial \mathbf{w}} = \frac{\partial}{\partial \mathbf{w}}\,.\,\tilde{\mathsf{B}} - \frac{m}{kT}\mathsf{B}\,.\,(\mathbf{w}-\mathbf{v}),$$

then (73.4) becomes

$$\mathrm{D}f = -\frac{\partial}{\partial \mathbf{w}}\,.\,(\mathbf{A}f) + \frac{1}{2}\frac{\partial^2}{\partial \mathbf{w}\,\partial \mathbf{w}}:(\tilde{\mathsf{B}}f), \tag{73.7}$$

of which the right-hand side is the Fokker–Planck collision term.

A special case of some importance is

$$\mathsf{B} = (2\nu kT/m)\mathbf{1}, \quad \text{for which} \quad \mathbf{A} = -\nu(\mathbf{w}-\mathbf{v})$$

and (73.7) becomes

$$\frac{\partial f}{\partial t} + \mathbf{w}\,.\,\frac{\partial f}{\partial \mathbf{r}} + \mathbf{F}\,.\,\frac{\partial f}{\partial \mathbf{w}} = \nu\frac{\partial}{\partial \mathbf{w}}\,.\left[(\mathbf{w}-\mathbf{v})f + \frac{kT}{m}\frac{\partial f}{\partial \mathbf{w}}\right]. \tag{73.8}$$

The collision term in (73.8) satisfies the conservation of number, momentum, and energy constraints given in § 71.4, and, furthermore, vanishes identically for a Maxwellian distribution.

The essential feature embodied in (73.7) is that only very slight changes in $\mathbf{w}$ are involved in the collision term, for not only have we imposed the constraint of 'locality' as in the BGK model but we have first employed an integration by parts to ensure that this constraint falls on the flux $\mathbf{J}$ of scattered particles rather than on their number. For apparent reasons the quantities $\mathbf{A}$ and B are called the 'friction' and 'diffusion' coefficients respectively, and a more detailed kinetic theory is required to obtain their values.

The model is especially appropriate for a plasma, as the long-range Coulomb forces acting between the charged particles cause a great number of small deflections at some distance and relatively few abrupt deflections from close encounters. For the extremely complicated form that $\mathbf{J}$ takes in a plasma, which is known as the Lenard–Balescu kinetic equation, the reader is referred to the text by Clemmow and Dougherty (1969).

73.3. *Fokker–Planck equation due to distant collisions*

Consider the motion of a charged particle Q moving through a cloud of similar particles, then the long-range Coulomb forces between the particles will cause Q to experience a sequence of 'distant collisions' with the other particles. These distant collisions will be far more numerous than close collisions, consequently almost all the changes in direction and speed experienced by Q will be small. Let Q have an initial velocity $\mathbf{w}$, then after a small time interval Δt, the distant collisions will generate a random-walk motion in Q producing a small cumulative change $\Delta\mathbf{w}$ satisfying $\Delta\mathbf{w} \ll \mathbf{w}$.

Let $P(\mathbf{w}|\Delta\mathbf{w})$ denote the transition probability density that Q experiences the change $\Delta\mathbf{w}$ in Δt, then $f(\mathbf{w}-\Delta\mathbf{w}, t-\Delta t)\, P(\mathbf{w}-\Delta\mathbf{w}|\Delta\mathbf{w})\, \mathrm{d}(\Delta\mathbf{w})$ is the number of particles like Q that are deflected from $(\mathbf{w}-\Delta\mathbf{w}, \mathbf{w})$ into the element $(\mathbf{w}, \mathbf{w}+\mathrm{d}\mathbf{w})$ owing to events occurring in $(t-\Delta t, t)$. These particles will contribute to the number $f(\mathbf{w}, t)\,\mathrm{d}\mathbf{w}$, and on the assumption that the process is Markovian, that is, that no earlier time intervals contribute to this number, we arrive at the Chapman–Kolmogorov equation

$$f(\mathbf{w}, t) = \int f(\mathbf{w}-\Delta\mathbf{w}, t-\Delta t) P(\mathbf{w}-\Delta\mathbf{w}|\Delta\mathbf{w})\, \mathrm{d}(\Delta\mathbf{w}). \tag{73.9}$$

Expanding (73.9) in a Taylor series to first order in Δt and to second-order in $\Delta\mathbf{w}$, we obtain the approximate form

$$f(\mathbf{w}, t) = \int \left\{(f-\Delta t \mathrm{D}f)P(\mathbf{w}|\Delta\mathbf{w}) - \left(\Delta\mathbf{w}\,.\frac{\partial}{\partial\mathbf{w}} - \tfrac{1}{2}\Delta\mathbf{w}\Delta\mathbf{w}\colon\frac{\partial^2}{\partial\mathbf{w}\,\partial\mathbf{w}}\right) \times \right.$$
$$\left. \times [f P(\mathbf{w}|\Delta\mathbf{w})]\right\} \mathrm{d}(\Delta\mathbf{w}).$$

As $\int P(\mathbf{w}|\Delta\mathbf{w})\, \mathrm{d}(\Delta\mathbf{w}) = 1$, the leading term containing f cancels with the left-hand side, and the expression remaining may be rearranged in the form of the Fokker–Planck equation (73.7), with

$$\mathbf{A} = \frac{\langle\Delta\mathbf{w}\rangle}{\Delta t} = \frac{1}{\Delta t}\int \Delta\mathbf{w} P(\mathbf{w}|\Delta\mathbf{w})\, \mathrm{d}(\Delta\mathbf{w})$$

and (73.10)

$$\mathsf{B} = \frac{\langle\Delta\mathbf{w}\,\Delta\mathbf{w}\rangle}{\Delta t} = \frac{1}{\Delta t}\int \Delta\mathbf{w}\,\Delta\mathbf{w} P(\mathbf{w}|\Delta\mathbf{w})\, \mathrm{d}(\Delta\mathbf{w}).$$

The problem is now reduced to that of obtaining a form for $P(\mathbf{w}|\Delta\mathbf{w})$. In plasma theory this is usually done by making the approximation of calculating $P(\mathbf{w}|\Delta\mathbf{w})$ for a test particle moving through a medium whose particles have a Maxwellian distribution. This topic is beyond the scope of the present text.

74. General kinetic equations

74.1. The BBGKY hierarchy of equations

For a system of N particles, $F_r(\mathbf{x}_1, \dots \mathbf{x}_r, t)$ is the marginal density giving the probability that particles $1, 2, \dots r$ lie in $\mathrm{d}\mathbf{x}_1, \mathrm{d}\mathbf{x}_2, \dots \mathrm{d}\mathbf{x}_r$ respectively, regardless of the location of the remaining particles. Thus

$$F_r(\mathbf{x}_1, \dots \mathbf{x}_r, t) = \int \dots \int F_N(\mathbf{x}_1, \dots \mathbf{x}_N, t)\, \mathrm{d}\mathbf{x}_{r+1} \dots \mathrm{d}\mathbf{x}_N, \tag{74.1}$$

and as the particles are identical, the functional form of $F_r(\mathbf{x}_1, \dots \mathbf{x}_r, t)$ is symmetric in $\mathbf{x}_1, \dots \mathbf{x}_r$ and independent of any particular choice for the r particles. Using this result, plus

$$\mathscr{H}_N = \mathscr{H}_r + \sum_{i=r+1}^{N} \left(\frac{\mathbf{p}_i}{m} \cdot \frac{\partial}{\partial \mathbf{q}_i} + \mathbf{K}_i \cdot \frac{\partial}{\partial \mathbf{p}_i}\right) - \sum_{r+1 \leqslant i < j \leqslant N} \Theta_{ij} - \sum_{i=1}^{r} \sum_{j=r+1}^{N} \Theta_{ij}, \tag{74.2}$$

which follows from eqn $(72.24)_2$, it is possible to replace the Liouville equation (72.27)

$$\left(\frac{\partial}{\partial t} + \mathscr{H}_N\right) F_N = 0, \tag{74.3}$$

by a system of equations for the densities F_r, $r = 1, 2, \dots N-1$, known as the BBGKY hierarchy.†

The integrals

$$\int \left(\frac{\mathbf{p}_i}{m} \cdot \frac{\partial}{\partial \mathbf{q}_i} + \mathbf{K}_i \cdot \frac{\partial}{\partial \mathbf{p}_i}\right) F_N\, \mathrm{d}\mathbf{x}_i, \iint \Theta_{ij} F_N\, \mathrm{d}\mathbf{x}_i\, \mathrm{d}\mathbf{x}_j \qquad \begin{pmatrix} i, j = 1, 2, \dots N \\ \mathrm{d}\mathbf{x}_i = \mathrm{d}\mathbf{p}_i\, \mathrm{d}\mathbf{q}_i \end{pmatrix}$$

may be evaluated directly, and constraining F_N to vanish at large values of $\mathbf{q}_i$ and $\mathbf{p}_i$ we find that their values are zero. Hence by (74.1), (74.2), (74.3), and the symmetry of F_N in its arguments

$$\begin{aligned} \left(\frac{\partial}{\partial t} + \mathscr{H}_r\right) F_r &= \sum_{i=1}^{r} \sum_{j=r+1}^{N} \int \dots \int \Theta_{ij} F_N\, \mathrm{d}\mathbf{x}_{r+1} \dots \mathrm{d}\mathbf{x}_N \\ &= (N-r) \sum_{i=1}^{r} \int \dots \int \Theta_{i,r+1} F_N\, \mathrm{d}\mathbf{x}_{r+1} \dots \mathrm{d}\mathbf{x}_N. \end{aligned}$$

† After Bogoliubov, Born, Green, Kirkwood, and Yvon. For these and other relevant references the reader could consult Ferziger and Kaper (1972).

The BBGKY hierarchy of equations now follows from another application of (74.1),

$$\left(\frac{\partial}{\partial t}+\mathscr{H}_r\right)F_r = (N-r)\int \sum_{i=1}^{r} \Theta_{i,r+1}F_{r+1}\,\mathrm{d}\mathbf{x}_{r+1}. \tag{74.4}$$

This equation may be viewed as being an r-particle Liouville equation, supplemented by a right-hand term representing the interaction of the r particles with the remaining particles.

In place of F_r it is convenient to use the reduced distribution functions $\mathscr{F}_r = V^r F_r$, for we can then eliminate wall effects by taking the thermodynamic limit $N \to \infty$, $V \to \infty$, $N/V = n$, and in place of (74.4) find the asymptotic form

$$\mathscr{D}_r\mathscr{F}_r = \left(\frac{\partial}{\partial t}+\mathscr{H}_r\right)\mathscr{F}_r = n\int \sum_{i=1}^{r} \Theta_{i,r+1}\mathscr{F}_{r+1}\,\mathrm{d}\mathbf{x}_{r+1}. \tag{74.5}$$

In particular (omitting the subscript 1 from $\mathscr{F}_1$ for convenience)

$$\mathscr{D}_1\mathscr{F} = \left(\frac{\partial}{\partial t}+\frac{\mathbf{p}_1}{m}\cdot\frac{\partial}{\partial \mathbf{q}_1}+\mathbf{K}_1\cdot\frac{\partial}{\partial \mathbf{p}_1}\right)\mathscr{F} = n\int \Theta_{12}\mathscr{F}_2\,\mathrm{d}\mathbf{x}_2, \tag{74.6}$$

and comparison with (71.3) gives

$$\left(\frac{\partial \mathscr{F}}{\partial t}\right)_{\mathrm{c}} = n\int \Theta_{12}\mathscr{F}_2\,\mathrm{d}\mathbf{x}_2.$$

We now have three different forms for the one-particle distribution function, namely, $f(\mathbf{r}, \mathbf{w}, t)$, $F_1(\mathbf{x}, t)$, and $\mathscr{F}(\mathbf{x}, t)$, where $\mathbf{x} = (\mathbf{q}, \mathbf{p})$. They are related by

$$f(\mathbf{r}, \mathbf{w}, t)\,\mathrm{d}\mathbf{w}\,\mathrm{d}\mathbf{r} = NF_1(\mathbf{x}, t)\,\mathrm{d}\mathbf{x} = n\mathscr{F}(\mathbf{x}, t)\,\mathrm{d}\mathbf{x}, \tag{74.7}$$

with

$$\mathrm{d}\mathbf{x} = \mathrm{d}\mathbf{q}\,\mathrm{d}\mathbf{p} = m^3\,\mathrm{d}\mathbf{r}\,\mathrm{d}\mathbf{w} \qquad (\text{if } \mathbf{q} = \mathbf{r}).$$

The right-hand side of (74.6) is an *exact* expression for the rate of change of $\mathscr{F}$ due to particle interactions of all kinds, but to keep it so we would need to solve the complete hierarchy of equations in (74.5). It is the prime object of kinetic theory to find a physically based means of truncating the hierarchy and thence to arrive at a closed equation for the one-particle distribution function $\mathscr{F}$. As the collision integral on the right-hand side of (74.6) has odd parity under motion-(or time-) reversal, the equation is reversible; it is in the closure of the hierarchy, or more precisely in the assignment of *initial* conditions for those distributions $\mathscr{F}_2, \dots \mathscr{F}_n$ remaining in the closed hierarchy, that this integral acquires the even parity required to render (74.6) irreversible. If it is supposed that the collision integral embodies more than one distinct process, then part of it could retain odd parity (see Klimontovich 1974), with the remainder constituting the irreversible element that must of necessity arise with less than complete knowledge of the state of the system.

74.2. The Vlasov equation

The method used in § 73.1 to deduce the kinetic equation for the BGK relaxation model was based on formula (72.10) for the rate of entropy production. Equivalently we could have started with (72.21),

$$\frac{\mathrm{d}S_1}{\mathrm{d}t} = -kN \int \mathscr{D}_1 F_1 \ln\left(\frac{F_1}{F_1^{\mathrm{eq}}}\right) \mathrm{d}\mathbf{x} \geqslant 0, \tag{74.8}$$

then by supposing $\mathscr{D}_1 F_1$ to be determined locally in $\mathbf{x}$-space, and that $|F_1 - F_1^{\mathrm{eq}}| \ll F_1^{\mathrm{eq}}$, arrived at the BGK equation (in terms of $\mathscr{F} = VF_1$)

$$\mathscr{D}_1 \mathscr{F} = (\mathscr{F}^{\mathrm{eq}} - \mathscr{F})/\tau_1, \tag{74.9}$$

where $\tau_1 \geqslant 0$ is the appropriate relaxation time, known as the *kinetic time*. It is about the average time between successive collisions for a given particle, that is, the mean free time.

We note the limiting cases:

$$\tau_1 = 0: \quad \mathscr{F} = \mathscr{F}^{\mathrm{eq}} \quad \text{(relaxed (Maxwellian) equilibrium)},$$

$$\tau_1 = \infty: \quad \mathscr{F} = \mathscr{F}^{0} \quad \text{(frozen (Vlasov) equilibrium)}.$$

$\mathscr{F}^0$ is the solution of $\mathscr{D}_1 \mathscr{F} = 0$. The distribution $\mathscr{F}^{\mathrm{eq}}$ need not be Maxwellian, other choices are possible. In the 'frozen' limit, the kinetic equation

$$\mathscr{D}_1 \mathscr{F}^0 = 0 \tag{74.10}$$

is sometimes termed the 'collisionless' Boltzmann equation, which is a little curious in view of the fact that his contribution hinged on the form of the collision integral.

Vlasov's kinetic equation may be cast into the same form as (74.10). It follows by taking the next simplest approximation to the collision integral after ignoring it completely as in (74.10). This is to adopt the product approximation

$$\mathscr{F}_2(\mathbf{q}_1, \mathbf{q}_2, \mathbf{p}_1, \mathbf{p}_2, t) = \mathscr{F}(\mathbf{q}_1, \mathbf{p}_1, t) \mathscr{F}(\mathbf{q}_2, \mathbf{p}_2, t) \tag{74.11}$$

in (74.6). Then as

$$\Theta_{12} = \frac{\partial \varphi}{\partial \mathbf{q}_1} \cdot \frac{\partial}{\partial \mathbf{p}_1} + \frac{\partial \varphi}{\partial \mathbf{q}_2} \cdot \frac{\partial}{\partial \mathbf{p}_2} \qquad \varphi = \varphi(|\mathbf{q}_1 - \mathbf{q}_2|),$$

we find that (74.6) may be written

$$\left(\mathscr{D}_1 - \frac{\partial U}{\partial \mathbf{q}_1} \cdot \frac{\partial}{\partial \mathbf{p}_1}\right) \mathscr{F} = 0, \tag{74.12}$$

where

$$U \equiv n \iint \varphi(|\mathbf{q}_1 - \mathbf{q}_2|)\mathscr{F}(\mathbf{q}_2, \mathbf{p}_2, t)\, \mathrm{d}\mathbf{q}_2\, \mathrm{d}\mathbf{p}_2.$$

If (74.12) is substituted into (74.8), the right-hand integral may be evaluated, leaving

$$\frac{\mathrm{d}S_1}{\mathrm{d}t} = 0, \tag{74.13}$$

which also holds for the collisionless Boltzmann equation. Both of these simple kinetic equations are reversible.

The Vlasov equation is particularly important in plasma physics, where the long-distance Coulomb collisions between particles means that the motion of each particle is determined by the cumulative action of *all* the particles in the plasma, rather than by a sequence of separate collisions. Consequently correlations between particles tend to be small, making (74.11) a good approximation. Irreversibility can be introduced by splitting the interaction force $-\partial U/\partial \mathbf{q}_1$ into rapidly fluctuating and slowly varying components. Most details of the fluctuations are suppressed by averaging over an ensemble, rather similar to the treatment of Brownian motion (§ 69.1), and an entropy

$$S_0 = -k \int \langle F_1 \rangle \ln \langle F_1 \rangle\, \mathrm{d}\mathbf{x} \geqslant S_1$$

is found to satisfy $\mathrm{d}S_0/\mathrm{d}t \geqslant 0$, where $\mathrm{d}/\mathrm{d}t$ is the rate of change on the slow time scale.

74.3. Use of time scales to effect closure

The procedure for finding kinetic equations just described suggests that for a more accurate kinetic equation one should start with the case $r = 2$ of (72.30), which is

$$\frac{\mathrm{d}S_2}{\mathrm{d}t} = -\tfrac{1}{2}kN \iint \mathscr{D}_2 F_2 \ln\left(\frac{F_2}{F_2^0}\right) \mathrm{d}\mathbf{x}_1\, \mathrm{d}\mathbf{x}_2 \geqslant 0 \qquad (\mathscr{D}_2 F_2^0 = 0),$$

and find instead of (74.9),

$$\mathscr{D}_2 \mathscr{F}_2 = (\mathscr{F}_2^0 - \mathscr{F}_2)/\tau_0, \tag{74.14}$$

where $\tau_0 \geqslant 0$ is called the *collision time* for a reason to be explained.

In a dilute neutral gas there exist three widely differing length scales: (1) the range r_0 of the interparticle potential, such that particles separated by more than r_0 are virtually non-interacting; (2) the mean free path $\lambda \sim (nr_0^2)^{-1}$; (3) the macroscopic length scale $L \sim |A/\nabla A|$, where A is a macroscopic variable. Associated with these lengths are three time scales: the collision time $\tau_0 = r_0/\bar{c}$, the kinetic time $\tau_1 = \lambda/\bar{c}$, and the macroscopic time $\mathscr{T} = L/u_s$, where $\bar{c}$ is the mean thermal speed of the particles and $u_s \sim \bar{c}$ is the speed of sound. Then, as $r_0 \ll \lambda \ll L$,

$$\tau_0 \ll \tau_1 \ll \mathscr{T}, \qquad \varepsilon \equiv \tau_0/\tau_1 = nr_0^3 \ll 1. \tag{74.15}$$

As (74.6) may be written in a dimensionless form with the small parameter ε multiplying the right-hand side, it follows that particle interactions cause $\mathscr{F}_2$ to change more rapidly than $\mathscr{F}$; similarly from (74.5) $\mathscr{F}_3$ changes more rapidly than $\mathscr{F}_2$, and so on, the ratio in each case being ε. Thus as $\mathscr{F}$ changes on the kinetic time scale τ_1, it follows that $\mathscr{F}_2$ changes on the collision time scale τ_0, which justifies our using τ_0 as the relaxation time in (74.14).

We are interested in relaxation processes occurring in the so-called *kinetic stage*, $\tau_0 < t < \tau_1$, for which (74.14) gives

$$\mathscr{F}_2 = \mathscr{F}_2^0 - \tau_0 \mathscr{D}_2 \mathscr{F}_2 = \mathscr{F}_2^0 - \varepsilon\tau_1 \mathscr{D}_2 \mathscr{F}_2,$$

so provided the operator $\tau_1 \mathscr{D}_2$ is of order unity when applied to $\mathscr{F}_2$, that is, *provided $\mathscr{F}_2$ changes on the kinetic time scale*, we may write

$$\mathscr{F}_2 = \mathscr{F}_2^0[1 + O(\varepsilon)]. \tag{74.16}$$

This basic assumption may be expressed by saying that, following an initial stage $0 < t < \tau_0$ during which $\mathscr{F}_2$ changes very rapidly, in the kinetic stage it becomes *synchronized* with $\mathscr{F}$. It allows us to write (74.6) as

$$\mathscr{D}_1 \mathscr{F} = n \int \Theta_{12} \mathscr{F}_2^0 \, \mathrm{d}\mathbf{x}_2 + O(\varepsilon^2), \tag{74.17}$$

in which the integral term is $O(\varepsilon)$. Further, as $\mathscr{D}_1 \mathscr{F} = O(\varepsilon)$, we are free to adopt 'Vlasov' approximation $\mathscr{D}_1 \mathscr{F}^0 = 0$ in the evaluation of $\mathscr{F}_2^0$, for which our equations now read

$$\begin{array}{cc} \mathscr{D}_2 \mathscr{F}_2^0 = 0 \quad (\tau_0 \approx 0), & \mathscr{D}_1 \mathscr{F}^0 = 0 \quad (\tau_1 \approx \infty). \\ \text{(relaxed)} & \text{(frozen)} \end{array} \tag{74.18}$$

For the stage $\varepsilon\tau_0 < t < \tau_0$, we use $\mathrm{d}S_3/\mathrm{d}t \geqslant 0$, and find by the same reasoning as above that (74.17) is replaced by

$$\mathscr{D}_1 \mathscr{F} = n \int \Theta_{12} \mathscr{F}_2 \, \mathrm{d}\mathbf{x}_2, \qquad \mathscr{D}_2 \mathscr{F}_2 = n \int (\Theta_{13} + \Theta_{23}) \mathscr{F}_3^0 \, \mathrm{d}\mathbf{x}_3 + O(\varepsilon^2),$$

and (74.18) by

$$\mathscr{D}_3 \mathscr{F}_3^0 = 0 \quad (\varepsilon\tau_0 \approx 0), \qquad \mathscr{D}_2 \mathscr{F}_2^0 = 0 \quad (\tau_0 \approx \infty), \qquad \mathscr{D}_1 \mathscr{F}^0 = 0 \quad (\tau_1 \approx \infty).$$

And so on to earlier stages, with closure being obtained at each stage; heavy algebra forces an early termination of the sequence.

There remains the task of solving (74.18) for $\mathscr{F}_2^0$, which we shall do in § 74.6 after first explaining the important contributions of Bogoliubov, Frieman, and Sandri.

74.4. *Bogoliubov's functional ansatz*

Bogoliubov (1946) gave the first new approach after Boltzmann to the problem of devising a kinetic equation for $\mathscr{F}$. He recognized the importance of the three basic time scales τ_0, τ_1, and $\mathscr{T}$, introduced above, and the fact that, whereas in the initial stage $\mathscr{F}, \mathscr{F}_2, \mathscr{F}_3, \ldots$ will change on time scales $\tau_1, \tau_0, \varepsilon\tau_0, \ldots$ respectively, in the kinetic stage it is intuitively reasonable to assume that all $\mathscr{F}_r$, $r \geqslant 2$, are in relaxed equilibrium and are synchronized with $\mathscr{F}$. We may indicate that the time dependence of $\mathscr{F}_r$ follows that of $\mathscr{F}$ by writing

$$\mathscr{F}_r(\mathbf{x}_1, \ldots \mathbf{x}_r, t) = \mathscr{F}_r(\mathbf{x}_1, \ldots \mathbf{x}_r | \mathscr{F}(t)) \qquad (\tau_0 < t < \tau_1),$$

which is known as Bogoliubov's *functional ansatz*. It enables us to evaluate $\partial\mathscr{F}_r/\partial t$ as a functional derivative, that is,

$$\frac{\partial \mathscr{F}_r}{\partial t} = \frac{\delta \mathscr{F}_r}{\delta \mathscr{F}} \frac{\partial \mathscr{F}}{\partial t} = \frac{\delta \mathscr{F}_r}{\delta \mathscr{F}}\left[-\mathscr{H}_1 \mathscr{F} + n \int \Theta_{12} \mathscr{F}_2 \, \mathrm{d}\mathbf{x}_2\right], \qquad (74.19)$$

by (74.6).

With the density n as an expansion parameter (equivalent to using ε) $\mathscr{F}_r$ may be expressed

$$\mathscr{F}_r(\mathbf{x}_1, \ldots \mathbf{x}_r | \mathscr{F}) = \mathscr{F}_r^{(0)}(\ldots|.) + n\mathscr{F}_r^{(1)}(\ldots|.) + n^2\mathscr{F}_r^{(2)}(\ldots|.) + \ldots, \qquad (74.20)$$

which is substituted in (74.5) after first using (74.19) to remove $\partial\mathscr{F}_r/\partial t$. Finally we equate equal powers of n to arrive at a chain of equations that are closed at whatever power of n they are terminated. Naturally this method of closure raises the question of convergence, and it is unfortunately the case that all terms in the series beyond n^2 diverge.

74.5. *Use of multiple time variables*

Frieman (1963) and Sandri (1963) used the techniques of multiple time variables to give greater clarity to the Bogoliubov theory and to generalize it in minor respects. In this technique the notion of time is extended by introducing a set $\tau_0, \tau_1, \tau_2, \ldots$ of disparate times, formally regarded as being independent variables and related to the original time variable by

$$\frac{\mathrm{d}\tau_n}{\mathrm{d}t} = \varepsilon^n \qquad (n = 0, 1, 2, \ldots,\ 0 < \varepsilon \ll 1), \qquad (74.21)$$

where the expansion parameter ε depends on the problem in question (it is given by (74.15) for a dilute neutral gas). We interpret (74.21) by describing τ_0 as a 'fast' time, τ_1 as a 'slower' time such that one unit in τ_1 is ε^{-1} times that in τ_0, and so on down the spectrum of slower and slower time variables. These variables are intended to correspond broadly with the various intrinsic time scales occurring in the system.

By increasing the number of time variables from one to several, we have in effect *embedded* the original problem in an extended space, a procedure that can be justified mathematically but which here we shall be content to accept as a heuristic move for solving the closure problem. The resulting freedom arising from this embedding is utilized to remove secularities (that is, unlimited increases with time in the magnitude of dependent variables) arising in higher-order terms.

From (74.21) it follows that

$$\frac{\partial}{\partial t} = \frac{\partial}{\partial \tau_0} + \varepsilon \frac{\partial}{\partial \tau_1} + \varepsilon^2 \frac{\partial}{\partial \tau_2} + \dots . \tag{74.22}$$

The expansion in (74.20) is replaced by

$$\mathscr{F}_r = \sum_{n=0} \varepsilon^n \mathscr{F}_r^{(n)}(\mathbf{x}_1, \dots \mathbf{x}_r, \tau_0, \tau_1, \tau_2, \dots),$$

so that

$$\frac{\partial \mathscr{F}_r}{\partial t} = \frac{\partial \mathscr{F}_r^{(0)}}{\partial \tau_0} + \varepsilon \left(\frac{\partial \mathscr{F}_r^{(1)}}{\partial \tau_0} + \frac{\partial \mathscr{F}_r^{(0)}}{\partial \tau_1} \right) + \varepsilon^2 \left(\frac{\partial \mathscr{F}_r^{(2)}}{\partial \tau_0} + \frac{\partial \mathscr{F}_r^{(1)}}{\partial \tau_1} + \frac{\partial \mathscr{F}_r^{(0)}}{\partial \tau_2} \right) + \dots .$$

These expansions are now used in the BBGKY hierarchy (74.5), in which the right-hand side is $O(\varepsilon)$ compared with the terms on the left-hand side. On equating equal powers of ε we obtain

$$\left.\begin{aligned}
\varepsilon^0: &\quad \frac{\partial \mathscr{F}_r^{(0)}}{\partial \tau_0} + \mathscr{H}_r \mathscr{F}_r^{(0)} = 0,\\
\varepsilon^1: &\quad \frac{\partial \mathscr{F}_r^{(1)}}{\partial \tau_0} + \mathscr{H}_r \mathscr{F}_r^{(1)} = -\frac{\partial \mathscr{F}_r^{(0)}}{\partial \tau_1} + \frac{n}{\varepsilon} \int \sum_{i=1}^{r} \Theta_{i,r+1} \mathscr{F}_{r+1}^{(0)}\, \mathrm{d}\mathbf{x}_{r+1},\\
\varepsilon^2: &\quad \frac{\partial \mathscr{F}_r^{(2)}}{\partial \tau_0} + \mathscr{H}_r \mathscr{F}_r^{(2)} = -\frac{\partial \mathscr{F}_r^{(0)}}{\partial \tau_2} - \frac{\partial \mathscr{F}_r^{(1)}}{\partial \tau_1} + \frac{n}{\varepsilon} \int \sum_{i=1}^{r} \Theta_{i,r+1} \mathscr{F}_{r+1}^{(1)}\, \mathrm{d}\mathbf{x}_{r+1},
\end{aligned}\right\} \tag{74.23}$$

and so forth. Given appropriate initial conditions for $\mathscr{F}_r^{(s)}$, these equations may be solved in turn and closure effected at any stage.

74.6. *Irreversibility via initial conditions*

Eqns (74.17) and (74.18) still possess the reversibility of the Liouville equation from which they have been derived, and our next task is to find a solution $\mathscr{F}_2^0$ that eliminates this feature.

Eqns (74.18) are special cases of the r-particle Liouville equation

$$\mathscr{D}_r \mathscr{F}_r^0 = \left(\frac{\partial}{\partial t} + \mathscr{H}_r \right) \mathscr{F}_r^0 = 0, \tag{74.24}$$

which shows that $\mathscr{F}_r^0$ is unchanged in magnitude in a frame convected with the phase-fluid moving through Γ_r-space (see § 30.2). Thus if $(\mathbf{x}_1', \ldots \mathbf{x}_r')$ is the phase of $\mathscr{F}_r^0$ at time t' and $(\mathbf{x}_1, \ldots \mathbf{x}_r)$ is its phase at t, (74.24) has the solution

$$\mathscr{F}_r^0(\mathbf{x}_1, \ldots \mathbf{x}_r, t) = \mathscr{F}_r^0(\mathbf{x}_1', \ldots \mathbf{x}_r', t').$$

But (74.24) also has the formal solution

$$\mathscr{F}_r^0(\mathbf{x}_1, \ldots \mathbf{x}_r, t) = \exp\left[-(t-t')\mathscr{H}_r(\mathbf{x}_1, \ldots \mathbf{x}_r)\right]\mathscr{F}_r^0(\mathbf{x}_1, \ldots \mathbf{x}_r, t'), \quad (74.25)$$

whence we may regard the operator $\exp\left[-(t-t')\mathscr{H}_r\right]$ as changing the phase from $(\mathbf{x}_1, \ldots \mathbf{x}_r)$ to $(\mathbf{x}_1', \ldots \mathbf{x}_r')$ as the time changes from t to t':

$$(\mathbf{x}_1', \ldots \mathbf{x}_r') = \exp(-\tau\mathscr{H}_r)(\mathbf{x}_1, \ldots \mathbf{x}_r) \qquad (\tau = t-t'). \quad (74.26)$$

Comparison of (74.5) and (74.24) shows that during this change the set of r particles do not interact with the remaining particles.

Take t' to be the *initial* time, and write $\mathscr{F}_2^0(\mathbf{x}_1', \mathbf{x}_2', t')$ in the form

$$\mathscr{F}_2^0(\mathbf{x}_1', \mathbf{x}_2', t') = \mathscr{F}^0(\mathbf{x}_1', t')\mathscr{F}^0(\mathbf{x}_2', t') + G_{12}(\mathbf{x}_1', \mathbf{x}_2', t').$$

We now introduce a key assumption, essential to the whole theory, namely, that any initial correlation G_{12} decays to *zero* on the collision time scale τ_0, which means that on the kinetic time scale we may choose as an initial state one in which G_{12} has already vanished. It is this specification of an *initial* condition that, distinguishing past from future, introduces the 'arrow of time', and, like Boltzmann's hypothesis of molecular chaos (§ 72.2), it confers irreversibility on the mathematical description of the system. It is less restrictive than Boltzmann's hypothesis, requiring as it does molecular chaos at only one instant in the past.

With $G_{12}(t') = 0$, eqns (74.25) for $r = 1, 2$ give

$$\mathscr{F}_2^0(\mathbf{x}_1, \mathbf{x}_2, t) = \mathscr{L}(\mathbf{x}_1, \mathbf{x}_2, \tau)\mathscr{F}^0(\mathbf{x}_1, \mathrm{t})\mathscr{F}^0(\mathbf{x}_2, t), \quad (74.27)$$

where $\mathscr{L}$ is the operator

$$\mathscr{L}(\mathbf{x}_1, \mathbf{x}_2, t) = \exp\{-t[\mathscr{H}_2(\mathbf{x}_1, \mathbf{x}_2) - \mathscr{H}_1(\mathbf{x}_1) - \mathscr{H}_1(\mathbf{x}_2)]\} = \exp(\tau\Theta_{12}).$$

In Fig. 17 is depicted the action of the operator $\mathscr{L}$ on the phases $\mathbf{x}_1 = (\mathbf{q}_1, \mathbf{p}_1)$, $\mathbf{x}_2 = (\mathbf{q}_2, \mathbf{p}_2)$. First by an application of (74.26)

$$\exp(-\tau\mathscr{H}_2)(\mathbf{x}_1, \mathbf{x}_2) = (\mathbf{x}_1', \mathbf{x}_2') = (\mathbf{q}_1', \mathbf{p}_1'; \mathbf{q}_2', \mathbf{p}_2')$$

traces the particles back to their initial phases at t', and secondly

$$\exp\{\tau[\mathscr{H}_1(\mathbf{x}_1) + \mathscr{H}_1(\mathbf{x}_2)]\}(\mathbf{x}_1', \mathbf{x}_2') = (\mathbf{Q}_1, \mathbf{P}_1; \mathbf{Q}_2, \mathbf{P}_2)$$

advances them to the phases they would have achieved at t had there been no interaction between them. If it may be assumed that the external forces

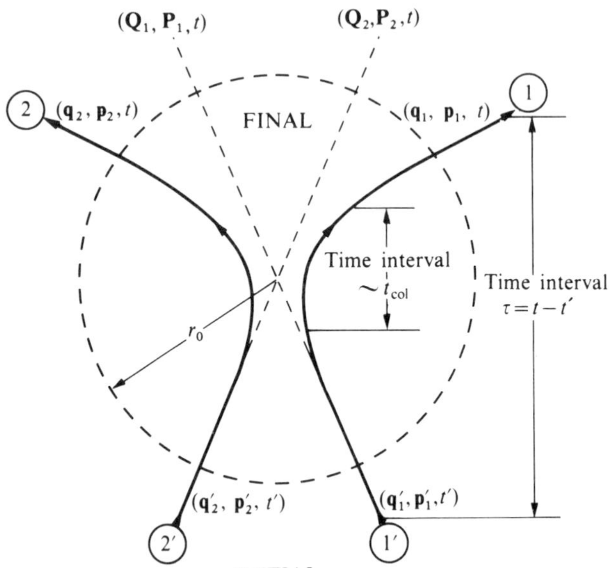

FIG. 17. Action of $\mathscr{L}$ on the phases of two particles.

have negligible effect during the short time of the interaction, then the momenta $\mathbf{P}_i$, $i = 1, 2$, will be the same as the initial momenta $\mathbf{p}_i'$ that yield the phases $\mathbf{x}_i$ at time t.

The superscript 0 may now be omitted and eqn (74.27) written as (cf. (74.11))

$$\mathscr{F}_2^0(\mathbf{x}_1, \mathbf{x}_2, t) = \mathscr{F}(\mathbf{Q}_1, \mathbf{P}_1, t)\mathscr{F}(\mathbf{Q}_2, \mathbf{P}_2, t) + O(\varepsilon),$$

where (74.28)

$$\mathbf{P}_i = P_i(\mathbf{x}_1, \mathbf{x}_2, -\tau) \qquad (i = 1, 2),$$

and similarly for $\mathbf{Q}_i$. Notice from the figure that provided $\tau \gg t_{col}$, where $t_{col} = r_0/\bar{c}$ is the collision time, $\mathbf{P}_i$ and $\mathbf{Q}_i$ will be independent of τ and we may replace $(74.28)_2$ by

$$\mathbf{P}_i = \mathbf{P}_i(\mathbf{x}_1, \mathbf{x}_2, -\infty). \tag{74.29}$$

74.7. *The generalized Boltzmann equation*

Bogoliubov's generalization of Boltzmann's kinetic equation now follows from (74.17) and (74.28),

$$\mathscr{D}_1\mathscr{F} = \left(\frac{\partial}{\partial t} + \frac{\mathbf{p}_1}{m} \cdot \frac{\partial}{\partial \mathbf{q}_1} + \mathbf{K}_1 \cdot \frac{\partial}{\partial \mathbf{p}_1}\right)\mathscr{F} = n \int \Theta_{12}\mathscr{F}(\mathbf{P}_1, \mathbf{Q}_1, t)\mathscr{F}(\mathbf{P}_2, \mathbf{Q}_2, t)\,\mathrm{d}\mathbf{x}_2; \tag{74.30}$$

the functions $P_i(\mathbf{x}_1, \mathbf{x}_2, -\tau)$ and $Q_i(\mathbf{x}_1, \mathbf{x}_2, -\tau)$ depends on the details of the two-particle intersection. In § 75.2 we shall derive Boltzmann's equation from it by introducing two approximations for the collisional process. Boltzmann's original derivation of his equation (§ 75.1) was quite different and shorter, being based more directly on physical intuition (none the worse for that!), but unlike BBGKY-based treatments it does not admit generalizations or extensions, at least by methods that are immediately obvious.

It remains to show how eqns (74.23) of the multiple-time-variable approach also lead to (74.30) in the first approximation. First, it is apparent from the first of (74.23) that $\mathscr{F}_2^{(0)}$ and $\mathscr{F}^{(0)}$ satisfy the same equations as $\mathscr{F}_2^0$ and $\mathscr{F}^0$ save that t is replaced by the fast time τ_0. Consequently, we can write down an equation like (74.28) for $\mathscr{F}_2^{(0)}$, in which t is replaced by τ_0 and $\mathscr{F}$ by $\mathscr{F}^{(0)}$. Secondly, the case $r = 1$ of $(74.23)_2$ gives

$$\frac{\partial \mathscr{F}^{(1)}}{\partial \tau_0} + \mathscr{H}_1 \mathscr{F}^{(1)} = -\frac{\partial \mathscr{F}^{(0)}}{\partial \tau_1} + \frac{n}{\varepsilon} \int \Theta_{12} \mathscr{F}^{(0)}(\mathbf{Q}_1, \mathbf{P}_1, \tau_0) \mathscr{F}^{(0)}(\mathbf{Q}_2, \mathbf{P}_2, \tau_0)\, \mathrm{d}\mathbf{x}_2,$$

on replacing $\mathscr{F}_2^{(0)}$ by the expression for it like (74.28). Let Φ denote the right-hand side of this equation, then its solution is easily verified to be

$$\mathscr{F}^{(1)}(\tau_0) = \exp(-\tau_0 \mathscr{H}_1) \mathscr{F}^{(1)}(0) + \int_0^{\tau_0} \exp[(t' - \tau_0)\mathscr{H}_1] \Phi(\mathbf{x}_1, t')\, \mathrm{d}t',$$

which increases without limit as $\tau_0 \to \infty$ unless Φ is zero. We are free to choose the derivative $\partial \mathscr{F}^{(0)}/\partial \tau_1$ to remove this secular behaviour in $\mathscr{F}^{(1)}$, thus

$$\varepsilon \frac{\partial \mathscr{F}^{(0)}}{\partial \tau_1} = n \int \Theta_{12} \mathscr{F}^{(0)}(\mathbf{Q}_1, \mathbf{P}_1, \tau_0, \tau_1, \dots) \mathscr{F}^{(0)}(\mathbf{Q}_2, \mathbf{P}_2, \tau_0, \tau_2, \dots)\, \mathrm{d}\mathbf{x}_2,$$

where we have made explicit the dependence on the remaining time variables $\tau_1, \tau_2, \dots$. If to this equation we add the case $r = 1$ of $(74.23)_1$, the left-hand side becomes

$$\left(\frac{\partial}{\partial \tau_0} + \varepsilon \frac{\partial}{\partial \tau_1} + \mathscr{H}_1\right) \mathscr{F}^{(0)} = \left(\frac{\partial}{\partial t} + \mathscr{H}_1\right) \mathscr{F} = \mathscr{D}_1 \mathscr{F},$$

on returning to the single variable t in (74.22) and ignoring the difference between $\mathscr{F}^{(0)}$ and $\mathscr{F}$. The final result now appears as in (74.30).

So far we have described only the first steps (to orders ε^0 and ε^1) of the multiple-time-variable treatment, and it is clear that the procedure may be continued to provide corrections to (74.30). However, just as Bogoliubov's synchronization method fails to give convergent higher-order corrections, so does the use of time-scaling in the 'simple' manner described above. For uniformly valid asymptotic solutions, a rather more complicated form of scaling is required to remove secularities from the higher-order terms (Ramanathan and Sandri 1969).

75. Boltzmann's equation

75.1. Classical derivation for a multi-component system

We shall return to the account in § 71 of a multi-component system, and obtain an explicit expression for the collision term $(\partial f_i/\partial t)_c$ following Boltzmann's original method.† The derivation is for a gas sufficiently dilute for only binary collisions to be significant. It is also assumed that walls and other boundaries are a sufficient number of mean free paths distant to have a negligible effect and that the external force $\mathbf{F}_i$ does not affect the collision cross-section. The most important assumption is that the velocity of a particle is uncorrelated with its position. This is the hypothesis of molecular chaos mentioned in § 72.2; it introduces irreversibility by ensuring positive parity under motion reversal for $(\partial f_i/\partial t)_c$, as will be seen. That it must be an approximation may be inferred by considering the relative motion of a pair of particles in a given volume element. If they are found to be approaching each other their positions and velocities will have a negligible probability of being correlated in a dilute gas, whereas if they are receding from each other, there is certainly some small probability that they have recently collided and therefore that their positions and velocities are correlated through the dynamics of the collision.

Let a particle P_i of the ith component collide with a particle P_j of the jth component, with relative velocities

$$\mathbf{g}_{ij} \equiv \mathbf{w}_i - \mathbf{w}_j = g_{ij}\mathbf{k}_{ij}; \qquad \mathbf{g}'_{ij} \equiv \mathbf{w}'_i - \mathbf{w}'_j = g'_{ij}\mathbf{k}'_{ij}, \tag{75.1}$$

(before collision) (after collision)

where $\mathbf{k}_{ij}$ and $\mathbf{k}'_{ij}$ are unit vectors; and for the velocities of the centre of mass before and after the collisions,

$$\mathbf{G}_{ij} = M_i\mathbf{w}_i + M_j\mathbf{w}_j, \qquad \mathbf{G}'_{ij} = M_i\mathbf{w}'_i + M_j\mathbf{w}'_j, \tag{75.2}$$

where $M_i = m_i/(m_i+m_j)$, $M_j \equiv m_j/(m_i+m_j)$. By conservation of momentum and energy it is readily shown that

$$\mathbf{G}_{ij} = \mathbf{G}'_{ij}, \qquad g_{ij} = g'_{ij}. \tag{75.3}$$

Shortly we shall require the values of the Jacobians

$$J \equiv \frac{\partial(\mathbf{G}_{ij}, \mathbf{g}_{ij})}{\partial(\mathbf{w}_j, \mathbf{w}_i)}, \qquad J' \equiv \frac{\partial(\mathbf{G}_{ij}, \mathbf{g}'_{ij})}{\partial(\mathbf{w}'_j, \mathbf{w}'_i)}.$$

† While this method does not fall into the 'entropy production rate' approach used to obtain constitutive relations in this text, it is such an important example of a constitutive relation arrived at intuitively, that it must be included for completeness. Further, it must be admitted that it is a somewhat easier treatment than the BBGKY approach described in § 74 and to be completed shortly in § 75.2. However, it does obscure the nature of the approximations involved.

By (75.1) and (75.2)

$$J = \frac{\partial(\mathbf{w}_i - M_j\mathbf{g}_{ij}, \mathbf{g}_{ij})}{\partial(\mathbf{w}_j, \mathbf{w}_i)} = \frac{\partial(\mathbf{w}_i, \mathbf{g}_{ij})}{\partial(\mathbf{w}_j, \mathbf{w}_i)} = -\frac{\partial(\mathbf{w}_i, \mathbf{w}_i - \mathbf{w}_j)}{\partial(\mathbf{w}_i, \mathbf{w}_j)} = \frac{\partial(\mathbf{w}_i, \mathbf{w}_j)}{\partial(\mathbf{w}_i, \mathbf{w}_j)} = 1.$$

Similarly, $J' = 1$.

Now the number of particles P_i arriving at a target P_j from a cone of solid angle $\mathrm{d}\mathbf{k}_{ij}$ (see Fig. 18) is $g_{ij}f_i\,\mathrm{d}\mathbf{w}_i$ per second where by f_i is meant $f_i(\mathbf{r}, \mathbf{w}_i, t)$,

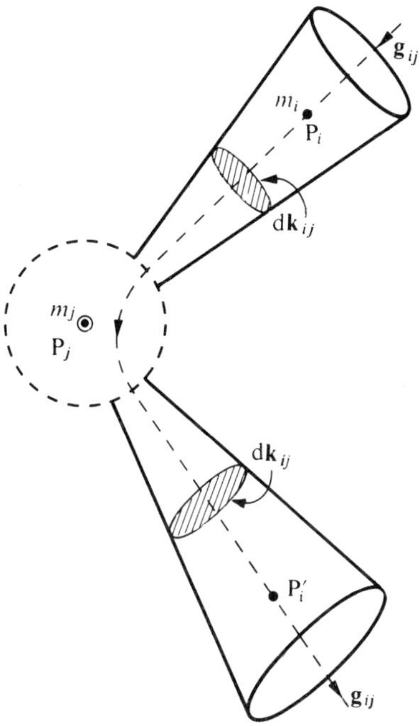

FIG. 18. Binary collision.

and as there are $f_j\,\mathrm{d}\mathbf{w}_j\,\mathrm{d}\mathbf{r}$ such targets in $\mathrm{d}w_j\,\mathrm{d}\mathbf{r}$, it follows from the assumption of molecular chaos that the collision rate is $f_if_jg_{ij}\,\mathrm{d}\mathbf{w}_i\,\mathrm{d}\mathbf{w}_j\,\mathrm{d}\mathbf{r}$. Let $W(\mathbf{k}_{ij}|\mathbf{k}_{ij}'; g_{ij}')\,\mathrm{d}\mathbf{k}_{ij}$ be the probability that the collision changes $\mathbf{k}_{ij}$ lying in $\mathrm{d}\mathbf{k}_{ij}$ into $\mathbf{k}_{ij}'$ lying in $\mathrm{d}\mathbf{k}_{ij}'$, then $\mathrm{d}\mathbf{w}_i\,\mathrm{d}\mathbf{r}[f_if_jg_{ij}W(\mathbf{k}_{ij}|\mathbf{k}_{ij}'; g_{ij})\,\mathrm{d}\mathbf{k}_{ij}'\,\mathrm{d}\mathbf{w}_j]$ is the number of particles lost from the element $\mathrm{d}\mathbf{w}_i\,\mathrm{d}\mathbf{r}$ owing to the particular set of encounters with P_j particles such that $\mathbf{w}_j$ and $\mathbf{k}_{ij}'$ lie in the ranges $\mathrm{d}\mathbf{w}_j$ and $\mathrm{d}\mathbf{k}_{ij}'$. The total loss for all values of $\mathbf{w}_j$ and $\mathbf{k}_{ij}'$ is thus

$$\mathrm{d}\mathbf{w}_i\,\mathrm{d}\mathbf{r}\iint f_if_jg_{ij}W(\mathbf{k}_{ij}|\mathbf{k}_{ij}'; g_{ij})\,\mathrm{d}\mathbf{k}_{ij}'\,\mathrm{d}\mathbf{w}_j \text{ per second.} \tag{75.4}$$

In addition to the (direct) collisions $\mathbf{w}_i \to \mathbf{w}_i'$ that scatter the particles P_i out of $\mathrm{d}\mathbf{w}_i\,\mathrm{d}\mathbf{r}$, there are others—namely, the inverse collisions $-\mathbf{w}_i' \to -\mathbf{w}_i$, that scatter particles P_i *into* $\mathbf{dw}_i\,\mathbf{dr}$. On reversing the direction of the collisions in (75.4), we find that the *gain* of particles due to such collisions is

$$\mathbf{dw}'_i\,\mathbf{dr}\iint f'_i f'_j g'_{ij} W(-\mathbf{k}'_{ij}|-\mathbf{k}_{ij};g_{ij})\,\mathbf{dk}_{ij}\,\mathbf{dw}'_j \text{ per second.} \tag{75.5}$$

Now $\mathbf{dw}_i\,\mathbf{dw}_j = |J^{-1}|\,\mathbf{dg}_{ij}\,\mathbf{dG}_{ij}$, where J is the Jacobian defined above. Hence, as $|J| = 1$,

$$\mathbf{dw}_i\,\mathbf{dw}_j = \mathbf{dg}_{ij}\,\mathbf{dG}_{ij} = g_{ij}^2\,\mathrm{d}g_{ij}\,\mathbf{dk}_{ij}\,\mathbf{dG}_{ij}, \tag{75.6}$$

on expressing the volume element $\mathbf{dg}_{ij}$ in spherical polar coordinates. From (75.3) and (75.6) it follows that

$$\mathbf{dk}'_{ij}\,\mathbf{dw}_i\,\mathbf{dw}_j = g_{ij}^2\,\mathrm{d}g_{ij}\,\mathbf{dk}_{ij}\,\mathbf{dk}'_{ij}\,\mathbf{dG}_{ij} = \mathbf{dk}_{ij}\,\mathbf{dw}'_i\,\mathbf{dw}'_j. \tag{75.7}$$

The particle equations of motion are invariant under time-reversal, with the consequence that the differential cross-section W must be unchanged by the transformation $\mathbf{k}_{ij} \to -\mathbf{k}'_{ij}$ (see Fig. 18). Further, as the interaction forces have central symmetry, W must also be invariant to an inversion of the coordinates. Hence

$$W(\mathbf{k}_{ij}|\mathbf{k}'_{ij};g_{ij}) = W(-\mathbf{k}'_{ij}|-\mathbf{k}_{ij};g_{ij}) = W(\mathbf{k}'_{ij}|\mathbf{k}_{ij};g_{ij}). \tag{75.8}$$

Eqns (75.7) and (75.8) enable us to combine (75.4) and (75.5) into the form $\mathbf{dw}_i\,\mathbf{dr}\,C(f_i, f_j)$, where $C(f_i, f_j)$ is the functional

$$C(f_i, f_j) \equiv \iint (f'_i f'_j - f_i f_j) g_{ij} W(\mathbf{k}_{ij}|\mathbf{k}'_{ij};g_{ij})\,\mathbf{dk}'_{ij}\,\mathbf{dw}_j, \tag{75.9}$$

known as Boltzmann's collision integral. This is the number written as C_{ij} in (71.4); thus

$$\left(\frac{\partial f_i}{\partial t}\right)_{\mathrm{c}} = \sum_j C(f_i, f_j), \tag{75.10}$$

from which follows Boltzmann's transport equation

$$D_i f_i = \frac{\partial f_i}{\partial t} + \mathbf{w}_i \cdot \frac{\partial f_i}{\partial \mathbf{r}} + \mathbf{F}_i \cdot \frac{\partial f_i}{\partial \mathbf{w}_i} = \sum_j C(f_i, f_j). \tag{75.11}$$

As each term in the integrand of the collision integral has even parity under time-reversal $(\partial f_i/\partial t)_{\mathrm{c}}$ has even parity, whereas $\mathrm{D}_i f_i$ has odd parity.

If there is only one component, the subscripts i and j used above stand for different classes of the same species. In this case, setting $i = 1$ and $j = 2$, Boltzmann's equation becomes

$$\begin{aligned} \mathrm{D}f_1 &= \frac{\partial f_1}{\partial t} + \mathbf{w} \cdot \frac{\partial f_1}{\partial \mathbf{r}} + \mathbf{F}_1 \cdot \frac{\partial f_1}{\partial \mathbf{w}} \\ &= \iint (f'_1 f'_2 - f_1 f_2) g_{12} W(\mathbf{k}_{12}|\mathbf{k}'_{12};g_{12})\,\mathbf{dk}'_{12}\,\mathbf{dw}_2, \end{aligned} \tag{75.12}$$

where

$$f_1 = f(\mathbf{r}, \mathbf{w}_1, t), \quad f_2 = f(\mathbf{r}, \mathbf{w}_2, t), \quad f_1' = f(\mathbf{r}, \mathbf{w}_1', t), \quad \text{and} \quad f_2' = f(\mathbf{r}, \mathbf{w}_2', t)$$

refer to the same distribution function.

75.2. Derivation from BBGKY hierarchy

We begin with Bogoliubov's result (74.30)

$$\mathscr{D}_1\mathscr{F} = n\iint \Theta_{12}\mathscr{F}(\mathbf{Q}_1, \mathbf{P}_1, t)\mathscr{F}(\mathbf{Q}_2, \mathbf{P}_2, t)\,\mathrm{d}\mathbf{q}_2\,\mathrm{d}\mathbf{p}_2, \tag{75.13}$$

where

$$\mathbf{P}_i = \mathbf{P}_i(\mathbf{q}_i, \mathbf{q}_2, \mathbf{p}_1, \mathbf{p}_2, \tau) \qquad (i = 1, 2),$$

and similarly for $\mathbf{Q}_i$. We shall show that with two simple approximations the collision integral in (75.13) may be reduced to a form like that appearing in (75.12).

The first approximation is to accept that the time interval τ between the instant of zero correlations and the present is much larger than the collision time t_{col}. As indicated in § 74.6, this permits us to replace τ by ∞, and hence to take $\mathbf{P}_i$ as being the (constant) momenta of the particles an infinite time in the past. The second approximation is to assume $\mathscr{F}$ to be spatially uniform over distances of the order of the range r_0 of the interparticle potential (see Fig. 17). This means that in the collision integral the *explicit* dependence of $\mathscr{F}$ on position may be neglected; of course its implicit dependence via $\mathbf{P}_i$ remains important.

As $\mathbf{P}_1$ and $\mathbf{P}_2$ are constants of the motion, independent of time, $\mathscr{D}_2\mathbf{P}_1 = \mathscr{H}_2\mathbf{P}_1 = 0$, and consequently

$$\mathscr{H}_2[\mathscr{F}(\mathbf{P}_1)\mathscr{F}(\mathbf{P}_2)] = 0, \tag{75.14}$$

where we have omitted position and time from the functional dependence of $\mathscr{F}$ as being merely parameters as far as the operator $\mathscr{H}_2$ is concerned. It follows from the definition of $\mathscr{H}_2$ given in (72.24) that (75.14) is equivalent to

$$\Theta_{12}\mathscr{F}(\mathbf{P}_1)\mathscr{F}(\mathbf{P}_2) = \frac{1}{m}\left(\mathbf{p}_1\cdot\frac{\partial}{\partial\mathbf{q}_1}+\mathbf{p}_2\cdot\frac{\partial}{\partial\mathbf{q}_2}\right)\mathscr{F}(\mathbf{P}_1)\mathscr{F}(\mathbf{P}_2).$$

Now $\mathbf{P}_1$ and $\mathbf{P}_2$ can depend on $\mathbf{q}_1$ and $\mathbf{q}_2$ only through the difference $(\mathbf{q}_1-\mathbf{q}_2)$, otherwise the collisional process would not be invariant to spatial translation. Hence

$$\Theta_{12}\mathscr{F}(\mathbf{P}_1)\mathscr{F}(\mathbf{P}_2) = \frac{1}{m}(\mathbf{p}_2-\mathbf{p}_1)\cdot\frac{\partial}{\partial\mathbf{q}_2}\mathscr{F}(\mathbf{P}_1)\mathscr{F}(\mathbf{P}_2). \tag{75.15}$$

We take a cylindrical coordinate system (ℓ, ε, ξ) for $\mathbf{q}_2$, with origin at particle 1 ($\mathbf{q}_1 = 0$) and axis parallel to $(\mathbf{p}_2 - \mathbf{p}_1)$, as shown in Fig. 19. Then

$$(\mathbf{p}_2 - \mathbf{p}_1) \cdot \frac{\partial}{\partial \mathbf{q}_2} = |\mathbf{p}_2 - \mathbf{p}_1| \cdot \frac{\partial}{\partial \xi},$$

$$\mathrm{d}\mathbf{q}_2 = \ell \, \mathrm{d}\ell \, \mathrm{d}\varepsilon \, \mathrm{d}\xi,$$

and (75.13) may be written

$$\mathscr{D}_1 \mathscr{F} = \frac{n}{m} \int (\mathbf{p}_2 - \mathbf{p}_1) \left\{ \int_0^\infty \int_0^{2\pi} [\mathscr{F}(\mathbf{P}_1)\mathscr{F}(\mathbf{P}_2)]_{\xi=-\infty}^{\xi=+\infty} \ell \, \mathrm{d}\ell \, \mathrm{d}\varepsilon \right\} \mathrm{d}\mathbf{p}_2 .$$

Recall that $\mathbf{P}_1$, $\mathbf{P}_2$ are the intial momenta with which particles 1 and 2 come from infinity and pass through the dynamical state $(\mathbf{q}_1, \mathbf{q}_2, \mathbf{p}_1, \mathbf{p}_2)$. Consider the dynamical state in which $\mathbf{q}_2 = (\ell, \varepsilon, -\infty)$. In this state the particles have not yet interacted, consequently the initial momenta are unchanged, that is, $\mathbf{P}_i = \mathbf{p}_i$. On the other hand, in the dynamical state $\mathbf{q}_2 = (\ell, \varepsilon, \infty)$ the particles have interacted, so that $\mathbf{P}_i$ cannot equal $\mathbf{p}_i$, but must have values $\mathbf{p}'_i$ such that $(\mathbf{p}'_1, \mathbf{p}'_2) \rightarrow (\mathbf{p}_1, \mathbf{p}_2)$ as a result of the collision. This is the inverse collision described after eqn (75.4) and shown in Fig. 19.

The kinetic equation now assumes the Boltzmann form

$$\mathscr{D}_1 \mathscr{F} = n \iiint [\mathscr{F}(\mathbf{p}'_1)\mathscr{F}(\mathbf{p}'_2) - \mathscr{F}(\mathbf{p}_1)\mathscr{F}(\mathbf{p}_2)] \frac{|\mathbf{p}_1 - \mathbf{p}_2|}{m} \ell \, \mathrm{d}\ell \, \mathrm{d}\varepsilon \, \mathrm{d}\mathbf{p}_2 . \quad (75.16)$$

In § 75.1 the differential cross-section $\ell \, \mathrm{d}\ell \, \mathrm{d}\varepsilon$ was expressed in the form $W \, \mathrm{d}\mathbf{k}_{12}$, so by (74.7) and (75.1) the identity of (75.12) and (75.16) immediately follows.

75.3. *The Maxwellian distribution function*

In § 24 we deduced the form for the equilibrium distribution function f^{eq} by an argument based on the functional dependence of the entropy. The usual treatment is more instructive and employs Boltzmann's kinetic equation.

First we derive a general, symmetric expression for the collisional rate of change of a transported property $\varphi(\mathbf{r}, \mathbf{w}, t)$ of a single component. Using the collision integral appearing in (75.12), we write the right-hand side of (71.8) as

$$\left(\frac{\partial \varphi}{\partial t}\right)_{\mathrm{c}} \equiv \int \varphi_1 \left(\frac{\partial f_1}{\partial t}\right)_{\mathrm{c}} \mathrm{d}\mathbf{w}_1$$

$$= \iiint \varphi_1 (f'_1 f'_2 - f_1 f_2) g_{12} W(\mathbf{k}_{12} | \mathbf{k}'_{12} ; g_{12}) \, \mathrm{d}\mathbf{k}'_{12} \, \mathrm{d}\mathbf{w}_2 \, \mathrm{d}\mathbf{w}_1 ,$$

where $\varphi_1 = \varphi(\mathbf{r}, \mathbf{w}_1, t)$ and we shall shortly introduce $\varphi_2 = \varphi(r, \mathbf{w}_2, t)$. Indeed as the result is clearly independent of whether class 1 or class 2

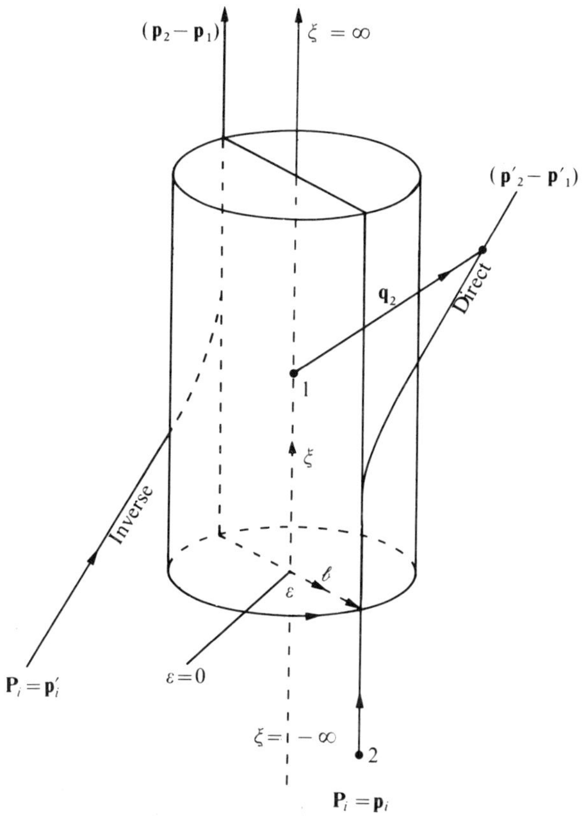

FIG. 19. Cylindrical coordinates.

particles are chosen in the integrand,

$$\left(\frac{\partial\varphi}{\partial t}\right)_c=\tfrac{1}{2}\iiiint(\varphi_1+\varphi_2)\times$$
$$\times(f_1'\,f_2'-f_1f_2)g_{12}^3W(\mathbf{k}_{12}|\mathbf{k}'_{12};g_{12})\,\mathrm{d}g_{12}\,\mathrm{d}\mathbf{G}_{12}\,\mathrm{d}\mathbf{k}_{12}\,\mathrm{d}\mathbf{k}'_{12}. \tag{75.17}$$

Let F, G, and φ be any functions of $\mathbf{r}$, $\mathbf{w}$, and t, then $\varphi_1F'_1G'_2$ in the integrand of

$$\iiiint\varphi_1F'_1G'_2g_{12}^3W(\mathbf{k}_{12}|\mathbf{k}'_{12};g_{12})\,\mathrm{d}g_{12}\,\mathrm{d}\mathbf{G}_{12}\,\mathrm{d}\mathbf{k}_{12}\,\mathrm{d}\mathbf{k}'_{12}$$

can be replaced by $\varphi'_1F_1G_2$, and similarly $\varphi_2F'_1G'_2$ can be replaced by $\varphi'_2F_1G_2$. To see this observe that the interchange of the roles of direct and inverse collisions cannot affect either the cross-section W because of (75.8) or the integration because of (75.7). When this result is applied to (75.17), the integral may be written in the symmetrical form

$$\left(\frac{\partial\varphi}{\partial t}\right)_{\mathrm{c}} = \frac{1}{4}\iiiint (\varphi_1+\varphi_2-\varphi'_1-\varphi'_2)\times$$
$$\times(f'_1 f'_2 - f_1 f_2)g_{12}^3 W(\mathbf{k}_{12}|\mathbf{k}'_{12};g_{12})\,\mathrm{d}g_{12}\,\mathrm{d}\mathbf{G}_{12}\,\mathrm{d}\mathbf{k}_{12}\,\mathrm{d}\mathbf{k}'_{12}. \qquad (75.18)$$

It follows that $(\partial\varphi/\partial t)_{\mathrm{c}}$ is zero if $\varphi_1+\varphi_2 = \varphi'_1+\varphi'_2$, and functions φ satisfying this relation are termed *summational invariants.* For the elastic smooth particles of the kinetic model the only summational invariants are the mass m, the momentum $m\mathbf{w}$, and the energy $\frac{1}{2}mw^2$, or linear combinations of these.

Turning now to the equilibrium distribution f^{eq}, we define this distribution to be that solution of the kinetic equation (75.12) that is independent of time. A *sufficient* condition for $f = f^{\mathrm{eq}}(\mathbf{r},\mathbf{w})$ to satisfy the kinetic equation is that $(\partial f/\partial t)_{\mathrm{c}}$ be zero, that is, that $f'_1 f'_2 = f_1 f_2$, or

$$\ln f(\mathbf{r},\mathbf{w}'_1)+\ln f(\mathbf{r},\mathbf{w}'_2) = \ln f(\mathbf{r},\mathbf{w}_1)+\ln f(\mathbf{r},\mathbf{w}_2).$$

Hence $\ln f^{\mathrm{eq}}$ must be a summational invariant, and from the remarks made in the last paragraph it follows that

$$\ln f^{\mathrm{eq}} = k_1+\mathbf{k}_2\,.\,\mathbf{w}+k_3 w^2,$$

where k_1, $\mathbf{k}_2$, and k_3 are independent of $\mathbf{w}$ but may depend on $\mathbf{r}$.

On replacing these coefficients by A, β, and $\mathbf{v}_0$, we have

$$f^{\mathrm{eq}} = A\exp[-\beta(\mathbf{w}-\mathbf{v}_0)^2],$$

where A, β, and $\mathbf{v}_0$ are to be determined from

$$n = \int f^{\mathrm{eq}}\,\mathrm{d}\mathbf{w}, \qquad \overline{\mathbf{w}} = \mathbf{v}, \qquad \overline{\tfrac{1}{2}m(\mathbf{w}-\mathbf{v})^2} = \tfrac{3}{2}kT.$$

The result is (see § 24.2)

$$f^{\mathrm{eq}} = n\left(\frac{m}{2\pi kT}\right)^{\frac{3}{2}}\exp[-\tfrac{1}{2}m(\mathbf{w}-\mathbf{v})^2/kT], \qquad (75.19)$$

where, in general, n, $\mathbf{v}$, and T depend on position (§ 75.5). To prove that (75.19) is a *necessary* condition for equilibrium we need Boltzmann's H-theorem.

75.4. Boltzmann's H-theorem

From (72.9), (75.9), and (75.10)

$$\sigma = -k\sum_{i,j}\iiint \ln f_i(f'_i f'_j - f_i f_j)g_{ij}W(\mathbf{k}_{ij}|\mathbf{k}'_{ij};g_{ij})\,\mathrm{d}\mathbf{k}'_{ij}\,\mathrm{d}\mathbf{w}_j\,\mathrm{d}\mathbf{w}_i$$
$$= -\tfrac{1}{4}k\sum_{i,j}\iiiint \ln\left(\frac{f_i f_j}{f'_i f'_j}\right)(f'_i f'_j - f_i f_j)g_{ij}^3 W(\ldots|\ldots)\,\mathrm{d}\mathbf{k}'_{ij}\,\mathrm{d}\mathbf{k}_{ij}\,\mathrm{d}g_{ij}\,\mathrm{d}\mathbf{G}_{ij}, \qquad (75.20)$$

where we have used an expression like (75.18) to symmetrize the integrand. As $\ln(f_i f_j / f'_i f'_j)$ is always opposite in sign to $(f'_i f'_j - f_i f_j)$, their product is always negative or zero, that is, $\sigma \geqslant 0$ and

$$\frac{\mathrm{d}H}{\mathrm{d}t} \equiv -\frac{1}{k}\int \sigma \,\mathrm{d}r \leqslant 0,$$

which is the original form of the famous H-theorem (see discussion in § 72.2).

The integral in (75.20) can vanish only if $f'_i f'_j = f_i f_j$, which is thus a necessary condition for equilibrium. It follows that Maxwell's distribution function is both necessary and sufficient for equilibrium.

75.5. Isentropic flow

Take f to be the Maxwellian distribution f^{eq} and consider a single component fluid. For this case (see §§ 24.2, 71.4),

$$\mathbf{p} = \rho\overline{\boldsymbol{cc}} = \rho(\overline{c_x^2}\mathbf{ii} + \overline{c_x c_y}\mathbf{ij} + \dots) = nkT(\mathbf{ii}+\mathbf{ij}+\mathbf{kk}),$$

that is, $\mathbf{p} = p\mathbf{1}$, so that the viscous stress tensor is zero. Similarly, $\mathbf{q} = \frac{1}{2}\rho\overline{\boldsymbol{c}c^2} = 0$. Then the basic equations for fluid motion given in § 38 reduce to

$$\frac{1}{\rho}\frac{\mathrm{d}\rho}{\mathrm{d}t} = \frac{3}{2}\frac{1}{T}\frac{\mathrm{d}T}{\mathrm{d}t} = -\nabla\,.\,\mathbf{v}, \qquad \frac{\mathrm{d}\mathbf{v}}{\mathrm{d}t} = -\frac{1}{\rho}\nabla p + \mathbf{F}, \qquad \sigma = 0, \quad (75.21)$$

where $\mathrm{d}/\mathrm{d}t = \partial/\partial t + \mathbf{v}\,.\,\nabla$ and we have used the perfect gas relation $u = 3kT/2m$. As the entropy production rate is zero, the flow is isentropic.

Next consider the kinetic equation for isentropic flow. The collision integral in (75.12) vanishes, leaving

$$\mathrm{D}f^{\mathrm{eq}} = \left(\frac{\partial}{\partial t} + \mathbf{w}\,.\,\nabla + \mathbf{F}\,.\,\frac{\partial}{\partial \mathbf{w}}\right)f^{\mathrm{eq}} = \left(\frac{\mathrm{d}}{\mathrm{d}t} + \boldsymbol{c}\,.\,\nabla + \mathbf{F}\,.\,\frac{\partial}{\partial \mathbf{w}}\right)f^{\mathrm{eq}} = 0. \quad (75.22)$$

By (75.19), $(75.21)_1$, and the fact that $\mathbf{r}, \mathbf{w}, t$ are independent variables,

$$\mathrm{D}f^{\mathrm{eq}} = -\frac{m}{2k}f^{\mathrm{eq}}\mathrm{D}\left(\frac{c^2}{T}\right) = \frac{m}{kT}f^{\mathrm{eq}}\left(\frac{c^2}{2T}\mathrm{D}T - \boldsymbol{c}\,.\,\mathrm{D}\boldsymbol{c}\right),$$

where

$$\mathrm{D}T = \frac{\mathrm{d}T}{\mathrm{d}t} + \boldsymbol{c}\,.\,\nabla T = -\tfrac{2}{3}T\nabla\,.\,\mathbf{v} + \boldsymbol{c}\,.\,\nabla T$$

and

$$\mathrm{D}\boldsymbol{c} = \mathrm{D}\mathbf{w} - \mathrm{D}\mathbf{v} = \mathbf{F} - \frac{\mathrm{d}\mathbf{v}}{\mathrm{d}t} - \boldsymbol{c}\,.\,\nabla\mathbf{v} = \frac{5}{2}\frac{k}{m}\nabla T - \boldsymbol{c}\,.\,\nabla\mathbf{v};$$

in the final step we have used $p = nkT \propto T^{\frac{5}{2}}$. Hence

$$\mathrm{D}f^{\mathrm{eq}} = f^{\mathrm{eq}}\left[\left(\frac{mc^2}{2kT} - \frac{5}{2}\right)\boldsymbol{c}\,.\,\nabla \ln T + \frac{m}{kT}(\boldsymbol{cc} - \tfrac{1}{3}c^2\mathbf{1}):\nabla\mathbf{v}\right]. \quad (75.23)$$

As $\boldsymbol{c}$ is an independent variable, (75.22) can be satisfied only if the coefficients of the various powers in $\boldsymbol{c}$ vanish, that is, from (75.23) only if ∇T is zero and $\nabla \mathbf{v}$ is an antisymmetric tensor (a consequence of (38.16)). The most general choice for $\nabla \mathbf{v}$ sets it equal to the tensor of some arbitrary axial vector $\boldsymbol{\omega}$, that is (see (48.4)) $\nabla \mathbf{v} = \boldsymbol{\omega} \times \mathbf{1}$. Thus we may write

$$\mathbf{v} = \mathbf{v}' + \boldsymbol{\omega} \times \mathbf{r}, \tag{75.24}$$

where $\mathbf{v}'$ is an arbitrary vector; the mean velocity of the gas is that of a rigid screw.

With zero temperature gradient and $\mathbf{v}$ given by (75.24), we may repeat the calculation to find that in steady flow

$$\begin{aligned} \mathrm{D}f^{\mathrm{eq}} &= f^{\mathrm{eq}}\left(\mathrm{D} \ln n - \frac{m}{kT}\boldsymbol{c} \,.\, \mathrm{D}\boldsymbol{c}\right) \\ &= f^{\mathrm{eq}}\left[\mathbf{w} \,.\, \nabla \ln n - \frac{m}{kT}(\mathbf{w} - \mathbf{v}) \,.\, (\mathbf{F} - \mathbf{v} \,.\, \nabla \mathbf{v})\right] = 0, \end{aligned}$$

which can be satisfied only if

$$\nabla(\ln n) - \frac{m}{kT}\mathbf{F} + \frac{m}{kT}\mathbf{v} \,.\, \nabla \mathbf{v} = 0, \qquad \mathbf{v} \,.\, \mathbf{F} = 0. \tag{75.25}$$

It follows from this that $\mathbf{F}$ must be the gradient of a scalar function $\Psi(\mathbf{r})$ satisfying $\mathbf{v} \,.\, \nabla \Psi = 0$, and (75.25) can be integrated to give

$$n = n_0 \exp\left[-m(\Psi + \Psi_0)/kT\right], \tag{75.26}$$

where $\Psi_0 = \frac{1}{2}v^2 = \frac{1}{2}(\mathbf{v}' + \boldsymbol{\omega} \times \mathbf{r})^2$ and n_0 is the number density at points where Ψ and Ψ_0 are zero.

From the above analysis it would appear that 'isentropic flow' is too degenerate to be of interest, but the relevance of the conclusions following from (75.23) depend on just how closely the two sides of the kinetic equation are coupled. As we shall see below, isentropic flow governed by (75.21), and without the constraints on T and $\mathbf{v}$ just derived, is a satisfactory model of fluid behaviour when the ratio of the kinetic time scale τ_1 to the macroscopic time scale $\mathscr{T}$ is vanishingly small.

76. On solving the kinetic equation

76.1. The hydrodynamic stage

For non-equilibrium states the best that can usually be achieved is to find the form of f for small disturbances from f^{eq}, which amounts to waiting until the transients of the kinetic state $\tau_0 < t < t_1$ have past. To this end we write

$$f = f^{\mathrm{eq}}(1+\varphi) \qquad (\varphi \ll 1) \tag{76.1}$$

and ignore quadratic terms in φ. By (72.2) and (72.3) φ is subject to the constraints

$$\int \Psi^{(i)} f^{\mathrm{eq}} \varphi \, d\mathbf{w} = 0 \qquad (i = 1, 2, 3),$$

where (76.2)

$$\Psi^{(i)} = (1, m\boldsymbol{c}, \tfrac{1}{2}mc^2).$$

After the kinetic stage comes the so-called *hydrodynamic stage* $\tau_1 < t < \mathscr{T}$, where $\mathscr{T}$ is the macroscopic time scale. Quantities like T and $\mathbf{v}$ change on this scale, and from (75.23) this is also true of f^{eq}. It is convenient to first discuss the BGK relaxation model. We start with (74.9) in the form $\mathrm{D}f = (f^{\mathrm{eq}} - f)/\tau_1$, $\tau_1 \geqslant 0$, or

$$f = f^{\mathrm{eq}} - \varepsilon\mathscr{T}\mathrm{D}f \qquad (\varepsilon = \tau_1/\mathscr{T} \ll 1). \tag{76.3}$$

In the kinetic stage f changes much more rapidly than f^{eq}, so we expect it to be almost completely relaxed during the hydrodynamic stage, with the operator $\mathscr{T}\mathrm{D}$ of order unity when applied to f. When f and f^{eq} are 'synchronized' in this way (cf. similar discussion of the kinetic stage in § 74.3), (76.3) gives

$$f = f^{\mathrm{eq}} - \tau_1 \mathrm{D} f^{\mathrm{eq}} + O(\varepsilon^2),$$

or (76.4)

$$\mathrm{D}f^{\mathrm{eq}} \approx (f^{\mathrm{eq}} - f)/\tau_1 = -f^{\mathrm{eq}}\varphi/\tau_1.$$

It follows from (76.1) and $(76.4)_1$ that the linearized form for φ is $\varphi = -\tau_1 \mathrm{D}(\ln f^{\mathrm{eq}})$. Hence by (75.23)

$$\varphi = -\tau_1 \left[\left(\frac{mc^2}{2kT} - \frac{5}{2} \right) \boldsymbol{c} \,.\, \nabla \ln T + \frac{m}{kT} \mathring{\boldsymbol{c}\boldsymbol{c}} \colon \nabla \mathbf{v} \right], \tag{76.5}$$

where $\mathring{\boldsymbol{c}\boldsymbol{c}}$ is the deviator of $\boldsymbol{c}\boldsymbol{c}$.

A similar argument applied to Boltzmann's kinetic equation (75.12) leads to the linearized form

$$\mathrm{D}f^{\mathrm{eq}} = \iint f_1^{\mathrm{eq}} f_2^{\mathrm{eq}} (\varphi_1' + \varphi_2' - \varphi_1 - \varphi_2) g_{12} W(\,..\,|\,..\,)\, d\mathbf{k}_{12}' \, d\mathbf{w}_2,$$

the leading right-hand term vanishing as $f_1^{\mathrm{eq}} f_2^{\mathrm{eq}} = f_1'^{\mathrm{eq}} f_2'^{\mathrm{eq}}$. By (75.23) this may be written

$$n^2 I(\varphi) = -f_1^{\mathrm{eq}} \left[\left(\frac{mc^2}{2kT} - \frac{5}{2} \right) \boldsymbol{c} \,.\, \nabla \ln T + \frac{m}{kT} \mathring{\boldsymbol{c}\boldsymbol{c}} \colon \nabla \mathbf{v} \right], \tag{76.6}$$

where the operator I is defined in

$$I(F) = \frac{1}{n^2}\iint f_1^{\text{eq}} f_2^{\text{eq}}(F_1+F_2-F_1'-F_2')g_{12}W(\ldots|\ldots)\,\mathrm{d}\mathbf{k}_{12}'\,\mathrm{d}\mathbf{w}_2. \tag{76.7}$$

We shall describe the solution of (76.6) in general terms shortly.

As the macroscopic gradients in $(76.4)_1$ are proportional to $L^{-1} = (\mathscr{T}u_s)^{-1}$, where $u_s \approx \bar{c}$ is the sound speed, in the limit as $\varepsilon = \tau_1/\mathscr{T}$ tends to zero $(76.4)_1$ reduces to $f = f^{\text{eq}}$ (isentropic flow), in which limit $\mathrm{D}f \approx \mathrm{D}f^{\text{eq}}$ is decoupled from $(\partial f/\partial t)_c$, as remarked at the end of § 75.5.

76.2. General form for φ

We shall now use the principles of process thermodynamics to find a general expression for φ, independent of the particular kinetic equation adopted. We return to the formula (72.11) for the entropy production σ; for one-component it reduces to

$$\sigma = -k\int \mathrm{D}f \ln(f/f^{\text{eq}})\,\mathrm{d}\mathbf{w} \geqslant 0.$$

During the hydrodynamic stage we may use the approximation in (76.3) to obtain

$$\sigma = -k\int \varphi \mathrm{D}f^{\text{eq}}\,\mathrm{d}\mathbf{w} \geqslant 0,$$

or by (75.23)

$$\sigma = -k\int f^{\text{eq}}\left[\varphi_1\left(\frac{mc^2}{2kT}-\frac{5}{2}\right)\boldsymbol{c}\,.\,\nabla \ln T + \varphi_2\frac{m}{kT}\mathring{\boldsymbol{c}\boldsymbol{c}}:\nabla\mathbf{v}\right]\mathrm{d}\mathbf{w} \geqslant 0, \tag{76.8}$$

where $\varphi_1 = \varphi_2 = \varphi$ is a linear dependency between the 'processes' φ_1 and φ_2. The second of eqns (76.2) allows us to remove the term involving $\frac{5}{2}$ from (76.8).

As a check on (76.8) we may compare it with (38.13), namely,

$$T\sigma = -\mathbf{q}\,.\,\nabla \ln T - \tilde{\boldsymbol{\pi}}:\nabla\mathbf{v},$$

and find

$$\mathbf{q} = \tfrac{1}{2}m\int f^{\text{eq}}\varphi \boldsymbol{c}c^2\,\mathrm{d}\mathbf{w} = \tfrac{1}{2}\rho\overline{\boldsymbol{c}c^2},$$

$$\tilde{\boldsymbol{\pi}} = m\int f^{\text{eq}}\varphi\mathring{\boldsymbol{c}\boldsymbol{c}}\,\mathrm{d}\mathbf{w} = \rho\overline{\boldsymbol{c}\boldsymbol{c}} - \tfrac{1}{3}\rho\overline{c^2}\mathbf{1} = \mathbf{p} - p\mathbf{1},$$

in agreement with the definitions given in § 71.4.

By the standard argument of process thermodynamics, we infer from (76.8) a linear constitutive relation of the form (cf. the more general theory given in § 66.1)

$$\varphi_1 = \varphi_2 = \varphi = -\frac{1}{n}\mathbf{A}\cdot\nabla\ln T - \frac{1}{n}\mathring{\mathbf{B}}:\nabla\mathbf{v}, \tag{76.10}$$

where $\mathbf{A}$ and $\mathring{\mathbf{B}}$ depend on $\boldsymbol{c}$ and on the thermodynamic state (n, T). That it is correct to use a deviator $\mathring{\mathbf{B}}$ here follows from the rule (§ 38.5) $\mathbf{C}:\mathring{\mathbf{D}} = \mathring{\mathbf{C}}:\mathring{\mathbf{D}}$ applied first to (76.8) and then to (76.10). The only vector that can be formed from the elements $\boldsymbol{c}$, n, and T is $\boldsymbol{c}$ itself multiplied by a scalar function of $c = |\boldsymbol{c}|$, n, and T; likewise the only deviator is $\mathring{\boldsymbol{cc}}$ times a similar scalar function. Thus

$$\varphi = -\frac{1}{n}A(c)\boldsymbol{c}\cdot\nabla\ln T - \frac{1}{n}B(c)\mathring{\boldsymbol{cc}}:\nabla\mathbf{v}, \tag{76.11}$$

where the scalar coefficients A, B also depend on (n, T). At first sight it appears that (76.11) could be inferred directly as φ_1, φ_2 are already multiplying the scalar affinities $\boldsymbol{c}\cdot\nabla\ln T$ and $\mathring{\boldsymbol{cc}}:\nabla\mathbf{v}$ in (76.8). However, φ_1 should be split into $\varphi_1^{(i)}$, $i = 1, 2, 3$, and distributed over the three scalar terms in $\boldsymbol{c}\cdot\nabla\ln T$, and similarly for φ_2, which means that, prior to imposing constraints from isotropy, (76.10) is the correct inference to take from (76.8).

By symmetry we find that for any function $F(c)$

$$0 = \int F(c)\boldsymbol{cc}\,\mathrm{d}\mathbf{w} - \tfrac{1}{3}\int F(c)c^2\,\mathrm{d}\mathbf{w}\,\mathbf{1} = \int F(c)\mathring{\boldsymbol{cc}}\,\mathrm{d}\mathbf{w}.$$

Using this and eliminating integrands that are odd functions of the components of $\boldsymbol{c}$, we find that (76.11) satisfies the three constraints in (76.2) if

$$\int f^{\mathrm{eq}}A(c)c^2\,\mathrm{d}\mathbf{w} = 0. \tag{76.12}$$

There is no restriction on B. The problem is now reduced to that of finding the scalars A and B subject to (76.12).

For the BGK model this task is immediately accomplished by comparing (76.5) and (76.11), and it is easily verified that (76.12) is satisfied for this case. For Boltzmann's model we substitute (76.11) into (76.6) and equate each of the coefficients of the independent affinities $\nabla\ln T$ and $\nabla\mathbf{v}$ to zero to arrive at the integral equations

$$nI(\boldsymbol{c}A) = f^{\mathrm{eq}}\left(\frac{mc^2}{2kT} - \frac{5}{2}\right), \qquad nI(\mathring{\boldsymbol{cc}}B) = f^{\mathrm{eq}}\frac{m}{kT}\mathring{\boldsymbol{cc}},$$

the (approximate) solution of which depends on the potential of the intermolecular forces, and involves tedious algebra not relevant to the purpose of this text. Only in the special case when the force between two molecules is proportional to minus the fifth power of the distance between them can an exact solution (due to Maxwell) be found.

From $(76.9)_1$ and (76.11)

$$\mathbf{q} = -\left[\frac{m}{6n}\int f^{\mathrm{eq}} c^4 A(c)\,\mathrm{d}\boldsymbol{c}\right]\nabla \ln T,$$

the various omitted integrals all having integrands odd in the components of $\boldsymbol{c}$. Similarly (see, for example, Chapman and Cowling (1970) or Ferziger and Kaper (1972)).

$$\boldsymbol{\pi} = -\left[\frac{m}{5n}\int f^{\mathrm{eq}}\mathring{\boldsymbol{c}}\boldsymbol{c}:\boldsymbol{c}\mathring{\boldsymbol{c}}B(c)\,\mathrm{d}\boldsymbol{c}\right]\mathring{\nabla}\mathbf{v}.$$

76.3. Systems with several components

For multi-component systems there are two distinct cases, depending on whether or not the kinetic time-scale can be split into several distinct kinetic time scales, as in the case of a plasma (§ 56.1), for example. Large mass differences between the particles of different components is a necessary condition for different kinetic time scales. Component equilibrium is then achieved much more rapidly than mixture equilibrium, consequently for each component the kinetic process proceeds in two stages: $f_i^{\mathrm{eq}} \to f_{\mathrm{ic}}^{\mathrm{eq}} \to f_{\mathrm{im}}^{\mathrm{eq}}$, with corresponding kinetic times $\tau_{ic} \ll \tau_{im}$. We shall leave this rather complicated type of multi-component system and give a brief account of the situation when one kinetic time serves for all components.†

The theory is based on (75.11) and follows the same development as for a single component. Thus with

$$f_i = f_{\mathrm{im}}^{\mathrm{eq}}(1+\varphi_i) \qquad (\varphi_i \ll 1), \tag{76.13}$$

corresponding to (76.6) we find the linearized form

$$\sum_j n_i n_j I_{ij}(\varphi) = -\mathrm{D}_i f_{\mathrm{im}}^{\mathrm{eq}}. \tag{76.14}$$

where

$$I_{ij}(F) \equiv \frac{1}{n_i n_j}\iint f_{\mathrm{im}}^{\mathrm{eq}} f_{\mathrm{im}}^{\mathrm{eq}}(F_i + F_j - F_i' - F_j')W(\ldots|\ldots)\,\mathrm{d}\mathbf{k}_{ij}'\,\mathrm{d}\mathbf{w}_i.$$

By eqns $(23.15)_2$ and (71.4)

$$\mathrm{D}_i f_{\mathrm{im}}^{\mathrm{eq}} = f_{\mathrm{im}}^{\mathrm{eq}}\left(\frac{\mathrm{d}}{\mathrm{d}t} + \mathscr{C}_i.\nabla + \mathbf{F}_i.\frac{\partial}{\partial \mathbf{w}_i}\right)\left[\frac{m}{kT}(g_i - \tfrac{1}{2}\mathscr{C}_i^2)\right] \quad \left(\frac{\mathrm{d}}{\mathrm{d}t} = \frac{\partial}{\partial t} + \mathbf{v}.\nabla\right).$$

† For accounts of kinetic processes in plasmas the works of Braginskii (1965) and of Shkarofsky, Johnston, and Bachynski (1966) may be consulted.

In isentropic flow

$$\frac{1}{n_i}\frac{dn_i}{dt} = -\nabla \cdot \mathbf{v} = \frac{3}{2}\frac{dT}{dt}, \qquad \frac{d\mathbf{v}}{dt} = -\frac{1}{\rho}\nabla p + \mathbf{F}, \qquad \left(\mathbf{F} = \frac{\sum_i \rho_i \mathbf{F}_i}{\rho}\right),$$

and as $m_i g_i/kT = \ln(n_i T^{-\frac{3}{2}}) + \text{constant}$, we find (use the result $T\nabla(g_i/T) = (\nabla g_i)_T - h_i \nabla \ln T$ given in § 53.4)

$$\mathrm{D}_i f^{\mathrm{eq}}_{\mathrm{im}} = f^{\mathrm{eq}}_{\mathrm{im}}\left[\frac{n}{n_i}\mathscr{C}_i \cdot \mathbf{d}_i + \left(\frac{m_i \mathscr{C}_i^2}{2kT} - \frac{5}{2}\right)\mathscr{C}_i \cdot \nabla \ln T + \frac{m_i}{kT}\overset{\circ}{\mathscr{C}_i\mathscr{C}_i} : \nabla \mathbf{v}\right], \quad (76.15)$$

where

$$\mathbf{d}_i \equiv \frac{\rho_i}{\rho}\left[(\nabla g_i)_T - \frac{1}{\rho}\nabla p + \mathbf{F} - \mathbf{F}_i\right] \qquad ((\nabla g_i)_T = kT\nabla n_i/\rho_i) \quad (76.16)$$

is an affinity driving the diffusion process. If $\mathbf{F} = \nabla p/\rho$, that is, inertia effects are dropped, $\mathbf{d}_i$ reduces to a form like that appearing in $(53.24)_2$.

The rate of entropy production in the hydrodynamic stage is

$$\sigma = -k\sum_i \int \varphi_i \mathrm{D}_i f^{\mathrm{eq}}_{\mathrm{im}}\, d\mathbf{w}_i \geqslant 0, \quad (76.17)$$

and by an argument similar to that leading to (76.11) we may infer from (76.15) and (76.17) that

$$\varphi_i = -\frac{1}{n}\sum_i \mathrm{D}_{ij}(\mathscr{C}_i)\mathscr{C}_i \cdot \mathbf{d}_j - \frac{1}{n}A_i(\mathscr{C}_i)\mathscr{C}_i \cdot \nabla \ln T - \frac{1}{n}B_i(\mathscr{C}_i)\overset{\circ}{\mathscr{C}_i\mathscr{C}_i} : \nabla \mathbf{v}, \quad (76.18)$$

where the coefficients D_{ij}, A_i, and B_i are determined from integral equations found by substituting (76.18) into (76.14) and equating the coefficients of corresponding terms. These coefficients must also satisfy constraints like that given in (76.12), but we shall not pursue the details here. What is worth noting is that the *form* of (76.18) is independent of the kinetic equation (76.14) and so is valid so long as Bogoliubov's synchronization hypothesis for the hydrodynamic stage holds good.

76.4. *Kinetic theory and magnetohydrodynamics*

As remarked in § 56.1, the presence of a magnetic field in a plasma introduces additional time scales into the problem, namely, the gyration times $\omega_i^{-1} = (m_i/e_i B)$, $i = 1, 2$, for the ions ($i = 1$) and electrons ($i = 2$) in the magnetic field. And these time scales have to be fitted somewhere in the Bogoliubov scheme of collisional, kinetic, and hydrodynamic time scales if the method of solving the kinetic equation described above is to be used. The ordering we shall consider here, although not particularly relevant in plasma physics, is the one originally chosen by Chapman and Cowling (1970), namely, ω_i^{-1}, $\tau_i \ll \mathscr{T}$, where τ_i, $i = 1, 2$, are the kinetic time scales; there is no particular constraint on the product $\omega_i\tau_i$.

The force acting per unit mass on the ion and electron components is

$$\mathbf{F}_i = \frac{e_i}{m_i}(\mathbf{E}+\mathbf{w}_i\times\mathbf{B}) = \mathscr{F}_i+\frac{e_i}{m_i}\mathscr{C}_i\times\mathbf{B} \qquad \left(\mathscr{F}_i \equiv \frac{e_i}{m_i}(\mathbf{E}+\mathbf{v}\times\mathbf{B}\right),$$

where $\mathscr{C}_i \equiv \mathbf{w}_i-\mathbf{v}$, $\mathbf{v}$ being the mass motion of the plasma as a whole. The averages $n_i\overline{\mathbf{F}_i\cdot\partial\varphi_i/\partial\mathbf{w}_i}$, $\varphi_i = (m_i, m_i\mathbf{w}_i, \frac{1}{2}m_i\mathbf{w}_i\cdot\mathbf{w}_i)$, required in § 71.4, yield terms $(0, n_ie_i(\mathbf{E}+\mathbf{v}_i\times\mathbf{B}), n_ie_i\mathbf{E}\cdot\mathbf{v}_i)$ to add to the right-hand sides of (71.19), (71.20), and (71.22), and we are thus led to the fluid equations described in the first paragraph of § 56.5.

With the definition

$$\mathscr{D}_if_i \equiv \left(\frac{\partial}{\partial t}+\mathbf{w}_i\cdot\frac{\partial}{\partial\mathbf{r}}+\mathscr{F}_i\cdot\frac{\partial}{\partial\mathbf{w}_i}\right)f_i,$$

(75.11) yields

$$\mathscr{D}_if_i = -\frac{e_i}{m_i}\mathscr{C}_i\times\mathbf{B}\cdot\frac{\partial f_i}{\partial\mathbf{w}_i}+\sum_j C(f_i^{\text{eq}}, f_j^{\text{eq}}), \tag{76.19}$$

of which the right-hand terms are proportional to ω_i and τ_i^{-1} respectively. In the hydrodynamic stage, that is, when f_i is synchronized with f_i^{eq}, the left-hand term is proportional to $\mathscr{T}^{-1}$. This to first order in the expansion (76.13), we find

$$\frac{e_i}{m_i}\mathscr{C}_i\times\mathbf{B}\cdot\frac{\partial f_i^{\text{eq}}}{\partial\mathbf{w}_i} = \sum_j C(f_i^{\text{eq}}, f_j^{\text{eq}}).$$

It is readily verified that this equation is satisfied by $f_i^{\text{eq}} = f_{\text{im}}^{\text{eq}}$ for $\partial f_{\text{im}}^{\text{eq}}/\partial\mathbf{w}_i$ is parallel to $\mathscr{C}_i$. To the next order we find that

$$\mathscr{D}_if_{\text{im}}^{\text{eq}} = -\frac{e_i}{m_i}\mathscr{C}_i\times\mathbf{B}\cdot\frac{\partial}{\partial\mathbf{r}}(f_{\text{im}}^{\text{eq}}\varphi_i)-\sum_j n_in_jI_{ij}(\varphi_i). \tag{76.20}$$

During the hydrodynamic stage the entropy is produced at the rate

$$\sigma = -k\sum_i\int\varphi_i\mathscr{D}_if_{\text{im}}^{\text{eq}}\,d\mathbf{w}_i \geqslant 0,$$

where $\mathscr{D}_if_{\text{im}}^{\text{eq}}$ is given by the right-hand side of (76.15), save that $\mathbf{d}_i$ is now

$$\mathbf{d}_i = \frac{\rho_i}{p}\left[(\nabla g_i)_T-\frac{1}{\rho}\nabla p-\left(\frac{e_i}{m_i}-\sum_j\frac{n_je_j}{\rho}\right)(\mathbf{E}+\mathbf{v}\times\mathbf{B})\right]. \tag{76.21}$$

Then φ_i is given, not by (76.18), but by the more general form

$$\varphi_i = -\frac{1}{n}\sum_j\mathbf{D}_{ij}\cdot\mathbf{d}_j-\frac{1}{n}\mathbf{A}_i\cdot\nabla\ln T-\frac{1}{n}\mathring{\mathbf{B}}_i:\nabla\mathbf{v}, \tag{76.22}$$

for now the coefficients $\mathbf{D}_{ij}$, $\mathbf{A}_i$, and $\mathring{\mathbf{B}}_i$ may depend not only on the vector $\mathscr{C}_i$, but also on the magnetic field vector $\mathbf{B} = \mathbf{b}B$.

Let $\mathbf{A}$, $\mathring{\mathbf{B}}$ denote typical coefficients in (76.22), then, whereas previously isotropy permitted us to write $\mathbf{A} = \mathbf{A}(\mathscr{C})\mathscr{C}$, $\mathring{\mathbf{B}} = \mathbf{B}(\mathscr{C})\mathring{\mathscr{C}\mathscr{C}}$, where A and B are scalars, now we require

$$\mathbf{A} = \boldsymbol{\mathsf{A}}(\mathscr{C}, \mathbf{B}) \,.\, \mathscr{C}, \qquad \mathring{\mathbf{B}} = \boldsymbol{\mathsf{B}}(\mathscr{C}, \mathbf{B}) : \mathring{\mathscr{C}\mathscr{C}},$$

where $\boldsymbol{\mathsf{A}}$ and $\boldsymbol{\mathsf{B}}$ are second and fourth-order tensors respectively, with lateral isotropy about $\mathbf{b}$. Such tensors are described in §§ 47.2 and 47.4. In the notation of (56.14) and (48.16) we obtain

$$\mathbf{A} = A_{\|}\mathscr{C}_{\|} + A_{\perp}\mathscr{C}_{\perp} + A_{\wedge}\mathbf{b}\times\mathscr{C}, \qquad \mathring{\mathbf{B}} = \sum_{j=1}^{s} B_j \boldsymbol{\mathsf{N}}_j : \mathring{\mathscr{C}\mathscr{C}},$$

the various scalar components of which depend on $(\mathscr{C}, B, n, T)$ and satisfy integral equations that are derived from (76.20) and (76.22).

REFERENCES AND FURTHER READING

The page on which the reference is cited is given in square brackets

Part I. Equilibrium thermodynamics

References cited in the text

BOGOLIUBOV, N. N. (1946). *Studies in statistical mechanics*, Vol. 1 (eds J. De Boer and G. E. Uhlenbeck). North-Holland, Amsterdam. [**89**]

CARATHÉODORY, C. (1909). *Math. Annln.* **67**, 355. [**29**]

GRAD, H. (1961). *Commun. Pure appl. Math.* **14**, 323–54. [**93**]

GROSSMAN, L. M. (1969). *Thermodynamics and statistical mechanics*, McGraw-Hill, New York. [**132**]

HUANG, K. (1963). *Statistical mechanics*, Wiley, New York. [**125**]

KHINCHINE, A. I. (1957). *Mathematical foundations of information theory*, Dover Publications, New York. [**136**]

OVERSETH, O. E. (1969). *Scient. Am.*, October, 89. [**70**]

STRATTON, J. A. (1941). *Electromagnetic theory*, McGraw-Hill, New York. [**38**]

TER HAAR, D. (1966). *Elements of Thermostatistics*, p. 78, Holt, Rinehart, and Winston, New York. [**90**]

ZEMANSKY, M. W. (1968). *Heat and thermodynamics* (5th edn). McGraw-Hill, New York. [**29**]

Further reading in equilibrium thermodynamics

BAZOROV, I. P. (1964). *Thermodynamics.* Pergamon Press, Oxford.

CALLEN, H. B. (1960). *Thermodynamics.* Wiley, New York.

GUGGENHEIM, E. A. (1967). *Thermodynamics* (3rd edn). North-Holland, Amsterdam.

LANDSBERG, P. T. (1961). *Thermodynamics.* Interscience, New York.

—— (ed) (1970). *Proceedings of the International Conference on Thermodynamics*, Butterworths, London.

MÜNSTER, A. (1970). *Classical thermodynamics.* Wiley–Interscience, New York.

PIPPARD, A. B. (1966). *The elements of classical thermodynamics.* Cambridge University Press.

Part II. Process thermodynamics

References cited in the text

ANDRONIKASHVILI, E. L. and MAMALADZE, YU. G. (1966). *Rev. mod. Phys.* **38**, 567–625. [**254**]

BATCHELOR, G. K. (1967). *Fluid dynamics*, p. 180. Cambridge University Press. [**249, 255**]

BHATNAGER, P. L., GROSS, E. P., and KROOK, M. (1954). *Phys. Rev.* **94**, 511–25. [**321**]

BOGOLIUBOV, N. N. (1946). *Studies in statistical mechanics*, Vol. 1 (eds J. De Boer and G. E. Uhlenbeck), North-Holland, Amsterdam. [**89**]

BOLTZMANN, L. (1872). *Sber. Akad. Wiss. Wien* **66**, 275–370. [**317**]

BRAGINSKII, S. I. (1965). *Rev. Plasma Phys.* **1**, 205–311. [**166, 230, 346**]

CALLEN, H. B., BARASCH, M. L., and JACKSON, J. L. (1952). *Phys. Rev.* **88**, 1382. [**288**]

—— and WELTON, T. A. (1951). *Phys. Rev.* **83**, 34. [**288**]

—— and GREENE, R. F. (1951). *Phys. Rev.* **83**, 1231. [**288**]

——, —— (1952). *Phys. Rev.* **86**, 702. [**288**]

CASIMIR, H. B. G. (1945). *Rev. mod. Phys.* **17**, 343. [**159**, **163**]
CHAPMAN, S. and COWLING, T. G. (1970). *The mathematical theory of non-uniform gases* (3rd ed). Cambridge University Press. [**199**, **346**]
CLEMMOW, P. C. and DOUGHERTY, J. P. (1969). *Electrodynamics of particles and plasmas*. Addison-Wesley, New York. [**323**]
COLEMAN, B. D. (1964). *Archs ration. Mech. Analysis* **17**, 1–46. [**156**]
CURIE, P. (1908). *Oeuvres*, p. 129. Paris. [**189**]
DE GROOT, S. R. and MAZUR, P. (1962). *Non-equilibrium thermodynamics*. North-Holland, Amsterdam. [(**vii**), **163**]
DONNELLY, R. J., HERMAN, R., and PRIGOGINE, I. (eds). (1966). *Non-equilibrium thermodynamics, variational techniques and stability*. Chicago University Press. [**222**]
EINSTEIN, A. (1905). *Annln Phys.* **17**, 891. [**293**]
ERINGEN, A. C. (1967). *Mechanics of continua*. Wiley, New York. [**156**]
FERZIGER, J. H. and KAPER, H. G. (1972). *Mathematical theory of transport procrsses in gases*. North-Holland, Amsterdam. [**324**]
FRIEMAN, E. A. (1963). *J. math. Phys.* **4**, 410. [**329**]
GLANSDORFF, P. and PRIGOGINE, I. (1971). *Thermodynamic theory of structure, stability and fluctuations*. Wiley–Interscience, New York. [**221**]
HAASE, R. (1968). *Thermodynamics of irreversible processes*. Addison-Wesley, New York. [**155**]
HALL, H. E. and VINEN, W. F. (1956). *Proc. R. Soc.* A**238**, 204. [**254**, **259**]
KHALATNIKOV, I. M. (1965). *An introduction to the theory of superfluidity*. Benjamin, New York. [**250**, **258**]
KLIMONTOVICH, YU. L. (1974). *Adv. plasma phys.* **5**, 35–49. [**325**]
KUBO, R. (1957). *J. phys. Soc. Japan* **12**, 570. [**288**]
—— and YOKOTA, M. (1957). *J. phys. Soc. Japan* **12**, 1203. [**288**]
LIN, C. C. (1963). In *Liquid helium* (ed. G. Careri), p. 93. Academic Press, New York. [**244**, **255**]
LONDON, F. (1954). *Superfluids*, Vol. 2. Wiley, New York. [**251**]
MEIXNER, J. (1941). *Annln Phys.* **39**, 333. [**163**]
—— (1942). *Annln Phys.* **41**, 409. [**163**]
—— (1943). *Annln Phys.* **43**, 1244. [**163**]
NYQUIST, H. (1928). *Phys. Rev.* **32**, 110. [**288**]
ONSAGER, L. (1931). *Phys. Rev.* **37**, 405. [**159**, **163**]
—— (1931). *Phys. Rev.* **38**, 2265. [**159**, **163**]
PAPOULIS, A. (1965). *Probability, random variables and stochastic processes*, p. 217. McGraw-Hill, New York. [**294**]
PRIGOGINE, I. (1945). *Classe Sci. Acad. R. Belg.* **31**, 600. [**220**, **222**]
RAMANATHAN, G. V. and SANDRI, G. (1969). *J. math. Phys.* **10**, 1763–73. [**333**]
SHKAROFSKY, I. P., JOHNSTON, T. W., and BACHYNSKI, M. P. (1966). *The particle kinetics of plasmas*, Addison-Wesley, New York. [**346**]
THOMSON, W. (1882). *Math. phys. Papers* **1**, 232. [**159**]
TRUESDELL, C. (1969). *Rational thermodynamics*. McGraw-Hill, New York. [**156**, **173**]
VAN DER POL, B. and BREMMER, H. (1950). *Operational calculus*, p. 118. Cambridge University Press. [**273**]
WILKS, J. (1967). *The properties of liquid and solid helium*. Oxford University Press. [**241**]
WOODS, L. C. (1971). In *Continuum physics* (ed. A. C. Eringen). Academic Press, New York. [**275**]
WU, T. Y. (1966). *Kinetic equations of gases and plasmas*. Addison-Wesley, New York. [**309**]

Further reading in process thermodynamics

LANDAU, L. D. and LIFSHITZ, E. M. (1959). *Statistical physics*. Pergamon Press, Oxford.

PRIGOGINE, I. (1967). *Introduction to the thermodynamics of irreversible processes* (3rd ed.). Wiley, New York.

STUART, E. B., GAL-OR, B., and BRAINARD, A. J. (1970). *A critical review of thermodynamics*. Mono Book Corp., New York.

YOURGRAU, W., VAN DER MERWE, A., and RAW, G. (1966). *Treatise on irreversible and statistical thermophysics*. Macmillan, New York.

INDEX